Electrochemistry for Materials Science

Electrochemistry for Materials Science

Waldfried Plieth

Technische Universität Dresden,
Fakultät Mathematik und Naturwissenschaften,
Fachrichtung Chemie und Lebensmittelchemie,
Physikalische Chemie und Elektrochemie,
D-01062, Dresden,
Germany

ELSEVIER

Amsterdam • Boston • Heidelberg • London • New York • Oxford
Paris • San Diego • San Francisco • Singapore • Sydney • Tokyo

Elsevier
Radarweg 29, PO Box 211, 1000 AE Amsterdam, The Netherlands
The Boulevard, Langford Lane, Kidlington, Oxford, OX5 1GB, UK

First edition 2008
Reprint 2008

British Library Cataloguing in Publication Data
A catalogue record for this book is available from the British Library

Library of Congress Cataloging-in-Publication Data
A catalog record for this book is available from the Library of Congress

ISBN: 978-0-444-52792-9

For information on all Elsevier publications
visit our website at www.elsevierdirect.com

Printed and bound in *Hungary*

08 09 10 10 9 8 7 6 5 4 3 2

Contents

List of Symbols

a	Lattice constant
a	Activity
$a_{\text{subscript}}$	Activity specified by the subscript
A	Surface area
A_e	Electron affinity
$\mathrm{A_n}$	Nucleation rate
$\boldsymbol{b}$	Burgers vector
B	Constant in the Kohlrausch equation (Eq. (1.7))
B_1	Constant in the Debye–Hueckel–Onsager equation (Eq. (1.19))
B_2	Constant in the Debye–Hueckel–Onsager equation (Eq. (1.19))
c	Concentration
$\bar{c}$	Bulk concentration
c_0	Total concentration
$c_{\text{subscript}}$	Concentration specified by the subscript
c_V	Volume percent of particles in the electrolyte (Eq. (12.4))
C	Constant of the extended Debye–Hueckel equation (Eq. (1.18))
C	Capacitance
C_{dl}	Double-layer capacitance
C_{sc}	Space-charge layer capacitance
C_{GC}	Capacitance of the Gouy–Chapman layer
C_H	Capacitance of the Helmholtz layer
C_M	Madelung constant
C_{SB}	Sauerbrey constant
d	Distance
$d_\pm$	Mean diameter of the ions of an electrolyte
d_{GC}	Thickness of the Gouy–Chapman layer
d_i	Diameter of an ion
d_p	Particle diameter
D	Density function
D	Diffusion coefficient
D_{ad}	Surface diffusion coefficient of ad-atoms
D_{ox}	Density of oxidized species in the electrolyte
D_{ox}	Diffusion coefficient of the oxidized substance
D_{o_2}	Diffusion coefficient of oxygen
D_{red}	Density of reduced species in the electrolyte
D_{red}	Diffusion coefficient of reduced substance
e_0	Elementary charge (1.6022×10^{-19} C)
E	Electrode potential
E_0	Nernst potential

$E_0^{\ominus}$	Standard Nernst potential
$E_{0,A}$	Nernst potential of alloy component A
$E_{0,B}$	Nernst potential of alloy component B
$E_{A_xB_{1-x}}$	Potential of alloy A_xB_{1-x}
$E_{0,A_xB_{1-x}}$	Nernst potential of alloy A_xB_{1-x}
$E^{\ominus}_{0,A_xB_{1-x}}$	Standard Nernst potential of alloy A_xB_{1-x}
E_a	Activation energy
$E_a^{\#}$	Anodic activation energy
$E_{0,a}^{\#}$	Anodic activation energy, Galvani potential difference $\Delta\varphi = 0$
$E_c^{\#}$	Cathodic activation energy
$E_{0,c}^{\#}$	Cathodic activation energy, Galvani potential difference $\Delta\varphi = 0$
E_{AA}	Activation energy of separation of atom A from kink site position AA^*
E_{AB}	Activation energy of separation of atom B from kink site position AB^*
E_{BA}	Activation energy of separation of atom A from kink site position BA^*
E_{BB}	Activation energy of separation of atom B from kink site position BB^*
E_{cor}	Free corrosion potential
E_{dif}	Diffusion potential
E_e	Electron energy
E_{iD}	Potential of deposition of a metal in an array of lower dimension on a substrate metal surface (bulk metal, E_0)
E_F	Fermi energy
$E_{F,ad}$	Fermi energy of the adsorption layer
$E_{F,el}$	Fermi energy of the electrolyte
$E_{F,m}$	Fermi energy of the metal
E_{Fl}	Flade potential
E_{fb}	Flat band energy
$E_{fb,n}$	Flat band energy, conduction band
$E_{fb,p}$	Flat band energy, valence band
E_g	Gap energy
E_L	Lattice energy
E_{mc}	Lower energy limit of electron mobility of the conduction band
E_{vc}	Upper energy limit of hole mobility of the valence band
E_{mg}	Mobility gap energy
E_p	Passivation potential
E_{ph}	Photopotential
$E_{ph,max}$	Photopotential maximum
E_{pzc}	Potential of zero charge
E_s	Sample potential
E_{SHE}	Potential versus standard hydrogen electrode
E_{SCE}	Potential versus saturated calomel electrode
E_t	Tip potential
f	Rotation frequency
f_0	Resonance frequency
F	Faraday constant (9.648×10^4 C·mol^{-1})
F_{adh}	Adhesion force

F_{frict}	Friction force
F_{shear}	Shear force
F_{stag}	Stagnation force
g	Constant in the Frumkin–Temkin isotherm
g_{AB}	Selectivity constant of kink site position A^* for electrolyte component B relative to component A
g_{AC}	Selectivity constant of kink site position A^* for electrolyte component C relative to component A
g_{BA}	Selectivity constant of kink site position B^* for electrolyte component A relative to component B
g_{BC}	Selectivity constant of kink site position B^* for electrolyte component C relative to component B
g_{CA}	Selectivity constant of kink site position C^* for electrolyte component A relative to component C
g_{CB}	Selectivity constant of kink site position C^* for electrolyte component B relative to component C
g^*_{AB}	Selectivity constant AB, extrapolated to current density zero and rotation rate infinity
g^*_{BA}	Selectivity constant BA, extrapolated to current density zero and rotation rate infinity
G	Gibbs energy
$\bar{G}$	Partial molar Gibbs energy
$G^\ominus$	Standard Gibbs energy
h	Planck constant (6.626×10^{-34} J·s)
H	Enthalpy
$\bar{H}$	Partial molar enthalpy
$H^\ominus$	Standard enthalpy
i	Current density
i_0	Exchange current density
$i_{subscript}$	Current density specified by the subscript
I	Ionic strength
I_p	Peak current
I_t	Tunneling current
I_0	Light intensity
k^*	Constant ($1/c_0 k_{diff}$)
k_0	Pre-exponential factor
k_0	Exchange rate constant of an electrode reaction, c_{red}, c_{ox} = 1 mol dm^{-3}
$k_0^\ominus$	Standard exchange rate constant of an electrode reaction, a_{red}, a_{ox} = 1
$k_{subscript}$	Rate constant specified by the subscript
k_B	Boltzmann constant (1.3807×10^{-23} J·K^{-1})
k_{dif}	Diffusion constant ($k_{dif} = 2FD/\delta$)
k_{dir}	Recombination constant, direct recombination
$k_{n,gs}$	Recombination constant of electrons, recombination via gap states
$k_{n,ss}$	Recombination constant of electrons, recombination via surface states
$k_{p,gs}$	Recombination constant of holes, recombination via gap states

$k_{p,ss}$	recombination constant of holes, recombination via surface states
$\bar{k}$	Wave vector
k_x	Wave vector, x component
K	Constant of stationary dissolution of an alloy
K_{ad}	Langmuir-type adsorption constant
$K_{subscript}$	Equilibrium constant specified by the subscript
l	Characteristic length (in Reynold's number)
l_{sc}	Diffusion length, space-charge layer
m	Molality
$m^{\ominus}$	Standard molality (1 mol kg^{-1})
m_e	Electron mass (9.110×10^{-31} kg)
M	Molar mass
M_{an}	Molar mass of anion
M_{app}	Apparent molar mass
M_{solv}	Molar mass of solvent molecules
n	Quantum number
n	Stoichiometric number of electrons in a redox reaction
n_i, n_j	Electrochemical reaction orders
$n_{ox,j}$	Electrochemical reaction orders, substances at the oxidized side of the net electrode reaction
$n_{red,i}$	Electrochemical reaction orders, substances at the reduced side of the net electrode reaction
n_s	Surface density of electrons in the conduction band
N_A	Avogadros number (6.022×10^{23} mol^{-1})
N_A	Number of monomer units A in a copolymer
N_{ad}	density of adsorbed molecules
N_B	Number of monomer units B in a copolymer
N_{cb}	Density of energy levels, conduction band
N_{cc}	Density of charge careers (donors or acceptors)
N_{max}	Maximum density of adsorption places
N_{oc}	Density of occupied electron energy levels
N_p	Number of particles passing the diffusion layer per area and time unit (Eq. (12.7))
$N_{p,ad}$	Density of adsorbed particles
$N_{p,co}$	Particle density in the co-deposition layer
$N_{p,el}$	Particle density in the electrolyte
$N_{p,max}$	Maximum density of adsorbed particles
N_{tr}	Transference number disc-ring
N_{un}	Density of unoccupied electron energy levels
N_{vb}	Density of energy levels, valence band
p	Vapor pressure
p^0	Vapor pressure, pure phase
p_{AA}	Probability of formation of kink site position AA^*
p_{AA}	Probability of reaction of monomer unit A^{+*} with polymer chain molecule $R{-}A^{+*}$

p_{AB}	Probability of formation of kink site position AB^{*}
p_{AB}	Probability of reaction of monomer unit B^{+*} with polymer chain molecule $R-A^{+*}$
p_{AC}	Probability of formation of kink site position AC^{*}
p_{BA}	Probability of formation of kink site position BA^{*}
p_{BA}	Probability of reaction of monomer unit A^{+*} with polymer chain molecule $R-A^{+*}$
p_{BB}	Probability of formation of kink site position BB^{*}
p_{BB}	Probability of reaction of monomer unit B^{+*} with polymer chain molecule $R-A^{+*}$
p_{BC}	Probability of formation of kink site position BC^{*}
p_{CA}	Probability of formation of kink site position CA^{*}
p_{CB}	Probability of formation of kink site position CB^{*}
p_{CC}	Probability of formation of kink site position CC^{*}
p_s	Surface density of holes in the valence band
P	Empirical factor (Eq. (12.7))
P	Transition matrix
$P(E)$	Probability of particle incorporation, potential dependent
P_m	Power maximum
q_1, q_j	Charge of ions in a lattice
q_{ad}	Density of charge of specifically adsorbed ions
q_{el}	Density of charge on the electrolytic side of the double layer
q_m	Density of charge on the metal side of the double layer
Q	Charge
r	Radius
r	Rotation rate
r_0	Distance between a central ion and nearest neighbors
r_1	Ratio k_{AA}/k_{AB} (Eq. (11.16))
r_2	Ratio k_{BB}/k_{BA} (Eq. (11.16))
$r_{1,j}$	Distance of ions in an ionic lattice
$r_{subscript}$	Rate of reaction specified by the subscript
r_B	Bjerrum radius
r_i	Distance between a central ion and ions i
r_j	Distance between a central ion and ions j
r_p	Particle radius
R	Universal gas constant ($8.314\ J \cdot K^{-1} \cdot mol^{-1}$)
P	Resistance
R_{ct}	Charge transfer resistance
R_d	Diffusion resistance
R_{el}	Electrolyte resistance
R_{sc}	Space-charge layer resistance
S	Overlap integral
S	Entropy
$\bar{S}$	Partial molar entropy
$S^{\ominus}$	Standard entropy

t	Time
t_-	Transference number (transport number) of anions
t_+	Transference number (transport number) of cations
T	Kelvin temperature
T_c	Critical temperature
T_g	Glass temperature
u_-	Mobility of anions
u_+	Mobility of cations
u_E	Electrophoretic mobility
$U, \Delta U$	Internal energies
U	Cell voltage
U_{ph}	Photocell voltage
$U_{ph,m}$	Photocell voltage at maximum power
$U_{ph,sc}$	Photocell voltage, short circuit
v	Abbreviation for x_A/x_B in the alloy
v	Ratio N_A/N_B (Eq. (11.16))
v	Flow rate
v	Scan rate
v_f	Free volume in a polymer
v_g	Free volume in a polymer at the glass temperature
V_A	Molar surface of the Wigner–Seitz cells of an element A
V_{bulk}	Molar volume of a bulk metal
V_{LJ}	Lennart–Jones potential
V_{upd}	Molar volume of an UPD metal
w	Abbreviation for c_A/c_B in the electrolyte
w_{sc}	Thickness, space-charge layer
x	Distance from electrode surface
$x_{subscript}$	Mole fraction specified by the subscript
x_{crit}	Critical ratio of particle diameter (Eq. (12.12))
x_m	Mass percent of particles in the metal matrix (Eq. (12.5))
x_V	Volume percent of deposited particles (Eq. (12.4))
z	Stoichiometric number of electrons, metal ion electrode
z	Number of electrons per monomer unit
z_-	Charge of an anion
z_+	Charge of a cation
z_{ad}	Charge of specifically adsorbed ions
z_i	Charge of an ion i
Z	Cell Constant
Z	General thermodynamic function
Z	Impedance
$\bar{Z}$	General partial molar thermodynamic function
$\|Z\|$	Modulus of impedance
$Z_{subscript}$	Impedance specified by the subscript
$[Ag_i^o]$	Density of interstitials in a metal lattice (e.g., silver)
$[V'_{Ag}]$	Density of vacancies in a metal lattice (e.g., silver)

$[A^*]$	Surface density of kink site positions A^*
$[AA^*]$	Surface density of kink site positions AA^*
$[AB^*]$	Surface density of kink site positions AB^*
$[B^*]$	Surface density of kink site positions B^*
$[BA^*]$	Surface density of kink site positions BA^*
$[BB^*]$	Surface density of kink site positions BB^*
$[ksp]$	Surface density of kink site positions
Δf	Frequency shift
Δf^*	Complex frequency shift
$\Delta G_{A_xB_{1-x}}$	Gibbs energy of alloy A_xB_{1-x}
$\Delta G^{\ominus}_{A_xB_{1-x}}$	Standard Gibbs energy, alloy A_xB_{1-x}
$\Delta \bar{G}_{upd}$	Partial molar Gibbs energy of a metal in its UPD modification
$\Delta_{for}G$	Gibbs energy of formation
$\Delta_{for}G_{upd}$	Gibbs energy of formation of an UPD layer
$\Delta_{for}G_{A_xB_{1-x}}$	Gibbs energy of formation of A_xB_{1-x}
$\Delta_{for}H$	Enthalpy of formation
$\Delta_{for}H_{upd}$	Enthalpy of formation of an UPD layer
$\Delta_{for}H_V$	Enthalpy of formation of vacancies
$\Delta_{fus}H$	Enthalpy of melting
$\Delta_{ion}H$	Enthalpy of ionization
$\Delta_{lat}H$	Lattice enthalpy
$\Delta_{solv}H$	Enthalpy of solvation
$\Delta_{sub}H$	Enthalpy of sublimation
$\Delta_{sub}H_{bulk}$	Sublimation enthalpy of bulk metal
$\Delta_{sub}H_{upd}$	Sublimation enthalpy of UPD layer
$\Delta_{vap}H$	Enthalpy of vaporization
$\Delta_{for}S$	Entropy of formation
$\Delta_{for}S_{upd}$	Entropy of formation of an UPD layer
Δm	Mass change
Δm_{Me}	Deposited metal mass per area and time unit (Eq. (12.5))
Δm_p	Deposited particle mass per area and time unit (Eq. (12.5))
Δn_{WS}	Difference of electron density on the surface of the Wigner–Seitz cells of elements A and B
Δw	Half width of resonance frequency
Δz_{ad}	Partial charge transfer of an ion in the adsorption process
Δz_i	Partial charge transfer in the *i*th partial reaction step of deposition
$\Delta \chi$	Difference of electronegativities
$\Delta \varphi$	Galvani potential difference
$\Delta \varphi_{dl}$	Potential drop across the double layer
$\Delta \varphi_{GC}$	Potential drop between Helmholtz plane and electrolyte
$\Delta \varphi_H$	Potential drop between metal and Helmholtz plane
$\Delta \varphi_{sc}$	Galvani potential difference, space-charge layer
$\Delta \varphi_{\underline{sc}}$	Potential drop in the space-charge layer
Δz_{sep}	Charge transfer in the separation reaction
Φ	Work function

Φ_s	Work function of the sample metal in tunneling spectroscopy
Φ_t	Work function of the tip in tunneling spectroscopy
Γ_{ad}	Surface concentration of specifically adsorbed ions
Γ_e	Surface concentration of electrons
Γ_i	Surface concentration of component *i*
$\Gamma_{i(s)}$	Relative surface excess of component *i*
$\Gamma_{i,a}$	Surface concentration of anion *i*
$\Gamma_{i,c}$	Surface concentration of cation *i*
Γ_{max}	Maximum of surface concentration
Γ_s	Surface concentration of solvent
Λ_-	Molar conductivity of anions
Λ_+	Molar conductivity of cations
Λ_∞	Molar conductivity at infinite dilution
Λ_m	Molar conductivity
Θ	Angle (small angle grain boundaries)
Θ	Angle between surface tension vectors σ_{mp}/σ_{ep}
Θ	Surface coverage
Θ_{ad}	Surface coverage, ad-atoms
T	Surface stress tensor
α	Degree of dissociation
α	Number of solvent molecules in the apparent mass (Eq. (11.4))
α_a	Charge transfer coefficient, anodic partial current density
α_c	Charge transfer coefficient, cathodic partial current density
$\alpha_{a,i}$	Charge transfer coefficient of the *i*th partial reaction step of dissolution
$\alpha_{c,i}$	Charge transfer coefficient of the *i*th partial reaction step of deposition
α_{des}	Charge transfer coefficient of a desorption reaction
α_{dis}	Charge transfer coefficient of a dissolution reaction
α_{sep}	Charge transfer coefficient of the separation reaction
α_f	Expansion coefficient of the free volume in a polymer
α_{sc}	Absorption coefficient, space-charge layer
β	A system constant (Eq. (12.4))
β	Constant of the exponential film growth with field strengths (Eq. (10.41))
β	Number of cations in the apparent mass (Eq. (11.4))
χ	Orbital function
χ	Surface (Volta) potential
δ	Thickness, general
δ_τ	Thickness growths of a metal layer during particle residence time (Eq. (12.4))
δ_{ad}	Average displacement of ad-atoms
δ_{crit}	Critical thickness of a metal layer for particle inclusion
δ_d	Thickness of the diffusion layer
ε	Relative dielectric permittivity
ε_0	Permittivity of the vacuum ($8.854 \times 10^{-12}\ J^{-1}C^2m^{-1}$)
ε_e	Elastic surface strain tensor
ε_p	Plastic surface strain tensor

ε_{tot}	Total surface strain tensor
ϕ	Phase shift
γ	Activity coefficient
γ	Electrosorption valency
$\gamma_{\pm}$	Mean activity coefficient
γ_{i}	Activity coefficient of an ion
γ_{S}	Function derived by Everett and Couchman, suggested name "surface tension"
η	Overpotential
η	Viscosity
η_{d}	Diffusion overpotential
$\eta_{\chi\tau}$	Charge transfer overpotential
φ	Galvani potential
$\varphi_{1/2}$	Enthalpy of an atom in a kink site position
φ_{AA}	Bond energy AA
φ_{AB}	Bond energy AB
φ_{BA}	Bond energy BA
φ_{BB}	Bond energy BB
φ_{H}	Galvani potential at the Helmholtz plane
φ_{m}	Galvani potential in the metal
$\varphi_{Me\text{–}Me}$	Bond energy of an atom in a kink site position to the nearest neighbor
φ_{x}	Galvani potential at a point x from the Helmholtz plane
$\varphi^{*}_{Me\text{–}Me}$	Bond energy of an atom in a kink site position to the nearest neighbor, corrected for long-range forces
κ	Constant describing the decay of the potential in front of an electrode surface (Gouy–Chapman) or with increasing distance from an ion (Debye)
κ_{2D}	Two-dimensional compressibility
λ	Partial charge transfer coefficient
λ	Thermal conductivity
λ	Wave length
μ	Dipole moment
μ	Shear modulus
$\mu_{1/2}$	Chemical potential of an atom in a kink site position
$\mu_{subscript}$	Chemical potential specified by the subscript
μ_{-}	Mobility of electrons
μ_{+}	Mobility of holes
$\tilde{\mu}$	Electrochemical potential
$\tilde{\mu}_{subscript}$	Electrochemical potential specified by the subscript
ν	Frequency
ν	Kinematic viscosity
ν_{0}	A system constant
ν_{at}	Atomic volume
$\nu_{subscript}$	Stoichiometric number specified by the subscript
ρ	Specific mass (density)
ρ	Specific resistivity
ρ_{M}	Specific mass of deposited metal

ρ_p	Specific mass of co-deposited particle
σ	Specific conductivity
σ_ε	Specific conductivity of electrons
σ_h	Specific conductivity of holes
σ	Gibbs surface energy (surface tension)
σ_{ep}	Surface tension between electrolyte and particle
σ_{max}	Surface tension at the potential of zero charge
σ_{me}	Surface tension between metal surface and electrolyte
σ_{mp}	Surface tension between metal surface and particle
τ	Transition time of diffusion
τ	Residence time
τ_{AA}	Residence time of atom A in kink site position AA^*
τ_{AB}	Residence time of atom B in kink site position AB^*
τ_{ad}	Residence time of ad-atoms
τ_{BA}	Residence time of atom A in kink site position BA^*
τ_{BB}	Residence time of atom B in kink site position BB^*
τ_p	Residence time of particles
ω	Angular frequency
ω	Half width
ψ	Inner potential
ψ	Wave function
ξ	Zeta potential

Preface

Electrochemistry, long-time corner stone for fundamental chemistry and physics, now plays an important role in many areas of applied science and technology. A very broad range of applications of electrochemical principles and technologies is found in materials science. Electrochemical deposition of metals and alloys, formation of oxide films and semi-conductors, corrosion and corrosion protection, new polymer materials that can switch between metallic conductivity and semi-conducting properties, and new applications in fast-evolving nanotechnologies are just some of the examples. To fully comprehend them, however, one should understand the basic electrochemical concepts. Therefore, these fundamentals are introduced and described in the first six chapters. The book begins with a look at the properties of electrolytes and electrodes whereby solid-state background, especially for electrode materials, is emphasized. Next, there is a discussion of the potential differences that develop when electrodes and electrolytes are combined. How thermodynamic functions of solids can be determined from potential measurements is explained as well. The properties of the electrochemical double layer are described on molecular and atomic levels. The kinetics of the two main reaction steps, mass transport and charge transfer, is also treated. The newcomer as well as the more experienced reader should find this introduction to the basic principles and subsequent discussion of the problems and examples in the first-half of the book, both readily comprehensible and informative.

The second-half of the book, which treats various applications of electrochemistry in several areas of materials science, introduces and discusses the principles and the special literature. The quickly evolving specific applications are mentioned but not treated in greater detail. The book's goal is to communicate a working knowledge of the fundamentals to people from all different areas and backgrounds and to guide the reader through an array of electrochemistry applications, strategies, and techniques in various scientific environments. The author's long teaching and research experience at a technical university, particularly in materials science, has also played a role in the book's conception and success.

The author would like to thank several people for their help in the preparation of the manuscript, in particular for their proofreading of individual chapters, valuable discussions, and critical comments. They are Dr. Andreas Bund, Dr. Hartmut Dietz, Dr. Karl Doblhofer, Prof. Dr. Ulrich Guth, Dr. Thomas Heine, Dr. Ursula Rammelt, and Dr. Holger Sahrhage. He would also like to thank Dr. Pamela Winchester for language polishing and editing the text.

Waldfried Plieth, Technische Universität Dresden, Germany

– 1 –

Electrolytes

Conductivity is based on the free mobility of electrons in a metal, but some substances can also transport electricity without free electrons. Instead free mobile ions transport the current and are therefore called ionic conductors. The first ionic conductors were aqueous electrolytes. Later on, polymer and solid electrolytes were discovered and the name electrolytes became a synonym for ionic conductors.

The physical chemistry of electrolytic solutions is a special area of physical chemistry with a large number of reference literatures. Classical descriptions are given in the books of Harned and Owen[1] or Robinson and Stokes.[2] A more recent advanced treatment is found in the book of Barthel, Krienke, and Kunz.[3] The special problems of ionic-conducting polymers and ionic solid electrolytes are described in various other reviews. Gray[4] described polymer electrolytes. A classical treatment of ionic solid electrolytes is the book by Rickert[5] or the Kudo and Fueki[6] compilation. Because these materials are used in batteries and fuel cells, there is much literature for this research field including such detailed reviews in the book by Julien and Nazri.[7] Another source for details and data compilations is the CRC *Handbook of Solid State Electrochemistry*.[8]

1.1 LIQUID ELECTROLYTE SOLUTIONS

The dissolution of a salt in water was described by Arrhenius[9] as a dissociation of the salt in positive and negative ions of free mobility. This was originally a difficult suggestion to accept because salts were very stable substances. Therefore, for a long time the theory was called Arrhenius' dissociation hypothesis. Later it became clear that the process was a reaction of the solvent water with the ions and that the ion–dipole interaction provided the necessary energy for the compensation of the strong lattice energies. For example, the process can be formulated for sodium chloride by the equation

$$NaCl + (m + n)H_2O \rightarrow Na^+_{mH_2O} + Cl^-_{nH_2O} \tag{1.1}$$

The enthalpy of solution connected with the solution process is relatively small because it is just the difference between the lattice enthalpy and the enthalpy of solvation of cations and anions. For several years the dissociation theory explained most of the experimental

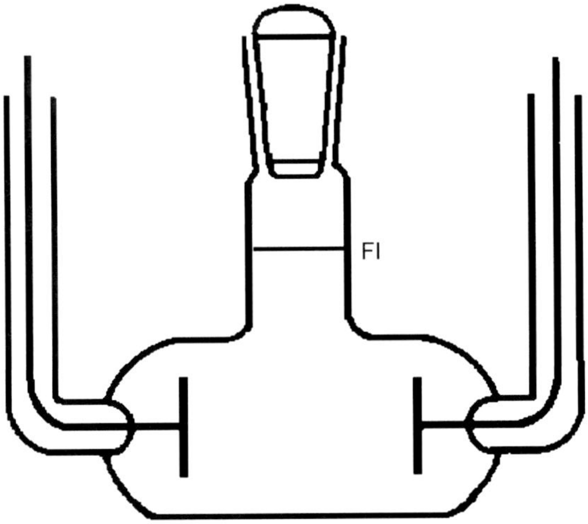

Figure 1.1 Conductivity cell and Fl filling level. The conductivity is usually measured with an alternating current to prevent polarization effects at the electrodes.

observations. The specific conductivity σ was defined in analogy to the conductivity of metals as the reciprocal value of the specific resistivity ρ. The specific resistivity ρ is usually measured in $\Omega \cdot$cm and the specific conductivity in $\Omega^{-1} \cdot \text{cm}^{-1} = \text{S} \cdot \text{cm}^{-1}$.

A typical cell for the measurement of the conductivity is shown in Figure 1.1. The cell is filled with a solution of an electrolyte with known conductivity, e.g., 0.1 mol $\cdot$ dm^{-3} KCl solution, $\sigma = 0.0129$ S $\cdot$ cm^{-1} up to a marked filling level. After the cell constant $\text{Z} = \text{R} \cdot \sigma$ is determined by the measured resistance of the cell, the electrolyte of the unknown conductivity is filled in. The specific conductivity of this electrolyte can be calculated with the newly measured resistance.

A molar conductivity is obtained if the specific conductivity σ is divided by the concentration c

$$\Lambda_m = \frac{\sigma}{c} \tag{1.2}$$

If σ is measured in S $\cdot$ cm^{-1}, c should be measured in mol $\cdot$ cm^{-3}. Then Λ_m is measured in S $\cdot$ cm^2 $\cdot$ mol^{-1}. The conductivity of aqueous electrolytes is related to the mobility of cations and anions, u_+ and u_-. Mobility is the rate of motion in a field of one V $\cdot$ cm^{-1}. The independent conductivity of anions and cations was postulated

$$\Lambda_+ = Fu_+ \tag{1.3}$$

$$\Lambda_- = Fu_- \tag{1.4}$$

(F is Faraday's constant). The molar conductivity is the sum of the ion conductivities

$$\Lambda_m = \Lambda_+ + \Lambda_- \tag{1.5}$$

The mobility was also used to define transference numbers (also called transport numbers) for anions and cations (Hittorf[10])

$$t_+ = \frac{u_+}{u_+ + u_-} \tag{1.6a}$$

$$t_- = \frac{u_-}{u_+ + u_-} = 1 - t_+ \tag{1.6b}$$

The transference numbers describe the contribution of cations and anions to the current.

For strong electrolytes (electrolytes with a nearly complete dissolution in cations and anions) Kohlrausch's law described the observed concentration dependence of the molar conductivity

$$\Lambda_m = \Lambda_\infty - B\sqrt{c} \tag{1.7}$$

Plotting Λ_m over $\sqrt{c}$ the molar conductivity at infinite dilution Λ_∞ can be determined.

The so-called weak electrolytes did not follow Kohlrausch's law. This could be partially explained by incomplete dissociation. The dissociation equilibrium of a salt CA (C cation, A anion) in a diluted electrolyte, where activities a can be approximately substituted by concentrations c, is described by the equation

$$\frac{a_C \cdot a_A}{a_{CA}} \approx \frac{c_C \cdot c_A}{c_{CA}} = K_c \tag{1.8}$$

With the introduction of the degree of dissociation α Eq. (1.8) can be written as

$$\frac{\alpha^2 \cdot c_0}{1 - \alpha} = K_c \tag{1.9}$$

The concentration c_0 is the total concentration of the salt. The degree of dissociation can be described by the equation

$$\alpha = \frac{\Lambda_m}{\Lambda_\infty} \tag{1.10}$$

Ostwald's law[11] is derived by combining Eqs. (1.9) and (1.10) and can be written in the following form:

$$\frac{1}{\Lambda_m} = \frac{1}{\Lambda_\infty} + \frac{1}{\Lambda_\infty^2 K_c} \cdot (c_0 \Lambda_m) \tag{1.11}$$

By plotting $1/\Lambda_m$ over $c_0\Lambda_m$ the characteristic constants Λ_∞ and K_c can be determined.

The reason for deviations of strong electrolytes from ideal behavior is ion–ion interaction. In 1923, Debye and Hückel[12] published a theoretical description of the interaction of ions in the electrolyte. The principal concept of this theory was the ionic atmosphere formed by charge of opposite sign surrounding each ion. The radius of this atmosphere was inversely dependent on a summarized ionic concentration and the ionic strength I was defined as

$$I = \frac{1}{2}\sum_i c_i z_i^2 \tag{1.12}$$

As a result, an approximate equation for the activity coefficient γ_i of an individual ion i was derived.

$$\log \gamma_i = -\frac{1.8246}{(\varepsilon T)^{3/2}} z_i^2 \sqrt{I} \tag{1.13}$$

This equation can be used up to an ionic strength of 10^{-2} mol · dm^{-3}. For water (relative dielectric permittivity $\varepsilon = 78.54$) at 25 °C the following equation can be used

$$\log \gamma_i = -0.509 z_i^2 \sqrt{I} \tag{1.14}$$

The dimension of the ionic strength in this equation is mol · dm^{-3}. The mean ionic activity coefficient is given by

$$\log \gamma_{\pm} = -0.509 \left| z_+ z_- \right| \sqrt{I} \tag{1.15}$$

A more advanced approximation that is valid for water at 25 °C up to ionic strength of 10^{-1} mol · dm^{-3} also takes into account the diameter of the ions d_i

$$\log \gamma_i = -\frac{0.509 z_i^2 \sqrt{I}}{1 + 3.29 d_i \sqrt{I}} \tag{1.16}$$

The dimension of the ionic strength in this equation is mol · dm^{-3}. The dimension of d_i is nm. For the mean ionic activity the following approximation with a mean ionic diameter $d_{\pm}$ can be used

$$\log \gamma_{\pm} = -\frac{0.509 \left| z_+ z_- \right| \sqrt{I}}{1 + 3.29 d_{\pm} \sqrt{I}} \tag{1.17}$$

For higher ionic strengths an additional term is added

$$\log \gamma_{\pm} = -\frac{0.509 \left| z_+ z_- \right| \sqrt{I}}{1 + 3.29 d_{\pm} \sqrt{I}} + C \cdot I \tag{1.18}$$

C and $d_{\pm}$ can be adapted to known activity data. An interpretation of the term C is based on salting out effects. A comparison of different equations for ion activities is shown in Figure 1.2.

In the extension of the Debye–Hückel theory by Onsager[13] an equation similar to Kohlrausch's law was derived

$$\Lambda = \Lambda_{\infty} - \left(B_1 \Lambda_{\infty} + B_2 \right) \sqrt{I} \tag{1.19}$$

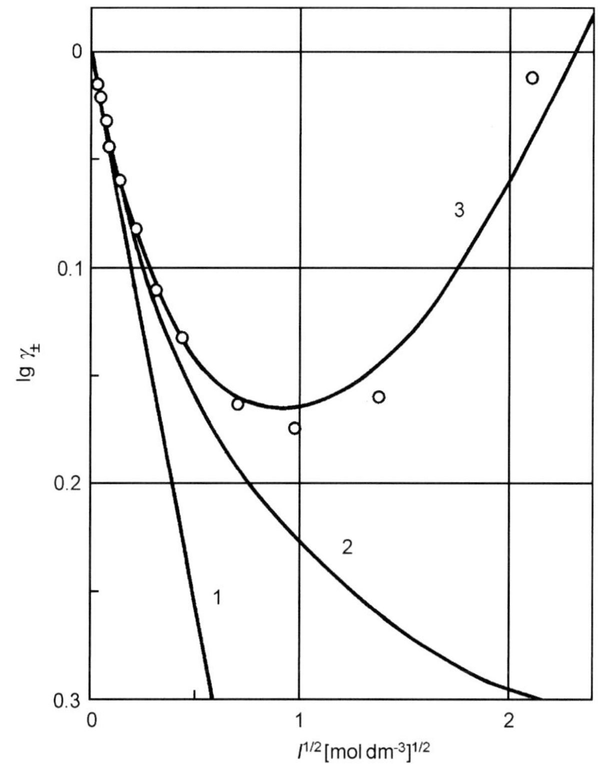

Figure 1.2 Comparison of experimental values of activity coefficients for NaCl with calculations using different approximations. Lg $\gamma_{\pm}$ is plotted versus the square root of the ionic strength *I*. (1) Eq. (1.15), (2) Eq. (1.17) with $d_{\pm} = 0.3$ nm, and (3) Eq. (1.18) with $d_{\pm} = 0.4$ nm and $C = 0.55$ $dm^3 \cdot mol^{-1}$.

The constants B_1 and B_2 for one/one valent electrolytes have the values

$$B_1 = \frac{8.204 \cdot 10^5}{(\varepsilon T)^{3/2}} \tag{1.20}$$

$$B_2 = \frac{82.5}{\eta(\varepsilon T)^{1/2}} \tag{1.21}$$

In aqueous electrolytes (relative dielectric permittivity $\varepsilon = 78.54$, viscosity $\eta = 0.8397$ P) and at 25 °C the following equation is obtained

$$\Lambda = \Lambda_{\infty} - \left(0.2290\Lambda_{\infty} + 60.32\right)\sqrt{I} \tag{1.22}$$

The two terms in the constant B of Kohlrausch's law represent the two correction effects in Onsager's approach: the electrophoretic effect where the ion moves against the flux in opposite direction of the ionic atmosphere and the relaxation effect in which the polarization of the ionic atmosphere creates a field that slows down the motion of the ion. Onsager's equation is a limiting law for dilute solutions ($<10^{-2}$ mol · dm^{-3}). It should be mentioned that deviations from the given equations were observed for very high electric fields ($>10^6$ V · m^{-1}) and for alternating electric fields at high frequencies ($>10^6$ Hz) (Wien and Debye-Falkenhagen effects, respectively).

It is remarkable that an approach (Bjerrum's association theory[14]), which is based on ion pair formation, was very successful for very concentrated solutions. The Bjerrum radius r_B

$$r_B = \frac{|z_+ z_-| e_0^2}{2\varepsilon k_B T} \tag{1.23}$$

describes a region around the central ion in which an association between the central ion and an ion of opposite sign is postulated, with e_0 the charge of the electron, ε the relative dielectric permittivity, and k_B the Boltzmann constant. Ions with distances smaller than the Bjerrum radius are considered neutral species.

1.2 IONIC MELTS

Separation of the ions in an ionic crystal is also possible by thermal energy. At higher temperatures the crystals melt which gives the ions a certain amount of free mobility. While pure salts have relatively high melting temperatures, mixtures of different salts form eutectic systems with considerably lower melting points. The melt of an ionic crystal is, however, a complicated system consisting of a variety of complexes. In classical ionic melts (alkali halide melts) these complexes are exclusively ionic, but the ions may also be embedded in a matrix of non-ionic molten components (glass forming molten salts). A new class of ionic melts with low melting points has come into focus for research, which is room temperature ionic melts or ionic liquids.

1.2.1 Alkali halide melts

After melting, alkali halides consist exclusively of metal cations and halide anions. It should be easy to give a description of the properties of such a simple system. Problems arise though because of the extraordinary number of complexes possible in these systems. Even in a pure salt melt like LiCl, NaCl, etc., several complexes can be formulated. The postulation that the preferential coordination number in the melt is four and the dominating species are MeX_4^{3-} and XMe_4^{3+} helps restrict the number of complexes (Smirnov *et al.*[15,16]). Furthermore, single ions can exist that are not bound to a specific complex. The first step of the transport mechanism is to separate a single ion from the complex so that it occupies a free space between the complexes. The next step is to move the single ion from place to place. The single ion disappears when it finds an unoccupied place in a complex. This mechanism can be compared with the transport in solids by interstitials (Section 1.4). This mobility model of ions in melts was also used to explain other properties like viscosity.

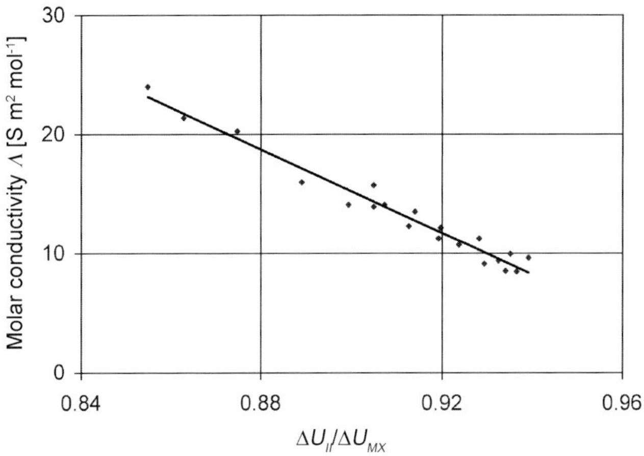

Figure 1.3 Molar conductivity of various alkali halides at T = 1100 K (fluorites at T = 1270 K) versus the ratio of the bond energy of individual and complex ions ΔU_{II} to the internal energy of the melt ΔU_{MX} (according to Khokhlov[17]). (Reproduced with permission from Ref. [18], © 1998 by Trans Tech Publications.)

The activation energy of the entire process is determined by the relative magnitude of the bond energy of individual and complex ions ΔU_{II} compared with the internal energy of the melt ΔU_{MX}. The molar conductivity versus the ratio of pair energy to melt energy is shown in Figure 1.3.

In mixed electrolytes like LiCl/KCl several more complexes should be considered but the general concept is a way to also understand the transport properties of such mixtures.

1.2.2 Glass forming molten salts

In the previous chapters liquid phases consisting exclusively of charged particles were considered. In some molten salts a separation into an ionic phase and a neutral glasslike matrix is observed. An example for this kind of ionic melts is a melt of 0.4 mole $Ca(NO_3)_2$ + 0.6 mole KNO_3 (CKN). Another example is CdF_2/LiF/AlF_3/PbF_2. Some glasses with metal salts as one component also belong, in the molten state, to this type of melts. Li^+ ion-containing glasses have been intensively investigated. An example is 0.6 mole LiCl/0.7 mole Li_2O/1.0 mole B_2O_3.

The structure of the liquid phase is the dominating property for determining the transport properties in these systems. As a result, structure sensitive methods are important for investigating these systems. Infrared and Raman spectroscopy and X-ray scattering or neutron scattering are some examples. High-frequency conductivity measurements have become exceptionally effective, because the different phases should show different time domains. Conductivity, shear viscosity, and other transport properties were investigated as a function of temperature with these methods. As an example, the frequency dependence of the conductivity is shown in Figure 1.4 as a function of frequency (after Funke *et al.*[18,19]). Considerable changes of the properties were observed in ultra high-frequency regions

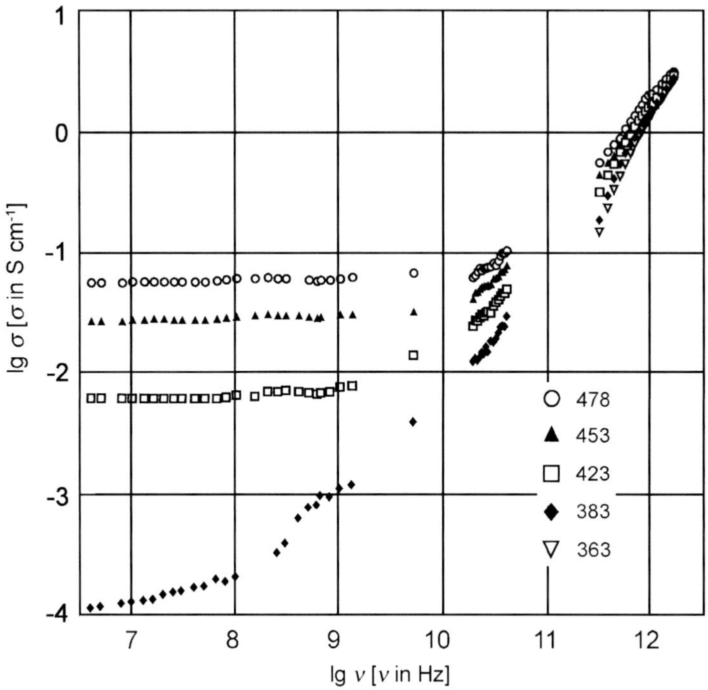

Figure 1.4 Frequency dependence of the real part of the complex conductivity of 0.4 mole $Ca(NO_3)_2$ + 0.6 mole KNO_3 (CKN) at different temperatures.[18,19] (Reproduced with permission from Ref. [17], © 1998 by Trans Tech Publications.)

(above 10^8 Hz). Ngai[20,21] has given theoretical descriptions of the dynamic phenomena in such systems at high frequencies. Reproduced with permission.

1.2.3 Ionic liquids

A new class of ionic melts is called room temperature melts or ionic liquids. They consist of organic ions or a mixture of organic and inorganic ions. A number of reviews have been published.[22,23] Examples of cations and anions are shown in Figures 1.5A and B, respectively.

Ionic liquids posses some remarkable properties:

- They are liquid at temperatures between room temperature and 100 °C.
- They are inflammable and have high thermal stability.
- Newly developed ionic liquids also have hydrophobic properties that allow handling in air.
- They have a negligible vapor pressure.
- Their large electrochemical potential window (Figure 1.6) allows electrochemical processes that are not possible in aqueous electrolytes such as deposition of metals that are not deposable in aqueous electrolytes.

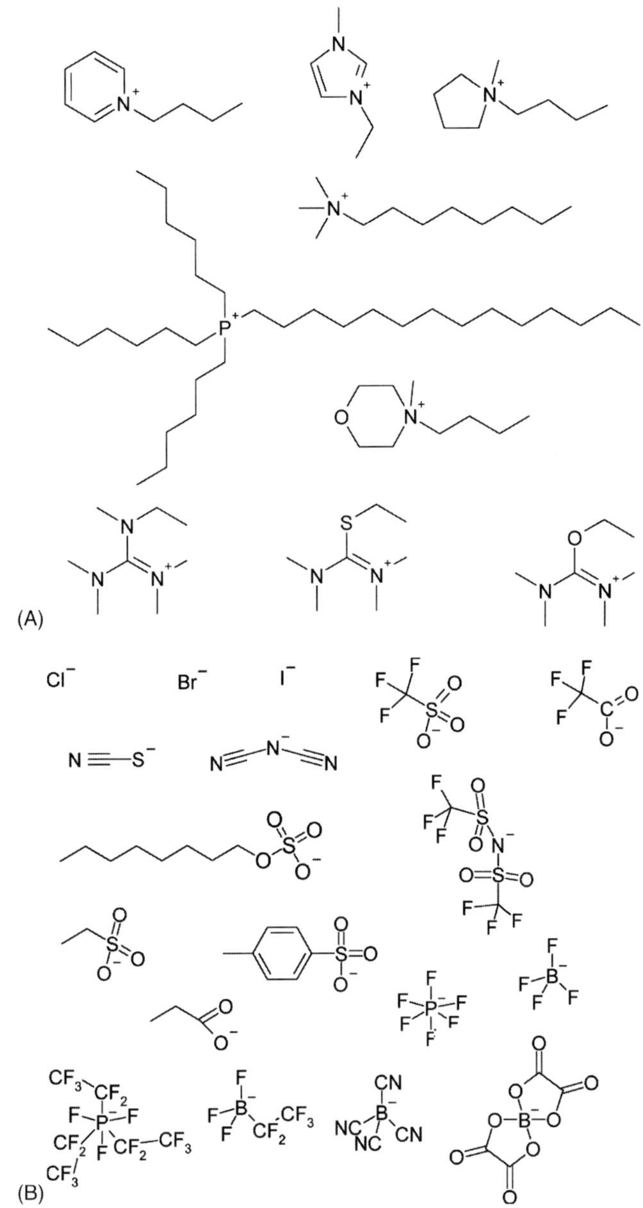

Figure 1.5 (A) Examples of cations in ionic liquids. (B) Examples of anions in ionic liquids. (Reproduced from Ref. [24] with permission of the author.)

The experimental simplicity of these materials has led to intensive investigation. An early electrochemical example was the deposition of aluminum and several aluminum alloys.[25] The addition of $AlCl_3$ leads to an acidic or basic character of the melt. This influences the potential window (Figure 1.7).

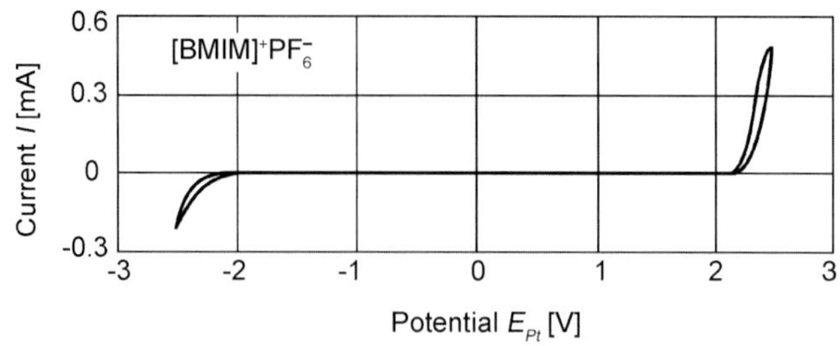

Figure 1.6 Electrochemical potential window of neutral 1-butyl-3-methyl imidazolium hexafluorophosphate ($[BMIM]^+PF_6^-$). (Reproduced with permission from Ref. [25], © 2002, Elsevier.)

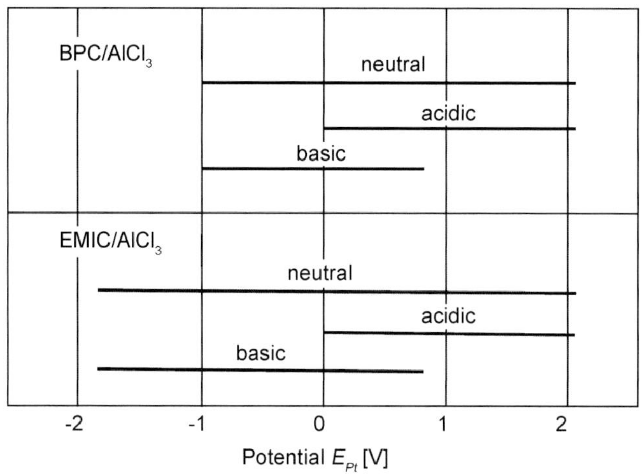

Figure 1.7 Electrochemical potential windows of *N*-butyl-pyridinium chloride (BPC) and 1-ethyl-3-methyl imidazolium chloride (EMIC) in the presence of $AlCl_3$ and depending on the acidity of the ionic melt, different potential windows exist. (Reproduced with permission from Ref. [25], © 2002, Elsevier.)

1.3 IONIC CONDUCTANCE IN POLYMERS

By mixing a polymer with a salt, one can form ionic-conducting polymers. The polymer can react with the ionic components partially dissociating the salt (polymer electrolytes). An ionic-conducting polymer can also be prepared by mixing a solution of an alkali salt in an organic solvent with a polymer, which forms a gel polymer electrolyte. The polymer itself can contain ionic groups in which case only the counter ion is mobile and the polymer has ionic exchange properties (ion exchangers). The application of these materials in lithium batteries has led to intensive interest in this class of materials.[26]

1.3.1 Polymer electrolytes

Some polymers with hetero-atoms carrying free electron pairs have solvating power to dissolve metal salts, e.g., alkali halides. An example is $LiClO_4$ dissolved in polyethylene oxide (PEO, Figure 1.8). The polymer has a helix like structure with oxygen atoms coordinated to Li cations. The anions stabilize the chain structure. The polymer and the $LiClO_4$ can form different complexes. In this context it is a common practice to describe the composition by the number of hetero-atoms per cation (PEO_nLiClO_4). The polymer has crystalline and amorphous regions. The conductivity increases with an increase in volume of the amorphous regions. A reference point is the glass transition temperature T_g (transition from crystalline structure to amorphous structure): the lower the glass transition temperature, the higher the conductivity at a given temperature.

The temperature dependence of the conductivity is described by the Vogel–Tamman–Fulcher equation.[27,28]

$$\sigma = \frac{A}{\sqrt{T}} \exp\left(-\frac{E_a}{k_B(T - T_g)} \right) \tag{1.24}$$

A is a constant related to the free carrier concentration, T_g the ideal glass transition temperature, E_a a pseudo activation energy. Examples of conductivity temperature relations are shown in Figure 1.9.

Several theoretical descriptions of the temperature dependence of the conductivity of these systems have been given. An example is the free-volume theory.[30] The conductivity is proportional to a "free volume", defined as the deviation of the real volume of the

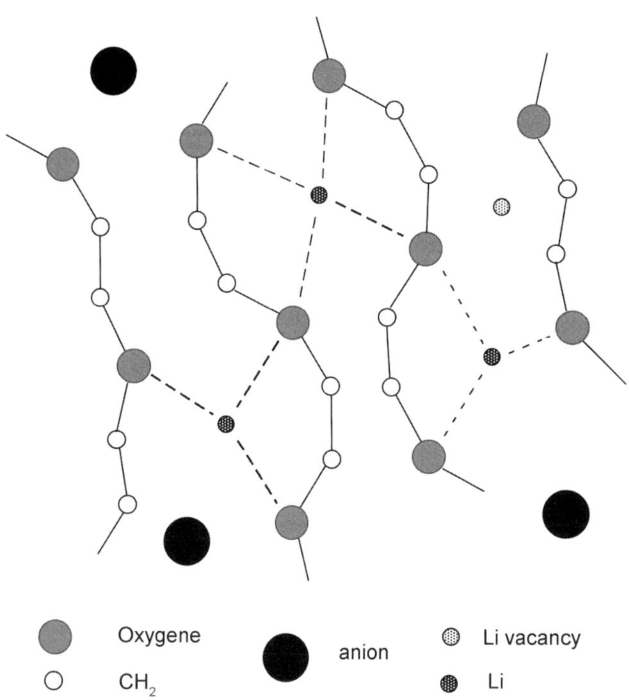

Figure 1.8 Polyethylene oxide/$LiClO_4$ polymer electrolyte structure.

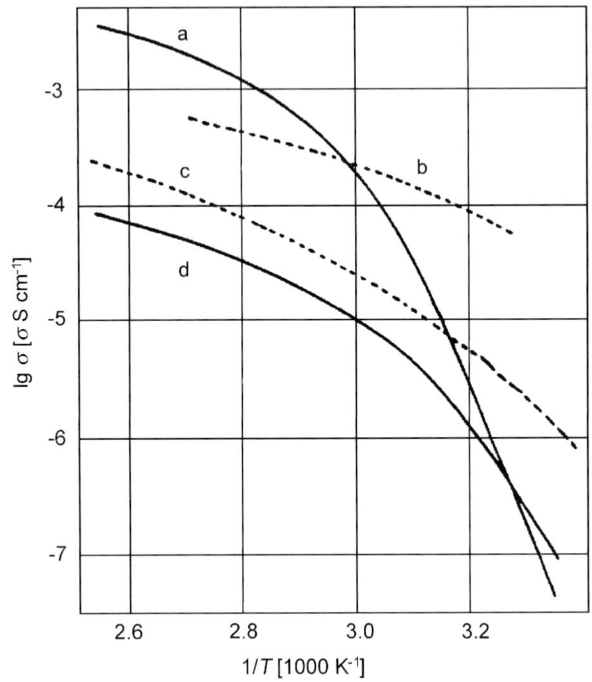

Figure 1.9 Temperature dependence of the conductivity of some examples of polymer electrolytes: (a) poly(ethylene oxide), $(\text{-}CH_2\text{-}CH_2\text{-}O\text{-})_n\text{-}LiClO_4$ (12:1), (b) poly(bis-methoxy-ethoxy-ethoxy) phosphazene $[(CH_3O\text{-}C_2H_4O\text{-}C_2H_4O)_2PN\text{-}]_n\text{-}LiCF_3SO_3$ (4:1), (c) poly(propylene oxide) $[\text{-}CH_2(CH_3)\text{-}CH_2\text{-}O\text{-}]_n\text{-}LiCF_3SO_3$ (9:1), and (d) poly(ethylene-amine $(\text{-}CH_2\text{-}CH_2\text{-}NH\text{-})_n\text{-}LiCF_3SO_3$ (9:1). (Reproduced with permission from Ref. [29], © 1994, Kluwer Academic Publishers.)

polymer from the theoretical volume (van der Waals volume). The free volume increases with temperature. An equation similar to the VTF-equation (Eq. (1.24)) could be derived.

Conductivity is increased by increasing the volume of the amorphous phase. This is achieved by using larger anions like $LiCF_3SO_3$ or by adding gel-forming components to the polymer electrolyte, e.g., polyethylene glycole.[31,32] This will be discussed in more detail in the following chapter.

1.3.2 Gel polymer electrolytes

The addition of gel-forming components (plasticizers) to polymer electrolytes (see the above) produces gel like structures. Therefore, this type of ion-conducting polymers can also be described as gel polymer electrolytes. Gel polymer electrolytes can also be prepared, if a solution of a salt in an organic solvent is added to a polymer matrix (polyvinyl chloride, polyvinyl fluoride). The solvent dissolves in the polymer matrix and forms a gel like structure. The conductivity as well as the current density and rate of diffusion, etc., are determined by the mobility of the solvated ions in the polymer matrix. The transport constants are again proportional to the free volume in the polymer.

1.3.3 Ion exchanging polymer electrolytes

The representative substance of this type of polymer ionic conductors is nafion. This polymer consists of a porous perfluorated carbon matrix with sulfonic acid groups. While the anion is fixed, the material shows high cation conductivity and is of high chemical stability. These materials are applied, for example, in membrane chlorine alkaline electrolysis and in polymer electrolyte membrane (PEM) fuel cells. Other types of fixed anions are carboxylic acid groups (weakly acid). Examples for ion exchange polymer electrolytes with fixed cations and anion conductivity are polymers with ternary amino groups (weakly basic) or quaternary amino groups (strongly basic).

1.4 IONIC CONDUCTANCE IN SOLIDS

The crystallographic structure in solids with fixed positions for atoms or ions restricts the free mobility of the ions. The ionic conductivity observed in solid electrolytes is based on defects and disorder in the crystallographic structure. One can distinguish

(i) *Crystal defects*—defects are formed during crystal growth.
(ii) *Intrinsic disorder*—defects are formed according to temperature and partial pressure and can be described by thermodynamics.
(iii) *Extrinsic disorder*—defects are formed by variation of the composition of a mixed phase or doping.
(iv) *Disorder in sub-lattices (quasi molten sub-lattices)*—these types of materials are called super ionic conductors or fast ionic conductors.

The value of the specific conductivity at room temperature is usually orders of magnitude smaller than in liquid electrolytes. For example, the ionic conductivity of NaCl at room temperature is of the order of 10^{-15} S $\cdot$ cm^{-1}. Conductivities with values almost as high as those for liquid electrolytes can nevertheless be achieved by increasing temperature. The transition from the region of low conductance at room temperature to super ionic conductance at higher temperature can occur continuously, as a transition of second-order, or as first-order transition, which is shown in Figure 1.10.

Different defects in solids will be described and some examples of super ionic conductors presented.

1.4.1 Crystal defects

During crystal growth defects in the ideal crystal lattice can occur. They are characterized as zero-, one-, or two-dimensional. A vacancy or an interstitial atom is a zero-dimensional defect due to a missing atom on a lattice place or an additional atom on an interstitial (Figure 1.11).

A zero-dimensional defect in the lattice creates varying distances between the neighboring atoms, which also generate tensions.

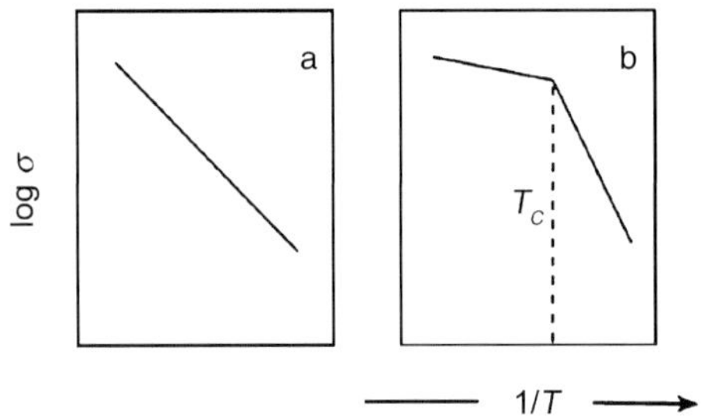

Figure 1.10 Transition from room temperature ionic conductance to super ionic conductance. (a) Continuous, (b) second-order transition, and (c) first-order transition, T_C, critical transition temperature.

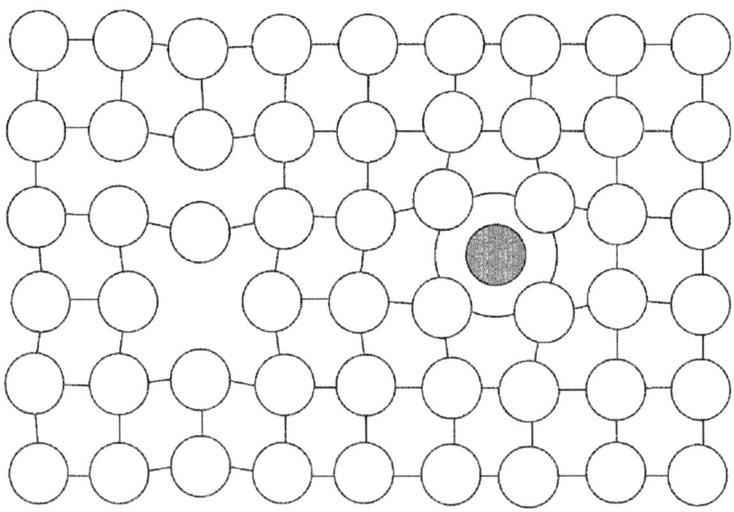

Figure 1.11 Vacancies and particles on interstitials in a lattice and tensions created by the defect.

Step dislocations (Figure 1.12) are one-dimensional defects. A step dislocation is generated if an additional lattice plane is inserted into the ideal lattice. A step dislocation marks a plane of deformation. The step dislocation is characterized by the Burgers vector ***b***. The construction of the Burgers vector is shown in Figure 1.13. The defect is circumvented on a trajectory along the neighboring lattice points beginning at S (starting point) and ending at E (end point). In the distorted lattice S and E are identical. The same process is then performed in an ideal lattice region. The distance between the endpoint in the first and second cases defines Burgers vector ***b***. The Burgers vector characterizes the distortion of the lattice by the inserted plane.

Another one-dimensional defect is a screw dislocation (Figure 1.14). It results if the planes of a lattice are twisted. Screw dislocations are centers of a permanent growth

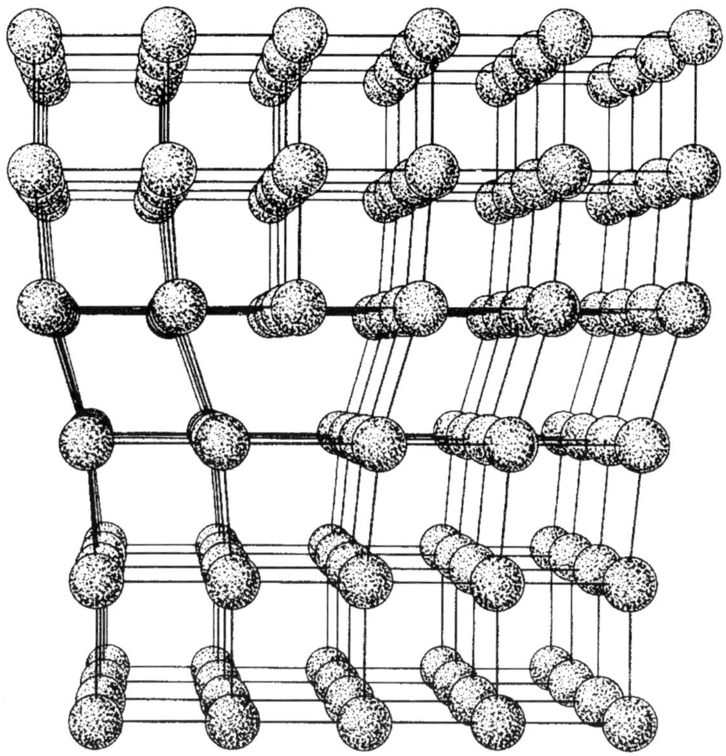

Figure 1.12 Step dislocation. An additional lattice plane is partially inserted into the ideal lattice (according to Kittel[33]).

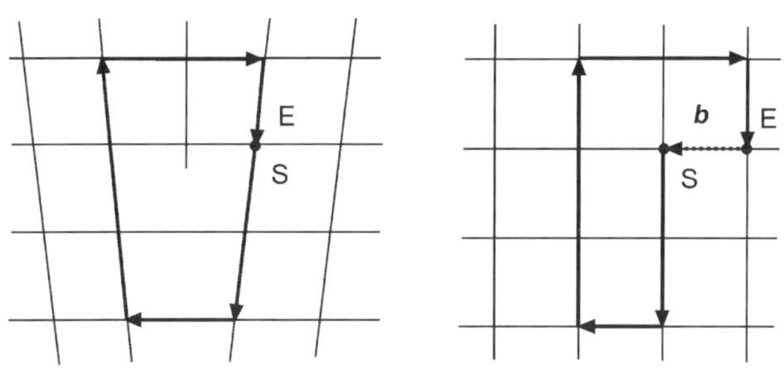

Figure 1.13 Construction of the Burgers vector, ***b*** to characterize a step dislocation. S, starting point, E, end point of circumvention of the step dislocation.

of crystals. Step dislocations and screw dislocations not only determine transport properties but also the mechanical properties of the crystal.

Grain boundaries, small angle grain boundaries, and all other phase separating borders in crystals are two-dimensional defects. As an example, a small angle grain boundary is shown in Figure1.15. Transport through the crystal can occur along these interfaces.

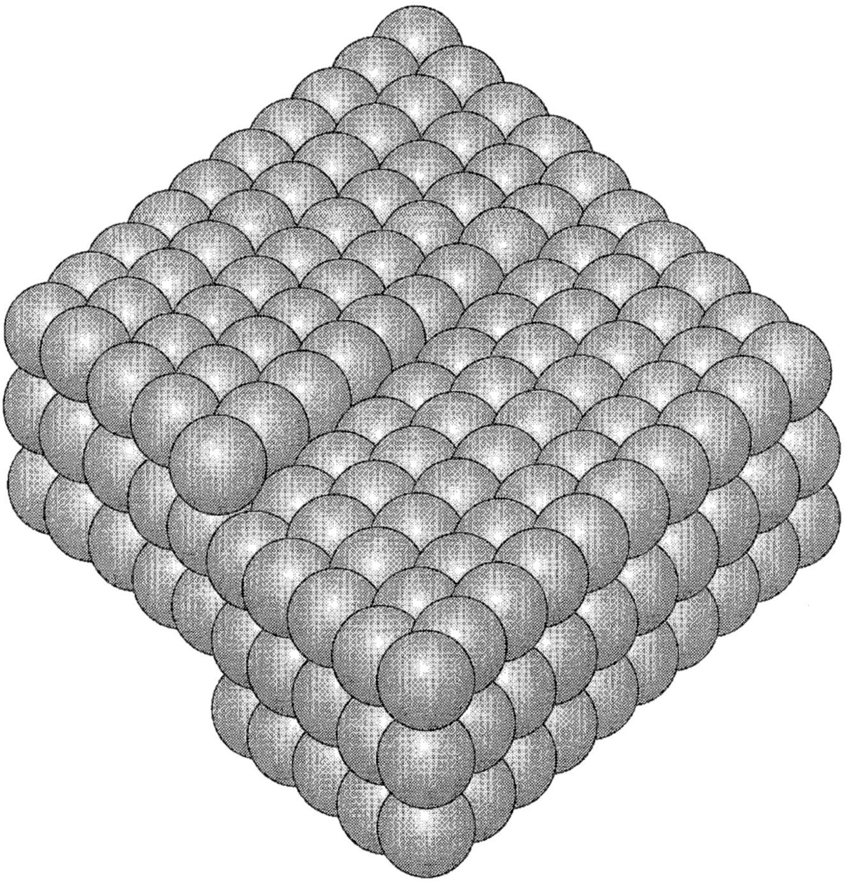

Figure 1.14 Screw dislocation. Screw dislocation results if the planes of a lattice are twisted. (Reproduced with permission from Ref. [34], © 2000, Hanser Verlag.)

1.4.2 Intrinsic disorder

Defects can be formed according to temperature and partial pressure. Eq. (1.25) describes the formation of a vacancy and an interstitial in a silver lattice according to the Kröger–Vink notation.

$$Ag_{Ag} \rightarrow V'_{Ag} + Ag_i^{o} \tag{1.25}$$

The thermodynamic description relates the density of vacancies $[V'_{Ag}]$ and of interstitials $[Ag_i^{\circ}]$ to the formation enthalpy $\Delta_{for}H_V$ (N_{Ag} number of Ag atoms per volume)

$$\left[V'_{Ag}\right] = \left[Ag_i^{o}\right] = N_{Ag}e^{-\frac{\Delta_{for}H_V}{RT}} \tag{1.26}$$

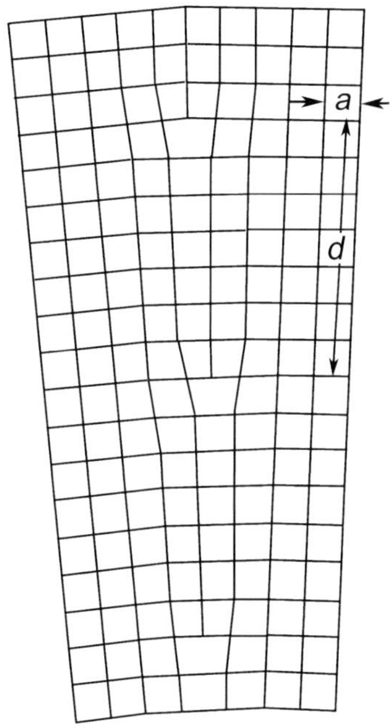

Figure 1.15 Small angle grain boundary. The defect can be considered as a two-dimensional array of step dislocations tilting the two crystals by an angle Θ. The defect is characterized by the distance d between the step dislocations and by the tilting angle $\Theta = a/d$ (a lattice constant according to Kittel[33]).

The formation enthalpy of a vacancy is of the order of 50–150 kJ · mol^{-1}, R is the gas constant and T the Kelvin temperature. Vacancies and interstitials can move through the lattice. The activation energy for this transport is approximately 20–30% smaller than the formation enthalpy of vacancies (Kanani[34]).

In ionic crystals, defects in one partial lattice must be compensated by defects in the second partial lattice to fulfill the law of electroneutrality. Therefore, one can distinguish (Hauffe[35]) among

- *Frenkel defects*: Vacancies in the cation lattice are compensated by cations on interstitials (Figure 1.16a).
- *Anti-Frenkel defects*: Vacancies in the anion lattice are compensated by anions on interstitials (Figure 1.16b).
- *Schottky defects*: Vacancies in the cation lattice are compensated by vacancies in the anion lattice (Figure 1.16c).
- *Anti-Schottky defects*: Cations on interstitials are compensated by anions on interstitials (Figure 1.16d).

The thermodynamic description of the formation of these defects is analogue to the description of vacancy-interstitial formation.

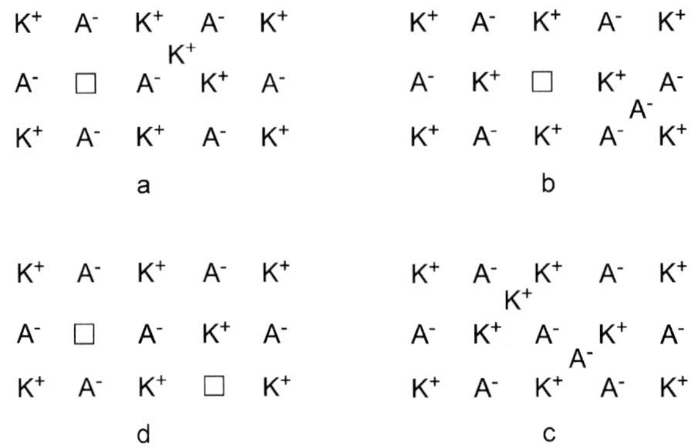

Figure 1.16 (a) Frenkel defect, vacancies in the cation lattice are compensated by cations on interstitials; (b) anti-Frenkel defect, vacancies in the anion lattice are compensated by anions on interstitials; (c) Schottky defect, vacancies in the cation lattice are compensated by vacancies in the anion lattice; and (d) anti-Schottky defect, cations on interstitials are compensated by anions on interstitials.

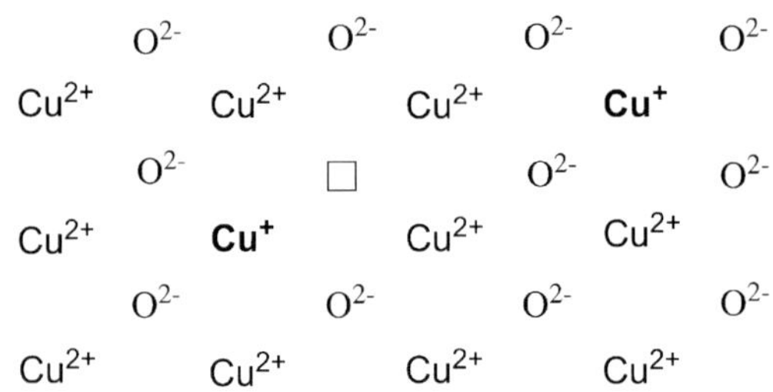

Figure 1.17 Non-stoichiometric Cu_2O crystal with charge compensation of missing Cu^+ ions by Cu^{2+} ions.

In compounds with ions of different oxidation states the compensation of ion vacancies or ions on interstitials is also possible by changing the oxidation state of the counter ion. For example, in non-stoichiometric Cu_2O with vacancies in the Cu^+ lattice, the negative excess charge of the oxygen ions in the oxygen lattice is compensated by oxidation of some Cu^+ ions to Cu^{2+} ions (Figure 1.17). The intrinsic disorder in an oxide lattice can be controlled by variation of the partial pressure of oxygen. Under electrochemical conditions the electrode potential is equivalent to the oxygen partial pressure.

1.4.3 Extrinsic disorder

Defects in mixed phases can be formed by doping or changing the stoichiometry of the mixed phase. An example for this type of disorder is yttria stabilized zirconia (YSZ).

The tetravalent zirconium ion is substituted by the three-valent yttrium ion. As a result twofold charged oxide ion vacancies are formed.

$$Y_2O_3 \rightarrow 2Y'_{Zr} + 3O''_O + V''_O \tag{1.27}$$

The highly conducting ZrO_2 crystallizes in the fluorite structure, which stabilizes by doping with Y_2O_3 or CaO. ZrO_2 is used in several electrochemical developments, e.g., as electrolyte in solid oxide fuel cells (SOFCs) or for an electrochemical oxygen sensor in the car industry (λ-probe).

1.4.4 Disorder in sub-lattices

These substances have more sites then ions. This led to high conductivity values, which led to the name super ionic conductors. The disorder can concern the cation lattice (cation conductors) or the anion lattice (anion conductors). The transfer numbers t_+ and t_- determine the classification with $t_+ \rightarrow 1$ for cation conductors or $t_- \rightarrow 1$ for anion conductors. Because of the usually small size of the cation compared to the anions, cation conductors are dominant.

The super conducting properties can extend over the entire experimental temperature range. One example is the Nasicon group of substance. These are compounds with the general formula $Na_{1+x}Zr_2Si_xP_{3-x}O_{12}$. Optimum conductivity is achieved with $1.8<x<2.2$. The stoichiometric compound $Na_3Zr_2Si_2PO_{12}$ is called nasicon, derived from the name natrium (sodium) super ionic conductor. It can be considered as a solid solution of three components, Na_2ZrO_3, SiO_2, and ZrP_2O_7. The composition triangle is shown in Figure 1.18.

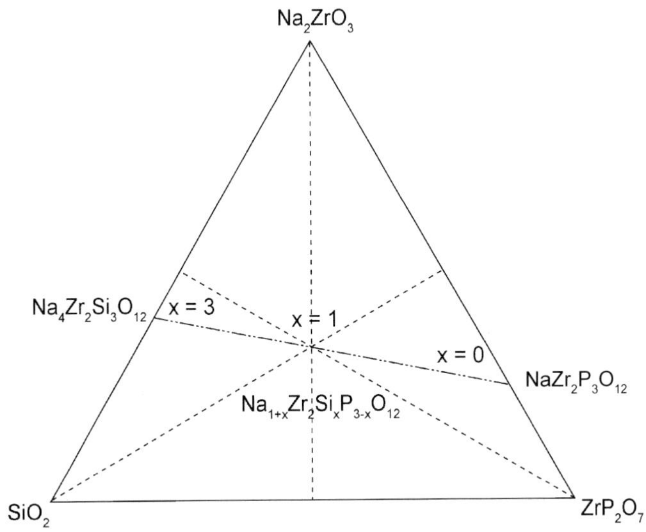

Figure 1.18 Composition triangle of nasicon. The original compound is represented in the center of the triangle. The experimental variation of the composition was preferentially made along the line for $x = 0$ (no Si) and $x = 3$ (no P).

Figure 1.19 Network of zirconium/oxygen octahedrons and Si_3PO tetrahedrons of nasicon (lantern structure, after Didishem *et al.*[36])

The material develops a structure of conduction channels within a network of zirconium/oxygen octahedrons and Si_3PO tetrahedrons (Figure 1.19). The conductivity of sodium ions in this network is of the order of 0.2 $S \cdot cm^{-1}$ at 300 °C. The temperature dependence is shown in Figure 1.20.

Another example is β-alumina, $Na_2O \cdot Al_2O_3$, and derivatives. It consists of aluminum oxide layers separated by intermediate layers of sodium and oxygen ions. The sodium ions are partially located on interstitials in channels with high mobility at ambient temperatures. The conductivity of sodium ions is of the order of 0.2 $S \cdot cm^{-1}$ at 300 °C. The conductivity in some derivatives can be 2 $S \cdot cm^{-1}$ at 300 °C. The temperature dependence is shown in Figure 1.20. In the development of a sodium-sulfur battery β-alumina was used as the electrolyte membrane separating sulfur and liquid sodium.

The super ionic conductivity can occur after a phase transformation at a characteristic transformation temperature. A representative example is AgI. The material has considerable conductivity already at room temperature. The large iodide ions form a lattice in which the much smaller Ag^+ ions can move via interstitials. At this temperature the stable form of silver iodide is the sphalerite lattice (see Chapter 2).

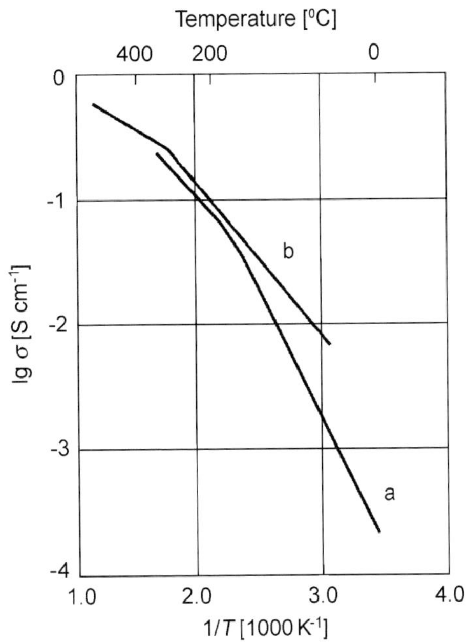

Figure 1.20 Arrhenius plots of conductivity-temperature dependence of (a) Nasicon and (b) β-alumina. (Reproduced with permission from Ref. [37], CRC Press.)

The transition to super ionic conductivity (γ-AgJ) is observed at 147 °C (Figure1.21). At 137 °C the wurtzite lattice is formed. The wurtzite lattice transfers into the γ-Form at 147 °C connected with a jump in conductivity from 10^{-3} to 1.3 S · cm^{-1}. This lattice consists primarily of a phase centered iodide lattice with a highly disordered Ag^+ sub-lattice with very mobile Ag^+ ions. The transition to super ionic conductivity is a first-order phase transition.

The high conductivity of AgI in the super ionic conducting state can be extended to lower temperatures by substitutions in the cation lattice. The best-known example is the $RbAg_4I_5$.

Similar properties were observed for Cu-ion conductors. An example of higher transition temperature is Bi_2O_3 with a transition temperature of 729 °C.

If the anions in the lattice are the dominant charge carrier, one speaks of anion conductors. There are two subclasses, halides and chalcogenides. In the class of halides the fluorites are the most intensively investigated anions. An example is the PbF_2 with a typical transition to super ionic conductivity between 250 and 350 °C (Figure 1.22). Another example of super ionic conductivity is LaF_3. By doping with Sr the conductivity of LaF_3 can be increased even more (Figure 1.22). More details of this class of materials can be found in the review by Schoonman.[38]

With the development of lithium batteries, lithium conducting solid electrolytes were intensively investigated. The representative compound is Li_3N. Another super ionic conductor is Li_3P while Li_3As shows metallic conductivity. Other Li ions conducting solid

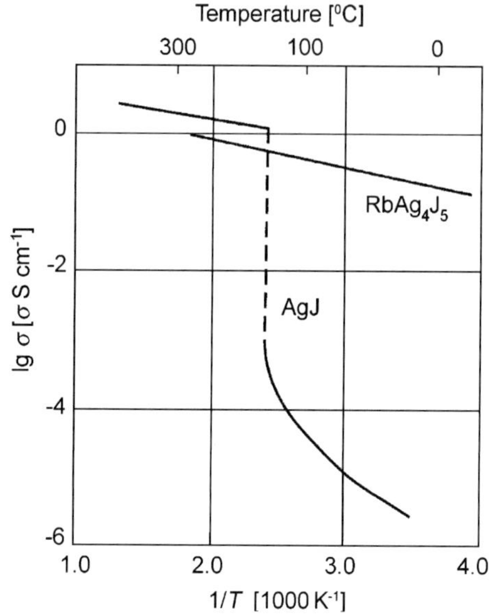

Figure 1.21 Conductivity of AgI. Transition to super ionic conductivity and comparison to $RbAg_4I_5$. (Reproduced with permission from Ref. [37], CRC Press.)

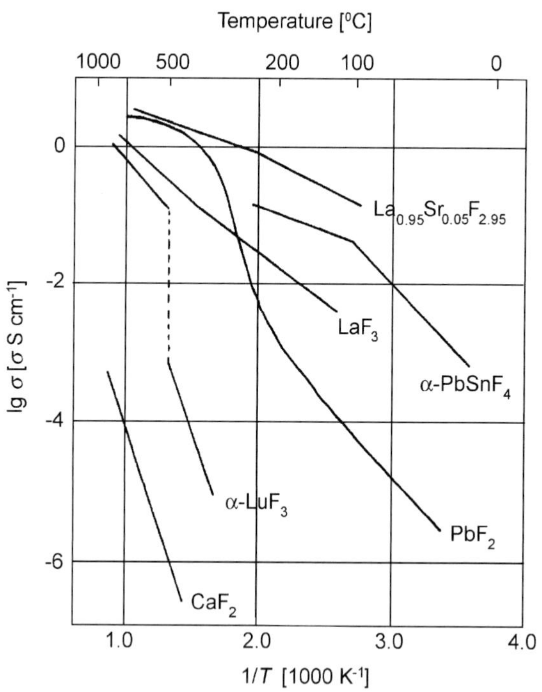

Figure 1.22 Conductivity of some fluorites (data from Julien and Nazri[7]). (Reproduced with permission from Ref. [29], © 1994, Kluwer Academic Publishers.)

electrolytes are composites consisting of a highly dispersed two-phase system, e.g., of a lithium salt with Al_2O_3. These materials go back to the work of Liang.[39]

1.4.5 Transport by defects

In metals the conductivity decreases with temperature according to the Wiedeman–Franz law.

$$\frac{\sigma}{\lambda} = \mathrm{const}\frac{1}{T} \tag{1.28}$$

The quotient of electrical conductivity σ and thermal conductivity λ is inversely proportional to the temperature. The ionic conductivity of solids depends on the lattice type and the type of the defects. The conductivity increases with temperature. This property is used to distinguish the ion conductor from the electron conductor. For vacancies and interstitials in the ion lattice, conductivity depends on the formation enthalpy for vacancy-interstitial pairs, $\Delta_{\mathrm{for}}H_{\mathrm{V}}$.

With increasing temperature the number of vacancies and atoms on interstitials will increase. Increasing temperature will also activate the motion of vacancies and interstitials through the crystal, the activation energy is E_{a}. The temperature dependence of the conductivity σ is given by the equation

$$\sigma = \sigma_0 \mathrm{e}^{\frac{-(\Delta_{\mathrm{for}}H_{\mathrm{V}}+E_{\mathrm{a}})}{RT}} \tag{1.29}$$

σ_0 is the conductivity for infinite temperature or $1/T = 0$.

1.4.6 Ion conducting glasses

Glasses are amorphous substances consisting of a network former like SiO_2, B_2O_3, and P_2O_5, which have a strong tendency to form solid polymeric networks below the glass temperature T_g instead of crystalline structures. In addition they usually contain a network modifier like Li_2O or Ag_2O and can contain other oxides like Al_2O_3. The conductivity is achieved by the addition of a doping salt, e.g., alkali halides.

Several models were developed to explain conductivity in the network.

- *Electrolyte models.* Assuming dissociation of the donor salt in the glass matrix, there is either a complete dissociation (strong electrolyte, Anderson–Stuart model[40]) or incomplete dissociation (weak electrolyte, Ravaine–Souquet[41] model) and the cations usually move in the matrix.
- *Defect model.* Analogue to crystalline ionic conductors, a transport via defects is discussed for glasses.
- *The coupling model.* This model, which was already discussed as an approach to understand conductivity in molten salts, could also be applied to glasses.

Depending on the type of ions, different ion conducting glasses are distinguished like lithium ion conducting glasses, sodium ion conducting glasses, as well as anion conducting glasses, e.g., fluorite glasses. The conductivity can reach values of 10^{-2} S · cm^{-1} (300 °C).

1.4.7 Mixed ionic and electronic conductance

Solids are mixed conductors that means electronic and ionic charge carriers show mobility in the lattice. One speaks of preferential ionic conductivity if the electronic transference number is $t_e < 0.01$. The electronic conductivity increases exponentially with the temperature and, for oxides, depends on the partial pressure of oxygen. Materials with preferential ionic conductivity can be found only in a certain temperature and pressure region. Materials with comparable ionic as well as electronic conductivity are called MIECs (mixed ionic electronic conductors). These materials have become especially interesting for applications. As an example, the ratio of electronic conductivity σ_e to ionic conductivity (hole conductivity σ_h) is shown in Figure 1.23 as a function of the partial pressure of oxygen for $Ce_{0.8}Gd_{0.2}O_{1.9}$.

The oxygen pressure is proportional to an electrochemical potential applied to the material. With increasing oxygen pressure the potential gets more anodic. The action on the $Ce_{0.8}Gd_{0.2}O_{1.9}$ is first a decrease of the electronic conductivity connected with a decrease of σ_e/σ_h. After passing a minimum decrease of whole conductivity causes the ratio to rise again.

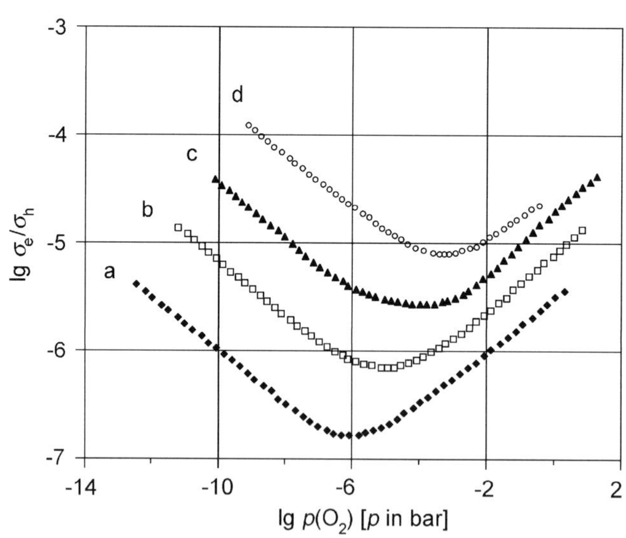

Figure 1.23 Ratio of electronic conductivity σ_e to ionic conductivity (hole conductivity σ_h) as function of the partial pressure of oxygen. $Ce_{0.8}Gd_{0.2}O_{1.9}$ at different temperatures: (a) 600°C, (b) 650°C, (c) 700°C, and (d) 750°C. (Reproduced from Ref. [42] with permission of the author.)

REFERENCES

1. H.S. Harned, The Physical Chemistry of Electrolyte Solutions, Reinhold, New York, 1950.
2. R.H. Robinson, R.H. Stokes, Electrolyte Solutions, Butterworths, London, 1959.
3. J.M.G. Barthel, H. Krienke, W. Kunz, Physical Chemistry of Electrolyte Solutions, in H. Baumgartel, E.U. Frank, W. Grunbein (Eds.), Topics in Physical Chemistry, Dr. Dietrich Steinkopf-Verlag, Darmstadt, Springer-Verlag, Berlin, 1998.
4. F.M. Gray, Solid Polymer Electrolytes, Wiley, New York, 1991; Polymer Electrolytes, Royal Society of Chemistry, Cambridge, 1997.
5. H. Rickert, Electrochemistry of Solids, Springer-Verlag, New York, 1982.
6. T. Kudo, K. Fueki (Eds.), Solid State Ionics, VCH, Weinheim, 1990.
7. C. Julien, G.A. Nazri, Solid State Batteries: Materials Design and Optimization, Kluwer Academic Publishers, Dordrecht, 1994.
8. P.J. Gelings. H.J.M. Bowmaster, CRC Handbook of Solid State Electrochemistry, CRC Press, Boca Raton, FL, 1997.
9. S. Arrhenius, Z. Physikal. Chem., *1*, 631 (1887); *2*, 491 (1888).
10. The work of Hittorf is described by W. Ostwald, Elektrochemie, Verlag Veit & Comp., Leipzig, 1896.
11. G. Kortüm, Z. Elektrochem., *57*, 874 (1953).
12. P. Debye, E. Hückel, Physik. Z., *24*, 311 (1923).
13. L. Onsager, Physik. Z., *27*, 388 (1926).
14. N. Bjerrum, Mat.-Fys. Medd. K. Dansk Videns Selsk, *7*, 1 (1926).
15. M.V. Smirnov, V.P. Stepanow, V.A. Khokhlov, Melts, *1*, 51 (1987).
16. M.V. Smirnov, K.A. Aleksandrov, V.A. Khokhlov, Electrochim. Acta, *22*, 543 (1977).
17. V.A. Khokhlov, Molten Salt Chemistry and Technology, in H. Wendt (Ed.), Molten Salt Forum, Vol. 5/6, Trans Tech. Publications, Switzerland, 1998, p. 202.
18. K.L. Ngai, Molten Salt Chemistry and Technology, in H. Wendt (Ed.), Molten Salt Forum, Vol. 5/6, Trans Tech. Publications, Switzerland, 1998, p. 13.
19. C. Cramer, K. Funke, T. Saatkamp, Phil. Mag. *B71*, 701 (1995); C. Cramer, K. Funke, M. Buscher, A. Happe, T. Saatkamp, Phil. Mag. *B71*, 713 (1995).
20. K.L. Ngai, J. Non-Cryst. Solids, *203*, 232 (1996).
21. K.L. Ngai, in R. Rickert, A. Blumen (Eds.), Effects of Disorder in Relaxational Processes, Transport and Relaxation in Glass Forming Molten Salts, Springer-Verlag, Berlin 1994, p. 89.
22. T. Welton, Chem. Rev., *99*, 2071 (1999).
23. P. Wasserscheid, W. Keim, Angew. Chem., *112*, 3926 (2000).
24. F. Endres, Workshop Ionic Liquids, FEM, Schwäbisch-Gmünd, 2006.
25. F. Endres, Chem. Phys. Chem., *3*, 144 (2002).
26. M.B. Arnaud, Solid State Ionics, *9/10*, 745 (1981).
27. G.F. Fulcher, J. Am. Ceram. Soc., *8*, 339 (1925).
28. C. Berthier, W. Gorecki, M. Minier, M.B. Armand, J.M. Chabagno, P. Rigaud, Solid State Ionics, *11*, 91 (1983).
29. C. Julien, G.A. Nazri, Solid State Batteries: Materials Design and Optimization, Kluwer Academic Publishers, Dordrecht, 1994, p. 357.
30. M.H. Cohen, D. Turnbull, J. Chem. Phys., *34*, 120 (1961); *52*, 3038 (1970).
31. I.E. Kelly, J.R. Owen, B.C.H. Steele, J. Power Sources, *14*, 13 (1985).
32. R. Spindler, D.M. Shriver, Macromolecule, *19*, 347 (1986).
33. C. Kittel, Introduction to Solid State Physics, 3rd Ed., Wiley, New York, 1966.
34. N. Kanani, Galvanotechnik, C. Hanser-Verlag, München-Wien, 2000, p. 31.
35. K. Hauffe, Reaktionen in und an festen Stoffen, Springer-Verlag, Berlin, 1955.

36. J.J. Didishem, E. Prince, B.J. Wuensch, Solid State Ionics, *19–19*, 981 (1986).
37. P.J. Gelings. H.J.M. Bowmaster, CRC Handbook of Solid State Electrochemistry, CRC Press, Boca Raton, FL, 1997.
38. J. Schoonman, Solid State Ionics, *5*, 71 (1981).
39. C.C. Liang, J. Electrochem. Soc., *120*, 1289 (1973).
40. O.L. Anderson, D.A. Stuart, J. Am. Ceram. Soc., *37*, 573 (1954).
41. D. Ravaine, J.L. Souquet, Phys. Chem. Glasses, *18*, 27 (1977); *19*, 115 (1978).
42. S. Lübke, H.-D. Wiemhöfer, Electronic Conductivity of Gadolinium Doped Ceria, Ber. Bunsenges. Phys. Chem., *102*, 642 (1998).

– 2 –

Structure and Bonding

In an electrochemical process leading to surface deposition or phase formation a special crystallographic structure is formed. The structure depends on several factors that are partially of geometric origin and partially determined by the bonding forces. A measure for the bonding forces is the bond energy or the cohesion energy. While bond energy is the energy of bonding between the closest neighbors, cohesion energy also depends on the long-range interaction including Coulomb and van der Waals forces and is the more general description for interactions in solids.

Insight into the nature of the chemical bond has been intensively developed since the publication of Pauling's[1] famous first review. Frenking and Shaik[2] gave a more recent review. More details can be found in Kittel's[3] classical treatment of solid state physics.

2.1 STRUCTURE FACTORS

Attempts were made in the literature to determine the factors, which are responsible for the formation of the structure of solids (metals, alloys, and intermetallic or other compounds). The results of these considerations were summarized in factors or, more precisely, factor groups. Combining two or more of these factor groups should allow prediction of the structure that will develop during the synthesis. Villars[4] was able to especially develop this concept and described five factors, which could be used for such predictions. These factors are

- Size
- Atomic number
- Cohesive energy
- Electrochemical
- Valence electron

In later publications a sixth factor, called the Mendeleev factor, was added that describes the position of an element in an empirical periodical table.

From the modern insight into the nature of the chemical bond, one easily understands that electronegativity and valence electron concentration are aspects of the electrostatic and chemical interaction of the components. Therefore, one can reduce the factors responsible for the structure of a solid phase in principal to three properties

- Size
- Coulomb forces by charge separation (ionic bonds)
- Bond forces by localized or extended orbitals (covalent bonding or metallic bonding).

This all depends, of course, on the chemical nature of the components that are represented by the position in the periodic table. Thus, these three factors in combination with the atomic number factor can lead to a successful classification and grouping of the elements. The results are given in structure tables, e.g., by Pettifor.[5]

In the following, a short overview of the most important structures and cohesion forces behind the structure formation will be given.

2.2 CLOSED PACKED STRUCTURES OF METALS

The main structure observed in metal deposits is a closed packing of identical spheres. A two-dimensional arrangement is shown in Figure 2.1. A layer of spheres is placed in the depression of the first layer to extend this arrangement into the third dimension. The arrangement of the third layer determines the type of crystal structure that will form. If the first layer is reproduced, the ABA arrangement forms the hexagonal closed packed structure, hcp (Mg type, Pearson symbol hP2, Figure 2.1a). If the third layer does not reproduce the first layer, the arrangement is ABC and the cubic closed packed structure ccp is formed (Cu type, Pearson symbol cF4, Figure 2.1b). Other packings are observed like ...ABAC... for lanthanum and ...ABABCBCAC for samarium.

The packing of layers is described either by the ABC notation or by looking at the neighboring layers (Jagodzinski and Pauling[6]). If a layer's neighbors are equal as in ABA for B, layer B has a hexagonal structure or h. If the neighboring layers are different like in ABC, layer B has a cubic structure or c. In the layer arrangement ABAC for lanthanum the notation is hc. The notation for samarium is chhchhchh or $(chh)_3$. Other notations are found in the literature.[7]

For both hcp and the ccp structures, the number of the next neighbors is 12, but the number of second and third next neighbors is different. The numbers are given in Table 2.1.

The structures obtained by closed packing of atoms in metals given in the literature are equilibrium structures. The arrangement of layers is a long-range order and is, especially

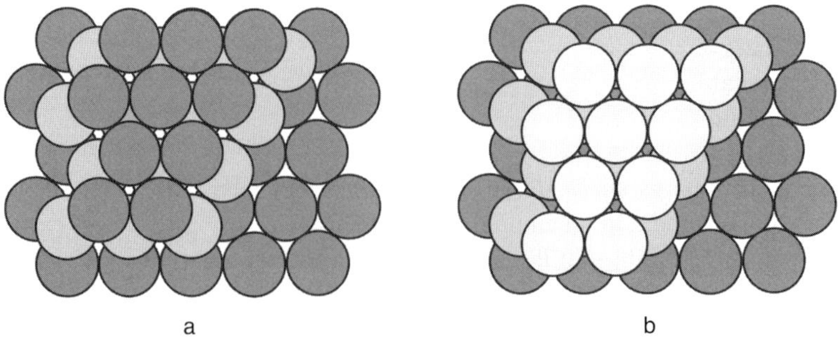

Figure 2.1 Closed packed structures. (a) hcp structure with ABAB packaging and (b) ccp or fcc structure with ABCABC packaging.

Table 2.1

Number of atoms and distances in coordination polyhedrons of closed packed structures[8]

Neighbors	Number of atoms	Distance	Number of atoms	Distance
	hcp	hcp	ccp	ccp
1	12	d	12	d
2	6	$\sqrt{2}d$	6	$\sqrt{2}d$
3	2	$\sqrt{\frac{8}{3}}d$	24	$\sqrt{3}d$
4	18	$\sqrt{3}d$	12	$2d$
5	12	$\sqrt{\frac{11}{3}}d$	24	$\sqrt{5}d$
6	6	$2d$	8	$\sqrt{6}d$
7	12	$\sqrt{5}d$	48	$\sqrt{7}d$

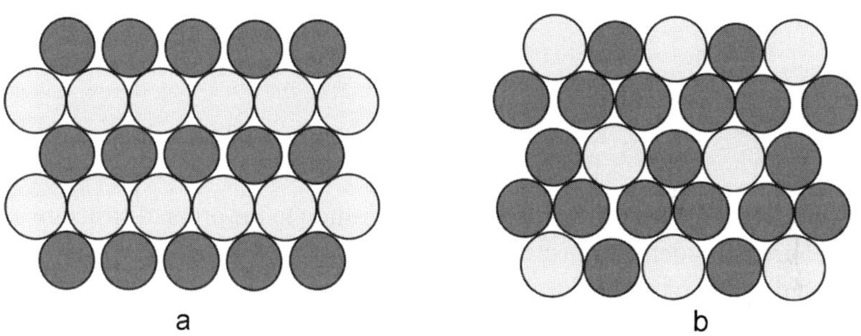

Figure 2.2 Structures of the alloy system $Au_{1-x}Cu_x$. (a) Layered structure of AuCu and (b) the $AuCu_3$ structure, white atoms for Au and dark atoms for Cu.

in more complex arrangements, sensitive to the conditions during crystallization. In electro deposition the formation of the equilibrium structure depends on many factors like potential, surface mobility, and the influence of the substrate. Copper, for example, usually forms a ccp structure. If deposited on platinum, the first layers (approximately nine layers) are deposited in the body centered cubic structure (Moffat[9]).

2.3 ALLOYS WITH CLOSED PACKED STRUCTURE

Alloys with similarly sized components and chemically similar atoms can also crystallize in closed packed structures. A statistic mixture of the different atoms is formed. An example is the system Au and Ag with atomic radii of $r_{Au} = 0.1442$ nm and $r_{Ag} = 0.1444$ nm. If the radius difference is larger, like in the system Au and Cu ($r_{Au} = 0.1442$ nm, $r_{Cu} = 0.1278$), complete miscibility is only observed at higher temperatures. At lower temperature the atoms form the structures AuCu and $AuCu_3$ as shown in Figure 2.2. The driving

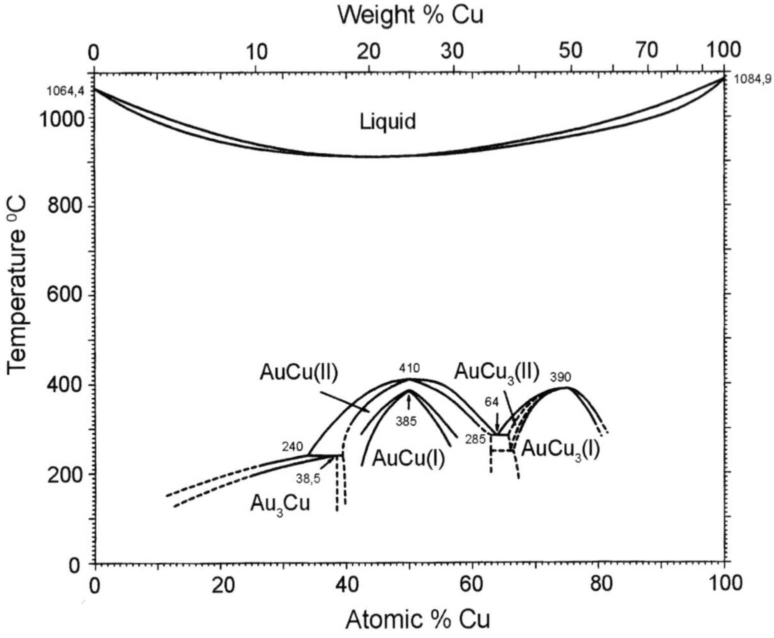

Figure 2.3 Phase diagram of the alloy AuCu.[10]

force is the reduction of the molar volume. The transition to the ordered structure is a phase transition of second order (λ-transition).

The phase diagram of $Au_{1-x}Cu_x$ is shown in Figure 2.3. At higher temperatures a complete statistical miscibility of Au and Cu in the solid phase is observed. This can be described as solid solution. The phase separating line shows an azeotropic minimum. The layered structure for $x = 0.5$ (AuCu) and the structure for $x = 0.75$ ($AuCu_3$) with copper surrounding the Au atoms (Figure 2.2) form at lower temperatures. The formation of AB and AB_3 structures is typical for these systems.

2.4 HUME-ROTHERY RULES FOR FORMATION OF SOLID SOLUTIONS

Hume-Rothery formulated some rules of thumb for the formation of solid solutions (Hume-Rothery,[11] p. 110) by comparing atomic diameters and electronegativities. The rules are summarized as follows:

- To form a solid solution over a broad range of composition the atomic diameter should not differ by more than 15%.
- One indicator for the formation of a solid solution is the difference in electronegativity. In such a case one can say that a small electronegativity difference is an indicator for extensive solid solutions, but it is difficult to express the influence of electronegativity quantitatively.

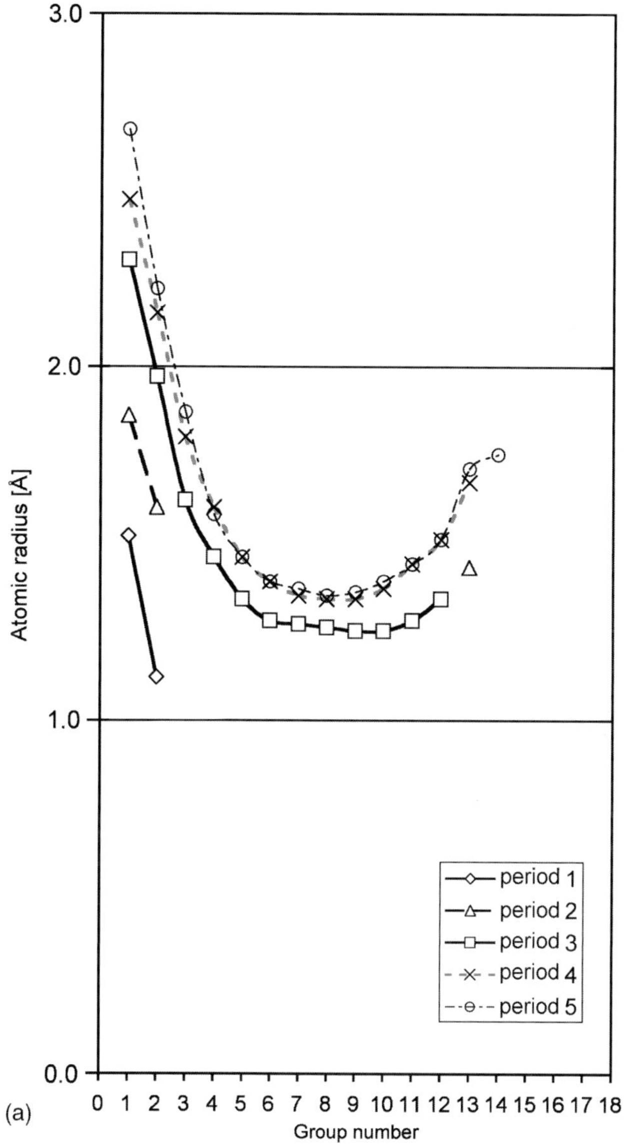

Figure 2.4 (a) Atomic radii (Ref. 7 p. 43) and (b) electronegativities[12] of metals plotted versus group numbers for the periods of the periodic table.

The compositional range for solid solutions is restricted, when the components tend to form a stable intermetallic compound. This tendency increases with increasing difference of electronegativities.

- A higher valent metal is more soluble in a lower valent metal than a lower valent metal in a higher valent metal.

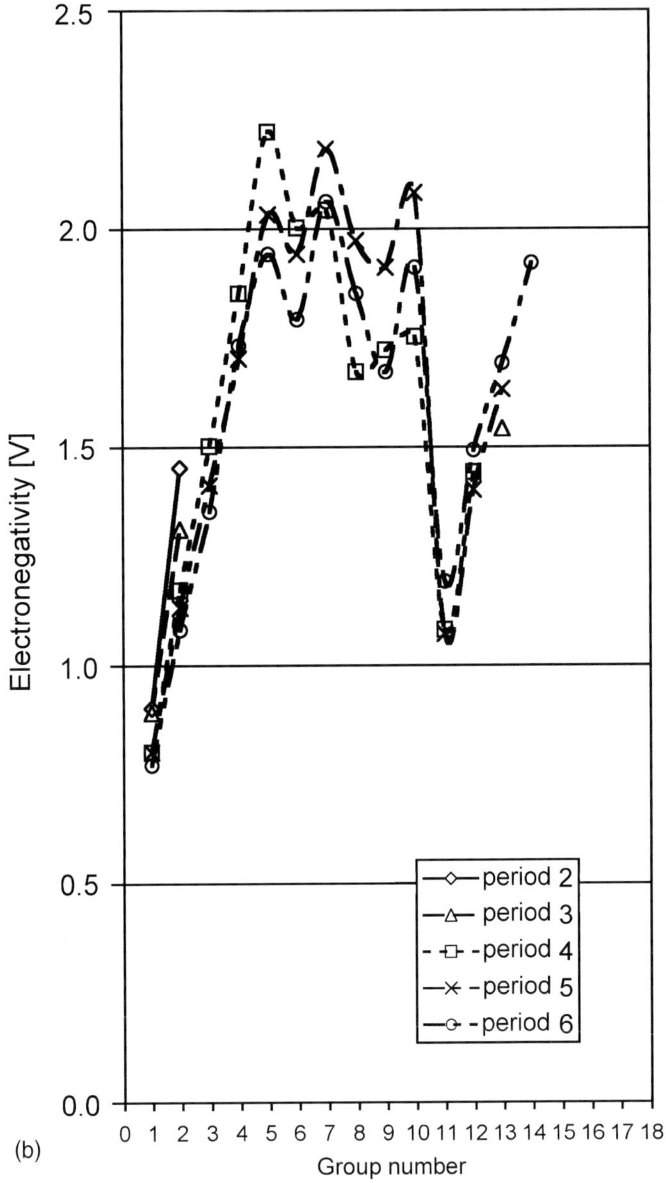

Figure 2.4 *(Continued)*

Plotting atomic radii and electronegativities of metals of the periodic table versus group numbers is a way to predict structure based on these rules (Figure 2.4). This is an example for the cooperative character of the structure factors introduced in Section 2.1.

Several improvements of these primary rules were suggested in the literature and have been summarized by Villars[34] or Alonso.[13]

2.5 BODY CENTERED CUBIC STRUCTURE

The coordination number of first nearest neighbors in closed packed structures is 12. Nevertheless, several metals crystallize in the body centered cubic structure (W type, Pearson symbol cI2, Figure 2.5). The coordination number is reduced to 8 but the difference of the distance between the first and the second nearest neighbors (white spheres) is small (15.5%). This structure is known as the CsCl structure.

Fifteen metals crystallize in this structure and many alloys prefer this structure for geometric reasons because of the different sizes of the components.

Of prime importance is the possibility to form superstructures with a twofold or even threefold transformation length by combining 8 or even 27 cells of CsCl type. In such systems different atoms can occupy the central position of each individual body centered cubic unit, providing a great variety of structures. The different lattice positions for a stack of 8 CsCl units are shown in Figure 2.6.

An example is the Heusler alloy $MnCu_2Al$ with a = Al, b = Mn, c = Cu, and d = Cu. Another example is the Zintl phase Li_2AgSb with a = Ag, b = Li, c = Sb, and d = Li.

2.6 HUME-ROTHERY PHASES

Inspecting the structure of copper-zinc alloys, Hume-Rothery observed that the transformation between different phases followed a change in the ratio of the number of valence electrons to the number of atoms in the Wigner–Seitz cell* (Hume-Rothery,[11]

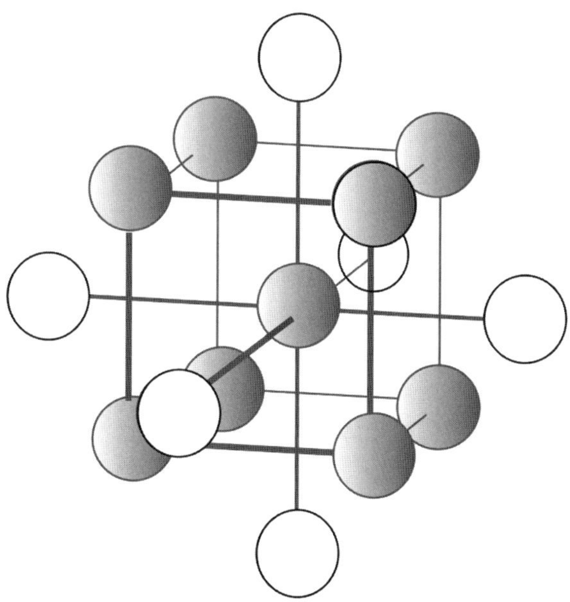

Figure 2.5 Body centered cubic structure (CsCl structure). The coordination number of the unit cell is eight. The second nearest neighbors are shown as white spheres. Its distance to the central atom is 15% larger than the distance of the first nearest neighbors.

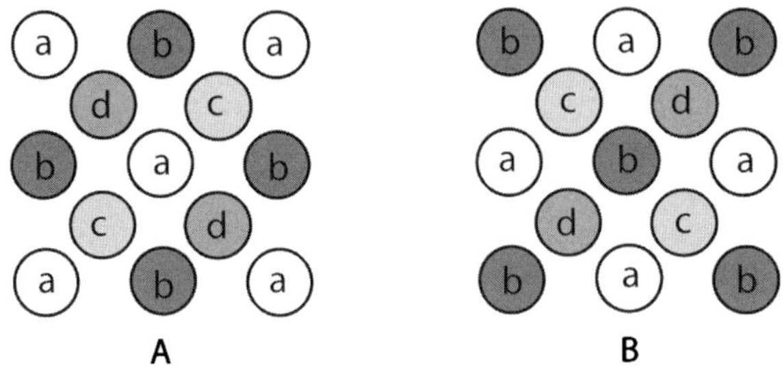

Figure 2.6 Different lattice positions in a superstructure of 8 CsCl units. (A) Lower half of unit cell and (B) upper half of unit cell.

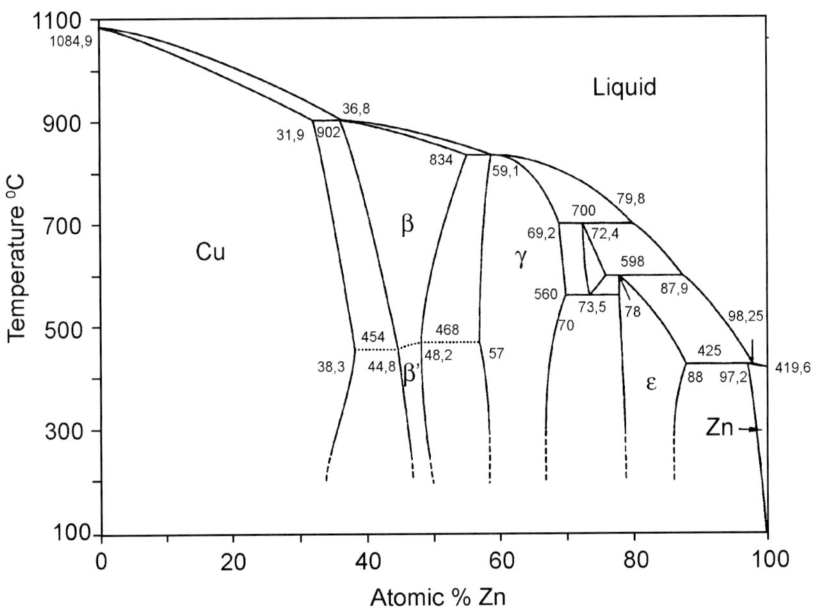

Figure 2.7 Phase diagram of CuZn.[10]

p. 203). An example is the CuZn alloy. The phase diagram of this system is shown in Figure. 2.7.

The formula $Cu_{(1-x)}Zn_x$ describes the composition. The structure of pure copper ($x = 0$) is the face centered cubic lattice (fcc, Pearson symbol cF4). Upon an increase in the zinc concentration, a solution of zinc in copper is observed (α-phase). The solubility extends up to $x = 0.38$ depending on the temperature. The zinc atoms are statistically distributed in the copper lattice.

For a copper/zinc ratio of 1:1 ($x = 0.5$, CuZn), the face centered cubic structure is no longer stable and transforms into a body centered cubic structure, the β-Phase (bcc, W type, Pearson symbol cI2). To explain the Hume-Rothery rule for this transformation, one has to compare the number of atoms in the Wigner–Seitz cell of the bcc structure (for a copper/zinc ratio of 1:1 the number is 2) with the number of valence electrons supplied by the two atoms (1 electron for copper, 2 electrons for zinc, which together is 3). The ratio 3:2 is 1.5, which marks the center of stability for the bcc structure. Copper and zinc atoms are still statistically distributed, at least at higher temperatures.

With a further increase of the zinc concentration at Cu_5Zn_8 and a valence electron/atom number ratio of 21:13 (1.62), a superstructure of the body centered cubic lattice is formed, the γ-phase (Cu_5Zn_8 type, Pearson symbol cI52). The superstructure consists of $3^3 = 27$ simple cells with 52 atoms and two vacancies compared to the simple cells. The atoms and vacancies are no longer statistically distributed but form an ordered structure.

The next phase change is observed at $CuZn_3$ and a valence electron/atom number ratio of 7:4 (1.75). At this ratio a hexagonal closed packed structure the ε-phase is formed (Mg type, Pearson symbol hP2).

Finally, from $x = 0.98$ until $x = 1.0$, a solution of copper in the zinc matrix is observed. The regions of homogeneous solutions are examples of Hume-Rothery's third rule. The higher valent atom (zinc) is more soluble in the lower valent solvent (copper) then vice versa.

There are several examples of alloys with Hume-Rothery faces, e.g., CuAl, CuSn, AgZn, and AgAl.

Jones[14] has given a theoretical interpretation of the Hume-Rothery phase formation. The change of structure occurs if the electrons in the Brillouin zone have to occupy anti-bonding levels. A structural change is then initiated with more bonding electron levels.

2.7 IONIC STRUCTURES

Metals and alloys in the solid state usually form metal bonds and preferentially crystallize in the described closed packed structures. Nevertheless, some metal like compounds have a salt character, e.g., Zintl phases.

If the components of a solid phase are stable cations and anions, the ions form an ionic lattice based on the following rules:

1. Ions of opposite sign try to be in contact or as near as possible to each other.
2. Ions of equal sign try to get as far apart as possible from each other.

The maximum coordination number of ions of similar radii and with direct contact between cations and anions is the CsCl structure, coordination number 8 (Figure 2.7a).

What happens if the radius of the cation gets smaller and the radius of the anion larger? Then the distance between the anions gets smaller. If the anions touch each other, the lattice structure gets instable; a change of the lattice type occurs. The stability of the CsCl lattice type is lost at a ratio of $r_C/r_A < 0.732$. The calculation of this ratio is explained in Figure 2.8b.

At this ratio the NaCl lattice is energetically more favorable (Figure 2.9). The coordination number is reduced to 6. At $r_C/r_A < 0.414$, as shown in Figure 2.9b, the NaCl lattice

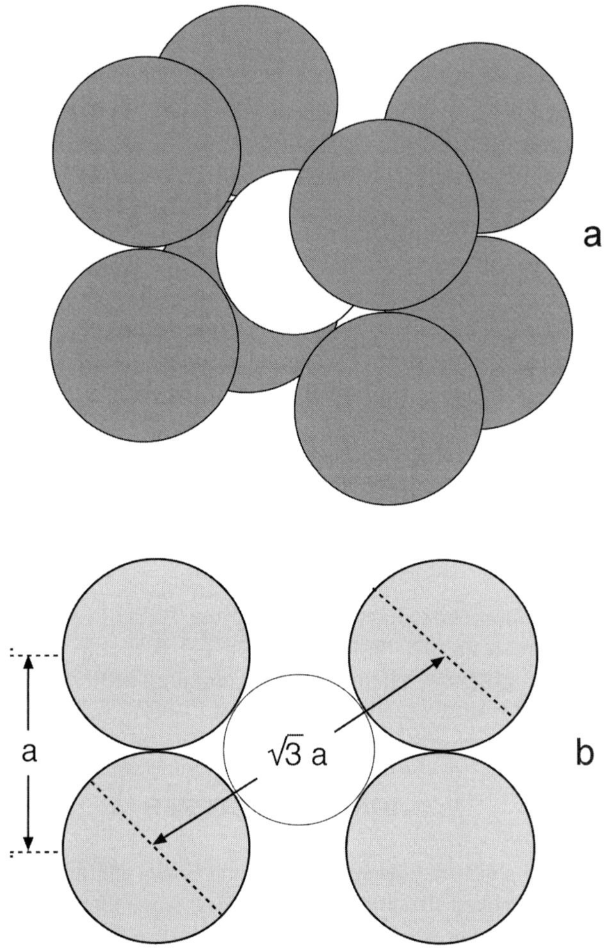

Figure 2.8 (a) CsCl octahedron with the critical ratio of cation/anion radius of 0.732 (white Cs^+, gray Cl^-) and (b) calculation of the critical ratio of cation/anion radius. With $a = 2\,r_A$ and the diagonal equal to $2(r_A + r_C)$ one gets $r_C/r_A = \sqrt{3}-1$.

is no longer stable and the sphalerite (ZnS) lattice is preferred (Figure 2.10). The coordination number is further reduced to 4.

The lattice energy E_L of a lattice of oppositely charged ions is determined by Coulomb forces and can be calculated using Coulomb's law

$$E_{lat} = \frac{1}{4\pi\varepsilon_0} \sum_j \frac{q_1 \cdot q_j}{r_{1,j}} \tag{2.1}$$

with q_1 the charge of the central ion and q_j the charge of the surrounding ions, $r_{1,j}$ the distance between ion 1 and ion j, and ε_0 the permittivity of the vacuum.

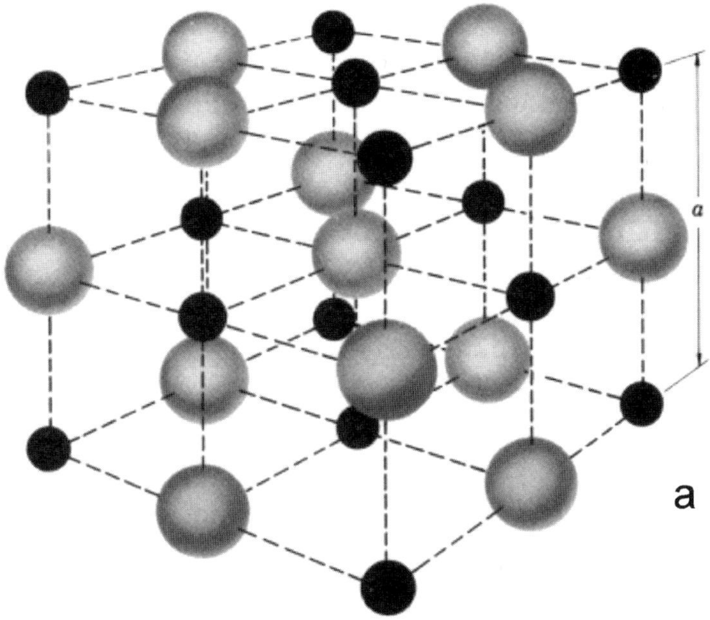

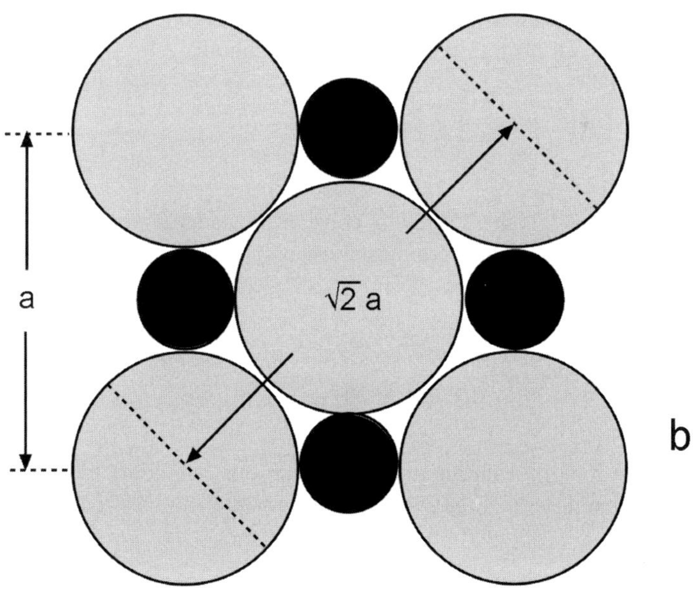

Figure 2.9 (a) NaCl lattice (black Na^+, gray Cl^-) and (b) calculation of the critical ratio of the cation/anion radius described in Figure 2.8.

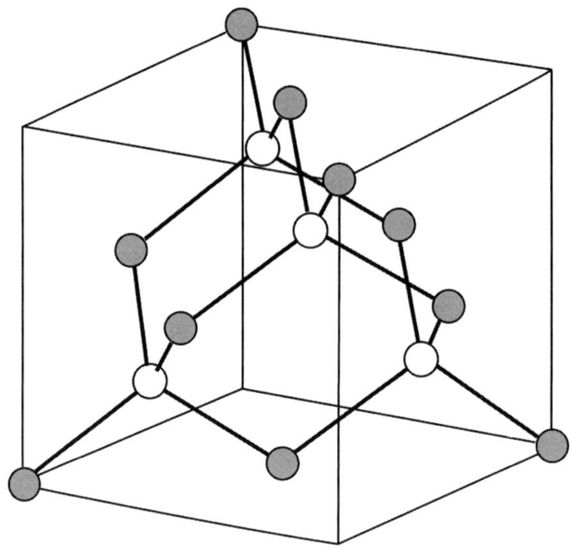

Figure 2.10 ZnS lattice (sphalerite).

The sum over the reciprocal distance values can be expressed as a convergent row, based on r_0, the distance between central ion and nearest neighbors.

$$C_M = \frac{1}{2} r_0 \left\{ \sum_i \frac{z_i / z_-}{r_i} + \sum_j \frac{z_j / z_+}{r_j} \right\} \tag{2.2}$$

The value of this row C_M is the Madelung constant.[15] Madelung constants depend on the lattice type. Some typical values are presented in Table 2.2.

The Madelung constant of CsCl is $C_M = 1.76267$. Then the lattice energy according to Eq. (2.1) is

$$E_{lat} = 1.76267 \frac{e_0^2}{4\pi\varepsilon_0} \frac{1}{r_0} \tag{2.3}$$

If the charges of the ions in a lattice are larger than one, shielding effects occur. This can be taken into account by a shielding constant. Tosi[16] has given a detailed treatment of this.

2.8 COORDINATION POLYHEDRONS OF MOLECULES

The repulsion between charges is a general phenomenon for the forces determining the structure of compounds. Gillespie and Nyholm[18] presented a very general rule for the influence of Coulomb interaction with the valence shell electron pair repulsion (VSEPR) theory.

Table 2.2

Madelung constants for some selected ionic lattice structures[17]

Compound	Lattice type	Madelung constant C_M
NaCl	fcc	1.74756
CsCl	bcc	1.76267
ZnS	Sphalerite (zinc blende)	1.63806
ZnS	Wurtzite	1.641
CaF_2	Fluorite	5.03878
TiO_2	Rutile	4.816
TiO_2	Anatase	4.800
Cu_2O	Cuprite	4.11552

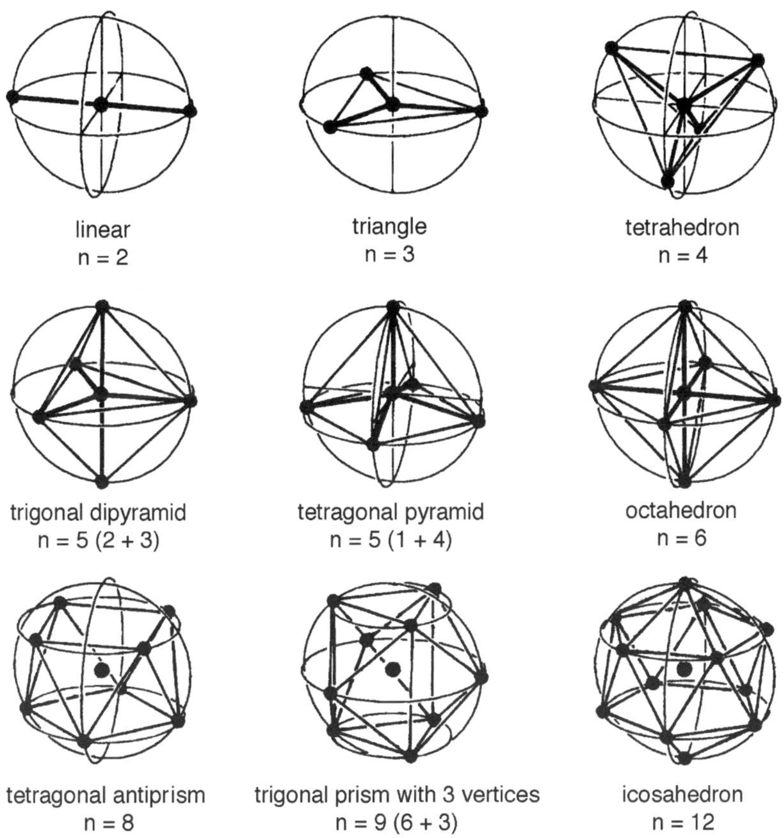

Figure 2.11 Arrangement of points on the surface of a sphere with minimal repulsion energy. *n* is the total number of points and in brackets are points of slightly different distance to the center. (Reproduced with permission from Ref. [7] p. 63, © 1992, Teubner Verlag.)

A detailed treatment is given by Heine *et al.*[19] The theory describes the distribution of valence electron pairs around a central atom under the influence of the repulsion between the negative charges. Counting the number of pairs of valence electrons a prediction about the coordination polyhedron can be made. The electron pairs take positions on a sphere around the central ion where they have the largest distance from each other and a minimum of repulsion energy. The different structures are shown in Figure 2.11.

The VSEPR theory was developed to explain molecular structures of main group elements with only bonding electron pairs. Molecular structures of subgroup elements must also take into account the influence of nonbonding electrons. The nonbonding electrons are preferentially d-electrons. The ligand field theory describes the interaction between d-electrons and bonding electrons (details are found, e.g., by Ballhausen[20]). The three-dimensional representation of d-electron orbitals is shown in Figure 2.12.

In the crystal, the d-electrons are either oriented diagonal to the *xy*, *xz*, and *yz* axes (t_{2g} orbitals) or point in the direction of the *x*, *y*, and *z* coordinates (e_g orbitals). The interaction of these d-orbitals with bonding orbitals leads to a splitting of the energy of the d-orbitals depending on the coordination polyhedron. This is shown in Figure 2.13.

VSEPR theory and ligand field theory show the great effect of electrostatic repulsion of the electrons in a molecular complex, but similar influences can also be expected at the formation of solid-state structures.

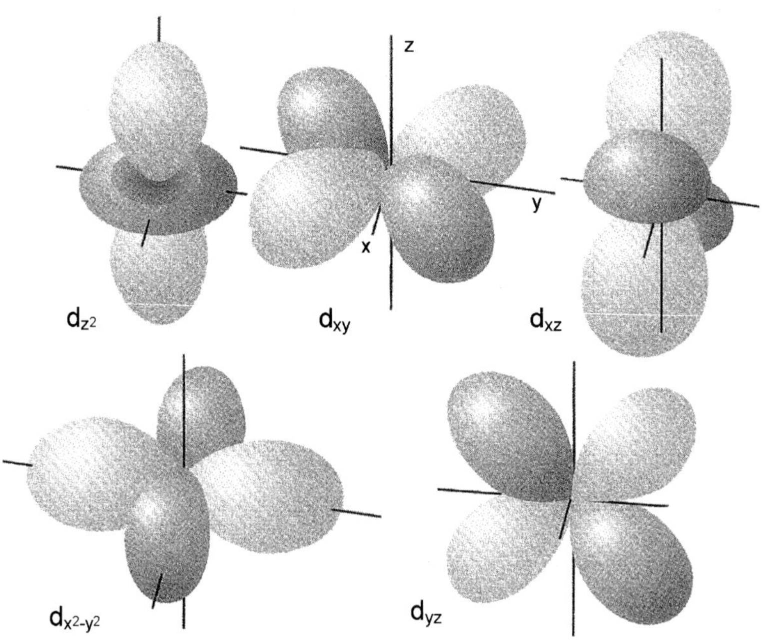

Figure 2.12 Three-dimensional representation of d-orbitals.

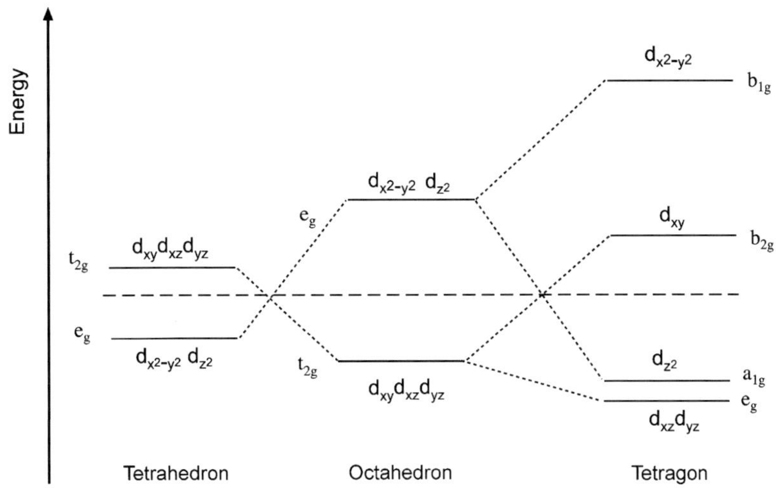

Figure 2.13 Splitting of the d-orbitals in a ligand field is shown for octahedron, tetrahedron, and tetragon.

2.9 THE BAND MODEL OF ELECTRONS IN SOLIDS

A simple description of electrons in a solid is the model of a free electron gas in the lattice of the ions as developed for the description of metals and metal clusters. The interaction of electrons and ions is restricted to Coulomb forces. This model is called a jellium model. Despite its simplicity, the model explains qualitatively several phenomena observed in the bulk and on the surface of metals. For a further development of the description of electrons in solids, the free electron gas can be treated by the rules of quantum mechanics. This treatment leads to the band model. Despite the complexity of the band model, Hoffmann[21] presented a simple description of bands in solids based on the molecular orbital theory of organic molecules that will also be discussed below.

2.9.1 Free electrons in a metal

The Schrödinger equation for a free electron in a box is the simplest model describing electrons in a metal. A comprehensive treatment is found in the book by Kittel.[3]

Looking at a row of atoms along an x-axis of length l (one-dimensional model), the Schrödinger equation is

$$-\frac{\hbar^2}{2m_e}\frac{\partial^2\psi}{\partial x^2} = E_e\psi \tag{2.4}$$

with $\hbar = h/2\pi$ and h Planck's constant, m_e the electron mass. Taking into account that the wave function is zero for $x \leqslant 0$ and $x \geq l$ (that is outside of the box, which is limited by potential walls of infinite height) the solution of this equation is a wave function

$$\psi_n = \sqrt{\frac{2}{l}}\sin\left(\frac{n\pi}{l}x\right) \tag{2.5}$$

The eigen values of the energy E_n are quadratic functions of the quantum number n

$$E_n = \frac{\hbar^2}{2m_e}\left(\frac{\pi n}{l}\right)^2 \tag{2.6}$$

One can introduce periodic boundary conditions. A wave function satisfying periodic boundary conditions has the form

$$\psi_n(x + l) = \psi_n(x) \tag{2.7}$$

A solution is a plane wave

$$\psi_n(x) = Ae^{ik_x \cdot r} \tag{2.8}$$

The wave number k_x (in the one-dimensional treatment) is defined as

$$k_x = \pm\frac{2\pi}{l}n \quad n = 0,\ 1,\ 2,... \tag{2.9}$$

In the three-dimensional case a wave vector must be defined with

$$\boldsymbol{k} = (k_x, k_y, k_z) \tag{2.10}$$

The energy can also be represented as a function of the wave vector (Figure 2.14a).

$$E_n = \frac{\hbar^2}{2m_e}(\bar{\boldsymbol{k}})^2 \tag{2.11}$$

The lattice of metal ions (distance between two ions a) restricts the free motion of an electron. The metal ions reflect the electrons. Applying Bragg's law of reflection, destructive

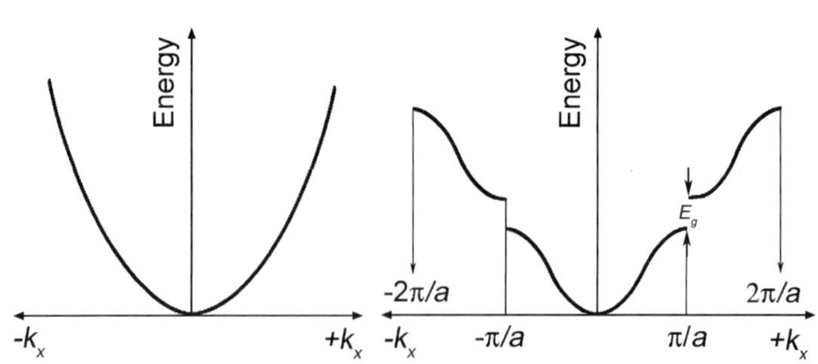

Figure 2.14 Free electron in a one-dimensional box of length l as function of wave number k_x. (a) Completely free electron model and (b) nearly free electron model, reflection of the electrons on the ions.

reflection in the one-dimensional case occurs for

$$k_x = \pm\frac{n\pi}{a} \tag{2.12}$$

This leads to energy gaps, the first observed for $n = 1$, $k_x = \pm\pi/a_x$ (Figure 2.14b).

The lattice of metal ions creates a periodic potential field. The potential energy of the electron as a function of the position $V(x)$ (the discussion is again restricted to one dimension) is periodic with respect to the atomic distances (lattice constant a).

$$V(x) = V(x + a) \tag{2.13}$$

The Schrödinger equation for the electron must now be solved in a periodic potential field. The result is that the eigen-functions of the Schrödinger equation for the electron in a periodic potential have the form

$$\psi_k(x) = \chi(x)e^{ik_x} \tag{2.14}$$

These functions are called Bloch functions.[22] The functions $\chi_k(x)$ are periodic functions of the wave vector $\boldsymbol{k}$.

2.9.2 Orbitals in solids

The structure of molecules is determined by the geometry of molecular orbitals formed by combinations of atomic orbitals. The characteristic geometries of atomic orbitals are obtained by the solution of the Schrödinger equation for the different quantum states of the electrons of an atom. The solutions provide the spherical model of s-orbitals, the orientation of p-orbitals along the main axis of the Cartesian coordinate system, and the more complex structures of d-orbitals, as was shown in Figure 2.12, or f-orbitals. In the molecular orbital theory, the bond between two atoms is described by a linear combination of the wave functions of the single atoms to two new wave functions, one lower in energy and one higher in energy. The new wave functions are molecular orbitals; the one with the lower energy is the bonding orbital, the other the anti-bonding orbital.

In a lattice, the set of atomic orbitals are ideal for the formation of Bloch functions. A band can be obtained by combining the atomic orbital functions. This model can be said to be composed of the orbital functions of the solid. Eq. (2.15) describes the combination for the one-dimensional case

$$\psi_k(x) = \sum_{n=0}^{N} e^{ik_x na}\chi_n(x) \tag{2.15}$$

a is the distance between the atoms and k_x the wave number. The number of atoms is counted from $n = 0$ to N. In the three-dimensional case a wave vector $\boldsymbol{k}$ must be defined. The wave vector $\boldsymbol{k}$ has several interpretations. The absolute value of $\boldsymbol{k}$ is the number of nods in the wave function ψ_k. $\boldsymbol{k}$ is also a translation vector in the reciprocal space.

If one looks at a chain of atoms (in one dimension) the region between $k_x = 0$ and $k_x = \pm\pi/a$ determines the smallest cell of the reciprocal space. This is called a Brillouin zone.

Each k_x represents one orbital of the solid. If the length of the chain reaches typical numbers of atoms in macroscopic dimensions, a very large number of crystal orbitals are formed. On the energy scale a single orbital is no longer distinguishable; a band is formed. A similar result was obtained with the nearly free electron model.

A short description of molecular orbitals in solids will be given in the following text. As was already mentioned, a comprehensive treatment of the orbital concept of bonding in solids was presented by R. Hoffmann.[21]

The combination of atomic s-orbitals is shown in Figure 2.15. The exponential term in Eq. (2.15) determines the sign of the combination. In Figure 2.15 the situation for $\boldsymbol{k} = 0$ (all signs positive, the energy of this molecular orbital is lowest), $\boldsymbol{k} = \pi/2a$ and $\boldsymbol{k} = \pi/a$ (border of the first Brillouin zone, alternating positive and negative sign, highest energy), is shown.

The energy of the wave functions plotted versus the absolute value of $\boldsymbol{k}$ in the first Brillouin zone is shown in Figure 2.16. As was shown in Figure 2.14, the first Brillouin zone in one dimension reaches from $-k_x$ to $+k_x$. The region for negative k_x-values in Figure 2.16 would be obtained by reflection on the energy axis. The width of the band (the dispersion) is determined by the intensity of the interaction of the atomic orbitals. The intensity of interaction depends on the distance a between the atoms. For a long distance a between the atoms, the interaction is weak and a small band is formed (Figure 2.16a). For a short distance a between the atoms, the interaction is strong and a broad band is formed (Figure 2.16b). The combination of s-orbitals can be demonstrated for a linear string of alkaline atoms, e.g., sodium atoms.

In the first example in Figure 2.15 s-orbitals were combined. What happens if p-orbitals are coupled? In this case the orientation of p_x, p_y, and p_z into the different axes of the coordination system must be taken into account. This is shown in Figure 2.17. If the axis of the chain is the x-direction, the orientation of the p_x orbital leads to a strong interaction,

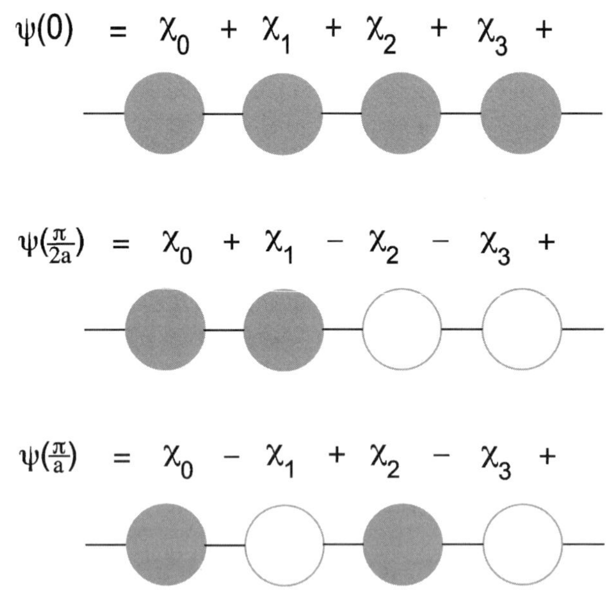

Figure 2.15 One-dimensional combination of atomic s-orbitals. $\boldsymbol{k} = 0$ (all signs positive, the energy of this molecular orbital is lowest), $\boldsymbol{k} = \pi/2a$, $\boldsymbol{k} = \pi/a$ (border of the first Brillouin zone, alternating positive and negative sign, and highest energy).

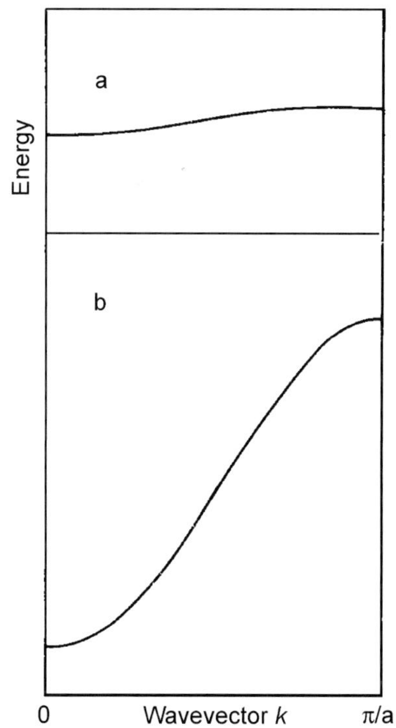

Figure 2.16 Energy band formed by the one-dimensional combination of atomic s-orbitals. (a) Weak interaction (long distance between the lattice points) and (b) strong interaction (short distance between the lattice points).

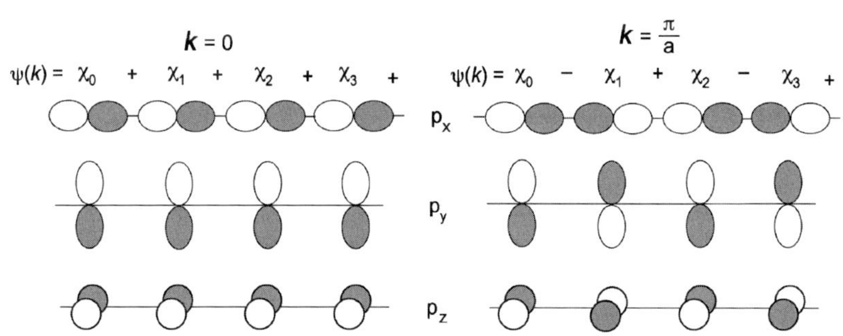

Figure 2.17 One-dimensional combination of atomic p-orbitals. Left site $\boldsymbol{k} = 0$, right site $\boldsymbol{k} = \pm\pi/a$.

but the direction of the energy change in the band is reversed. Wave function $\psi(\boldsymbol{k} = 0)$ has the highest energy; wave function $\psi(\boldsymbol{k} = \pi/\mathrm{a})$ the lowest energy. While for a linear combination of s-orbitals the energy increases with k, for a linear combination of p_x orbitals in x-direction energy decreases with k. (Figure 2.18) The energy of p_y and p_z orbitals increases again with k but the band width is small because the overlap is small (large distance).

When several bands are formed in a solid, the bands can overlap or can be separated by a band gap (Figure 2.19). An example for overlapping s- and p-bands are the earth alkaline metals.

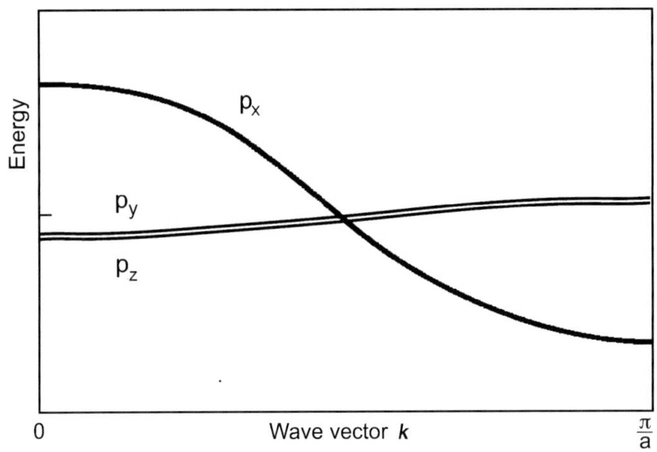

Figure 2.18 Energy as a function of wave vector $\boldsymbol{k}$ for atomic p-orbitals. For the p_x orbitals, energy decreases with $\boldsymbol{k}$ and the energy of p_y and p_z orbitals increases with $\boldsymbol{k}$. The different band widths (strong for p_x orbitals and small for p_y and p_z orbitals) represent the different intensities of interaction.

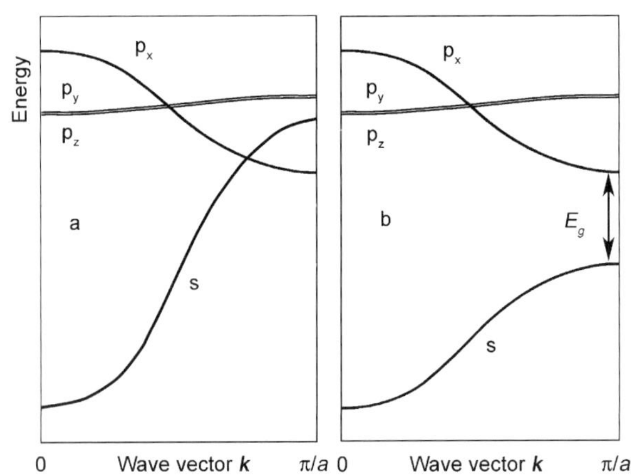

Figure 2.19 s- and p-bands in a solid. (a) Overlapping and (b) forming a band gap. The energy gap is E_g.

2.9.3 Density of states (DOS)

A band contains as many states (orbitals) as there are atoms in the macroscopic solid. The density of orbitals called DOS is a characteristic property of a band. The density of orbitals is the number of orbitals (atoms), $\mathrm{d}N$ in an interval of electron energy, $\mathrm{d}E_e$, which is usually divided by the total number of orbitals (atoms)

$$D(E_e) = \frac{1}{N}\frac{\mathrm{d}N}{\mathrm{d}E_e} \tag{2.16}$$

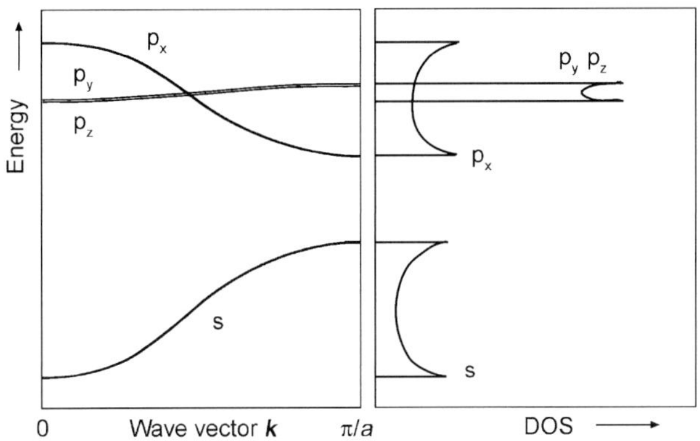

Figure 2.20 Energy bands and DOS versus electron energy for s- and p-bands.

Each band contains the same number of energy levels given by the number of atoms or molecular units in the chain. If a small band is considered, this band has a high DOS. However, a broad band with distribution of the same number of orbitals over a larger energy interval has a lower DOS. But the DOS differs even within a band. For the bands described in Figures 2.16 and 2.18 the DOS increases at the bottom and top of the band. As a result, schematic DOS-energy diagrams can be constructed (Figure 2.20).

2.9.4 Filling up with electrons; Fermi energy

The orbitals are defined as the energy levels of electrons in a solid. This is independent of the number of electrons. To construct an orbital diagram the electrons must be filled into the orbitals. According to the rules two electrons (spin up and spin down) can be filled in one orbital. The filling goes on from the bottom to top of the energy scale until all free electrons have found an orbital (band state). The energy of the highest occupied orbital (at T = 0 K) is called Fermi energy (E_F).

In Figure 2.21, the filling into the orbital diagram of a free electron model is shown with the Fermi energy as the highest occupied orbital (dark area).

In the case of sodium N electrons can be filled into the σ-band. The σ-band is half filled with electrons (see Figure 2.23).

What happens for $T > 0$ K? In this case the electrons can take up additional thermal energy. Some orbitals above the Fermi energy get occupied. The same number of orbitals below the Fermi energy becomes unoccupied. The overlap of occupied and unoccupied levels is given by the Fermi–Dirac function[3]

$$f(E_e) \approx \frac{1}{1 + \exp\left(\dfrac{E_e - E_F}{kT}\right)} \tag{2.17}$$

This function is shown in Figure. 2.22. In an exact derivation the Fermi energy in this equation must be substituted by the chemical potential of the electrons. At experimental

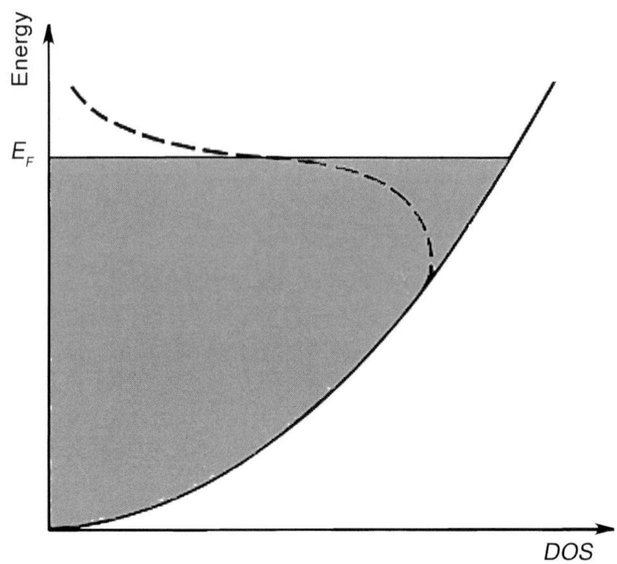

Figure 2.21 Electrons filled into the energy diagram of the free electron model. The dark area is the situation for $T = 0$ K. The energy of the highest occupied orbital is the Fermi energy E_F. The hatched line is $T > 0$ K. Some electrons are excited into orbitals above E_F, which creates unoccupied levels below E_F. The Fermi energy is now the point of the equal number of filled and unfilled energy levels.

temperatures ($T << 10^5$ K) Fermi energy and chemical potential have approximately the same value.

In Figure 2.22 the function $f(E_e)$ is shown for a high temperature. For relevant temperatures only a small region around E_F is concerned.

2.9.5 Crystal orbital overlap population: the formation of bonds

The final description of the binding forces in the orbital model of solids is the distribution of electrons to bonding and anti-bonding states. In the description of the bond between the two atoms A and B, the bond energy is associated with the overlap integral

$$S = \int (\chi_A \cdot \chi_B) \mathrm{d}v \tag{2.18}$$

Hoffmann[21] introduced a similar description for the band model of solids with the calculation of the crystal orbital overlap population (COOP). In this description the overlap integral in a lattice of atoms or molecules is calculated while taking into account the overlap between all the neighbors. This is shown for a chain of s-orbitals in Figure 2.23. Positive COOP values represent bonding contributions; negative values anti-bonding contributions. The sign obtained by integration up to the Fermi level gives a value that is characteristic for the bond order.

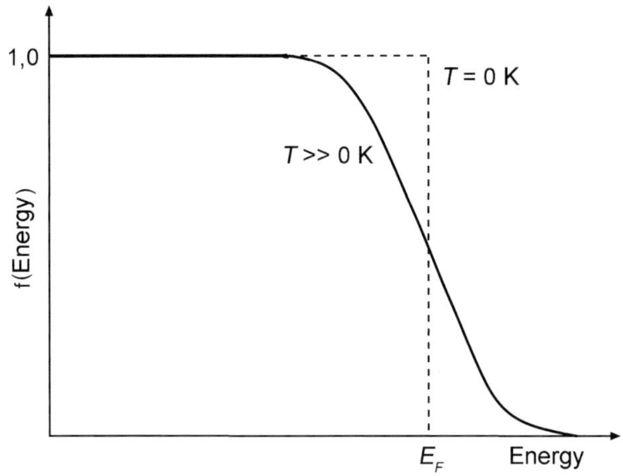

Figure 2.22 Fermi–Dirac function (after Kittel[3]); the widths of the distribution around E_F is k_BT.

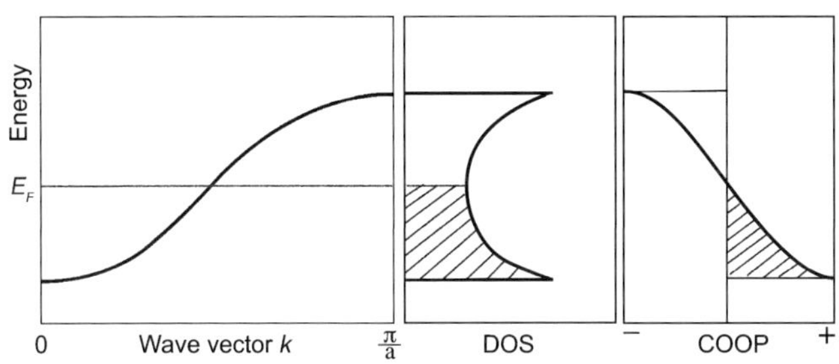

Figure 2.23 Band, DOS, and COOP for a one-dimensional s-band. The Fermi level is shown, for example, Na atoms with a half-filled band.

In Figure 2.23 only first nearest neighbors were taken into account. In a two- and three-dimensional lattice contributions from second and higher nearest neighbors can also be considered.

2.9.6 Extension to more dimensions

The principals of an orbital concept of solids can be extended to more dimensions. The band construction for the first Brillouin zone follows the same principles as described for the linear chain. The wave vector $\boldsymbol{k}$ must now be set up, e.g. for two dimensions k_x and k_y. The combination of s-atomic orbitals in two dimensions is shown in Figure 2.24.

The energy is usually represented along a line through the area of the first Brillouin zone. This is shown for a quadratic two-dimensional lattice in Figure 2.25.

In Figure 2.26 the energy of the linear combinations of s-orbitals along the line Γ-X-M-Γ is shown.

Three-dimensional treatment is possible in the same way. In Figure 2.27 the Brillouin zone of a primitive cubic lattice and characteristic points are shown.

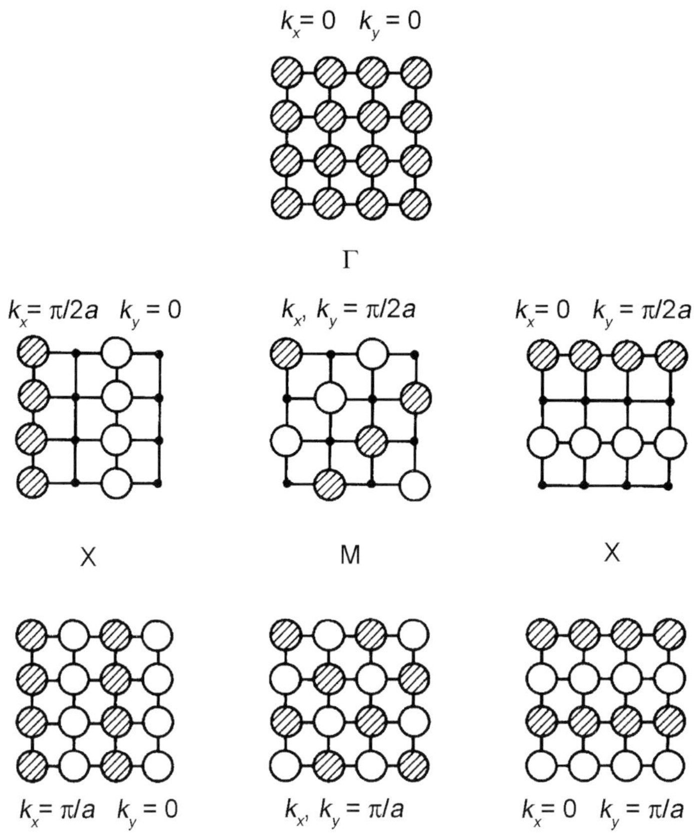

Figure 2.24 Combination of s-atomic orbitals in two-dimensions (according to Hoffmann[21]).

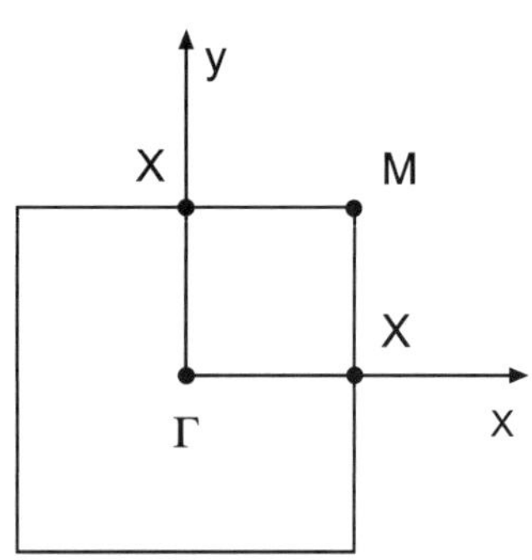

Figure 2.25 Brillouin zone of a quadratic two-dimensional lattice and characteristic points in the Brillouin zone. The energy-***k***-dependence is represented along a line in the Brillouin zone through these points.

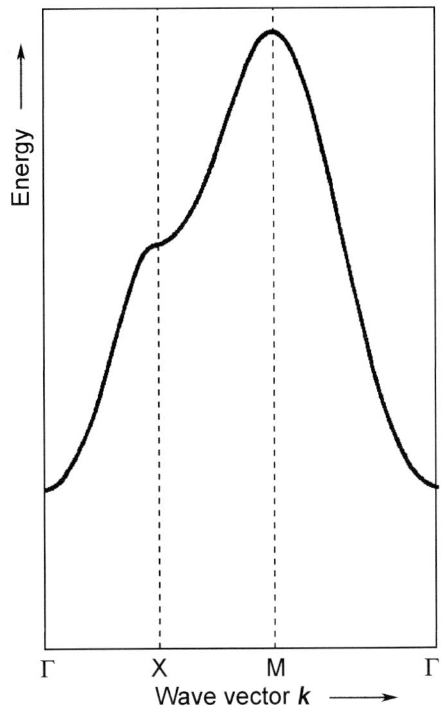

Figure 2.26 The s-band of a two-dimensional lattice is represented along a line Γ-X-M-Γ.

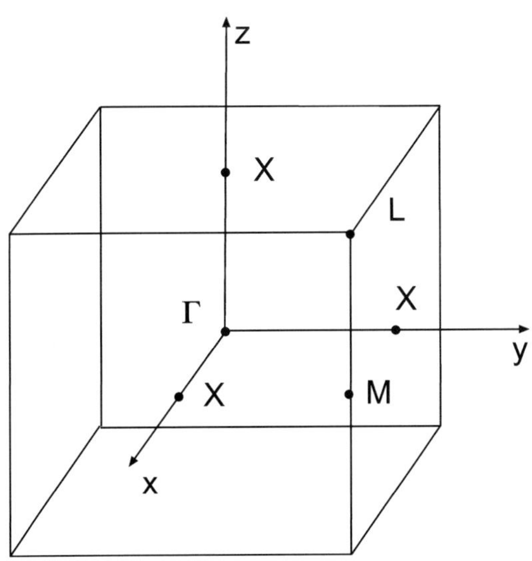

Figure 2.27 Brillouin zone of a cubic lattice and characteristic points in the Brillouin zone.

2.9.7 Band structure of d-metals

The described method can be applied to d-metals. A schematic plot of the expected DOS and COOP is shown in Figure 2.28.

A calculation by Saillard and Hoffmann[23] for nickel is shown in Figure 2.29.

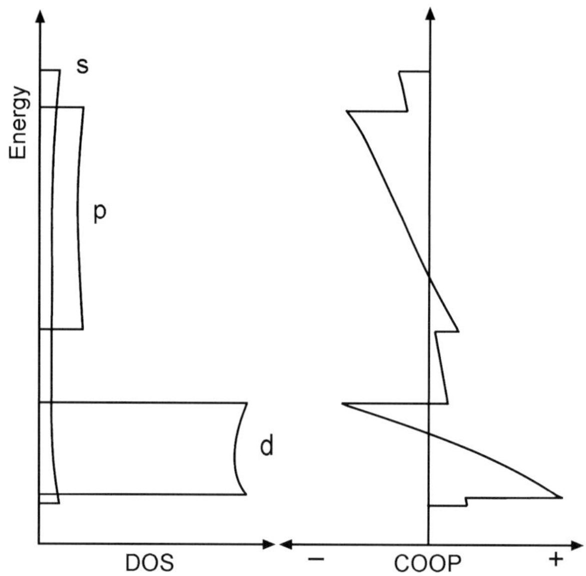

Figure 2.28 Schematic plot of DOS and COOP versus energy for d-metals.

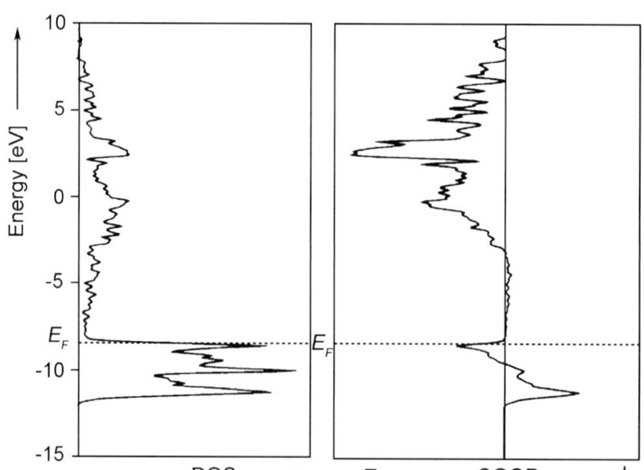

Figure 2.29 Calculated band structure, DOS and COOP of Ni (Saillard and Hoffmann[23]); the position of the Fermi energy of Ni shows that Ni has metallic character. (Reproduced with permission from Ref. [21], © 1988, Wiley-VCH.)

2.9.8 Semiconductors: example TiO_2

Part of the crystal structure of TiO_2 (rutile) is shown in Figure 2.30. The characteristic feature is the oxygen octahedron. A calculation of orbital structure, DOS- and COOP-functions for the orbitals of the components, in this case Ti-atoms and O-octahedrons, is shown in Figure 2.31.

The characteristic feature of this band structure is that the properties of the orbitals of the atoms and atomic groups are approximately preserved. The lowest band is primarily an O 2s-band. The next band is primarily an O 2p-band. These bands are fully occupied valence bands. The next band is primarily of t_{2g} character; the following band primarily of e_g character. These bands are unoccupied and are called conduction bands. The Fermi level in the ideal crystal is between the highest occupied and the lowest unoccupied bands. The band gap of 3,2 eV makes TiO_2 an insulator.

Usually the crude band sketch of Figure 2.32 describes this situation. The intrinsic semiconductor is shown in Figure 2.32i. In the case of TiO_2 donor states are available just below the conduction band. The doping by these donor states makes TiO_2 an n-semiconductor (Figure 2.32n). The doping by acceptor states, which creates holes in the valence band, as was observed for NiO, is shown in Figure 2.32p.

2.9.9 Peierls distortion

The combination of s-orbitals could also be demonstrated for a linear string of hydrogen atoms. Otherwise, hydrogen does not form a stable metallic solid, neither a one- nor three-dimensional one. The chain separates into individual hydrogen molecules. A similar

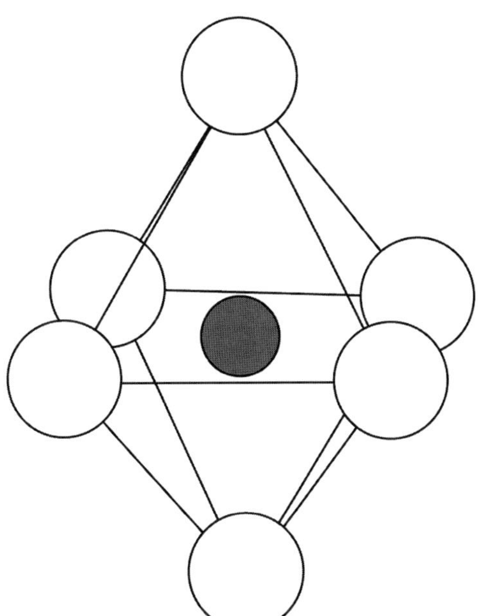

Figure 2.30 Crystal structure of TiO_2 (rutile) with the characteristic feature, a Ti atom in an oxygen octahedron.

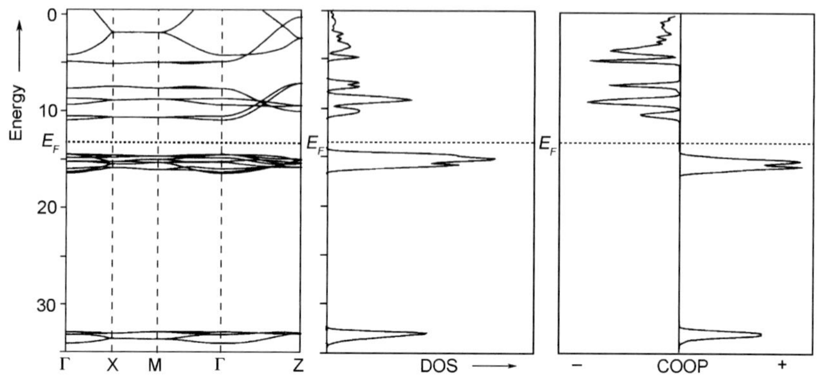

Figure 2.31 Calculated band structure, DOS and COOP of TiO_2 (rutile, according to Hoffmann[21] with reference to Refs. 24 and 25), position of the Fermi energy shows, TiO_2 is an insulator; by doping it behaves as a semiconductor. (Reproduced with permission from Ref. [21], © 1988, Wiley-VCH.)

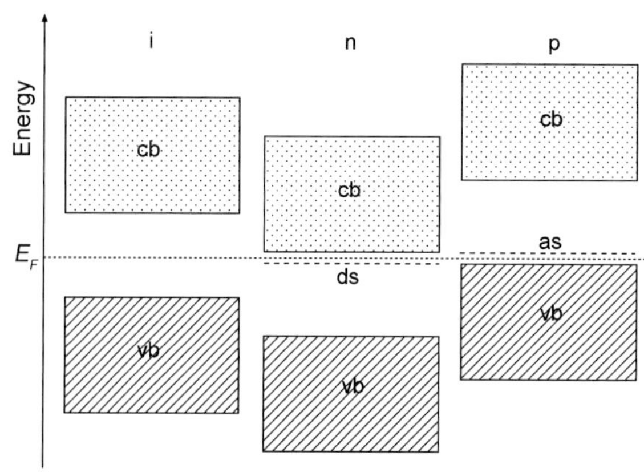

Figure 2.32 Schematic band sketch of intrinsic (i), n-doped (n), and p-doped (p) semiconductor.

situation is observed for a linear chain of CH strings. A chain of alternating single and double bonds is the chemical description of these strings.

The reason for the bond separation is called Peierls[26] distortion. In the case of hydrogen atoms, this leads to a complete decay of the chain into individual hydrogen molecules under normal conditions. Metallic hydrogen is only possible under extreme conditions, e.g., on the planet Jupiter.[27] If the chains consist of –CH– units (poly-acetylene), Peierls distortion leads to an asymmetry of the neighboring bonds.

Poly-acetylene is an insulator, but it can be transferred by oxidation into an electronic conductor. These intrinsically conducting polymers will be described in Chapter 11.

2.9.10 Energy bands in electrolytes

The concept of energy bands was extended to electrolytes. The principal concept is based on the difference of solvation of ions in the oxidized and reduced state. The concepts were developed by Gurney[28] and Essin[29] and later extended by Gerischer.[30] The concentration of ions in the reduced state, e.g., an Fe^{2+} ion, can be considered as a density of occupied energy states of electrons (the occupied energy band) in the electrolyte. The concentration of oxidized ions, e.g., Fe^{3+} ions, is therefore the density of unoccupied states (the unoccupied energy band) in the electrolyte. Gaussian distribution functions describe approximately the DOS of the solvated ions as function of energy. The width of the two Gaussian distribution functions is given by the distribution of the energy of the solvation states of the ions, approximately 0.3–0.6 eV.

Because the energy of a solvated ion is given by the ion–dipole interaction, the energy states of the oxidized ions are higher than the energy states of the reduced ions. The distribution functions of energy states of electrons for reduced and oxidized ions are shown in Figure 2.33.

The two distributions overlap at the Fermi energy. As will be shown in Chapter 3, this energy is equal to the equilibrium redox potential and marks the electrochemical potential of the electrons in the electrolyte.

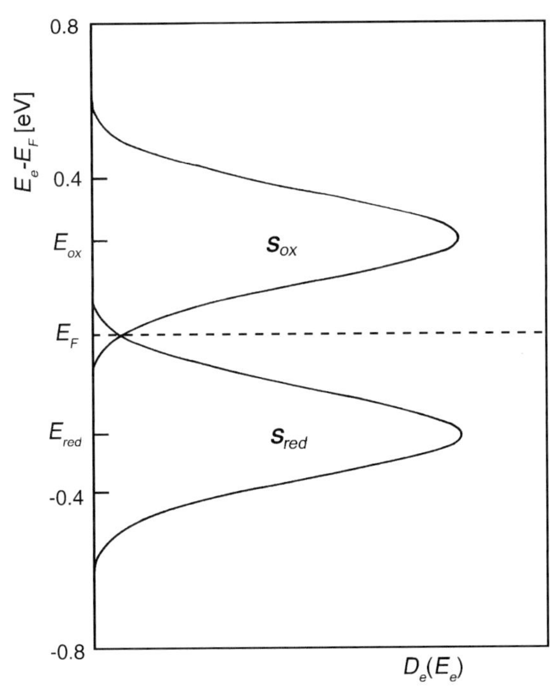

Figure 2.33 DOS of electrons for reduced and oxidized ions as function of energy in an electrolyte. Equal concentrations of oxidized and reduced ions are assumed; the reference point for the energy is the Fermi energy, defined as energy of equal DOS of occupied and reduced states. The Fermi energy is equal to the equilibrium redox potential (Chapter 3).

2.10 COHESION IN SOLIDS

2.10.1 Lattice enthalpy

The cohesion of the atoms of a crystal lattice is determined by the energy that holds the atoms together and which is released when the crystal is disassembled into its atomic, molecular, or ionic components. It will be called the lattice energy or lattice enthalpy $\Delta_{lat}H$.

Forces of different origin contribute to the lattice energy and depend on the distances in the lattice.

- Attractive forces, which are called dispersion forces or London forces.[31]

$$\Delta_{dis}H \approx -\frac{C_{dis}}{r^6} \tag{2.19}$$

- Repulsive forces, which can also be described by an exponential law.

$$\Delta_{rep}H \approx +\frac{C_{rep}}{r^{12}} \tag{2.20}$$

The exponent 12 is often used in this equation but smaller values are also found in the literature. The combination of the two equations is called the Lennard–Jones potential.

$$V_{LJ} = -4\varepsilon\left\{\left(\frac{r_0}{r}\right)^6 - \left(\frac{r_0}{r}\right)^{12}\right\} \tag{2.21}$$

- Dipole forces or multi-pole forces, if the lattice components have dipoles or multipoles. Dipoles can be either intrinsic or induced. Depending on the orientation, the dipole–dipole interaction energy for linear-orientation (Figure 2.34a) is

$$\Delta H = -\frac{2\mu_1\mu_2}{4\pi\varepsilon_0 r^3} \tag{2.22}$$

For a parallel orientation (Figure 2.34a) one gets

$$\Delta H = -\frac{\mu_1\mu_2}{4\pi\varepsilon_0 r^3} \tag{2.23}$$

- When ions form the lattice Coulomb forces as described in Section 2.7.
- Orbital interaction energies as described in Section 2.9.

The lattice enthalpy is the sum of all interactions in the lattice. Sometimes it is necessary to separate it into contributions from interactions between two atoms, molecules, or ions, mostly the nearest neighbors. This will be called the bond energy φ_{i-j} between components i and j.

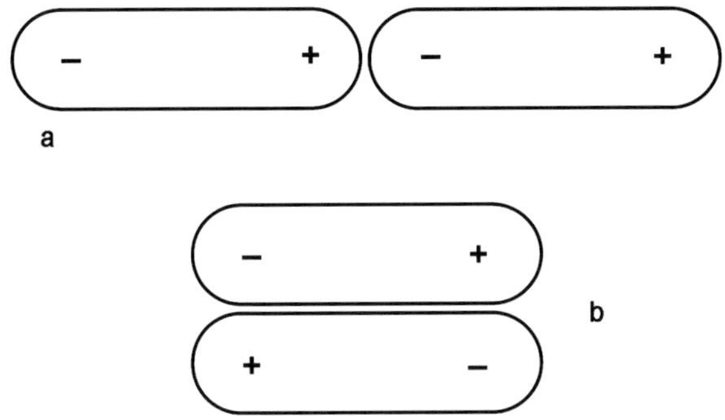

Figure 2.34 Different dipole orientations. (a) Linear orientation and (b) parallel orientation.

2.10.2 Sublimation enthalpy

Experimentally one can determine the energy to separate an atom or molecule from the lattice. This is called the sublimation enthalpy $\Delta_{sub}H$. The sublimation enthalpy is the energy to separate an atom from an ideal kink site position on the surface of the lattice. In this ideal kink site position the atom or molecule is bound with just half of the bonds as in the bulk. Thus, the sublimation enthalpy is just one half of the lattice enthalpy:

$$\Delta_{sub} H = \frac{1}{2} \Delta_{lat} H \qquad (2.24)$$

On a real surface the atoms undergo reorientation processes, which lead to changes in the distribution of energies between different bonds. The bond of an atom in a kink site position (the calculation will be described in the next chapter) can differ from the ideal value.

An approximate value of the sublimation enthalpy is the sum of enthalpy of fusion $\Delta_{fus}H$ and the enthalpy of vaporization $\Delta_{vap}H$.

$$\Delta_{sub} H \cong \Delta_{fus} H + \Delta_{vap} H \qquad (2.25)$$

Values of the sublimation enthalpy of main group and transition metals are shown in Table 2.3[32]

2.10.3 Bond energies of metals

The bond energy φ_{i-j} of main group and transition metals of an atom in an ideal kink site position can be calculated from the sublimation enthalpy $\Delta_{sub}H$.

In the process of sublimation the atom is separated from a special position on the surface, the kink site position (half crystal position). In the ideal kink site position the

Table 2.3

Sublimation enthalpies, energies of atoms in ideal kink positions and bond energies between the two nearest atoms, main group and transition metals[32]

Metal	Structure	$\Delta_{sub}H$ (kJ mol^{-1}) 298.2 K	± (kJ mol^{-1})	φ_{ksp} (eV) 298.2 K	φ_{Me-Me} (eV) 298.2 K
Li	bcc	159.3	± 1.0	1.65	0.27
Na	bcc	107.5	± 0.7	1.11	0.18
K	bcc	89.0	± 0.8	0.92	0.15
Rb	bcc	80.9	± 0.8	0.84	0.14
Cs	bcc	76.5	± 1.0	0.79	0.13
Be	h	324	± 5	3.36	0.49
Mg	h	147.1	± 0.8	1.52	0.22
Ca	c	111.8	± 0.2	1.16	0.17
Sr	c	164.4		1.70	0.25
Ba	c	180.0		1.87	0.31
Sc	h	377.8		3.92	0.57
Y	h	421.3		4.37	0.63
La	hc	431.0		4.47	0.65
Ti	h	473	± 3	4.90	0.71
Zr	h	608.8		6.31	0.91
Hf	h	619.2		6.42	0.93
V	bcc	514.2		5.33	0.87
Nb	bcc	725.9		7.52	1.23
Ta	bcc	782.0		8.10	1.33
Cr	bcc	396.6		4.11	0.67
Mo	bcc	658.1		6.82	1.12
W	bcc	849.4		8.80	1.44
Mn		280.7		2.91	0.48
Tc	h	678.0		7.03	1.02
Re	h	769.9		7.98	1.16
Fe	bcc	416.3		4.31	0.71
Co	h	424.7		4.40	0.64
Ni	c	429.7		4.45	0.65
Ru	h	642.7		6.66	0.97
Rh	c	556.9		5.77	0.84
Pd	c	378.2		3.92	0.57
Os	h	791.0		8.20	1.19
Ir	c	665.3		6.90	1.00
Pt	c	565.3		5.86	0.85
Cu	c	337.4	± 1.2	3.50	0.51
Ag	c	284.9	± 0.8	2.95	0.43
Au	c	366.1		3.79	0.55
Zn	h	130.4	± 0.4	1.35	0.20
Cd	h	111.8	± 0.2	1.16	0.17
Hg	c	61.38	± 0.04	0.64	0.09
Al	h	330	± 4	3.42	0.50
Ga		272.0		2.82	
In	c	243.3		2.52	0.37

Table 2.3 *(Continued)*

Metal	Structure	$\Delta_{sub}H$ (kJ mol^{-1}) 298.2 K	± (kJ mol^{-1})	φ_{ksp} (eV) 298.2 K	φ_{Me-Me} (eV) 298.2 K
Tl	h	182.2		1.89	0.27
Ge		372	± 3	3.86	
Sn		301.2	± 1.5	3.12	
Pb	c	195.2	± 0.8	2.02	0.29
Sb		262.3		2.72	
Bi		207.1		2.15	
Ce	c	423.0		4.38	0.64
Pr	hc	355.6		3.69	0.53
Nd	hc	327.6		3.40	0.49
Pm	hc				
Sm	hhc	206.7		2.14	0.31
Eu	bcc	175.3		1.82	0.26
Gd	h	397.5		4.12	0.60
Tb	h	388.7		4.03	0.58
Dy	h	290.4		3.01	0.44
Ho	h	300.8		3.12	0.45
Er	h	317.1		3.29	0.48
Tm	h	232.2		2.41	0.35
Yb	c	152.3		1.58	0.23
Lu	h	427.6		4.43	0.64
Th	c	602.0		6.24	0.90
U		533	± 8	5.52	0.80

atom is bound by exactly one half of its lattice bonds. On a real surface, the atoms undergo reorientation processes, which lead to changes in the distribution of energies between different bonds. The bond energy of an atom in a kink site position can differ from the ideal value.

The Gibbs energy of an atom in the ideal kink site position $\mu_{1/2}$ is the characteristic value for the molecular interaction of the atom with the bulk matrix. It takes into account bond strengths among first, second, and further neighbors. It is exactly one half of the lattice energy.

The Gibbs energy of an atom in the kink position (chemical potential) can be derived from the sublimation enthalpy (Volmer[33]); a discussion is also found in the book of Budevski, Staikov and Lorenz.[34]

$$\mu_{1/2} = \frac{\Delta_{sub} H}{N_A} + \frac{1}{2} k_B T \tag{2.26}$$

N_A is Avogadros number. In the second term, k_B is the Boltzmann constant and T the temperature, the second term (entropic term) is usually small ($<1\%$). It contains only one degree of freedom because of the critical oscillation of the atom leading to the bond breaking.

The enthalpy of an atom in an ideal kink site position is

$$\varphi_{1/2} \approx \frac{\Delta_{\text{sub}} H}{N_A} \tag{2.27}$$

No strict distinction between free energy and enthalpy is made. Values for enthalpies of atoms in kink positions for main group and transition metals are listed in Table 2.3.

Using an appropriate lattice model, one can calculate the bond energy between two atoms. For instance, in a closed packed hexagonal or cubic lattice with 12 nearest neighbors and 6 nearest neighbors in the kink position, one can approximately write

$$\varphi_{\text{Me-Me}} \approx \frac{\varphi_{1/2}}{6} \tag{2.28}$$

Second and further nearest neighbors are neglected in this first approach. Second and higher nearest neighbors can be taken into account by a suitable assumption for the degree of decay of the interaction energies with the radius. One approximation is based on Eq. (2.19) for the radius dependence. Taking into account all neighbors in a radius of $2r_0$ (r_0 is the radius of the closed packed layer) gives for the ccp and the hcp lattice for the corrected bond energy

$$\varphi^*_{\text{Me-Me}} \approx \frac{\varphi_{1/2}}{6.9} \tag{2.29}$$

In the case of a body centered cubic lattice, the number of next nearest neighbors in the lattice is eight, but the difference between the distance of first and second nearest neighbors is smaller, and the calculation of the bond energy by an equation taking into account only nearest neighbors should be avoided. For calculation of a value of the corrected bond energy Eq. (2.30) could be used

$$\varphi^*_{\text{Me-Me}} \approx \frac{\varphi_{1/2}}{6} \tag{2.30}$$

Values of energies of atoms in ideal kink site positions and bond energies between two metal atoms were summarized for transition and main group metals in Table 2.3. On the real surface, the bond energies in kink site positions and the bond energies between two neighbors can differ from the ideal model values because of relaxation processes on the surface.

2.10.4 Bond energies of alloys

The situation for alloys is more complex. In theoretical calculations of lattice energies one has to take into account the different contributions from Coulomb interactions (ionic bonds), localized orbitals (covalent bonds), and delocalized orbitals (metal bonds). Therefore an *ab initio* calculation is difficult. Solutions of the problem are under development and one can expect that in the near future reliable data will be presented.

An empirical approach to calculate bond energies between different components of an alloy was presented by Miedema.[35] The basic idea of this approach is to use the interaction at the surface of the Wigner–Seitz cell of the component A and the surrounding

Wigner–Seitz cells of the component B for an empirical calculation. The result is an enthalpy value of the interaction of A surrounded by B (or B surrounded by A).

The interaction was described by two terms with different signs. The first term took into account the ionic part of the interaction. It was proportional to the difference of the electronegativities. Only values of the electronegativities were chosen, which took into account empirical corrections. This term was principally a negative term. The second contribution came from the interaction of the electrons, which is, in principal, proportional to the electron density at the interface between the Wigner–Seitz cells.

The Wigner–Seitz cell is the smallest cell of the crystal lattice. While the unit cell is chosen for reasons of symmetry, the Wigner–Seitz cell represents the full periodicity of the electron shell. The Wigner–Seitz cell is an example of the construction of a primitive elementary cell. The characteristic feature of the primitive cell is that it contains only one lattice point. The rules for constructing a Wigner–Seitz cell are:

- A central lattice point must be connected with the surrounding lattice points by straight lines.
- A plane is laid in the middle point of each connecting line.
- The smallest volume set up by the planes is the Wigner–Seitz cell.

This is shown in Figure 2.35 for a two-dimensional lattice.

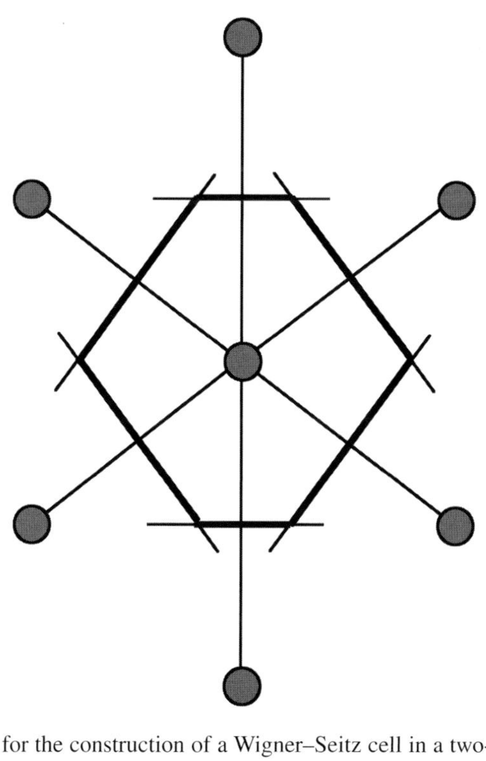

Figure 2.35 Example for the construction of a Wigner–Seitz cell in a two-dimensional lattice.

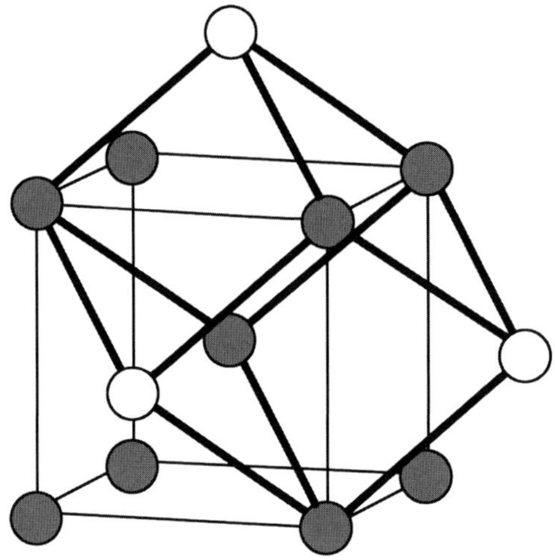

Figure 2.36 Unit cell and Wigner–Seitz cell (thick lines) of the body centered cubic lattice.

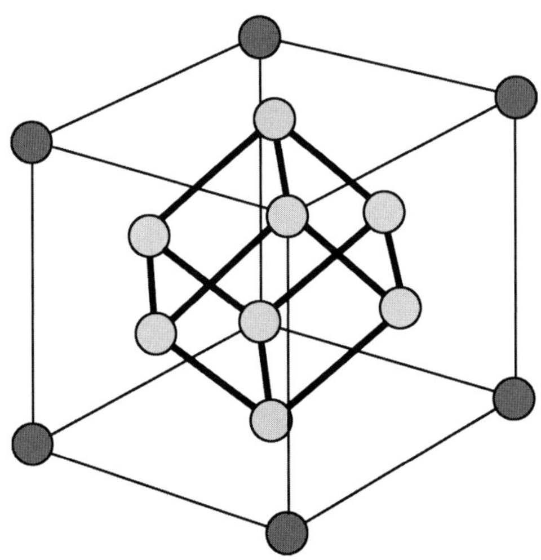

Figure 2.37 Unit cell and Wigner–Seitz cell (thick lines) of the face centered cubic lattice.

Figure 2.36 shows the body centered cubic unit cell and the associated Wigner–Seitz cell. The face centered cubic unit cell and associated Wigner–Seitz cell can be seen in Figure 2.37.

An experimental way to find the smallest primitive cell is the atomic environment concept. The distances of all lattice points around a central lattice are put into a histogram. In the histogram the largest gap between different lattice points is used as the guiding rule to define the coordination polyhedron (Figures 2.38a and b).

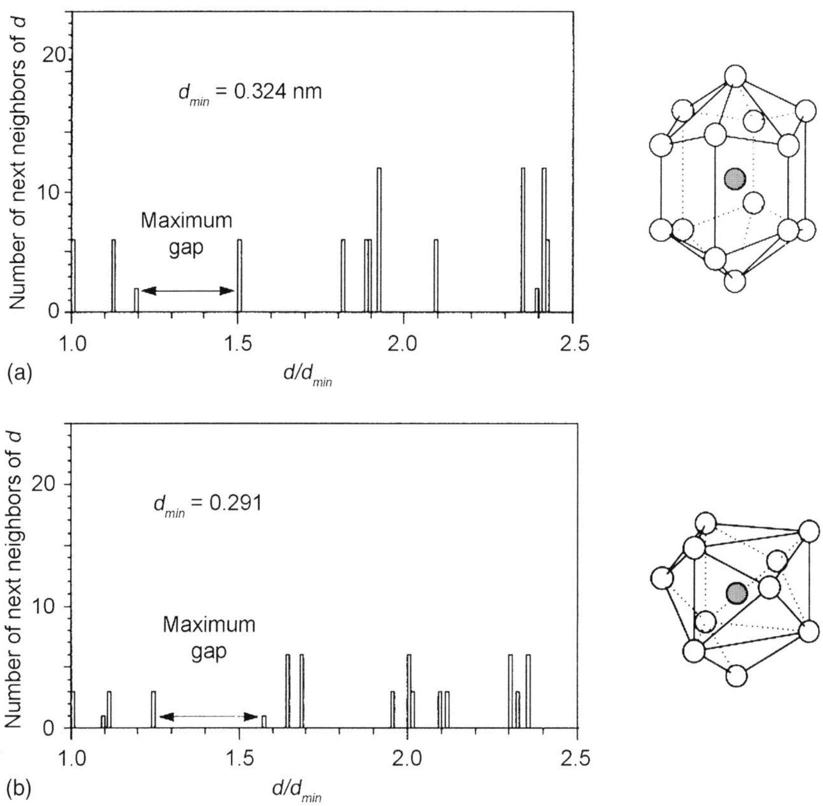

Figure 2.38 Determination of the coordination polyhedron of $CaIn_2$. (a) For Ca as central atom and (b) for In as central atom. (Reproduced with permission from Ref. [36], © 2000, J. Wiley.)

At the interface between the Wigner–Seitz cells of two different components the electron density must be equilibrate. This can be done by a reorganization of the orbital structure and the orbital occupation at the interface. This is, in principal, a positive contribution. This was multiplied by the molar surface area of the Wigner–Seitz cell of component A (for A in B) $V_A^{2/3}$ (V_A molar volume of the Wigner–Seitz cell of component A) and divided by an average of the electron density of the adjacent Wigner–Seitz cells, $(n_{WS}^{1/3})_{av}$ leading to the final equation

$$\Delta H_{\text{A in B}}^{\text{interface}} = V_A^{2/3} \frac{-P(\Delta\chi)^2 + Q(\Delta n_{WS}^{1/3})^2}{(\eta_{WS}^{1/3})_{av}} \tag{2.31}$$

The constants P and Q had to be chosen according to the kind of metals in the alloy, e.g., transition group metals. The empirical selection of the constants was based on the borderline between alloys with formation of intermetallic compounds ($\Delta H_{\text{A in B}}^{\text{interface}}$ negative) and alloys with the tendency to form solid solutions ($\Delta H^{\text{interface}}_{\text{A in B}}$ around zero) or segregating into the components ($\Delta H^{\text{interface}}_{\text{A in B}}$ with larger positive values). For transition metals the following values were selected:

$P = 14.2$ for two metals with valence values larger than two
$P = 10.7$ for mono- and di-valent atoms
$Q = 9.4P$

Enthalpies of interactions $\Delta H^{\text{interface}}_{\text{A in B}}$ calculated by this approach are presented in Table 2.4a. Values $\Delta H^{\text{interface}}_{\text{B in A}}$ could be calculated with the equation

$$\Delta H^{\text{interface}}_{\text{B in A}} = \frac{V_{\text{B}}^{2/3}}{V_{\text{A}}^{2/3}} \Delta H^{\text{interface}}_{\text{A in B}} \tag{2.32}$$

The approach was extended to alloys between transition metals and elements of main groups. Considerable empiric assumptions were necessary to force the different components into this frame.

The values given by Miedema are the formation enthalpies of solid solutions of A in B or vice versa. The solved metal is embedded into the matrix of the solvent. Using a Haber–Born cycle as shown in Figure 2.39 the sublimation enthalpy of the solved metal embedded in the matrix of the solvent can be calculated. Results of sublimation enthalpies of the solved metal embedded in the matrix of the solvent are shown in Table 2.4b.

Bond energies φ_{A-B} can then be calculated with the model of the solved metal A embedded in the matrix of the solvent B (Table 2.4c).

The empirical treatment of Miedema *et al.* is based on some experimental information on the magnitude of the expected cohesion energies. Values of $\Delta H^{\text{interface}}_{\text{A in B}}$ of this first approach are a guide for the magnitude of the expected interactions.

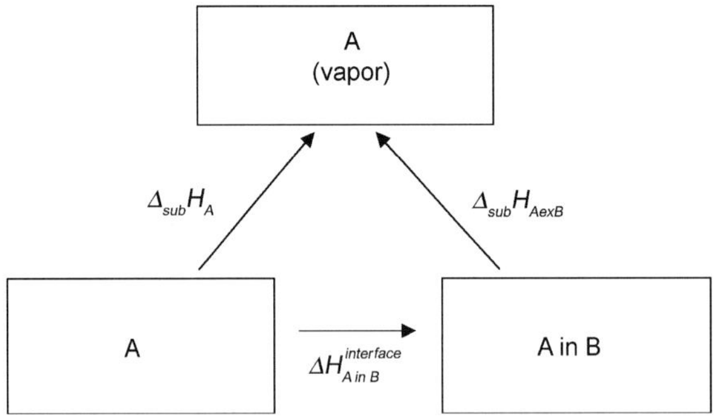

Figure 2.39 Born–Haber cycle for calculation of sublimation enthalpies from enthalpies $\Delta H^{\text{interface}}_{\text{A in B}}$.

Table 2.4a

Solution enthalpies $\Delta H^{\text{interface}}_{\text{A in B}}$ calculated by Miedema for alloys of transition metals

B \ A	Sc	Ti	V	Cr	Mn	Fe	Co	Ni	Y	Zr	Nb	Mo	Tc	Ru	Rh	Pd	La	Hf	Ta	W	Re	Os	Ir	Pt
Sc		28	25	2	−28	−39	−100	−130	4	16	66	38	−141	−157	−221	−319	10	20	59	34	−63	−141	−227	−333
Ti	33		−6	−27	−30	−62	−103	−126	73	−1	8	−14	−153	−169	−204	−260	99	1	6	−23	−100	−160	−227	−304
V	33	−7		−8	−3	−28	−54	−69	85	−17	−5	0	−89	−103	−123	−151	118	−10	−4	−3	−54	−97	−146	−198
Cr	3	−33	−8		8	−6	−18	−27	59	−58	−32	2	−39	−49	−58	−65	93	−44	−30	4	−19	−46	−78	−110
Mn	−40	−36	−3	9		1	−21	−33	−8	−74	−17	22	−37	−46	−70	−100	14	−57	−17	28	−2	−38	−80	−128
Fe	−53	−74	−29	−6	1		−2	−6	−6	−118	−70	−9	−13	−20	−23	−19	25	−98	−67	0	−1	−17	−38	−59
Co	−139	−126	−58	−18	−21	−2		−1	−112	−197	−111	−22	−1	−3	−9	−7	−93	−168	−109	−6	9	0	−15	−33
Ni	−180	−154	−75	−27	−33	−6	−1		−162	−237	−136	−32	3	2	−4	0	−146	−204	−133	−14	10	6	−7	−22
Y	3	51	54	35	−5	−4	−67	−97		34	102	81	−96	−112	−181	−286	1	41	93	81	−14	−93	−180	−289
Zr	16	−1	−13	−41	−52	−85	−137	−165	42		15	−23	−195	−212	−261	−339	60	−1	10	−34	−120	−202	-281	−377
Nb	76	8	−4	−26	−13	−57	−88	−107	139	17		−22	−139	−156	−176	−208	177	17	0	−33	−100	−152	−207	−267
Mo	45	−15	0	1	18	−7	−18	−27	114	−27	−23		−45	−56	−59	−59	156	−17	−20	−1	−27	−55	−86	−114
Tc	−162	−155	−80	−33	−31	−11	−1	2	−132	−232	−147	−45		−1	1	17	−111	−201	−142	−28	1	1	−6	−13
Ru	−185	−176	−95	−41	−40	−17	−3	2	−158	−260	−169	−58	−1		5	25	−137	−227	−164	−40	−4	0	−2	−4
Rh	−257	−211	−113	−50	−60	−20	−8	−3	−253	−316	−188	−61	1	5		8	−245	−277	−185	−37	4	8	3	−7
Pd	−352	−256	−132	−53	−82	−16	−5	0	−379	−391	−212	−57	17	24	7		−388	−342	−210	−26	26	31	24	8
La	7	64	69	52	8	14	−51	−81	1	45	120	103	−75	−90	−162	−270		53	111	104	9	−70	−158	−267
Hf	21	1	−8	−32	−41	−71	−119	−145	51	−1	15	−15	−172	−188	−232	−301	72		11	−228	−111	−178	−253	−340
Ta	67	5	−4	−24	−14	−54	−86	−105	127	12	0	−19	−135	−152	−173	−206	164	12		−29	−96	−147	−203	−263
W	39	−23	−3	4	23	0	−5	−11	111	−39	−34	−1	−28	−38	−36	−26	154	−27	−30		−16	−38	−62	−82
Re	−72	−101	−48	−16	−2	−1	8	8	−20	−153	−105	−27	1	−4	4	26	14	−129	−100	−17		−4	−13	−18
Os	−162	−164	−88	−39	−32	−15	0	5	−128	−241	−161	−56	1	0	8	32	−104	−210	−156	−39	−4		−2	−2
Ir	−235	−228	−130	−65	−66	−32	−13	−5	−242	−331	−216	−85	−6	−2	2	24	−230	−293	−211	−63	−13	−2		2
Pt	−354	−290	−167	−87	−101	−47	−26	−17	−367	−421	−264	−108	−12	−4	−6	8	−368	−374	−200	−79	−17	−2	1	
Th	0	28	28	7	−24	−33	−93	−122	5	15	67	43	−132	−147	−210	−306	11	20	60	40	−55	−131	−216	−320
U	14	−1	3	−9	−18	−38	−77	−98	47	−13	15	8	−116	−130	−167	−222	70	−7	12	4	−64	−121	−184	−257
Pt	−3	6	16	6	−13	−22	−66	−88	18	−13	34	31	−98	−112	−157	−224	34	−4	30	31	−39	−100	−167	−247
Cu	−115	−40	21	51	15	53	26	14	−117	−110	12	82	37	31	−11	−62	−115	−81	9	101	82	46	1	−56
Ag	−119	−6	63	98	46	103	69	56	−136	−87	66	147	96	91	37	−29	−143	−53	61	172	151	111	62	−3
Au	−291	−180	−69	−1	−39	28	25	25	−318	−303	−124	13	52	56	27	0	−328	−257	125	44	76	68	48	17

Table 2.4b

Sublimation enthalpies calculated with the Born–Haber cycle in Figure 2.39 for transition metals A in B

A	Sc	Ti	V	Cr	Mn	Fe	Co	Ni	Y	Zr	Nb	Mo	Tc	Ru	Rh	Pd	La	Hf	Ta	W	Re	Os	Ir	Pt
$\Delta_{sub}H$	378	473	514	397	281	416	425	430	421	609	726	658	678	643	557	378	431	619	782	849	770	791	665	565
B																								
Sc		445	489	395	309	455	525	560	417	593	660	620	819	800	778	697	421	599	723	815	833	932	892	898
Ti	345		520	424	311	478	528	556	348	610	718	672	831	812	761	638	332	618	776	872	870	951	892	869
V	345	480		405	284	444	479	499	336	626	731	658	767	746	680	529	313	629	786	852	824	888	811	763
Cr	375	506	522		273	422	443	457	362	667	758	656	717	692	615	443	338	663	812	845	789	837	743	675
Mn	418	509	517	388		415	446	463	429	683	743	636	715	689	627	478	417	676	799	821	772	829	745	693
Fe	431	547	543	403	280		427	436	427	727	796	667	691	663	580	397	406	717	849	849	771	808	703	624
Co	517	599	572	415	302	418		431	533	806	837	680	679	646	566	385	524	787	891	855	761	791	680	598
Ni	558	627	589	424	314	422	426		583	846	862	690	675	641	561	378	577	823	915	863	760	785	672	587
Y	375	422	460	362	286	420	492	527		575	624	577	774	755	738	664	430	578	689	768	784	884	845	854
Zr	362	474	527	438	333	501	562	595	379		711	681	873	855	818	717	371	620	772	883	890	993	946	942
Nb	302	465	518	423	294	473	513	537	282	592		680	817	799	733	586	254	602	782	882	870	943	872	832
Mo	333	488	514	396	263	423	443	457	307	636	749		723	699	616	437	275	636	802	850	797	846	751	679
Tc	540	628	594	430	312	427	426	428	553	841	873	703		644	556	361	542	820	924	877	769	790	671	578
Ru	563	649	609	438	321	433	428	428	579	869	895	716	679		552	353	568	846	946	889	774	791	667	569
Rh	635	684	627	447	341	436	433	433	674	925	914	719	677	638		370	676	896	967	886	766	783	662	572
Pd	730	729	646	450	363	432	430	430	800	1000	938	715	661	619	550		819	961	992	875	744	760	641	557
La	371	409	445	345	273	402	476	511	420	564	606	555	753	733	719	648		566	671	745	761	861	823	832
Hf	357	472	522	429	322	487	544	575	370	610	711	673	850	831	789	679	359		771	1077	881	969	918	905
Ta	311	468	518	421	295	470	511	535	294	597	726	677	813	795	730	584	267	607		878	866	938	868	828
W	339	496	517	393	258	416	430	441	310	648	760	659	706	681	593	404	277	646	812		786	829	727	647
Re	450	574	562	413	283	417	417	422	441	762	831	685	677	647	553	352	417	748	882	866		795	678	583
Os	540	637	602	436	313	431	425	425	549	850	887	714	677	643	549	346	535	829	938	888	774		667	567
Ir	613	701	644	462	347	448	438	435	663	940	942	743	684	645	555	354	661	912	993	912	783	793		563
Pt	732	763	681	484	382	463	451	447	788	1030	990	766	690	647	563	370	799	993	982	928	787	793	664	
Th	378	445	486	390	305	449	518	552	416	594	659	615	810	790	767	684	420	599	722	809	825	922	881	885
U	364	474	511	406	299	454	502	528	374	622	711	650	794	773	724	600	361	626	770	845	834	912	849	822
Pu	381	467	498	391	294	438	491	518	403	622	692	627	776	755	714	602	397	623	752	818	809	891	832	812
Cu	493	513	493	346	266	363	399	416	538	719	714	576	641	612	568	440	546	700	773	748	688	745	664	621
Ag	497	479	451	299	235	313	356	374	557	696	660	511	582	552	520	407	574	672	721	677	619	680	603	568
Au	669	653	583	398	320	388	400	405	739	912	850	645	626	587	530	378	759	876	907	805	694	723	617	548

Table 2.4c

Bond energies φ_{A-B} calculated from the sublimation enthalpies for alloys of transition metals

B	A CN	Sc	Ti	V	Cr	Mn	Fe	Co	Ni	Y	Zr	Nb	Mo	Tc	Ru	Rh	Pd	La	Hf	Ta	W	Re	Os	Ir	Pt
Sc	hcp		0.77	0.85	0.68	0.53	0.79	0.91	0.97	0.72	1.02	1.14	1.07	1.41	1.38	1.34	1.20	0.73	1.04	1.25	1.41	1.44	1.61	1.54	1.55
Ti	hcp	0.60		0.90	0.73	0.54	0.83	0.91	0.96	0.60	1.05	1.24	1.16	1.44	1.40	1.31	1.10	0.57	1.07	1.34	1.51	1.50	1.64	1.54	1.50
V	bcc	0.60	0.83		0.70	0.49	0.77	0.83	0.86	0.58	1.08	1.26	1.14	1.32	1.29	1.17	0.91	0.54	1.09	1.36	1.47	1.42	1.53	1.40	1.32
Cr	bcc	0.65	0.87	0.90		0.47	0.73	0.76	0.79	0.63	1.15	1.31	1.13	1.24	1.19	1.06	0.77	0.58	1.15	1.40	1.46	1.36	1.45	1.28	1.17
Mn		0.72	0.88	0.89	0.67		0.72	0.77	0.80	0.74	1.18	1.28	1.10	1.24	1.19	1.08	0.83	0.72	1.17	1.38	1.42	1.33	1.43	1.29	1.20
Fe	bcc	0.74	0.94	0.94	0.70	0.48		0.74	0.75	0.74	1.26	1.37	1.15	1.19	1.14	1.00	0.69	0.70	1.24	1.47	1.47	1.33	1.40	1.21	1.08
Co	hcp	0.89	1.03	0.99	0.72	0.52	0.72		0.74	0.92	1.39	1.45	1.17	1.17	1.12	0.98	0.67	0.91	1.36	1.54	1.48	1.31	1.37	1.18	1.03
Ni	hcp	0.96	1.08	1.02	0.73	0.54	0.73	0.74		1.01	1.46	1.49	1.19	1.17	1.11	0.97	0.65	1.00	1.42	1.58	1.49	1.31	1.36	1.16	1.01
Y	hcp	0.65	0.73	0.79	0.62	0.49	0.73	0.85	0.91		0.99	1.08	1.00	1.34	1.30	1.27	1.15	0.74	1.00	1.19	1.33	1.35	1.53	1.46	1.48
Zr	hcp	0.62	0.82	0.91	0.76	0.57	0.87	0.97	1.03	0.66		1.23	1.18	1.51	1.48	1.41	1.24	0.64	1.07	1.33	1.53	1.54	1.72	1.63	1.63
Nb	bcc	0.52	0.80	0.90	0.73	0.51	0.82	0.89	0.93	0.49	1.02		1.17	1.41	1.38	1.27	1.01	0.44	1.04	1.35	1.52	1.50	1.63	1.51	1.44
Mo	bcc	0.57	0.84	0.89	0.68	0.45	0.73	0.76	0.79	0.53	1.10	1.29			1.21	1.06	0.76	0.48	1.10	1.39	1.47	1.38	1.46	1.30	1.17
Tc	hcp	0.93	1.08	1.03	0.74	0.54	0.74	0.74	0.74	0.96	1.45	1.51	1.21		1.11	0.96	0.62	0.94	1.42	1.60	1.52	1.33	1.36	1.16	1.00
Ru	hcp	0.97	1.12	1.05	0.76	0.55	0.75	0.74	0.74	1.00	1.50	1.55	1.24	1.17		0.95	0.61	0.98	1.46	1.63	1.54	1.34	1.37	1.15	0.98
Rh	ccp	1.10	1.18	1.08	0.77	0.59	0.75	0.75	0.75	1.16	1.60	1.58	1.24	1.17	1.10		0.64	1.17	1.55	1.67	1.53	1.32	1.35	1.14	0.99
Pd	ccp	1.26	1.26	1.12	0.78	0.63	0.75	0.74	0.74	1.38	1.73	1.62	1.24	1.14	1.07	0.95		1.41	1.66	1.71	1.51	1.28	1.31	1.11	0.96
La	hc	0.64	0.71	0.77	0.60	0.47	0.69	0.82	0.88	0.73	0.97	1.05	0.96	1.30	1.27	1.24	1.12		0.98	1.16	1.29	1.31	1.49	1.42	1.44
Hf	hcp	0.62	0.82	0.90	0.74	0.56	0.84	0.94	0.99	0.64	1.05	1.23	1.16	1.47	1.43	1.36	1.17	0.62		1.33	1.86	1.52	1.67	1.59	1.56
Ta	bcc	0.54	0.81	0.90	0.73	0.51	0.81	0.88	0.92	0.51	1.03	1.25	1.17	1.40	1.37	1.26	1.01	0.46	1.05		1.52	1.50	1.62	1.50	1.43
W	bcc	0.59	0.86	0.89	0.68	0.45	0.72	0.74	0.76	0.54	1.12	1.31	1.14	1.22	1.18	1.02	0.70	0.48	1.12	1.40		1.36	1.43	1.26	1.12
Re	hcp	0.78	0.99	0.97	0.71	0.49	0.72	0.72	0.73	0.76	1.32	1.44	1.18	1.17	1.12	0.96	0.61	0.72	1.29	1.52	1.50		1.37	1.17	1.01
Os	hcp	0.93	1.10	1.04	0.75	0.54	0.74	0.73	0.73	0.95	1.47	1.53	1.23	1.17	1.11	0.95	0.60	0.92	1.43	1.62	1.53	1.34		1.15	0.98
Ir	ccp	1.06	1.21	1.11	0.80	0.60	0.77	0.76	0.75	1.15	1.62	1.63	1.28	1.18	1.11	0.96	0.61	1.14	1.58	1.72	1.58	1.35	1.37		0.97
Pt	ccp	1.26	1.32	1.18	0.84	0.66	0.80	0.78	0.77	1.36	1.78	1.71	1.32	1.19	1.12	0.97	0.64	1.38	1.72	1.70	1.60	1.36	1.37	1.15	
Th	ccp	0.65	0.77	0.84	0.67	0.53	0.78	0.89	0.95	0.72	1.03	1.14	1.06	1.40	1.36	1.32	1.18	0.73	1.04	1.25	1.40	1.42	1.59	1.52	1.53
U		0.63	0.82	0.88	0.70	0.52	0.78	0.87	0.91	0.65	1.07	1.23	1.12	1.37	1.33	1.25	1.04	0.62	1.08	1.33	1.46	1.44	1.58	1.47	1.42
Pu		0.66	0.81	0.86	0.67	0.51	0.76	0.85	0.89	0.70	1.07	1.20	1.08	1.34	1.30	1.23	1.04	0.69	1.08	1.30	1.41	1.40	1.54	1.44	1.40
Cu	ccp	0.85	0.89	0.85	0.60	0.46	0.63	0.69	0.72	0.93	1.24	1.23	1.00	1.11	1.06	0.98	0.76	0.94	1.21	1.34	1.29	1.19	1.29	1.15	1.07
Ag	ccp	0.86	0.83	0.78	0.52	0.41	0.54	0.61	0.65	0.96	1.20	1.14	0.88	1.01	0.95	0.90	0.70	0.99	1.16	1.25	1.17	1.07	1.17	1.04	0.98
Au	ccp	1.16	1.13	1.01	0.69	0.55	0.67	0.69	0.70	1.28	1.57	1.47	1.11	1.08	1.01	0.92	0.65	1.31	1.51	1.57	1.39	1.20	1.25	1.07	0.95

END NOTE

*The construction of the Wigner–Seitz cells is explained in Section 2.10.4.

REFERENCES

1. L. Pauling, The Nature of the Chemical Bond, 2nd Ed., Cornell University Press, Ithaca, NY, 1948.
2. G. Frenking, S Shaik (Eds.), 90 Years of Chemical Bonding, J. Comput. Chem., *28*, 1–466 (2007).
3. C. Kittel, Introduction into Solid State Physics, Wiley, New York, 1966.
4. P. Villars, Crystal Structures of Intermetallic Compounds, Factors Governing Crystal Structures, in J.H. Westbrook, R.L. Fleischer (Eds.), Wiley, Chichester, 2000, p. 1.
5. D.G. Pettifor, Crystal Structures of Intermetallic Compounds, Structure Mapping, in J.H. Westbrook, R.L. Fleischer (Eds.), Wiley, Chichester, 2000, p. 195.
6. See: S. Mardix, Acta Crystallogr., A *46*, 1 (1990).
7. Müller, Anorganische Strukturchemie, 2nd Ed., Teubner Verlag, Stuttgart, 1992.
8. J. Hauck, K, Micka, Crystal Structures of Intermetallic Compounds, in J.H. Westbrook, R.L. Fleischer (Eds.), Wiley, Chichester, 2000, p. 55.
9. T.P. Moffat, Electrochemical Nanotechnology, STM Studies of Electrodeposition of Strained-Layer Metallic Superlattices, in W.J. Lorenz, W. Plieth (Eds.), Wiley, Weinheim, 1998, p. 171.
10. T.B. Massalski, H. Okamoto, P.R. Subramanian, L. Kacprzak, Binary Alloy Phase Diagrams, 2nd Ed., ASM International, Metals Park, OH, 1991.
11. W. Hume-Rothery, R.E. Smallman, C.W. Haworth, The Structure of Metals and Alloys, The Institute of Metals, London, 1969.
12. A.J. Martyanov, S.S. Batsanov, Russ. J. Inorg. Chem., *25*, 1737 (1980).
13. J.A. Alonso, Rev. Latinoam. Metal. Mater., *5*, 3 (1985).
14. H. Jones, The Theory of Brillouin Zones and Electronic States in Crystals, North Holland, Amsterdam, 1962.
15. E. Madelung, Physik. Z., *19*, 524 (1918).
16. M.P. Tosi, Solid State Phys., *16*, 1 (1964).
17. G. Kortüm, Elektrochemie, VCH, Weinheim, 1957, p. 97.
18. See: R.J. Gillespie, Molekülgeometrie, VCH, Weinheim, 1975.
19. G. Merino, A. Vela, T. Heine., Chem. Rev., *105*, 3812 (2005).
20. C.J. Ballhausen, Ligand Field Theory, MacGraw-Hill, New York, 1962.
21. R. Hoffmann, Solids and Surfaces: A Chemists View on Bonding in Extended Structures, VCH, Weinheim, 1988.
22. C. Kittel, Quantum Theory of Solids, 4th Ed., Wiley, New York, 1967.
23. J.Y. Saillard, R.J. Hoffmann, J. Am. Chem. Soc., *106*, 2006 (1984).
24. J.K. Burdett, T. Hughbanks, Inorg. Chem., *24*, 1741 (1985).
25. J.K. Burdett, Inorg. Chem., *24*, 2244 (1985).
26. R.E. Peierls, Quantum Theory of Solids, Oxford University Press, Oxford, 1972.
27. N.W. Ashcroft, Nature, *419*, 569 (2002).
28. R.W. Gurney, Proc. Royal Soc. (London), A*134*, 137 (1931).
29. O. Essin, Acta Physicochim. USSR, *13*, 123 (1940).
30. H. Gerischer, Z. Physikal. Chem. NF, *26*, 223, 325 (1960).
31. F. London, Z. Physik, *63*, 245 (1930); Trans. Faraday Soc., *33*, 8 (1937).

32. D.R. Lide (Ed.), Handbook of Chemistry and Physics, Taylor and Francis, 2006.
33. M. Volmer, Kinetik der Phasenbildung, Theodor Steinkopf-Verlag, Dresden and Leipzig, 1939.
34. E. Budevski, G. Staikov, J.W. Lorenz, Electrochemical Phase Formation and Growth, VCH, Weinheim, 1996.
35. F.R. de Boer, R. Boom, W. Mattens, A.R. Miedema, A.K. Niessen, Cohesion in Metals: Transition Metal Alloys, in Cohesion and Structure, in F.R. de Boer, D.G. Petifor (Eds.), Vol. 1, North Holland, Amsterdam, 1998.
36. J.L.C. Daams, Crystal Structures of Intermetallic Compounds, Atomic Environment in some related Intermetallic Compounds, in J.H. Westbrook, R.L. Fleischer (Eds.), Wiley, Chichester, 2000, p. 139.

– 3 –

Electrode Potentials

3.1 PURE METALS

3.1.1 Equilibrium between a metal phase and an electrolyte phase

If a metal is immersed in an electrolyte the equilibrium of the metal atoms and the hydrated metal ions in an aqueous electrolyte is described by the equation

$$Me^{z+} \cdot nH_2O + ze^- \leftrightarrow Me + nH_2O \tag{3.1}$$

According to the jellium model the metal can be considered as metal ions embedded in the electron plasma. The thermodynamic condition for electronic equilibrium between both phases is, from a chemist's point of view, equal values of the electrochemical potentials $\tilde{\mu}_e$ of the electrons.

$$\tilde{\mu}_{e,\mathrm{m}} = \tilde{\mu}_{e,\mathrm{el}} \tag{3.2}$$

From a physicist's point of view, the condition for electronic equilibrium is equal values of the Fermi energy E_F. Electronic equilibrium concerns all charged particles and might also be formulated for the metal ions. The equilibrium contact between a metal phase and an electrolyte phase is shown in Figure 3.1.

The equilibrium is achieved by the build-up of a potential difference, which can be expressed by the equation

$$\mu_{e,\mathrm{m}} - F\varphi_{\mathrm{m}} = \mu_{e,\mathrm{el}} - F\varphi_{\mathrm{el}} \tag{3.3}$$

where μ is the chemical potential, φ the Galvani potential, and F the Faraday constant. The Galvani potential describes the electrical condition of a phase, the sum of inner potential and surface (Volta) potential.

$$\varphi = \psi + \chi \tag{3.4}$$

By controlling the surface charge, the Volta potential can be controlled. The inner and Galvani potentials can neither be controlled nor measured. Only differences of the Galvani potential between two phases of equal chemical composition can be measured. Trassatti and Parsons[1] have given a detailed description.

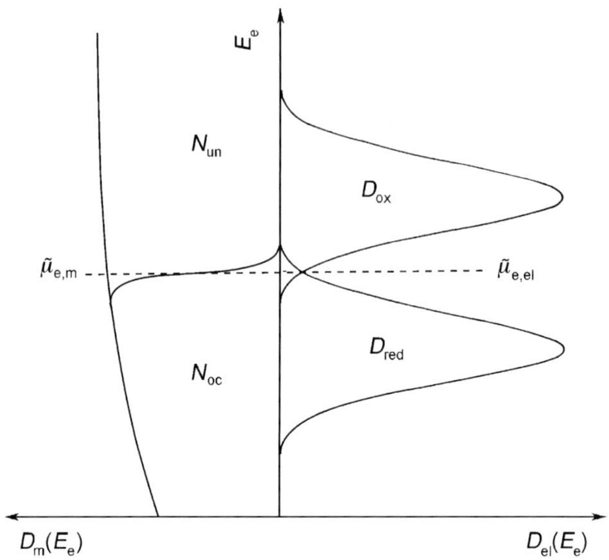

Figure 3.1 Electronic equilibrium between a metallic phase and an electrolyte phase. The electronic energy states in the metal are described by the energy band (Section 2.9). The occupied states are N_{oc} and D_{red}. The density of states of electrons in the electrolyte are the energy distribution functions of the reduced and oxidized components of a redox system, e.g., Fe^{2+} and Fe^{3+} ions (Section 2.9.10). The equilibrium condition is equal values of the electrochemical potentials $\tilde{\mu}$ of the electrons in both phases. An alternative condition is equal values of the Fermi energy E_F in both phases.

Eq. (3.3) can be reformated to obtain an equation for the Galvani potential difference

$$\varphi_m - \varphi_{el} = -\frac{\mu_{e,m} - \mu_{e,el}}{F} \tag{3.5}$$

By connecting the metal phase with a reference electrode of constant Galvani potential, variations of $\varphi_m - \varphi_{el}$ with the experimental conditions can be measured as electrode potential E, which is the difference of the inner potential of the Cu electrode and the connecting Cu wire. This is shown in the potential diagram in Figure 3.2.

3.1.2 Standard electrode potentials

After obtaining a reversible system and inserting the concentration dependence of the chemical potential of the electrons in the electrolyte into Eq. (3.5), the Nernst equation for the equilibrium potential E_0 (Nernst potential) of the metal/metal ion equilibrium is derived.

$$E_0 = E_0^{\ominus} + \frac{RT}{zF} \ln a_{Me^{z+}} \tag{3.6}$$

$E_0^{\ominus}$ is the standard potential, defined for ambient pressure of $p=1$ atm (1.01325 bar) and an activity of $a=1$. In recent definitions for standard conditions $p=1$ bar was suggested.

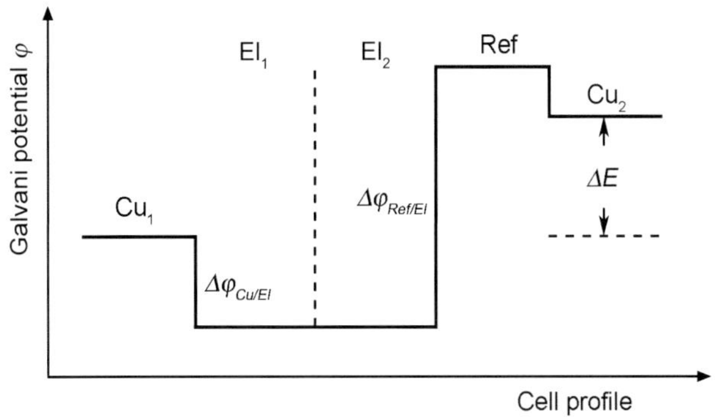

Figure 3.2 Potential diagram of an electrochemical cell with a Cu working electrode with unknown Galvani potential difference connected to a reference electrode of constant Galvani potential difference. Changes of the unknown Galvani potential difference of the working electrode can be measured as difference of the inner potentials of the Cu electrode Cu_1 and the connecting Cu wire Cu_2. The diffusion potential between electrolytes El_1 and El_2 is neglected.

The activity is defined by the equation

$$a_{Me^{z+}} = \frac{m_{Me^{z+}}}{m^{\ominus}} \gamma_{Me^{z+}} \tag{3.7}$$

where $m_{Me^{z+}}$ the molality of metal ions, $m^{\ominus}$ the molality for standard conditions equal to $1\ \text{mol} \cdot \text{kg}^{-1}$, and $\gamma_{Me^{z+}}$ the activity coefficient.

In these definitions the molar masses of the metal ions are measured in molalities ($\text{mol} \cdot \text{kg}^{-1}$). A more common way to measure molar masses is the concentration c, $\text{mol} \cdot \text{dm}^{-3}$ or $\text{mol} \cdot \text{l}^{-1}$ electrolyte (*molarity*). In this case the molar mass for standard conditions is $1\ \text{mol} \cdot \text{l}^{-1}$. The molar masses (mole fractions) of the metal atoms in a pure metal and the molar masses of the solvent molecules of a diluted electrolyte (usually water) are one and therefore do not occur in Eq. (3.7).

In an electrolyte, the activity coefficient of the metal ions $\gamma_{Me^{z+}}$ cannot be neglected and deviates considerably from one. The activity coefficient depends on the ionic strength I, defined in Eq. (1.12). For diluted solutions ($c<10^{-2}\ \text{mol} \cdot \text{kg}^{-1}$), it can be calculated by the Debye–Hückel theory in the first approximation (Eq. (1.15)). In practice, if working with an excess of supporting electrolytes, the activity coefficient is approximately constant. The variation of the equilibrium potential E_0 with the concentration is determined by the Nernst factor $2.303\ RT/F = 59.2$ mV at 25 °C for $z_+ = 1$ (decadal logarithm).

To measure the potential, the half-cell is connected to a suitable second half-cell. For example, a copper electrode (Cu metal in $CuSO_4$) can be coupled with a zinc electrode (Zn metal in $ZnSO_4$). A diffusion potential E_{dif} usually occurs then at the interface of the two electrolytes. The measured potential is given by the equation

$$\Delta E = E_{Cu} - E_{Zn} + E_{dif} \tag{3.8}$$

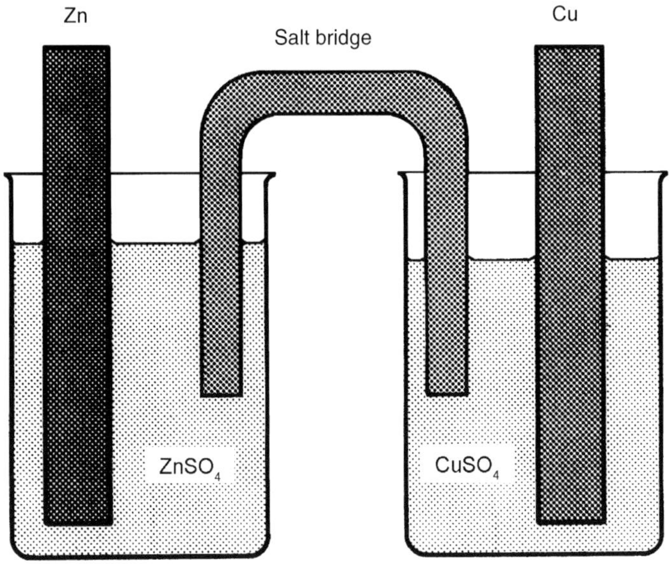

Figure 3.3 Combination of two half-cells (Cu electrode in a $CuSO_4$ electrolyte and Zn electrode in a $ZnSO_4$ electrolyte) joined by a salt bridge. The salt bridge suppresses the diffusion potential.

The diffusion potential is approximately given by the Henderson equation

$$E_{dif} = -\frac{u_+ - u_-}{u_+ + u_-}\frac{RT}{F}\ln\frac{a_\pm^{Zn}}{a_\pm^{Cu}} \tag{3.9}$$

where u_+ and u_- being the mobility of ions and $a_\pm$ the mean ion activity. A way to suppress the diffusion potential is to join the two half-cells by a salt bridge, an electrolytic connection with a saturated solution of, e.g., potassium chloride or potassium nitrate in agar jelly (Figure 3.3). The selected salt should have a similar mobility of cations and anions so that the diffusion potential is brought down to the usual measurement accuracy of 1 mV.

The copper electrode can be coupled to a standard hydrogen electrode (SHE) as shown in Figure 3.4.

If both half-cells contain sulfuric acid of a concentration much larger than the copper sulfate concentration, the diffusion potential can be neglected. The cell reaction is given by the equation

$$Cu^{2+} + H_2 \rightarrow Cu + 2H^+ \tag{3.10}$$

The cell voltage is given by the equation

$$\Delta E = \Delta E_0 + \frac{RT}{2F}\ln\frac{a_{Cu^{2+}}\,p_{H_2}}{a_{H^+}^2} \tag{3.11}$$

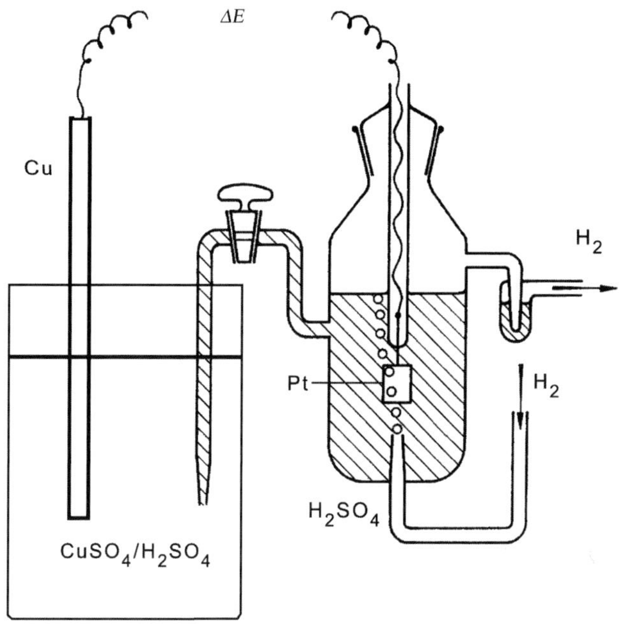

Figure 3.4 Combination of the Cu electrode with a standard hydrogen electrode (SHE). The SHE is operated at standard conditions (ambient pressure of 1 atm (1.01325 bar) and proton activity $a=1$) and at the temperature of 25 °C.

For standard conditions (ambient pressure of 1 atm (1.01325 bar), proton activity $a = 1$) ΔE_0 is equal to the standard potential of the copper electrode $E_0^{\ominus}$. The dependence on the decadic logarithm of the copper ion activity at 25 °C is

$$E_0 = E_0^{\ominus} + \frac{0.059}{2} \log a_{Cu^{2+}} \tag{3.12}$$

Values of standard electrode potentials of metals in aqueous electrolytes are given in Table 3.1.

For unknown activities the measurement of the standard electrode potential is more complicated. The standard electrode potential is defined at $m^{\ominus} = 1$ mol · kg^{-1} with the hypothetical activity coefficient of $\gamma = 1$ (ideal diluted solution). First principal experimental determinations of standard potentials may only be made by extrapolation to this hypothetical value. For measurements, selected cell arrangements are used with complete elimination of the diffusion potential and with diluted electrolytes. For the correction of the activity the Debye–Hückel approximation (Eq. (1.15)) may be used, for example, for the Harned cell Ag/AgCl, HCl ($m_{\pm}$)/Pt(H_2).[3] A concentration corrected potential value is plotted versus the square root of the molality. The extrapolation to $m_{\pm} = 0$ gives the standard potential of the Ag/AgCl electrode (Figure 3.5). Using this electrode as reference electrode other standard potentials can be determined.

Table 3.1

Standard electrode potentials[2] of metal ion electrodes versus
(a) the standard hydrogen electrode (SHE) at 25 °C and (b) on the physical scale with $E_{SHE} = -4{,}44V$

Electrode process	Potential versus SCE (V)	Physical scale (eV)	Electrode process	Potential versus SCE (V)	Physical scale (eV)
$Li^+ + e^- \leftrightarrow Li$	−3.045	−1.40	$Ga^{+++} + 3e^- \leftrightarrow Ga$	−0.52	−3.92
$Rb^+ + e^- \leftrightarrow Rb$	−2.925	−1.51	$S^{--} \leftrightarrow S + 2e^-$	−0.51	−3.93
$Cs^+ + e^- \leftrightarrow Cs$	−2.92	−1.52	$Fe^{++} + 2e^- \leftrightarrow Fe$	−0.441	−4.00
$K^+ + e^- \leftrightarrow K$	−2.922	−1.52	$Cd^{++} + 2e^- \leftrightarrow Cd$	−0.400	−4.04
$Ba^{++} + 2e^- \leftrightarrow Ba$	−2.92	−1.52	$In^{+++} + 3e^- \leftrightarrow In$	−0.34	−4.10
$Sr^{++} + 2e^- \leftrightarrow Sr$	−2.89	−1.55	$Tl^{++} + e^- \leftrightarrow Tl$	−0.336	−4.10
$Ca^{++} + 2e^- \leftrightarrow Ca$	−2.76	−1.68	$Co^{++} + 2e^- \leftrightarrow Co$	−0.283	−4.16
$Na^+ + e^- \leftrightarrow Na$	−2.712	−1.73	$Ni^{++} + 2e^- \leftrightarrow Ni$	−0.236	−4.20
$La^{+++} + 3e^- \leftrightarrow La$	−2.4	−2.04	$Mo^{+++} + 3e^- \leftrightarrow Mo$	−0.20	−4.24
$Mg^{+++} + 2e^- \leftrightarrow Mg$	−2.375	−2.06	$Sn^{++} + 2e^- \leftrightarrow Sn$	−0.136	−4.30
$Th^{++++} + 4e^- \leftrightarrow Th$	−2.1	−2.34	$Pb^{++} + 2e^- \leftrightarrow Pb$	−0.126	−4.31
$Ti^{++} + 4e^- \leftrightarrow Ti$	−1.63	−2.81	$Fe^{+++} + 3e^- \leftrightarrow Fe$	−0.045	−4.40
$HfO^{++} + 2H + 4e^- \leftrightarrow Hf + H_2O$	−1.7	−2.74	$2H^+ + 2e^- \leftrightarrow H_2$	−0.000	−4.44
$Be^{++} + 2e^- \leftrightarrow Be$	−1.70	−2.74	$BiO^- + 2H^+ + e^- \leftrightarrow Bi + H_2O$	+0.32	−4.76
$Al^{+++} + 3e^- \leftrightarrow Al$	−1.706	−2.73	$Cu^{++} + 2e^- \leftrightarrow Cu$	+0.340	−4.79
$ZrO^{++} + 2H^+ + 4e^- \leftrightarrow Zr + H_2O$	−1.5	−2.94	$Cu^+ + e^- \leftrightarrow Cu$	+0.521	−4.96
$V^{++} + 2e^- \leftrightarrow V$	−1.5	−2.94	$Te^{++++} + 4e^- \leftrightarrow Te$	+0.56	−5.00
$WO_4^{--} + 4H_2O^+ + 6e^- \leftrightarrow W + 8OH^-$	−1.1	−3.34	$Po^{+++} + 3e^- \leftrightarrow Po$	+0.56	−5.00
$Mn^{++} + 2e^- \leftrightarrow Mn$	−1.05	−3.39	$Rh^{++} + 2e^- \leftrightarrow Rh$	+0.6	−5.04
$Te^{--} \leftrightarrow Te + 2e^-$	−0.92	−3.52	$Hg_2^{++} + 2e^- \leftrightarrow 2Hg$	+0.796	−5.24
$UO_2^{++} + 4H^+ + 6e^- \leftrightarrow U + H_2O$	−0.82	−3.62	$Ag^+ + e^- \leftrightarrow Ag$	+0.799	−5.24
$Se^{--} \leftrightarrow Se + 2e^-$	−0.78	−3.66	$Ir^{+++} + 3e^- \leftrightarrow Ir$	+1.0	−5.44
$Zn^{++} + 2e^- \leftrightarrow Zn$	−0.763	−3.68	$Pt^{++} + 2e^- \leftrightarrow Pt$	+1.2	−5.64
$H_3BO_3 + 3H^+ + 3e^- \leftrightarrow B + 3H_2O$	−0.73	−3.71	$Au^{+++} + 3e^- \leftrightarrow Au$	+1.42	−5.86
$Cr^{+++} + 3e^- \leftrightarrow Cr$	−0.71	−3.73	$Au^+ + e^- \leftrightarrow Au$	+1.68	−6.12
$SbO_2^- + 2H_2O + 3e^- \leftrightarrow Sb + 4OH^-$	−0.67	−3.77	$HAsO_2 + 3H^+ + 3e^- \leftrightarrow As + 2H_2O$	+0.25	−4.69
$Cr^{++} + 2e^- \leftrightarrow Cr$	−0.56	−3.88	$Pd^{++} + 2e^- \leftrightarrow Pd$	+0.987	−5.43

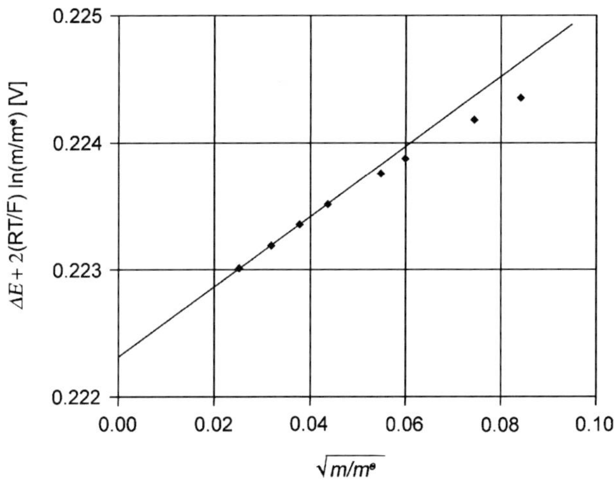

Figure 3.5 Procedure of extrapolation to determine the standard potential of the Ag/AgCl electrode.[3]

Table 3.2

Reference electrode potentials versus the SHE at 25 °C[4]

Reference electrode	Ion concentration	Potential versus SHE (V)	Physical scale (eV)
SHE (H_2/H^+)	$a = 1, p = 1.01325$ bar	0.000	−4.440
$Ag/AgCl/Cl^-$	$a = 1$	0.222	−4.662
	$c = 1$ mol l^{-1}	0.237	−4.677
	sat	0.198	−4.638
$Ag/AgBr/Br^-$	$a = 1$	0.071	−4.511
$Ag/AgJ/J^-$	$a = 1$	−0.152	−4.288
$Hg/Hg_2Cl_2/Cl^-$	$a = 1$	0.268	−4.708
	$c = 1$ mol l^{-1}	0.281	−4.721
	sat	0.241	−4.681
$Hg/Hg_2Br_2/Br^-$	$a = 1$	0.140	−4.580
$Hg/Hg_2J_2/J^-$	$a = 1$	−0.040	−4.400
$Hg/Hg_2SO_4/SO_4^{2-}$	$a = 1$	0.614	−5.054
	$c = 0.5$ mol l^{-1}	0.682	−5.122
$Hg/HgO/OH^-$	$a = 1$	0.097	−4.537
	$c = 1$ mol l^{-1}	0.140	−4.580
$Pb/PbSO_4/SO_4^{2-}$	$a = 1$	−0.276	−4.164

Under experimental conditions the SHE is rarely used. Reference electrodes of a second kind are used instead, which are simpler to handle and are commercially available. The Ag/AgCl electrode was already mentioned. Other examples are the calomel electrode based on $Hg/Hg_2Cl_2/KCl$ (for instance, as saturated calomel electrode (SCE)), the mercury sulfate electrode $Hg/Hg_2SO_4/H_2SO_4$ (0.5 mol · l^{-1}), and the mercury oxide electrode Hg/HgO/NaOH (1 mol · l^{-1}). Potentials of some reference electrodes versus the SHE are shown in Table 3.2.

Any cell reaction can be considered to be an electron transfer between two coupled half-cells. The measured potential corresponds to the difference of the electron energy. The arbitrary definition of a reference electrode raises the question of whether the electrochemical potential scale can be correlated with energy scales of electrons in surface physics. If measuring work functions or electron affinities, the reference value is the free electron in vacuum. Mehl and Lohmann[5] calculated for the electron affinity of a hydrogen electrode −4.5 eV using the following Born–Haber process

$$\mathrm{Ag} \rightarrow \mathrm{Ag_{vac}} \text{ evaporation enthalpy } \Delta_{\mathrm{vap}} H \quad (3.13a)$$

$$\mathrm{Ag_{vac}} \rightarrow \mathrm{Ag^+} + e_{\mathrm{vac}} \text{ ionization enthalpy } \Delta_{\mathrm{ion}} H \quad (3.13b)$$

$$\mathrm{Ag^+} + n\mathrm{H_2O} \rightarrow \mathrm{Ag^+_{solv}} \text{ solvation enthalpy } \Delta_{\mathrm{solv}} H \quad (3.13c)$$

The value neglects the transfer of the ion through a dipole layer at the water surface. Taking into account the most appropriate model for this dipole layer, Trassatti[6] recommended a value of −4.44 eV for the hydrogen electrode on the physical scale. The potentials on the physical scale are also listed in Table 3.2.

The Gibbs energy of the cell reaction (Eq. (3.10)) can be calculated from the standard potential. For this calculation the following procedure is applied. The potential of the copper electrode (I) is referred to the potential of the hydrogen electrode (II).

$$E_0 = E_{\mathrm{I}} - E_{\mathrm{II}} \quad (3.14)$$

A cathodic current is then passed through the copper electrode according to the direction of the arrow in Eq. (3.10). The cathodic current leads to copper deposition and hydrogen ionization. One mole turnover defines the Gibbs energy of the cell reaction (Eq. (3.10)). The standard value of the Gibbs energy (ion activity of one) is obtained by the equation

$$\Delta G^{\ominus}_{\mathrm{Cu^{2+}/Cu}} = -zFE^{\ominus}_0 \quad (3.15)$$

The sign in this equation stands for the definition[7,8] of the voltage measurement: (i) electrode I minus electrode II and (ii) direction of the cell reaction in Eq. (3.10). The sign changes if the conditions of the voltage measurements are altered.

The change in the direction of the cell reaction (Eq. (3.10)) means the substitution of one mole of hydrogen ions by one mole of metal ions. This is called the Gibbs energy of ion formation and is calculated by the equation

$$\Delta_{\mathrm{for}} G^{\ominus}_{\mathrm{Me^{z+}}} = +zFE^{\ominus}_0 \quad (3.16)$$

The reference value for the Gibbs energy of ion formation is the hydrogen ion for standard conditions. The temperature dependence of the potential gives the corresponding value of the ion formation entropy. The reference value is again the hydrogen ion.

$$\Delta_{\mathrm{for}} S^{\ominus}_{\mathrm{Me^{z+}}} = -zF\left(\frac{\partial E^{\ominus}_0}{\partial T}\right)_p \quad (3.17)$$

Then the enthalpy of formation of metal ions is calculated using the Gibbs–Helmholtz equation.

$$\Delta_{for} H^{\ominus}_{Me^{z+}} = zF\left[E_0^{\ominus} - T\left(\frac{\partial E_0^{\ominus}}{\partial T}\right)_p\right] \quad (3.18)$$

Values for some metal ions are shown in Table 3.3.[9]

In non-aqueous electrolytes, the different properties of the solvated metal ions lead to different equilibrium and standard potentials. For comparing standard potentials, electrode reactions should be defined as reference systems with similar values in different solvents. Koepp, Wendt, and Strehlow[10] suggested ferrocene/ferrocinium and cobaltocene/cobaltocenium redox systems. The redox systems are bis-pentadienyl complexes of Fe^{2+}/Fe^{3+} and Co^{2+}/Co^{3+}, respectively. Gritzner and Kuta[11] recommended ferrocene/ferrocinium and bis(biphenyl)Cr(I)/bis(biphenyl)Cr(0). Salt bridges with conventional cells should be avoided. Similar to aqueous electrolytes a reference to the physical potential scale is possible.[6] Similar considerations hold for ionic melts and molten and solid electrolytes.

3.1.3 Standard electrode potentials of metal complexes

Different equilibrium values and potentials are obtained if chemicals are added to the electrolyte, forming complexes of higher stability than the water complex. The complex

Table 3.3

Examples of Gibbs energy, entropy, and enthalpy of ion formation in reference to the hydrogen ion[9]

Metal ion (aq)	$\Delta_{for}G^{\ominus}$ (kJ mol^{-1})	$\Delta_{for}S^{\ominus}$ (J K^{-1} mol^{-1})	$\Delta_{for}H^{\ominus}$ (kJ mol^{-1})
H^+	0	0	0
Al^{3+}	−485	−321.7	−531
Ba^{2+}	−560.77	9.6	−537.64
Pb^{2+}	−24.43	10.5	−1.7
Br^-	−103.96	82.4	−121.55
Cd^{2+}	−77.61	73.2	−75.90
Ca^{2+}	−553.58	−53.1	−542.83
Cl^-	−131.23	56.5	−167.16
CrO_4^{2-}	−727.75	50.21	−881.15
Fe^{2+}	−78.90	−137.7	−89.1
F^-	−278.79	−13.8	−332.63
J^-	−51.57	111.3	−55.19
K^+	−283.27	102.5	−252.38
Cu^{2+}	65.49	−99.6	−64.77
Li^+	−293.31	13.4	−278.49
Mg^{2+}	−454.8	−138.1	−466.85
Na^+	−261.91	59.0	−240.12
Hg_2^{2+}	153.52	84.5	172.4
OH^-	−157.24	−10.75	−229.99
Ag^+	77.11	72.68	105.58
Zn^{2+}	−147.06	−112.1	−153.89
Sn^{2+}	−27.2	−17	−8.8

formation reaction is

$$Me^{z+} + vL^- \leftrightarrow (MeL_v)^{(v-z)-} \tag{3.19}$$

The free metal ion is still the hydrated ion as described in Eq. (3.1).

The complex concentration is constant, approximately equal to the initial metal ion concentration, and is part of the complex constant K_c

$$K_c = c_{Me^{z+}} \cdot c_{L^-}^v \tag{3.20}$$

Inserting Eq. (3.20) into Eq. (3.6) gives

$$E_{c,0} = E_{c,0}^{\ominus} - \frac{RT}{zF} \ln c_{L^-}^v \tag{3.21}$$

$$E_{c,0}^{\ominus} = E_0^{\ominus} + \frac{RT}{zF} \ln K_c \tag{3.22}$$

An example is the formation of the silver cyanide complex given by

$$Ag^+ + 2(CN)^- \leftrightarrow Ag(CN)_2^- \tag{3.23}$$

The complex constant is

$$K_c = c_{Ag^+} \cdot c_{CN^-}^2 \tag{3.24}$$

The complex constant has a value of $K_c = 8.00 \times 10^{-22}$ (mol $\cdot$ l^{-1}).[3] The standard electrode potential shifts from $E_0^{\ominus} = +0.799$ V to $E_{c,0}^{\ominus} = -0.446$ V.

Complex constants and standard potentials of some metal ion complex electrodes with various complex agents are given in Table 3.4.[12]

3.2 ALLOYS

3.2.1 Partial molar Gibbs energies

For an alloy one has to describe the mixture of two or more metals in one phase. This can be done by the introduction of partial molar thermodynamic functions for each component. This will be described for a binary alloy $A_x B_y$. The partial molar Gibbs energy for component A is defined for the gas phase.

$$\Delta \overline{G}_{A,g} = RT \ln \frac{p_A}{p_A^0} \tag{3.25}$$

This is the difference of the Gibbs energy of the component A in the gas phase over the alloy (pressure p_A) and the Gibbs energy of the gas phase over the pure metal phase A

Table 3.4

Complex constants and standard potentials of some metal ion complex electrodes[12]

	z	$E_0^{\ominus}$ (V)	Complex	K_c	log K_c	(0.059/z)log K_c	$E_0^{\ominus}$ (V)
Ag/Ag^+	1	0.7991	$Ag(NH_3)_2^+$	9.31E-08	−7.03	−0.415	0.384
			Ag_2Br^+	2.00E-10	−9.70	−0.572	0.227
			$AgBr_4^{3-}$	6.20E-10	−9.21	−0.543	0.256
			Ag_2Cl^+	2.00E-07	−6.70	−0.395	0.404
			$AgCl_4^{3-}$	4.80E-06	−5.32	−0.314	0.485
			$Ag(CN)_2^-$	8.00E-22	−21.10	−1.245	−0.446
			$Ag(CN)_4^{3-}$	2.10E-21	−20.68	−1.220	−0.421
			AgOH	5.00E-03	−2.30	−0.136	0.663
			Ag_3I^{2+}	8.00E-15	−14.10	−0.832	−0.033
			AgI_4^{3-}	1.80E-14	−13.74	−0.811	−0.012
			$Ag(CNS)_4^{3-}$	9.30E-04	−3.03	−0.179	0.620
			$Ag(SeCN)_3^{2-}$	1.32E-13	−12.88	−0.760	0.039
			$Ag(SO_4)_2^{3-}$	5.90E-01	−0.23	−0.014	0.786
			$Ag(SO_3)_2^{3-}$	4.50E-08	−7.35	−0.433	0.366
			$Ag(S_2O_3)_2^{3-}$	3.50E-14	−13.46	−0.794	0.005
Cu/Cu^{2+}	2	0.345	$Cu(NH_3)_4^{2+}$	2.14E-13	−12.67	−0.374	−0.029
			$CuBr^+$	5.00E-01	−0.30	−0.009	0.336
			$CuCl^+$	7.70E-01	−0.11	−0.003	0.342
			$CuOH^+$	3.40E-07	−6.47	−0.191	0.154
			$Cu(OH)_4^{2-}$	7.60E-17	−16.12	−0.476	−0.131
			$Cu(P_2O_7)_2^{6-}$	1.00E-09	−9.00	−0.266	0.080
			$Cu(CNS)_3^-$	6.50E-06	−5.19	−0.153	0.192
			$CuSO_4$	4.50E-03	−2.35	−0.069	0.276
			$Cu(SO_3)_3^{5-}$	6.50E-10	−9.19	−0.271	0.074
			$Cu(P_4O_{12})_2^{6-}$	2.30E-05	−4.64	−0.137	0.208
Ni/Ni^{2+}	2	−0.25	$Ni(NH_3)_6^{2+}$	1.86E-09	−8.73	−0.258	−0.508
			$Ni(N_2H_4)_6^{2+}$	1.10E-12	−11.96	−0.353	−0.603
			$NiOH^+$	2.50E-05	−4.60	−0.136	−0.386
			$Ni(P_2O_7)_2^{6-}$	6.50E-08	−7.19	−0.212	−0.462
			$Ni(CNS)_3^-$	1.55E-02	−1.81	−0.053	−0.303
			$NiSO_4$	4.00E-03	−2.40	−0.071	−0.321
			$NiP_4O_{12}^{2-}$	1.12E-05	−4.95	−0.146	−0.396
			NiS_2O_3	8.70E-03	−2.06	−0.061	−0.311
			$NiP_3O_9^-$	6.00E-04	−3.22	−0.095	−0.345
Pb/Pb^{2+}	2	−0.126	$PbBr_4^{2-}$	1.00E-03	−3.00	−0.089	−0.215
			$PbCl_3^-$	1.40E-02	−1.85	−0.055	−0.181
			$PbOH^+$	6.00E-07	−6.22	−0.184	−0.310
			PbI_3^-	2.22E-05	−4.65	−0.137	−0.263
			$PbNO_3^+$	3.30E-01	−0.48	−0.014	−0.140
			$Pb(P_2O_7)_2^{6-}$	4.74E-06	−5.32	−0.157	−0.283
			$Pb(S_2O_3)_3^{4-}$	4.48E-07	−6.35	−0.187	−0.313
Sn/Sn^{2+}	2	−0.14	$SnBr_3^-$	4.50E-02	−1.35	−0.040	−0.180
			$SnCl_3^-$	2.10E-02	−1.68	−0.049	−0.189
			$SnOH^+$	2.00E-10	−9.70	−0.286	−0.426

(pressure p_A^0). In the condensed phase the partial molar Gibbs energy for component A is defined by the activity in Eq. (3.26)

$$\Delta\overline{G}_{A,c} = RT \ln a_A \tag{3.26}$$

In equilibrium, the partial molar Gibbs energy for component A of the gas phase is equal to the partial molar Gibbs energy for component A in the condensed phase

$$\Delta\overline{G}_{A,c} = \Delta\overline{G}_{A,g} \tag{3.27}$$

Comparison of Eqs. (3.25) and (3.26) leads to Eq. (3.28)

$$a_A = \frac{p_A}{p_A^0} \tag{3.28}$$

The activity of component A in the alloy phase is equal to the ratio of the partial pressures of A in the gas phase over the alloy and the pure component A.

In the principal form, corresponding equations also exist for component B. But experimentally, the vapor pressure of only one component, the more volatile component, can be determined. The vapor pressure of the other component can be calculated by the Gibbs–Duhem or the Duhem–Margules equation, as will be explained in detail in Section 3.2.3.

The measurement of partial pressures over a liquid or solid mixture of two metals is not as simple. Mostly, it is restricted to higher temperatures or even to the molten phase. The direct measurement can be done, for example, in high or ultra high vacuum, using a Knudsen cell and a mass spectrometer for selective pressure determination. Dynamic measurements were developed, e.g., transportation methods. A steady stream of a reactive gas is passed over the sample transporting the reactive component to a cooled region of the apparatus. From the measured mass of the transported metal and the flow rate the vapor pressure can be calculated. Kubaschewski *et al.*[13] have given a detailed description of the experimental possibilities.

Examples of the activity of a component A (equal to the ratio of the partial pressure p_A over the alloy and the pressure p_A^0 over the pure component A) as function of the mole fraction x_A are shown in Figure 3.6.

Figure 3.6a shows the ideal case of a linear dependence of the activity on the mole fraction

$$a_A = x_A \tag{3.29}$$

This expression is called Raoult's law. More complex behavior is shown in Figure 3.6b–d. Raoult's law is observed as a limiting case for $x_A \rightarrow 1$. Another limiting law is found for $x_A \rightarrow 0$ that is called Henry's law

$$a_A = K_A x_A \tag{3.30}$$

where K_A is the equilibrium constant of the distribution of component A between the gas phase and the condensed phase. The importance of these limiting laws is that experimental

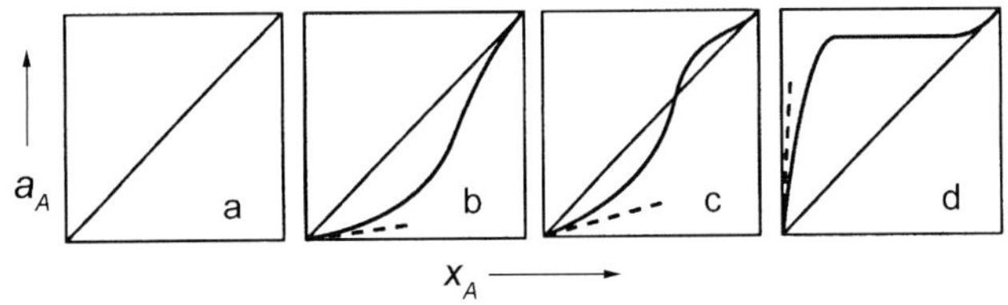

Figure 3.6 Activity of a_A of component A in the alloy versus the mole fraction x_A. (a) represents the ideal case (Raoult's law), (b) and (c) the different forms of interaction between the components, and (d) the case of a miscibility gap (dashed line Henry's law).[13]

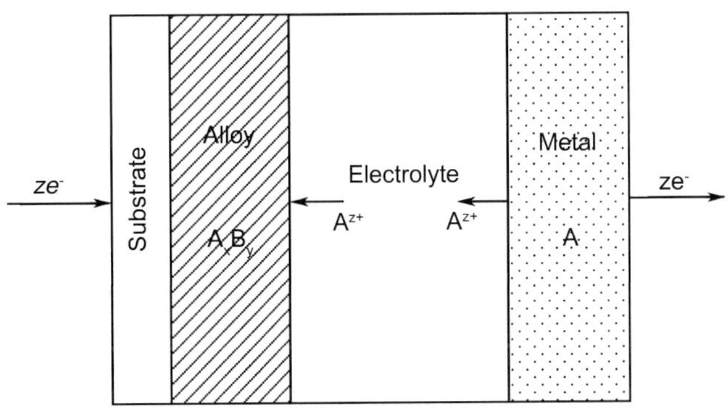

Figure 3.7 Combination of a bulk metal electrode and an alloy electrode. The alloy is shown as a layer deposited on a substrate. The cell reaction is the transformation of one mole of pure metal into the alloy of composition A_xB_{1-x}.

data at small concentrations can only be determined with a large error. Using the limiting laws one can extrapolate measurements down to $x_A \to 0$ or $x_A \to 1$.

3.2.2 Electrochemical measurements of partial molar functions

A second method to determine partial molar Gibbs energies and activities of alloy components in an alloy is the measurement of the cell voltage of the cell shown in Figure 3.7. The alloy can be formed by metallurgical alloying or, as is shown in Figure 3.7, be deposited on a substrate either electrochemically or by vapor deposition methods.

The cell reaction is described by the following equations:

$$A \rightarrow A^{z+} + ze^- \tag{3.31a}$$

$$A_xB_y + A^{z+} + ze^- \rightarrow A_{x+dx}B_y \tag{3.31b}$$

$$A + A_xB_y \rightarrow A_{x+dx}B_y \tag{3.31c}$$

The equilibrium potential of this cell ΔE_A is the difference between the potential of the metal in the alloy state E_A minus the potential of the pure metal E_A^0

$$\Delta E_A = E_A - E_A^0 \tag{3.32}$$

It is related to the partial molar Gibbs energy of the component A by the equation

$$\Delta \bar{G}_A = -z_A F \Delta E_A \tag{3.33}$$

The sign is defined by the procedure of Section 3.1.2, Eqs. (3.14) and (3.15). The alloy electrode (electrode I) is measured versus the pure metal electrode (electrode II) and a cathodic current is assumed through the alloy electrode determining the direction of the cell reaction.

To measure the equilibrium potential of such a cell several experimental restrictions must be observed. Some examples are given in the following listing:

- At the interface alloy/electrolyte the concentration of the alloy components must be the same as in the bulk of the alloy. In aqueous electrolytes or in molten salts the less noble component might dissolve in the electrolyte and thus lead to concentration failures. This could be avoided by keeping up a small cathodic current preventing dissolution. The current may not distort the potential measurement.
- The equilibrium of the ion exchange at the alloy/electrolyte interface must be established. A slow mobility of metal atoms in the alloy could prevent equilibrium conditions.
- If a solid electrolyte is used the conductivity of the electrolyte might be a problem. The conductivity necessary for the measurement may only be achieved at sufficiently high temperatures.

For these reasons, potential measurements of alloys are frequently made at higher temperatures with solid electrolytes or by using molten salt electrolytes. Measurements in aqueous electrolytes and at room temperatures were mostly restricted to mercury alloy (amalgam) electrodes.

From Eq. (3.33), one can derive relations between the activity a_A and the cell voltage ΔE_A and between the logarithm of the activity coefficient γ_A and the cell voltage ΔE_A of A in the alloy

$$a_A = \exp\left(-\frac{z_A F \Delta E_A}{RT}\right) \tag{3.34}$$

$$\ln \gamma_A = -\frac{z_A F \Delta E_A}{RT} - \ln x_A \tag{3.35}$$

If ΔE_A and its temperature dependence are measured, the partial molar entropy of component A can be determined

$$\Delta \bar{S}_A = +z_A F \left(\frac{\partial \Delta E_A}{\partial T}\right)_p \tag{3.36}$$

Then the partial molar enthalpy of A in the alloy can be calculated using the equation

$$\Delta\overline{H}_{A} = -z_{A}F\left[\Delta E_{A} - T\left(\frac{\partial \Delta E_{A}}{\partial T}\right)_{a}\right] \tag{3.37}$$

As an example, the alloy system AgAu will be described in Section 3.2.3.

3.2.3 Ag_xAu_y—example of a solid solution

Ag and Au form a system of complete miscibility in the solid state. The atomic diameter of Ag is $d = 0.2890$ nm, the diameter of Au is 0.2884 nm. The electronegativity of Ag (after the calculation of Martynov and Batsanov) is $\chi = 1.08$ for Ag and $\chi = 1.19$ for Au.[14,15] As a result of this geometrical and chemical similarity, the atoms in the crystal lattice can substitute each other.

Kubaschewski and Huchler[16] measured the cell voltage as a function of temperature. The cell used for such experiments is shown in Figure 3.8. Figure 3.8a shows the ceramic cup in which the Ag_xAu_y/Ag electrode couple separated by a thin glass electrolyte film was placed. The ceramic cup was placed in a larger cell (Figure 3.8b) with thermocouple and facilities for measuring in an inert gas atmosphere (nitrogen). Other cell types for high temperature measurements can be found in the literature.[13]

In the alloy Ag_xAu_y, silver is the electronegative component (the less noble one) and gold is the electropositive component (the more noble one). The electrolyte in the given example was a special glass with Ag ion conductivity. The glass was melted on the silver electrode. The electrolyte film had a thickness of 0.1 mm and the resistance was of the order of 2000 Ω. The cell reaction was described in the previous chapter. The electrode process (Eq. (3.31)) consisted of the transfer of metal atoms from the pure Ag into the alloy environment Ag_xAu_y keeping the alloy composition constant. Eqs. (3.33)–(3.37) could then be applied to calculate the partial molar functions of the Ag_xAu_y system.

The potential difference $\Delta E_{Ag} = E_I - E_{II}$ was measured as function of temperature in the temperature range of 350–650 °C. Linear plots were obtained and average values of ΔE_{Ag} could be determined for selected temperatures. In Figure 3.9, values for 400, 500, and 600 °C are presented as a function of the mole fraction of Ag in the alloy. Larger values of the voltage were measured at a higher gold content. The potential differences approached zero for silver-rich alloys.

Corresponding partial molar Gibbs energies for 500 °C were calculated using Eq. (3.33). From the temperature dependence of the potentials partial molar entropies were calculated using Eq. (3.36). Finally, partial molar enthalpies were obtained using Eq. (3.37). Values of the partial molar functions of Ag as a function of composition are summarized in Table 3.5.

The activity of Ag in the alloy can be calculated with Eq. (3.34) as the function of the mole fraction. Results for 500 °C are shown in Figure 3.10.

A comparison of Figures 3.10 and 3.6 shows that the Ag_xAu_y alloy represents the type of alloy shown in Figure 3.6b. For the determination of the linear extrapolation according to Henry's law the region of low mole fractions is shown in Figure 3.11.

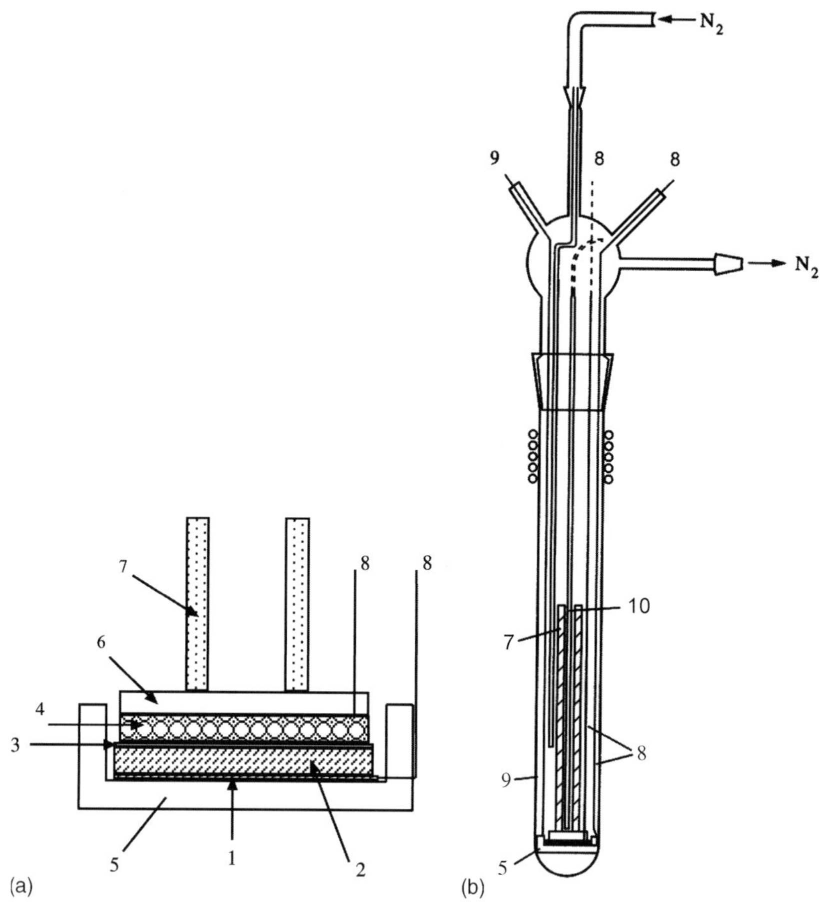

Figure 3.8 (a) Ceramic cup for potential measurements in the alloy system Ag_xAu_y. (1) Ag plate with connecting Pt wire (8), (2) Ag electrode, (3) glass film melted on the Ag electrode (thickness ca. 0.1 mm), (4) Ag_xAu_y alloy electrode with platinum contact wire (8) in a ceramic cup, (5) pressed on the electrolyte layer by the cover of the ceramic cup (6), and an alumina tube (7).[16] (b) Apparatus used for potential measurements in the alloy system Ag_xAu_y with the ceramic cup (5) shown in detail in Figure 3.8a. (7) alumina tube, (8) platinum wire connections to the electrodes, (9) wire to hold the ceramic cup, and (10) thermocouple.[16]

The equation determined for $Ag_x Au_y$ is

$$a_{Ag} = 0.16 \cdot x_{Ag} \tag{3.38}$$

Extrapolations to higher mole fractions can be made by the data fit in Figure 3.10 and by Raoult's law. The possibility for such data fits provides a way for smoothing experimental data and to obtain data in regions of limited accuracy, typically for $x_A \rightarrow 0$ or $x_A \rightarrow 1$. For reasons of transparency, only the experimental values of Kubaschewski and Huchler[16] were used in the following calculations.

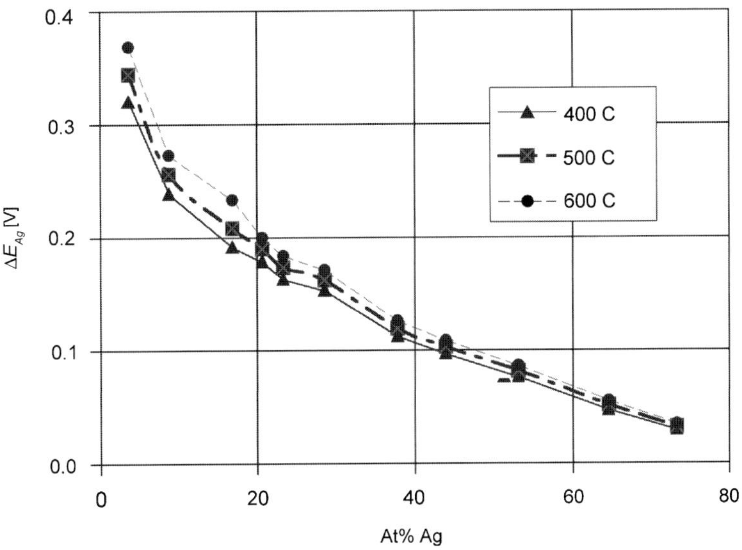

Figure 3.9 Potential difference between a Ag_xAu_y alloy electrode and a pure Ag electrode at three different temperatures.[16]

Table 3.5

Values of partial molar thermodynamic functions of silver and gold in the Ag_xAu_y alloy as a function of composition for 500 °C. The values for gold were obtained by integration of the Gibbs-Duhem equation (Section 3.2.4, data for $x_{Ag}=0.169$ were omitted for the integration of the entropy values)

x_{Ag}	$\Delta\bar{G}$ (Ag) (kJ mol^{-1})	$\Delta\bar{S}$ (Ag) (JK^{-1} mol^{-1})	$\Delta\bar{H}$ (Ag) (kJ mol^{-1})	$\Delta\bar{G}$ (Au) (kJ mol^{-1})	$\Delta\bar{S}$ (Au) (JK^{-1} mol^{-1})	$\Delta\bar{H}$ (Au) (kJ mol^{-1})
0.000				0.00	0.00	0.00
0.037	−33.16	23.21	−15.22	−0.26	0.00	−0.26
0.088	−24.63	16.50	−11.87	−0.79	0.46	−1.15
0.169	−20.01	19.93	−4.60	−1.46		
0.208	−18.27	10.37	−10.25	−1.86	1.56	−3.07
0.234	−16.69	9.79	−9.12	−2.31	1.73	−3.65
0.287	−15.60	8.73	−8.85	−2.69	2.10	−4.31
0.379	−11.46	6.66	−6.31	−4.76	3.15	−7.20
0.440	−9.88	5.60	−5.55	−5.86	3.89	−8.87
0.532	−7.82	4.83	−4.08	−7.82	4.63	−11.40
0.647	−4.52	3.83	−1.55	−12.08	6.10	−16.80
0.733	−3.07	2.55	−1.10	−16.24	9.03	−23.22
1.000	0.00	0.00	0.00			

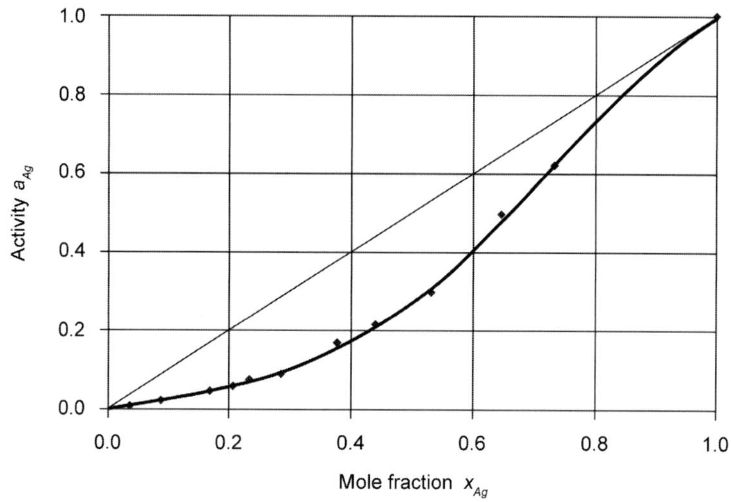

Figure 3.10 Activity of Ag in the alloy Ag_xAu_y as function of the mole fraction of Ag, temperature 500 °C.

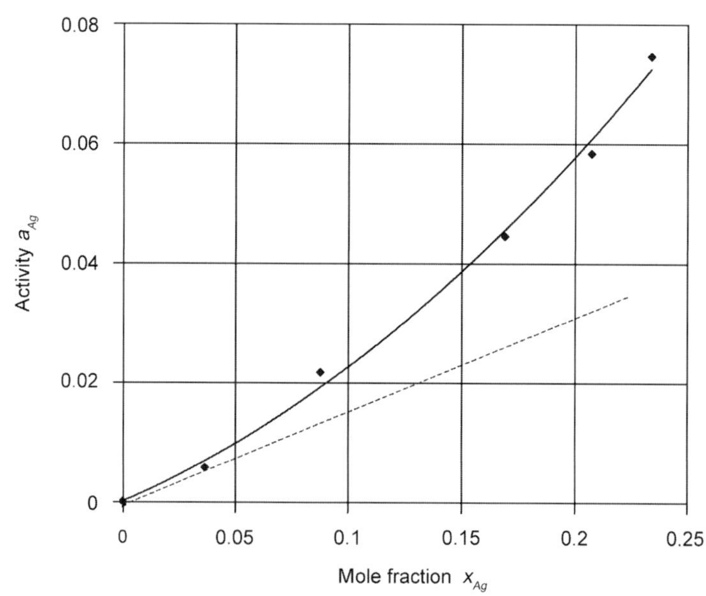

Figure 3.11 Activity of Ag in the alloy Ag_xAu_y at low mole fractions of Ag, temperature 500 °C. The solid line is a polynomial fit of the experimental data and the constant in Henry's law can be determined from the linear fit.

3.2.4 Partial molar functions of component B

The partial molar functions of the component B can be calculated using the Gibbs-Duhem equation. The Gibbs–Duhem equation for any partial molar function $\overline{Z}$ ($\overline{Z}$ equal to $\Delta\overline{G}$, $\Delta\overline{S}$, $\Delta\overline{H}$, etc.) has the general form

$$\sum_i x_i d\overline{Z}_i = 0 \tag{3.39}$$

Applied to the Gibbs function of a binary alloy one obtains

$$d\Delta\overline{G}_{\mathrm{B}} = -\frac{x_{\mathrm{A}}}{x_{\mathrm{B}}} d\Delta\overline{G}_{\mathrm{A}} \tag{3.40}$$

$$d\ln a_{\mathrm{B}} = -\frac{x_{\mathrm{A}}}{x_{\mathrm{B}}} d\ln a_{\mathrm{A}} \tag{3.41}$$

The integration of these equations gives the partial molar functions of component B. For the integration the integration constants must be known. Because the potential difference was measured between the alloy and the pure metal, the partial molar functions are relative values referring to the pure metal as a reference state. Therefore, integration is carried out between $x_{\mathrm{B}}=1$, $x_{\mathrm{A}}=0$ and x_{B}, and $x_{\mathrm{A}}=1-x_{\mathrm{B}}$.

Integrating these equations, one sometimes extrapolates to $x_{\mathrm{A}}\rightarrow 0$. For $x_{\mathrm{A}}\rightarrow 0$, a_{A} also goes to zero when the activity coefficient γ_{A} approaches a constant value. Therefore, a better way is to integrate the activity coefficients

$$d\ln\gamma_{\mathrm{B}} = -\frac{x_{\mathrm{A}}}{x_{\mathrm{B}}} d\ln\gamma_{\mathrm{A}} \tag{3.42}$$

In a similar manner one can reduce $\Delta\overline{G}_{\mathrm{A}}$ and $\Delta\overline{S}_{\mathrm{A}}$ to the non-ideal part with the equations

$$\Delta\overline{G}_{\mathrm{A}}^{\mathrm{ex}} = \Delta\overline{G}_{\mathrm{A}} - RT\ln x_{\mathrm{A}} \tag{3.43}$$

$$\Delta\overline{S}_{\mathrm{A}}^{\mathrm{ex}} = \Delta\overline{S}_{\mathrm{A}} + RT\ln x_{\mathrm{A}} \tag{3.44}$$

These functions are called excess functions.

Integration of Eq. (3.42) from $x_{\mathrm{A}} = 0$, $x_{\mathrm{B}} = 1$ ($\gamma_{\mathrm{B}} = 1$, $\ln\gamma_{\mathrm{B}} = 0$) to x_{A}, $x_{\mathrm{B}} = 1-x_{\mathrm{A}}$ gives the equation

$$\ln\gamma_{\mathrm{B}} = -\int_0^{\ln\gamma_{\mathrm{A}}} \frac{x_{\mathrm{A}}}{x_{\mathrm{B}}} d\ln\gamma_{\mathrm{A}} \tag{3.45}$$

For the integration of Eq. (3.45) one can plot $x_{\mathrm{A}}/x_{\mathrm{B}}$ versus $\ln\gamma_{\mathrm{A}}$ as shown in Figure 3.12 for the $Ag_x Au_y$ system.

Integration of the data gives the logarithm of the activity coefficients of gold, $\ln\gamma_{\mathrm{Au}}$. Using these data, the activities of gold on the Ag_xAu_y alloy can be calculated. The results plotted versus the molar fraction of Ag are shown in Figure 3.13.

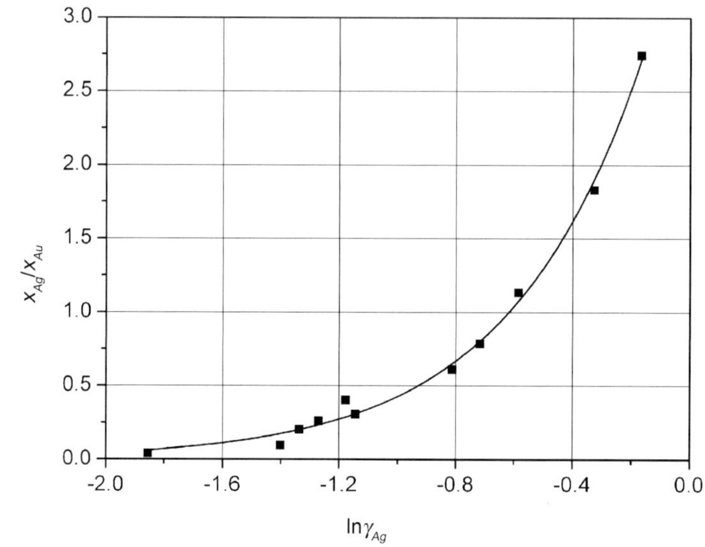

Figure 3.12 Plot of x_A/x_B versus $\ln \gamma_A$ for the Ag_xAu_y system.

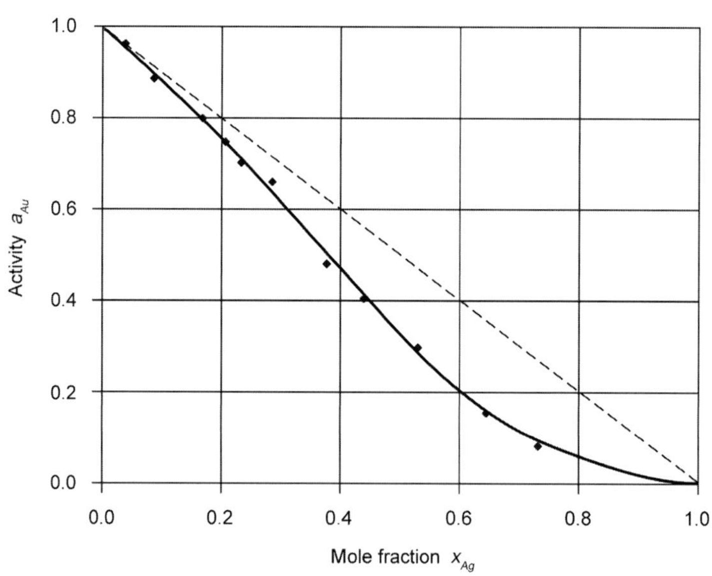

Figure 3.13 Activity of Au in the alloy Ag_xAu_y versus the mole fraction of Ag, temperature 500 °C. The experimental data can be represented by a curve similar to the data fit in Figure 3.10.

The data shown in Figure 3.13 can be compared with the data for silver shown in Figure 3.10. In Figure 3.10 the data for $x_{Ag}>0.8$ could be fitted by the solid curve. Calculations with such extended data files provide a way to also obtain values for gold in this region and to support the data fit in this region in Figure 3.13.

With the data in Figure 3.13 the partial molar Gibbs energy of gold was calculated. With the same integration procedure the partial molar entropy of gold was obtained using excess entropy values. Finally, the partial molar enthalpy was calculated. The values for the complete set of thermodynamic functions for gold in the alloy system as function of composition are shown in Table 3.5.

3.2.5 From partial molar functions to integral functions

With the partial molar functions $\Delta\bar{Z}_A$ and $\Delta\bar{Z}_B$ one can calculate the integral functions. One obtains for an alloy of composition x_A and x_B the thermodynamic functions of formation

$$\Delta_{for} Z = x_A \Delta\bar{Z}_A + x_B \Delta\bar{Z}_B \tag{3.46}$$

Another way to obtain integral thermodynamic functions is provided by the Duhem–Margules equation. The partial molar function $\Delta\bar{Z}_A$ is plotted versus y (the composition of the alloy is represented by the formula A_yB) and then is integrated from $y=0$ to y

$$\Delta_{for} Z = -x_B \int_0^y \Delta\bar{Z}_A \mathrm{d}y \tag{3.47}$$

This procedure provides a second way to obtain partial molar functions of component B. The partial molar functions of A, $\Delta\bar{Z}_A$, are known from potential measurements. One can then calculate the partial molar functions of component B from $\Delta_{for} Z$ subtracting the partial molar function $\Delta\bar{Z}_A$ according to the equation

$$\Delta\bar{Z}_B = \frac{1}{x_B}(\Delta_{for} Z - x_A \Delta\bar{Z}_A) \tag{3.48}$$

The procedure is demonstrated for the example Ag_xAu_y. The excess function of the partial molar Gibbs energy calculated with Eq. (3.43) is shown as function of the stoichiometric number y in the formula Ag_yAu (y is equal to the ratio x_A/x_B) in Figure 3.14. For the integration between $y=0$ and y, $\Delta\bar{G}_{ex,Ag}$ must be extrapolated to $y=0$. A logarithmic data fit was used for the extrapolation.

The excess function of the Gibbs energy of formation, $\Delta_{for}G_{ex}$, calculated with Eq. (3.48), is shown in Figure 3.15, where the values of $\Delta_{for}G_{ex}$ are also represented by a Gaussian fit. This is a further possibility of data smoothing and data extrapolation.

From $\Delta_{for}G_{ex}$, first the excess function and then the partial molar Gibbs energy of gold can be calculated. The calculated values and values from Table 3.5, which were obtained by the integration of the Gibbs equation described in the previous chapter, are shown in Figure 3.16.

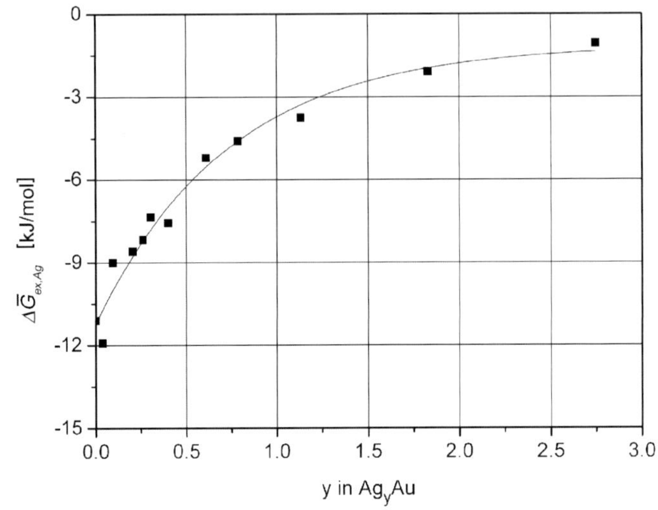

Figure 3.14 Excess function of the partial molar Gibbs energy of silver in the silver–gold alloy plotted as function of y in Ag_yAu. The solid line is a logarithmic fit of the experimental data.

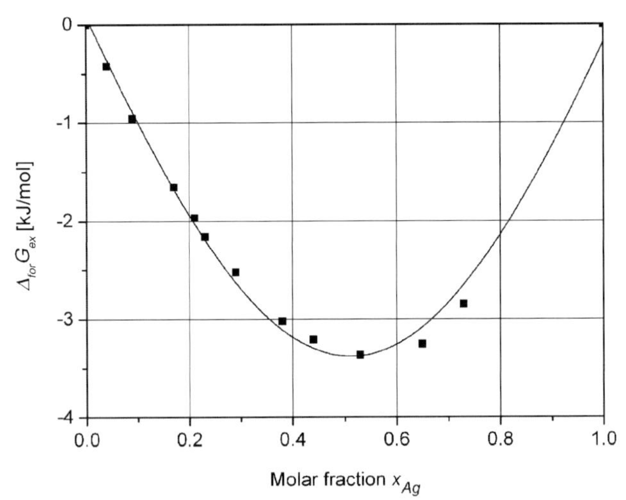

Figure 3.15 Excess function of the Gibbs energy of formation, $\Delta_{\text{for}}G_{ex}$, as function of the mole fraction of Ag and with Gaussian fitted data.

3.3 INTERMETALLIC PHASES AND COMPOUNDS

3.3.1 Potential versus mole fraction diagrams

In the example above, the Ag_xAu_y system was seen to consist of two metals (either with complete miscibility or just a miscibility gap) and to only form solid solutions. In many other systems the chemical similarity is much less pronounced. Solubility is limited and new

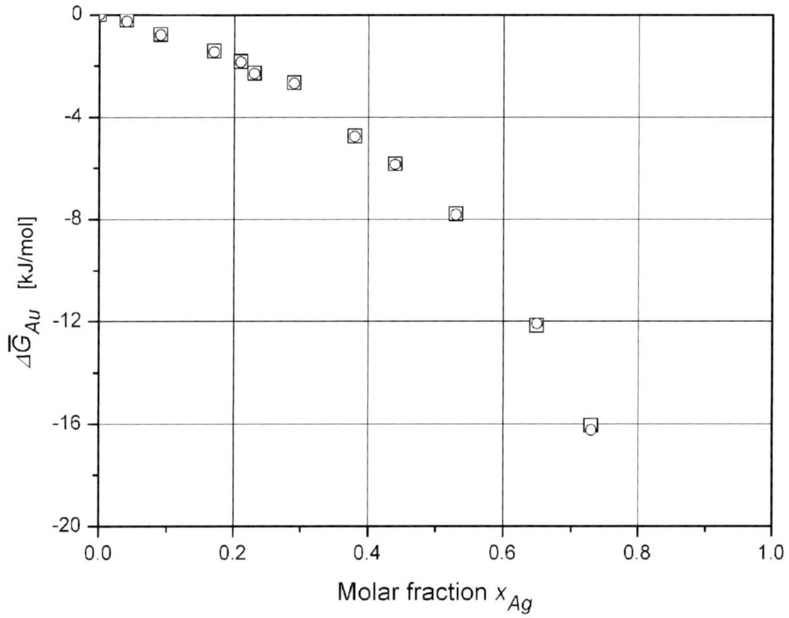

Figure 3.16 Partial molar Gibbs energy of Au in the alloy Ag_xAu_y as function of the mole fraction of Ag. Two sets of values are shown: The first set (squares) was calculated with the Gibbs-Duhem equation, as described in Section 3.2.3, and the second set of data (circles) was calculated with the Duhem–Margules equation, temperature 500 °C.

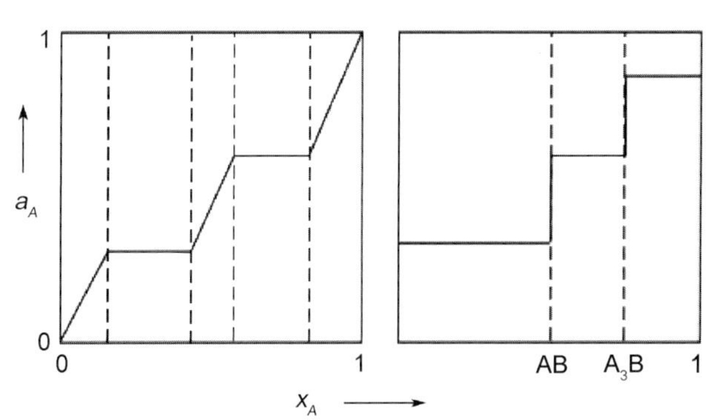

Figure 3.17 Activity diagrams for (a) systems with a non-stoichiometric phase and (b) systems forming intermetallic compounds.

crystallographic structures (phases) or even new compounds (intermetallic compounds) are formed. Information about compound or phase formation is obtained from the same experimental parameters as introduced in Section 3.2.1. The ratio of the partial pressure of component A is above the condensed phase and above the pure phase. This is equal to the activity of A in the condensed phase.

The activity can also be obtained from voltage measurements versus the pure phase A as the reference electrode. But in the case of the formation of intermetallic phases or compounds, the dependence of the activity on the mole fraction shows the characteristic slopes and plateaus schematically shown in Figure 3.17.

The more complex diagrams have led to the development of some modifications of the methods to determine the characteristic thermodynamic functions. This will be described in the following chapters.

3.3.2 Coulometric titration

The measurements of metallurgically prepared alloys, which were used as alloy electrodes in cells as in Figure 3.7, are limited in their number of data points. A big advantage that was introduced by Wagner[17] is the coulometric titration. The cell is either a two-electrode cell or a three-electrode cell with a solid electrolyte or a molten salt electrolyte. The alloy is not preformed but is formed instead by a slow electrochemical deposition of component A into the component B (current I). The mole number of A deposited is determined by the equation

$$n_A = \frac{1}{zF}\int_0^t I\mathrm{d}t \tag{3.49}$$

After deposition, the system must achieve an equilibrium distribution of the deposited metal. This is only possible if the metal A has enough mobility within the matrix of B, which is only possible at higher temperatures. This also limits the thickness of an alloy sample so that either a thin wire or a thin film electrode is used. With ultra thin metal films or with metal clusters, new applications of this method seem possible at lower temperatures.

3.3.3 Coulometric titration: the system LiAl

Weppner and Huggins[18–20] extensively applied the coulometric titration method. One example of the formation of intermetallic phases or compounds is the LiAl system. Here the electrolyte was a eutectic mixture of LiCl and KCl.[18] Li was deposited into Al by coulometric titration. The Li content in the Al film electrode was increased in small steps. The potential was measured after equilibration of the alloy electrode. In Figure 3.18 the potential of an LiAl electrode versus an Li electrode is shown as function of y, the stoichiometric coefficient in the formula Li_yAl.

The interpretation of the diagram is the following:

- At low Li contents Li is dissolved in aluminum (α-phase). The activity of Li is very low but increases fast until the LiAl phase (β-phase) begins to form at 9.2 at% ($y = 0.10$ in Figure 3.18).
- The well-defined plateau marks the equilibrium between Li-saturated aluminum and the LiAl phase. The equilibrium is described by the equation

$$\mathrm{Li} + \mathrm{Al} \leftrightarrow \mathrm{LiAl} \tag{3.50}$$

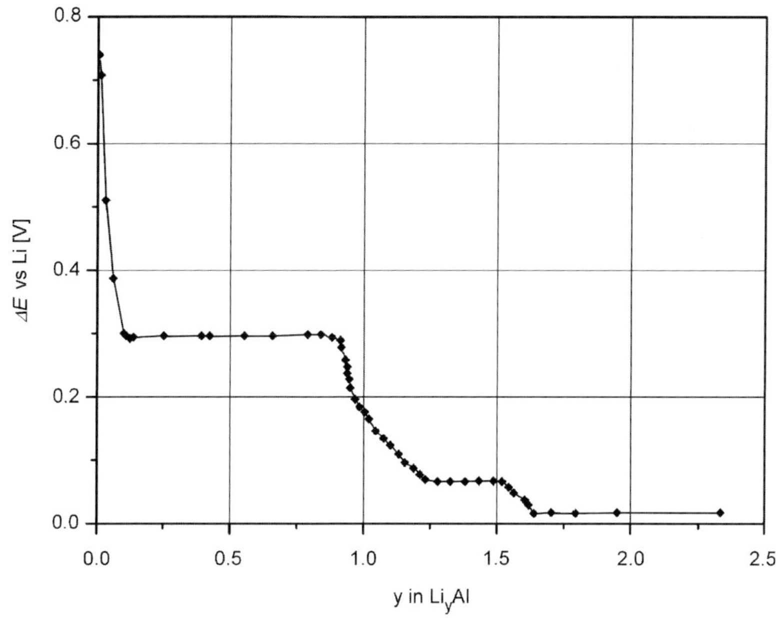

Figure 3.18 Potential of an Li_yAl electrode with increasing Li content versus a pure Li electrode. (Reproduced with permission from Ref. [18], © 1979, The Electrochem. Soc.)

Potential and activity of Li stay constant until at 46.8 at% Li ($y = 0.88$ in Figure 3.18), where the lower existence limit of the LiAl β-phase is reached.

- At first the activity of Li in this new compound is of the vacancies and then above 50 at% Li ($y = 1.00$ in Figure 3.18) of interstitial Li atoms whereupon the slope of the curve in the potential diagram changes.
- At 56 at% Li ($y = 1.27$ in Figure 3.18) equilibrium to a third phase (γ-phase) is achieved with the average stoichiometry of Li_3Al_2. This is described by the equation

$$Li + 2LiAl \leftrightarrow Li_3Al_2 \tag{3.51}$$

- The γ-phase Li_3Al_2 exists in a smaller region between 60.3 and 61.7 at% Li ($y = 1.52$ and $y = 1.61$ in Figure 3.18).
- At higher Li concentrations a solution of Al in Li is formed as a fourth phase.

Measurements at Li concentrations above 60 at% Li do not hint of other LiAl phases, but the measurements at these Li concentrations are difficult and are no real proof for the non-existence of such phases.

The evaluation of thermodynamic data follows the same principles as described in the previous chapters but the existence of intermetallic phases or compounds must be taken into account. The extrapolation to well-defined limiting values is restricted. While for solid solutions thermodynamic functions are mostly defined for a composition A_xB_{1-x},

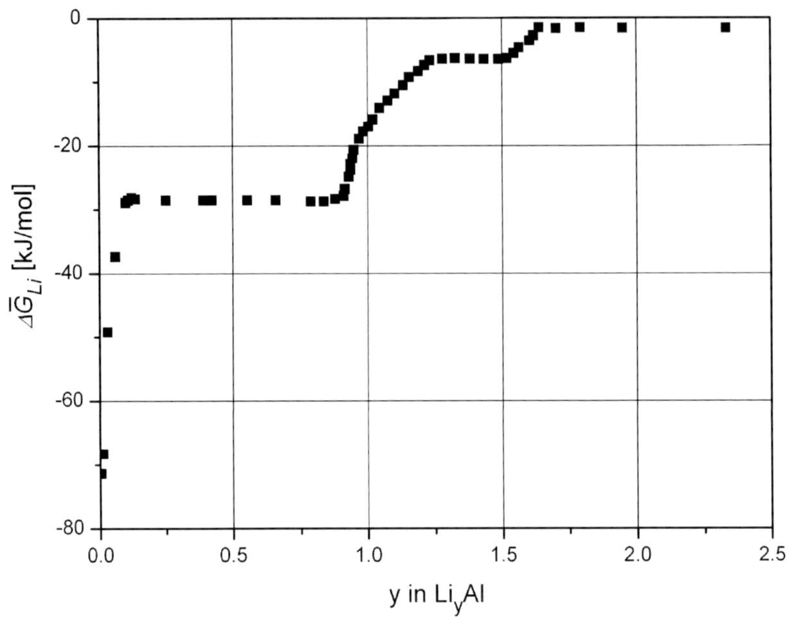

Figure 3.19 Partial molar Gibbs energies $\Delta\bar{G}_{\mathrm{Li}}$ in an LiAl electrode with increasing Li content, which is represented by y in $\mathrm{Li}_y\mathrm{Al}$.

the stoichiometric formula is used as an alternative for an intermetallic compound. For calculation of thermodynamic functions, the data are plotted as function of y (y the stoichiometric coefficient in $\mathrm{A}_y\mathrm{B}$).

The first data calculated from the potential values are the partial molar Gibbs energies of Li and the activities of Li in the alloy phases. The partial molar Gibbs energies are shown versus y in $\mathrm{Li}_y\mathrm{Al}$ in Figure 3.19.

Using the Duhem–Margules equation,

$$\Delta_{\mathrm{for}} G_{\mathrm{A}_{x_\mathrm{A}}\mathrm{B}_{x_\mathrm{B}}} = \frac{1}{y+1}\int_0^y \Delta\bar{G}\mathrm{d}y \tag{3.52}$$

one obtains by integration of the partial molar Gibbs energies the Gibbs energies of formation of $\mathrm{A}_{x_\mathrm{A}}\mathrm{B}_{x_\mathrm{B}}$. The values for $\mathrm{Li}_x\mathrm{Al}_{1-x}$ are shown in Figure 3.20. The activity and activity coefficient of component B (Al) can be obtained from the energy of formation $\Delta_{\mathrm{for}}G(\mathrm{A}_{x_\mathrm{A}}\mathrm{B}_{x_\mathrm{B}})$ by the previously described procedure.

The Gibbs energy of formation is defined as

$$\Delta_{\mathrm{for}} G_{\mathrm{A}_{x_\mathrm{A}}\mathrm{B}_{x_\mathrm{B}}} = x_\mathrm{A}\,\Delta\bar{G}_\mathrm{A} + x_\mathrm{B}\,\Delta\bar{G}_\mathrm{B} \tag{3.53}$$

$$\Delta_{\mathrm{for}} G_{\mathrm{A}_{x_\mathrm{A}}\mathrm{B}_{x_\mathrm{B}}} = x_\mathrm{A} RT\ln a_\mathrm{A} + x_\mathrm{B} RT\ln a_\mathrm{B} \tag{3.54}$$

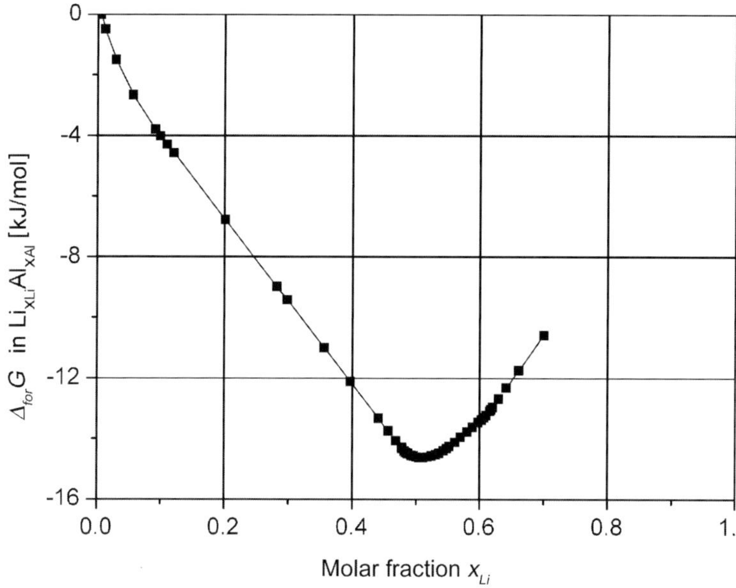

Figure 3.20 Gibbs energy of formation of $\Delta_{for}G$ of $Li_{x_{Li}}Al_{x_{Al}}$ as function of the mole fraction of Li, x_{Li}.

The values of ln a_B can then be calculated by the equation

$$\ln a_B = \frac{1}{x_B RT} \Delta_{for} G_{A_yB} - \frac{x_A}{x_B} \ln a_A \tag{3.55}$$

For phases of mean composition, e.g., $Li_{0.5}Al_{0.5}$, another presentation of thermodynamic functions is possible. The value of a thermodynamic function relative to the value of stoichiometric composition is given. This is extensively described by Weppner *et al.*[18,19]

3.3.4 Intermetallic compounds: the system LiSb

In Figure 3.21 potential measurements in the system LiSb are shown.[19] While the borders between intermetallic phases and intermetallic compounds in the LiAl system were fluctuant, the system in Figure 3.21 shows two intermetallic compounds Li_2Sb and Li_3Sb which are of nearly ideal stoichiometry.

The Gibbs energy of formation is calculated from the plateau potential of Li_2Sb by the equation

$$\Delta_{for}G_{Li_2Sb} = -F\Delta E\int_0^2 dy_{Li} \tag{3.56}$$

A value of (-176.0 ± 1) kJ · mol^{-1} was obtained.

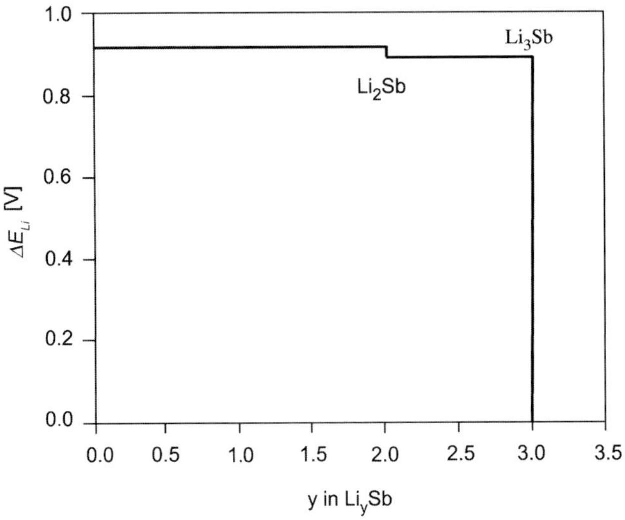

Figure 3.21 Coulometric titration plot of deposition of Li into Sb. Potentials versus a pure Li electrode, temperature 360°C. (Reproduced with permission from Ref. [19], © 1978, The Electrochem. Soc.)

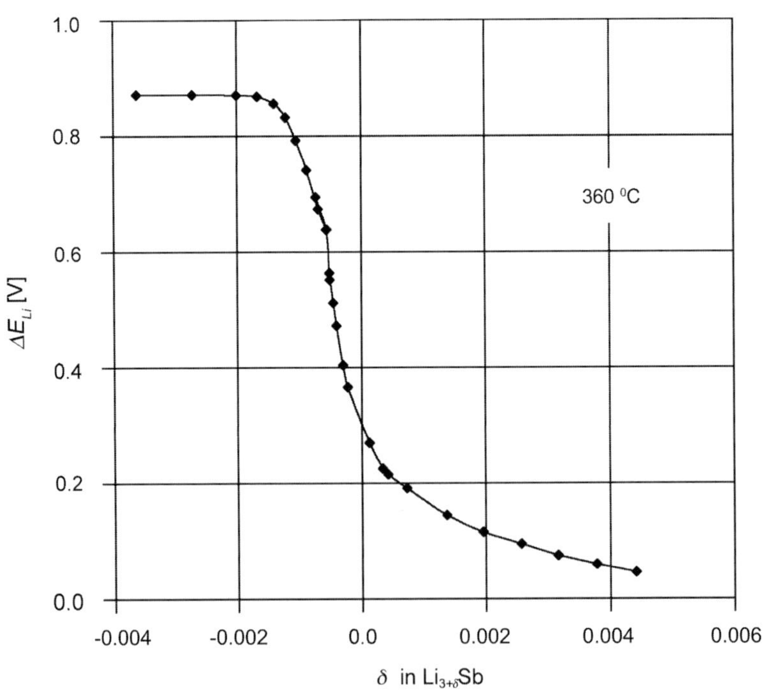

Figure 3.22 Coulometric titration of Li_3Sb. Region of variable stoichiometry, temperature 360°C. (Reproduced with permission from Ref. [19], © 1978, The Electrochem. Soc.)

The authors also measured the temperature dependence of the plateau potentials in the range of 355–600 °C and then calculated the entropy of formation and the enthalpy of formation of Li_2Sb using Eqs. (3.57) and (3.58).

$$\Delta_{for} S_{Li_2Sb} = -F \int_0^y \frac{\partial \Delta E}{\partial T} dy \tag{3.57}$$

$$\Delta_{for} H_{Li_2Sb} = -F \left\{ \int_0^y \Delta E \bullet dy - T \int_0^y \frac{\partial \Delta E}{\partial T} dy \right\} \tag{3.58}$$

The values obtained by the calculation are $(-66 \pm 1)\ J \cdot mol^{-1} \cdot K^{-1}$ for the entropy of formation and $(-220.4 \pm 0.7)\ kJ \cdot mol^{-1}$ for the enthalpy of formation.

In Figure 3.21 the second intermetallic compound Li_3Sb also looks like a stoichiometric compound. The high accuracy of the coulometric titration method reveals that this compound has a very small region of variable stoichiometry of $Li_{3\pm\delta}Sb$ with $\delta = \pm 10^{-3}$. This is shown in Figure 3.22.

Figure 3.22 demonstrates the high precision of coulometric titration and the accuracy achievable for thermodynamic data by electrochemical potential measurements.

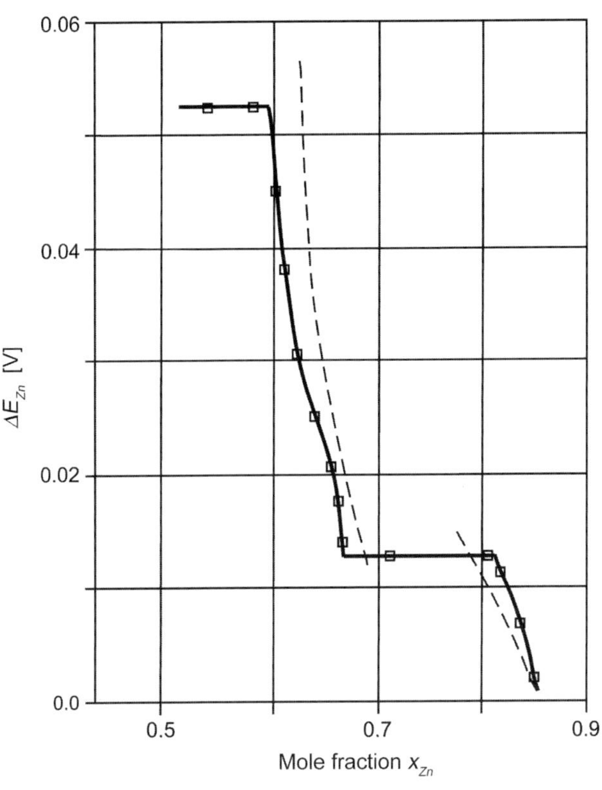

Figure 3.23 Potential measurements of CuZn alloys at room temperatures.[21] The bracketed line was calculated with the data of Ölander.[22]

3.3.5 Measurements at room temperatures: CuZn

So far all the measurements were made at higher temperatures. Measurements of alloy potentials, however, can also be made at room temperature. For this purpose, Schwitzgebel[21] developed a cell where the surface of an alloy electrode could be grinded in the electrolyte. On the fresh surface, reproducible potential measurements were possible. One system investigated was the classical brass alloy CuZn. The first potential measurements at higher temperatures using a molten salt electrolyte were published by Ölander.[22] The present phase diagram is still based on these measurements. The new measurements at 25 °C are shown in Figure 3.23.

The results are in excellent agreement with Ölander's work and extend the phase information down to room temperature.

REFERENCES

1. S. Trassatti, R. Parsons, Pure Appl. Chem., *58*, 437 (1986).
2. A.J. Bard, R. Parsons, J. Jordan (Eds.), Standard Potentials in Aqueous Solutions, IUPAC, Marcel Dekker, New York, 1985.
3. R.G. Bates, V.E. Bower, J. Res. Nat. Bur. Stand., *53*, 283 (1954).
4. D.J.G. Ives, J. Janz, Reference Electrodes, Academic Press, New York, 1961.
5. W. Mehl, G. Lohmann, Private Communication, 1975.
6. S. Trassatti, Pure Appl. Chem., *58*, 955 (1986).
7. K.J. Vetter, Elektrochemische Kinetik, Springer, Heidelberg, 1961.
8. I. Mills, T. Cvitas, K. Homann, N. Kallay, K. Kuchitsu, Quantities, Units and Symbols in Physical Chemistry, 2nd Ed., IUPAC, Blackwell Science, Oxford, 1993, p. 58.
9. P.W. Atkins, Physikalische Chemie, 2nd Ed., VCH, Weinheim, 1996, p. 1036.
10. Z.M. Koepp, H. Wendt, H. Strehlow, Z. Elektrochem., *64*, 483 (1960).
11. G. Gritzner, J. Kuta, Pure Appl. Chem., *56*, 461 (1984).
12. S. Wehner, Doctoral Thesis, Technische Universität Dresden, 2005.
13. O. Kubaschewski, C.B. Alcock, P.J. Spencer, Materials Thermochemistry, 6th Ed., Pergamon Press, Oxford, 1993.
14. A.J. Martyanov, S.S. Batsanov, Russ. J. Inorg. Chem., *25*, 1737 (1980).
15. P. Villars, Factors Governing Crystal Structures, in J.H. Westbrook, R.L. Fleischer (Eds.), Intermetallic Compounds, Crystal Structures of Intermetallic Compounds, Wiley, Chichester, 2000, p. 1.
16. O. Kubaschewski, O. Huchler, Z. Elektrochem., *52*, 170 (1948).
17. C. Wagner, J. Chem. Phys., *21*, 1819 (1953).
18. C. John Wen, B.A. Boukamp, R.A. Huggins, W. Weppner, J. Electrochem. Soc., *126*, 2258 (1979).
19. W. Weppner, R. Huggins, J. Electrochem. Soc., *125*, 7 (1978).
20. W. Weppner, Solid State Electrochemical Methods for the Characterization of the Kinetics, Thermodynamics and Phase Equilibria of Lithium Battery Materials, in C. Julien, Z. Stoynov (Eds.), Materials for Lithium-Ion Batteries, Kluwer Academic Publishers, Dordrecht, NL, 2000, p. 451.
21. G. Schwitzgebel, Z. Phys. Chem. NF, *95*, 15 (1975).
22. A. Ölander, Z. Phys. Chem. Abt. A, *164*, 428 (1933).

– 4 –

Ad-Atoms and Underpotential Deposition

In the previous chapter, the equilibrium between the solid and the liquid phase was described. The interface between the two bulk phases is the mediator and any processes between the two neighboring phases will be influenced by the properties of this interface. In this chapter the properties of this interface and methods of investigation will be described.

4.1 THE THERMODYNAMIC DESCRIPTION OF THE INTERPHASE

4.1.1 The electrochemical double layer

In Figure 4.1 a model is shown of the interface of an electrode in contact with an electrolyte that is otherwise known as the electrochemical double layer.

The description of the surface region distinguishes several layers.[1] These layers locate the different charges on both sides of the interface, q_m on the solid side and q_{el} on the electrolytic side, that build up the potential difference between the sides. If the solid phase is a metal, the charge q_m is located in the Thomas–Fermi layer which is less than an Angstrom (0.1 nm) thickness. The charge for a semiconductor, on the other hand, is located in the space-charge region which can have a thickness of several hundred nanometers depending on the doping level (Figure 4.2). On the electrolytic side of the double layer, ions can be specifically adsorbed on the electrode forming the inner Helmholtz layer. In the adsorption process the ions can lose part of the solvation shell and part of their charge. Also uncharged molecules can be adsorbed in the inner Helmholtz layer. If only electrostatic forces are acting, the ions will just approach the electrode to a distance determined by the thickness of the solvation layer at the electrode surface and of the solvation shell of the ions. This layer is called "the outer Helmholtz layer." The contribution of the thermal forces can distribute the charge in the electrolyte. This layer is called the "Gouy–Chapman layer" or "diffuse double layer." Its contribution to the surface properties is most pronounced when there is a small surface charge and at low electrolyte concentrations.

The crudest description of the electrical properties is the application of the model of a plate condenser to the double layer. The capacitance is given by the equation

$$C = \varepsilon\varepsilon_0 \frac{A}{d} \tag{4.1}$$

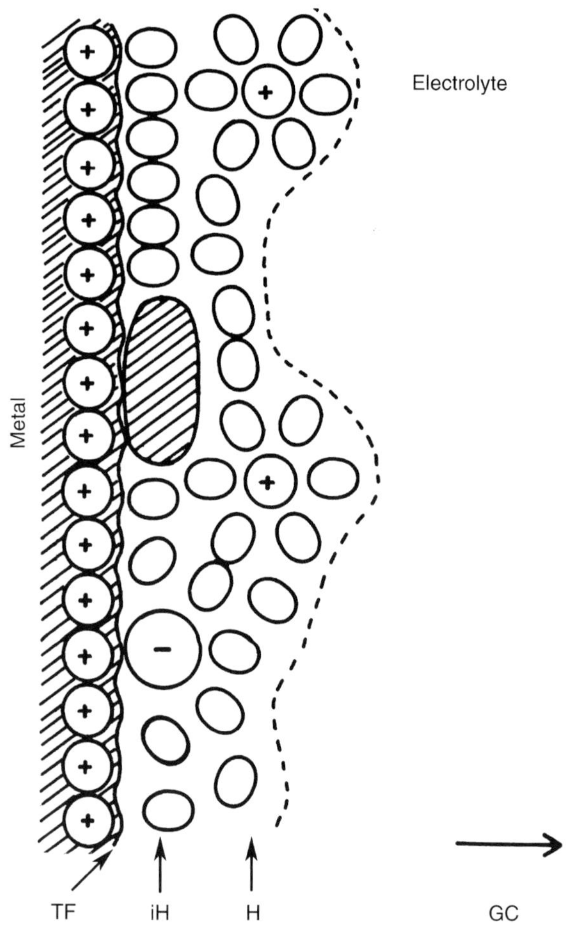

Figure 4.1 Model of the electrochemical double layer on a metal electrode. TF Thomas–Fermi layer, iH layer of specific adsorption (inner Helmholtz layer), H Helmholtz layer (outer Helmholtz layer), and GC Gouy–Chapman layer.

where C is the capacitance, ε the permittivity of the medium between the condenser plates, and ε_0 the permittivity of the vacuum. A is the electrode area and d the distance between the charged surface planes. This equation is an approximation for the outer Helmholtz layer in the case of the absence of specific adsorption. The potential difference of the inner or Galvani potential across this layer between the metal and the Helmholtz plane $\Delta\varphi_{\mathrm{H}} = \varphi_{\mathrm{m}} - \varphi_{\mathrm{H}}$ is proportional to the surface charge density q_{m} of the metal side of the double layer. It can be described by the equation

$$\Delta\varphi_{\mathrm{H}} = \frac{d}{\varepsilon\varepsilon_0} q_{\mathrm{m}} \tag{4.2}$$

Gouy and Chapman developed the theory for the diffuse part of the double layer in front of the electrode.[2] In a first approximation and for a z/z valent electrolyte, an exponential potential decay with increasing distance x from the Helmholtz plane was derived.

$$\varphi_x = \varphi_{\mathrm{H}} \mathrm{e}^{-\kappa \cdot x} \tag{4.3}$$

The potential φ_{H} is the potential in the plane of closest approach for the solvated ions to the electrode surface, which is the Helmholtz plane. This plane can be different for anions and cations. The constant κ of the exponential function is given by the equation

$$\kappa = \sqrt{\frac{F^2(z_+^2 c_+ + z_-^2 c_-)}{\varepsilon RT}} \tag{4.4}$$

The reciprocal value is a measure for the thickness of the Gouy–Chapman layer.

$$\frac{1}{\kappa} = d_{\mathrm{GC}} \tag{4.5}$$

On the surface of a metal the surface charge concentrates in the very thin Thomas–Fermi layer. The charge for a semiconductor is distributed in a layer similar to the Gouy–Chapman layer called "the space-charge layer." A theory similar to the Gouy–Chapman theory can be developed for the space-charge layer.[3] If one combines the potential variation in the different layers of the electrochemical double layer, one gets the description shown in Figure 4.2.

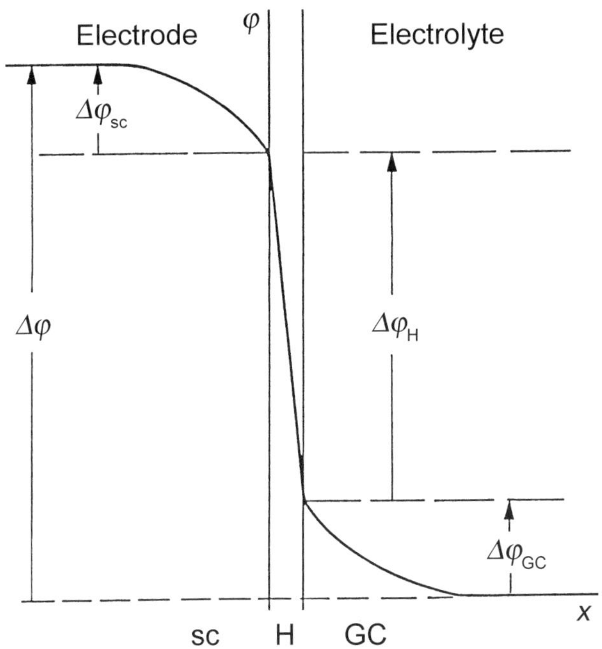

Figure 4.2 Potential changes in the different parts of the double layer. $\Delta\varphi_{\mathrm{sc}}$ potential drop in the space-charge layer, $\Delta\varphi_{\mathrm{H}}$ potential drop in the Helmholtz layer, and $\Delta\varphi_{\mathrm{GC}}$ potential drop in the Gouy–Chapman layer (Bard, Memming. Miller[4]).

The potential change between the electrode and the electrolyte is the sum of the different potential drops

$$\Delta\varphi_{dl} = \Delta\varphi_{sc} + \Delta\varphi_{H} + \Delta\varphi_{GC} \tag{4.6}$$

The total capacitance of the double layer is given by a linear combination of the reciprocal capacitances of the different layers, that is, for a metal electrode

$$\frac{1}{C_{dl}} = \frac{1}{C_{H}} + \frac{1}{C_{GC}} \tag{4.7}$$

For a semiconductor electrode a third term, the capacitance of the space-charge layer, must be added

$$\frac{1}{C_{dl}} = \frac{1}{C_{sc}} + \frac{1}{C_{H}} + \frac{1}{C_{GC}} \tag{4.8}$$

Usually the capacitance of the Helmholtz layer and at higher electrolyte concentrations the capacitance of the Gouy–Chapman layer are much larger than the capacitance of the space-charge layer. Therefore, the reciprocal term can be neglected. The space-charge layer is the dominant element and represents the properties of the double layer for semiconductor electrodes

$$\frac{1}{C_{dl}} \approx \frac{1}{C_{sc}} \tag{4.9}$$

The adsorption layer (inner Helmholtz layer) is also important. In the case of the specific adsorption of a selected ion, the adsorbed mass is given by the surface concentration Γ_{ad} (mol m^{-2}) and the corresponding charge is $q_{ad} = z_{ad} F\Gamma_{ad}$ (C m^{-2}). In a more specific description in which only metal electrodes will be discussed, the following relation can be derived for the relation between the measured capacitance to the capacitance values of the different layers

$$\frac{1}{C_{dl}} = \frac{1}{C_{H}} + \frac{1}{C_{GC}}\left[1 - \left(\frac{\partial q_{ad}}{\partial q_{m}}\right)_{\mu}\right] \tag{4.10}$$

The partial derivation $\partial q_{ad}/\partial q_{m}$ is made at constant concentration that means constant chemical potential μ.

The charge on the metal surface is q_{m}. Substituting the charge q_{ad} gives

$$\frac{1}{C_{dl}} = \frac{1}{C_{H}} + \frac{1}{C_{GC}}\left[1 - z_{ad}F\left(\frac{\partial \Gamma_{ad}}{\partial q_{m}}\right)_{\mu}\right] \tag{4.11}$$

A general description of the different capacitances of electrified interfaces was given by Trasatti and Parsons.[5]

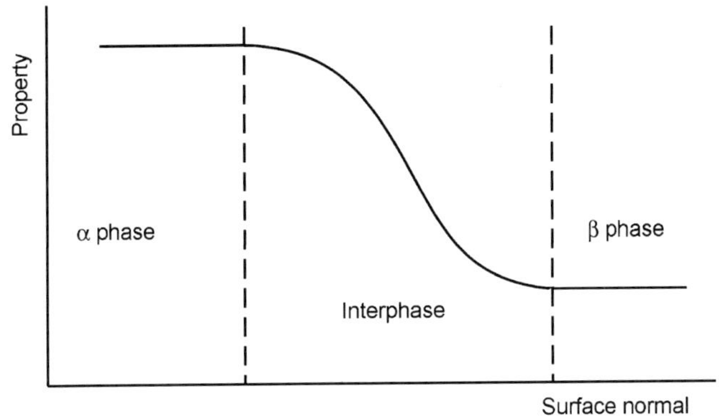

Figure 4.3 The concept of an interphase to describe the surface properties between two electrified phases, α-phase and β-phase.

4.1.2 Ideally polarizable electrodes

The thermodynamic description of the boundary layer between electrode and electrolyte is usually based on the application of Gibbs's fundamental equation to the double layer. For this application the surface region is considered as a separate phase called the "interphase" (Figure 4.3). In the following description it is assumed that no Faradaic current between the electrode and the electrolyte is possible, which means no reduction or oxidation can occur. This is called an "ideally polarizable electrode."

The interphase region is limited on both sides by vanishing differences of the main intensive parameters to the bulk phase. At constant temperature and constant pressure the Gibbs–Duhem equation for the interphase has the form[6]

$$-\mathrm{d}\sigma = q_{\mathrm{m}}\mathrm{d}E + \sum_i \Gamma_i \mathrm{d}\mu_i \tag{4.12}$$

where σ is the Gibbs surface energy (for liquid electrodes equal to the surface tension), q_{m} is the surface charge density of the metal side of the interphase (in C cm^{-2}), E the electrode potential versus a constant reference electrode, Γ_i the surface concentration of a component in the interphase (in mol·cm^{-2}), and μ_i the chemical potential of the component i in the electrolyte.*

4.1.3 Electrocapillary curves

Eq. (4.12) leads to two quantities, which can be experimentally determined. The charge on the metal side of the interphase is obtained by the dependence of the surface energy on the potential at constant activities

$$q_{\mathrm{m}} = -\left(\frac{\partial \sigma}{\partial E}\right)_{a_i} \tag{4.13}$$

This equation is called Lippmann equation.[7]

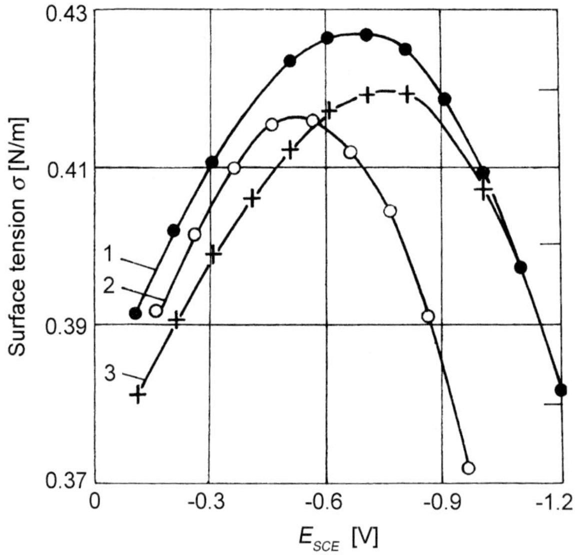

Figure 4.4 Dependence of the surface energy (surface tension) of a mercury electrode on the potential. (1) 0.55 mol·dm^{-3} H_2SO_4, (2) 1.1 mol·dm^{-3} HBr (Br^- adsorption), and (3) 0.55 mol·dm^{-3} H_2SO_4 + 0.1 mol·dm^{-3} $C_6H_{11}NH_2$ ($C_6H_{11}NH_3^+$ adsorption).[8]

For the application of this equation one has to measure surface energy (surface tension) and its dependence on electrode potential at constant activities of the different components. Precise measurements are restricted to liquid metals like mercury or gallium and their alloys. Classical experiments were made with the Lippmann electrocapillary meter. The measurement of the drop time or the drop frequency of a dropping mercury electrode is easier.

A typical measurement for mercury in an inert electrolyte is shown in Figure 4.4. The charge of the interphase is given by the derivative of the surface tension versus the electrode potential. The maximum of the capillary curve represents a point with charge zero. This is called the potential of zero charge (pzc). On the cathodic side of the pzc the electrode is charged negatively and on the anodic side it is charged positively. If substances are added, which are adsorbed on the electrode surface, the pzc either shifts to negative potentials with anion adsorption (curve 2) or to positive potentials with cation adsorption (curve 3).

The charge on each side of the interphase is the sum of the charge of the different charged components. For a metal electrode and on the metal side, using the simple jellium model, this is the difference between the positive metal kernels and the electrons. This difference represents the asymmetry of the electron density on the metal surface.

$$q_{\mathrm{m}} = F\sum_i z_i \Gamma_{i,+} - F\Gamma_e \tag{4.14}$$

The charge of the electrode side q_{m} must be balanced by the charge of the electrolytic side $q_{\mathrm{el}} = -q_{\mathrm{m}}$, which is the sum of the charge of the different ionic components in the double layer, the diffuse as well as the Helmholtz double layer. The model was shown in Figure 4.1. The sum is given by the equation

$$q_{\mathrm{el}} = F \sum_{i} z_i \Gamma_i \tag{4.15}$$

The surface concentration Γ stands for the different ions and their different charges z_i (cations $z_i > 0$ and anions $z_i < 0$).

Much less accurate than the surface tension measurements on liquid electrodes are surface energy or surface tension measurements on solid electrodes. The special properties of metal electrodes will be treated in Section 4.1.7, but the measurement of the capacitance can be an alternative even for liquid electrodes. The differential capacitance is related to the surface energy by the equation

$$C_{\mathrm{dl}} = \frac{\mathrm{d}q_{\mathrm{m}}}{\mathrm{d}E} = -\frac{\mathrm{d}^2\sigma}{\mathrm{d}E^2} \tag{4.16}$$

As an example, the capacitance of a mercury electrode in NaF on the potential is shown for different electrolyte concentrations (Figure 4.5). The pzc is obtained from these measurements at low electrolyte concentrations. By integration of the differential capacitance versus potential curves the surface charge is obtained.

Capacitance measurements are extensively used for investigation of semiconductor electrodes (Chapter 9).

4.1.4 Adsorption isotherms

The second quantity obtained from the Gibbs–Duhem equation is the relative surface excess $\Gamma_{i(s)}$, which is the difference of the amount of a substance in the interphase and the relative amount of solvent (s) expressed by the ratio of the mole fractions x_i and x_s.

$$\Gamma_{i(s)} = \Gamma_i - \frac{x_i}{x_s}\Gamma_s \tag{4.17}$$

The relative surface excess is a result of the restriction that only i−1 chemical potentials are independent quantities. The meaning of the expression is explained in Figure 4.6.

The relative surface excess is obtained from the dependence of the surface energy on the activity of the independent variable components.

$$\Gamma_{i(s)} = -\frac{1}{RT}\left(\frac{\partial \sigma}{\partial \ln a_i}\right)_{a_{j\neq i},E} \tag{4.18}$$

This is called the Gibbs adsorption isotherm. The equation represents the most general and pure thermodynamic description of enrichment of a substance on a surface. The description of the surface concentration of ions on the electrode is also possible by special

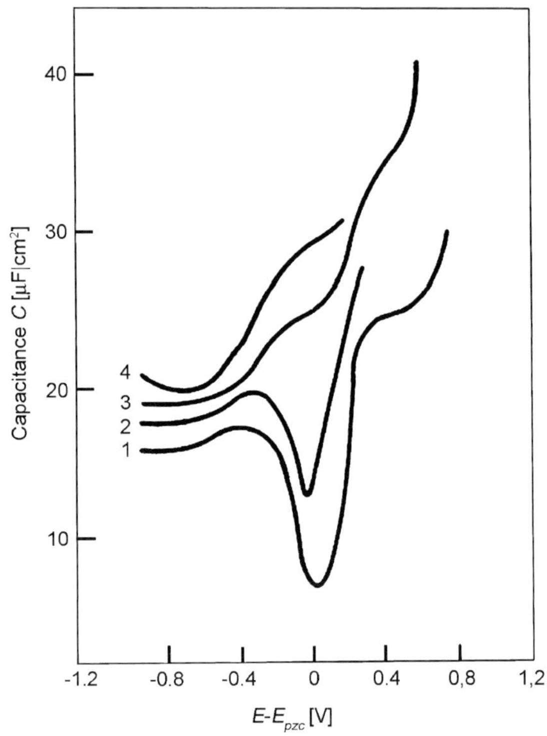

Figure 4.5 Dependence of the differential capacitance of a mercury electrode on the potential. (1) 0.001 mol·dm^{-3} NaF, (2) 0.01 mol·dm^{-3} NaF, (3) 0.1 mol·dm^{-3} NaF, and (4) 0.9 mol·dm^{-3} NaF. The representation is based on measurements by Grahame.[9]

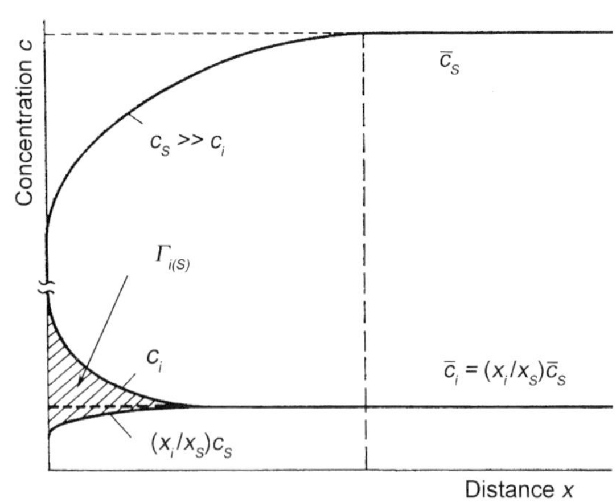

Figure 4.6 Schematic presentation of the relative surface excess $\Gamma_{i(s)}$. $c_i(x)$ and $c_s(x)$ are the concentration of substance i and solvent s in front of the surface. Usually large excess of the solvent in the electrolyte, represented by the ratio x_i/x_s, reduces the concentration of the solvent to the concentration of the substance i. The relative surface excess is the hatched area in the diagram. If $(x_i/x_s)c_s \approx \bar{c}_i$ the relative surface excess is approximately the surface excess of the adsorbed substance over its bulk concentration.

models, which lead to adsorption isotherms that describe the dependence of the surface concentration Γ on the electrolyte concentration c with model parameters. For convenience the surface concentration is usually expressed by the surface coverage $\Theta = \Gamma/\Gamma_{max}$, the ratio between the surface concentration and the maximum surface concentration. The Langmuir isotherm is the simplest example.

$$\Theta = \frac{K_{ad}c}{K_{ad}c + 1} \tag{4.19}$$

The adsorption constant K_{ad} is characteristic for the surface enrichment of a substance. For the description of ion adsorption the strong electrostatic forces between the ions must be taken into account. The first equation applicable to these problems was the Frumkin isotherm.[10]

$$\frac{\Theta}{1-\Theta} = K_{ad}c\exp\left(\frac{g\Theta}{RT}\right) \tag{4.20}$$

The constant g determines the ion or molecule interaction. $g < 0$ implies repulsion and $g > 0$ attraction. Temkin[11] derived a similar isotherm for rough surfaces. Several more models of adsorption isotherms are described in the literature.

4.1.5 Reversible electrodes

The thermodynamic description of the electrode–electrolyte interface of a reversible electrode differs from the simple case of the polarizable electrode.[12] The concentrations of the redox components in the electrolyte and the potential are related by the Nernst equation. This means dE must be substituted by

$$\mathrm{d}E = \frac{RT}{zF}\mathrm{d}\ln\frac{a_{ox}}{a_{red}} \tag{4.21}$$

No longer an independent variable, the electrode potential is determined by the redox concentrations. Accordingly, the surface excess of redox components and charge q_m are also not independently determinable. The expressions for the redox components are shown in the following equation

$$-\mathrm{d}\sigma = \left(\Gamma_{ox(s)} + \frac{q_m}{zF}\right)RT\mathrm{d}\ln a_{ox} + \left(\Gamma_{red(s)} - \frac{q_m}{zF}\right)RT\mathrm{d}\ln a_{red} \tag{4.22}$$

As an example, the surface excess of the quinon–hydroquinone redox components on mercury is shown in Figure 4.7.

While the surface excess of quinon or hydroquinone cannot be determined without addition or substraction of q_m/zF, the charge term is eliminated if the sum of the relative surface excess values of quinon and hydroquinone is formed.

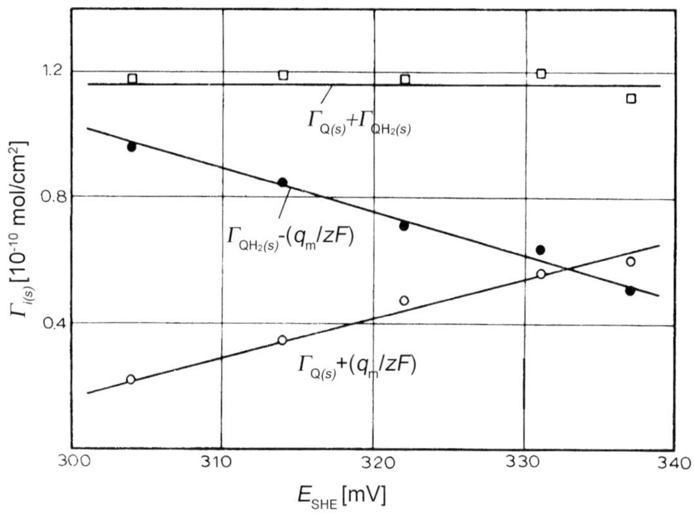

Figure 4.7 Relative surface excess of the redox components quinon and hydroquinone on a mercury electrode.[13]

For an example of a metal electrode in equilibrium with a solution of metal ions, the equation for the relative surface excess of the metal ions is

$$-\mathrm{d}\sigma = \left(\Gamma_{\mathrm{Me}^{z+}(s)} + \frac{q_\mathrm{m}}{zF}\right) RT\mathrm{d}\ln a_{\mathrm{Me}^{z+}} \tag{4.23}$$

Surface charge and charge of adsorbed ions can be considerably different. For complete coverage one can calculate

$$q_\mathrm{m} << zF\Gamma_{\mathrm{Me}^{z+}(s)} \tag{4.24}$$

Therefore, if the charge $zF\Gamma_{\mathrm{Me}^{z+}(s)} + q_\mathrm{m}$ in Eq. (4.23) is much larger than 10^{-5} C cm^{-2} it is mostly determined by adsorbed metal ions. The same argument can be applied to the surface excess of the redox pair in Figure 4.7. Under similar conditions the contribution of the surface charge to the surface excess values of quinon and hydroquinone can be neglected.

4.1.6 Partial charge and electrosorption valency

For molecules and ions adsorbed at the interphase the possibility of partial discharge is discussed in the literature. Metal deposition, for example, can occur in two steps with a partial charge transfer

$$\mathrm{Me}^{z+} + \Delta z_\mathrm{ad}\mathrm{e}^- \rightarrow \mathrm{Me}_\mathrm{ad}^{z_\mathrm{ad}} \tag{4.25a}$$

$$\mathrm{Me}_\mathrm{ad}^{z_\mathrm{ad}} + z_\mathrm{ad}\mathrm{e}^- \rightarrow \mathrm{Me} \tag{4.25b}$$

The intermediate adsorbed ion is called ad-atom (Section 4.3). It can have a charge between the charge of the ion in the solution and the discharged metal atom. Lorenz and Salié[14,15] and Plieth and Vetter[16] independently described the concept of partial charge $z_{ad} = \lambda z$. In Lorenz and Salié's description the partial charge is the consequence of a bond formation between the adsorbed ion and the metal surface forming a molecular orbital with an asymmetric electron distribution. Plieth and Vetter discussed a redox reaction and gave a statistical interpretation based on the fast electron transfer between the adsorbed redox system and the metallic substrate.

As was explained in Section 2.9.10, the reduced and oxidized ions of a redox couple interact with the solvent dipoles by ion–dipole interaction. This influences the energy of the electronic states. The fluctuation of the solvent molecules around the ion with only a statistical equilibrium solvation leads to a distribution of the electron energies around a central value of Gaussian form. Two energy distribution functions describe the energy distribution, one for the reduced ions (the occupied states) and the other for the oxidized ions (the unoccupied states). This was shown in Figure 2.33. The development of two different distribution functions is based on stable oxidation states. In each state the ion–dipole interaction can achieve a quasi equilibrium distribution.

If the ion is specifically adsorbed, the clear distinction between reduced and oxidized states is no longer possible as a consequence of the formation of bonding orbitals and the fast electron fluctuation between the metal surface and the adsorbed ion. The solvation shell cannot achieve a stable configuration for oxidized and reduced states of the adsorbed ion. Instead, an intermediate solvent distribution develops for a mean residence probability of an electron on the adsorbed ion.

Only one distribution function exists for a metal ion in the electrolyte with a fixed charge z. But in the specifically adsorbed state, the situation is the same as for a redox couple. Electrons are in a fast exchange process between the energy bands of the metal electrode and the energy levels of the adsorbed ion. The adsorbed metal ions become partially discharged. The charge z_{ad} is the fraction $\lambda \cdot z$ of the charge of the metal ion in the electrolyte. The electron energy distribution functions of the metal ions in the electrolyte and the partially discharged adsorbed metal ions, the so-called ad-atoms, are shown in Figure 4.8 (together with the electron energy levels in the metal).

Discharge of a metal ion during the deposition process is often assumed for two-valent metal ions like Zn^{2+} or Cu^{2+} thus forming a one-valent ad-atom Zn^{+} or Cu^{+}. The concept of partial discharge is a generalization of this concept. Lorenz and Salié have investigated several systems.[17,18,19]

The experimental verification of the partial charge transfer and determination of the partial charge of ions adsorbed on the electrode surface is connected with a general controversy about the pure thermodynamic interpretation of the partial charge and attempts to interpret measurements by models allowing an approximate view on the charge distribution in the double layer.

The thermodynamic interpretation is based on the derivative of the Lippmann equation defining the electrosorption valency γ:

$$\gamma = -\frac{1}{F}\left(\frac{\partial q_{\mathrm{m}}}{\partial \Gamma_i}\right)_E = \frac{1}{F}\left(\frac{\partial \mu}{\partial E}\right)_\Gamma = z_i + \sum_{j \neq i} \frac{\partial \Gamma_j}{\partial \Gamma_i} \tag{4.26}$$

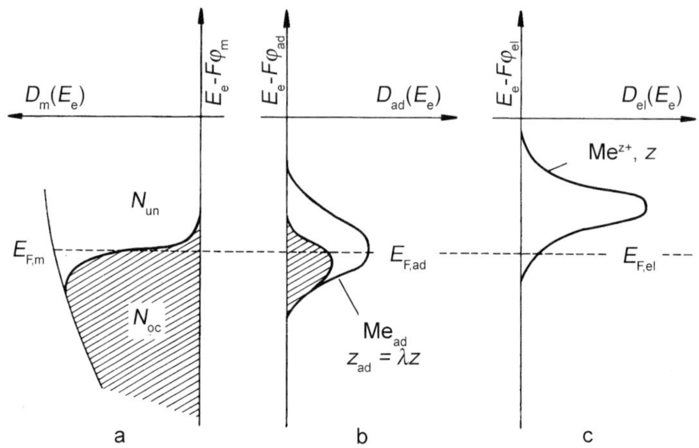

Figure 4.8 Electron energy distribution function $D(E_e)$ as function of the energy $E_e - F\varphi$. (a) In the metal, potential φ_m, (b) for an adsorbed ion in the adsorption layer (inner Helmholtz layer), potential φ_{ad}, and (c) for the metal ion in the electrolyte, potential φ_{el}. E_F is the Fermi energy. In equilibrium is $E_{F,m} = E_{F,ad} = E_{F,el}$. The Fermi energy of the electrolyte is equal to the equilibrium redox potential (Nernst potential).[16]

Eq. (4.26) was probably first derived by Schmidt.[20] Vetter and Schultze[21] introduced the expression "electrosorption valency".

Thermodynamically it is the total charge balance of the adsorbed ion and co-adsorption of counter ions. The value can be obtained by independent measurements of adsorbed charge and adsorbed mass. The charge is usually determined by chronoamperometry and integration of potential scans either in the negative direction (adsorption) or the positive one (desorption and anodic stripping method). The latter method is usually preferred because the adsorption film reaches a stable condition. The mass can be determined by several methods. The experimental techniques and examples will be described in Section 4.2.

A second coefficient can be derived from the Gibbs–Duhem equation if, instead of the potential, the chemical potential is kept constant

$$\gamma^* = -\frac{1}{F}\left(\frac{\partial q_m}{\partial \Gamma_i}\right)_\mu = \frac{1}{F}\left(\frac{\partial \mu}{\partial E}\right)_{q_m} \tag{4.27}$$

This value is the reciprocal value of the Esin–Markov coefficient.[22]

The partial charge can only be determined from the electrosorption valency by introduction of specific models about geometric structure and charge distribution in the double layer. The basic Gibbs adsorption isotherm does not give any information on the charge separation in the interphase. One model has already been presented by Grahame.[23,24] He gave a picture of a variable position for the adsorbed ion in the double layer. The model of partial charge appointed a fixed position to the adsorbed ion in the double layer (potential φ_{ad}) and a partial charge $z_{ad} = z\lambda$. Parsons and Sass[25] and Schmickler and Guidelli[26] connected the electrosorption valency with a dipole moment of the adsorbed ion. While the dipole moment concept might be sufficient for a description of a polarized bond between adsorbed ions and the metal, as was pointed out by Lipkowski in his investigations of anion adsorption,[22] the concept of partial charge and partial charge transfer is something one has to take into account in kinetics.[15,27,28] Rate equations for ion desorption or equations for surface diffusion, for example, can depend on the charge of the adsorbed ion. Examples will be described in Chapters 4 (Section 4.3) and 10.

If the concept of partial charge is introduced into the electrocapillary equation one gets the following result for the relative surface excess of the adsorbed ion[29]

$$-\mathrm{d}\sigma = \left(\lambda\Gamma_{\mathrm{ad}(s)} + \frac{q_M}{zF}\right)RT\mathrm{d}\ln a_{\mathrm{Me}^{z+}} \tag{4.28}$$

The ad-atom is divided into a charged part still belonging to the electrolytic side of the interphase and a discharged part already part of the metallic phase. Again this is not a thermodynamic result but the thermodynamic reflection of the model of charge separation characteristic for the description of the electrochemical double layer.

4.1.7 Thermodynamics of solid electrolyte interfaces

So far the description has been focused on interfaces between liquid metals and electrolytes. These systems served preferentially for fundamental experiments because of the easy access to the free surface energy by just measuring surface tension. The situation becomes more complicated if a solid electrode is considered. For a solid electrode one must distinguish the work connected with the creation of a new interface, e.g., by cleavage from the work required to increase the surface area by stretching.

In the first case the atom density in the surface remains constant at least as long as no restructuring occurs. The transfer of N_A atoms from the bulk into the newly created surface increases the surface area. As the process is the same as for liquid electrodes, this may be termed surface energy σ. In the review of Linford[30] this contribution is called "superficial work."

The second contribution to the surface energy is new because it is based on an extension of the atomic distances in the surface area by a force called surface stress. The symbol used for the surface stress in this book is T as in the review of Linford.[30] The surface stress is a tensor with the tensor components T_{xx}, T_{xy}, T_{xx}, and T_{yy}. The tensor is only independent in its direction for an isotropic solid. The surface stress causes an elastic deformation of the surface described by the surface strain tensor ε_{e}.

Otherwise, a deformation can have two components: the above-mentioned elastic deformation ε_{e} and a plastic deformation by replacing atoms from the bulk into the surface ε_{p}. Both parts of the surface strain were combined to a new function derived by Everett and Couchman.[31]

$$\gamma_s = \frac{\mathrm{d}\varepsilon_{\mathrm{p}}}{\mathrm{d}\varepsilon_{\mathrm{tot}}}\sigma + \frac{\mathrm{d}\varepsilon_{\mathrm{e}}}{\mathrm{d}\varepsilon_{\mathrm{tot}}}\mathrm{T} \tag{4.29}$$

A good name for this function is surface tension. In the review of Linford[30] this contribution is called "surface energy."

The classical Gibbs–Duhem equation must be modified to describe thermodynamic properties of solids. If temperature and pressure are kept constant, one obtains by the function of Everett and Couchman[31]

$$-\mathrm{d}\sigma = q_{\mathrm{m}}\mathrm{d}E + \sum_i \Gamma_i \mathrm{d}\mu_i + (\sigma - \gamma_s)\,\mathrm{d}\varepsilon_{\mathrm{tot}} \tag{4.30}$$

With Eq. (4.29) one gets

$$-\mathrm{d}\sigma = q_{\mathrm{m}}\mathrm{d}E + \sum_i \Gamma_i \mathrm{d}\mu_i + (\sigma - \mathrm{T})\,\mathrm{d}\varepsilon_{\mathrm{e}} \tag{4.31}$$

From Eq. (4.31) follows the Shuttleworth equation[32]

$$-\left(\frac{\partial \sigma}{\partial \varepsilon_{\mathrm{e}}}\right)_{E,\mu_i} = \sigma - \mathrm{T} \tag{4.32}$$

Furthermore, a modified Lippmann equation is obtained

$$-\left(\frac{\partial \sigma}{\partial E}\right)_{\mu_i} = q_{\mathrm{m}} + (\sigma - \mathrm{T})\left(\frac{\partial \varepsilon_{\mathrm{e}}}{\partial E}\right)_{\mu_i} \tag{4.33}$$

Gokhshtein derived a modified form of this equation.[33]

The question of how to describe a solid surface is still being developed. To give some examples, Grafov recently derived a generalized Shuttleworth equation.[34] The present concept was critically discussed by Lang and Heusler[35] and Guidelli.[36] Experimental efforts to verify the thermodynamic approach were published, e.g., by Haiss, Nichols, Sass, and Charlé.[37]

4.2 PRINCIPAL METHODS FOR THE INVESTIGATION OF THE ELECTROCHEMICAL DOUBLE LAYER

The first verification of the fundamental equations of Gibbs and Lippmann was based on surface tension (surface energy) measurements. The investigation of solid electrodes with this method is restricted. In the following paragraph a description is given of some of the modern methods, which are used for investigation of the electrochemical double layer.

4.2.1 Measurement of capacitance

Capacitance measurements have become an important method in electrochemistry. Combinations of resistance and capacitance elements, the so-called equivalent circuits, describe the electrochemical properties of the double layer. In the case of an ideally polarizable electrode, the equivalent circuit is a linear combination of a double layer capacitance and an Ohmic electrolyte resistance (Figure 4.9a). The equivalent circuit of an

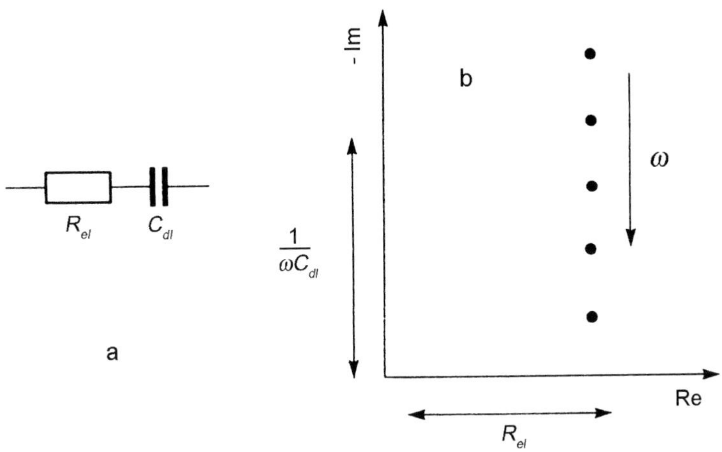

Figure 4.9 (a) Equivalent circuit of an ideally polarizable electrode, R_{el} electrolyte resistance and C_{dl} capacitance of the electrode. (b) Representation of the complex impedance as function of frequency, ω in a Nyquist diagram.

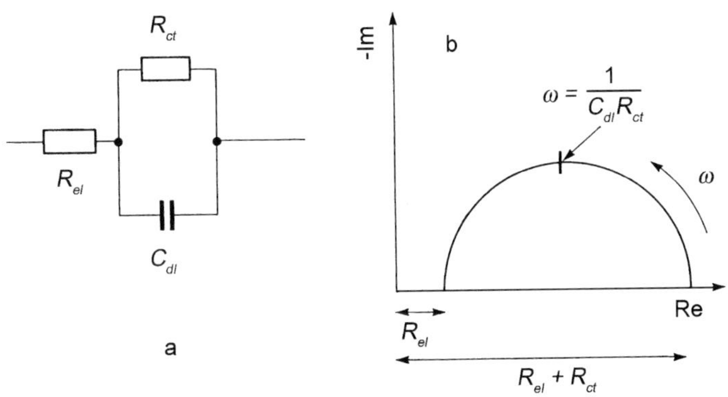

Figure 4.10 (a) Equivalent circuit of an electrode with charge transfer. R_{el} electrolyte resistance, C_{dl} capacitance of the electrode, and R_{ct} charge transfer resistance. (b) Representation of the complex impedance as function of frequency, ω in a Nyquist diagram.

electrode with an additional charge transfer resistance across the double layer is shown in Figure 4.10a. More complex equivalent circuits are necessary if more elements must be taken into account. The Warburg impedance, for example, represents diffusion limitation. A detailed description is given in Chapter 5. In the most general approach a transfer function can be defined. The determination of the elements of an equivalent circuit is of prime importance in electrochemical investigations.

The result of the combination of a resistance and a capacitance is the complex impedance

$$Z = Z_{re} + iZ_{im} \tag{4.34}$$

Any impedance point can be represented in the complex plane (Figure 4.11). For a capacitance, the imaginary part of the impedance has a negative sign.

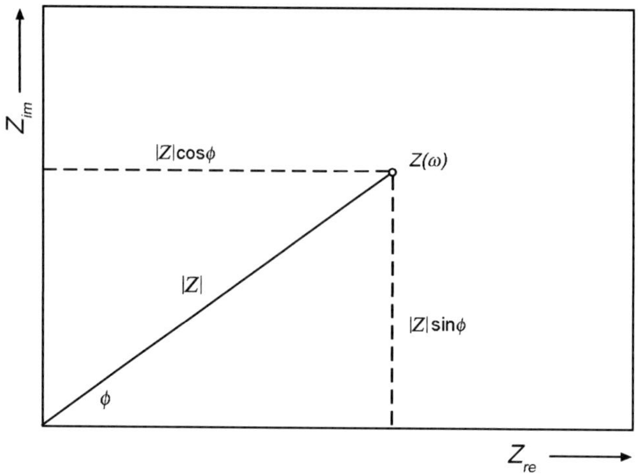

Figure 4.11 Representation of an impedance value $Z(\omega)$ in the complex plane.

The impedance can also be expressed in an exponential form

$$Z = |Z| e^{i\phi} \tag{4.35}$$

with |Z| the modulus of the impedance and ϕ the phase shift between output and input. The following equations relate the two representations:

$$|Z| = \sqrt{Z_{\mathrm{re}}^2 + Z_{\mathrm{im}}^2} \tag{4.36}$$

$$\tan\phi = \frac{Z_{\mathrm{re}}}{Z_{\mathrm{im}}} \tag{4.37}$$

The impedance of an Ohmic resistance is a real quantity

$$Z_R = R \tag{4.38}$$

The impedance of a capacitance is an imaginary quantity

$$Z_C = -i\frac{1}{\omega C} \tag{4.39}$$

The serial combination of a resistance and a capacitance has the impedance

$$Z_{\mathrm{s}} = R - i\frac{1}{\omega C} \tag{4.40}$$

The impedance of a parallel combination of a resistance and a capacitance is given by

$$\frac{1}{Z_{\mathrm{p}}} = \frac{1}{R} + i \cdot \omega C \tag{4.41}$$

The impedance of the equivalent circuit in Figure 4.10a is obtained by the reciprocal combination of Z_R and Z_C. The result is linear (serial) combined with $Z_{el} = R_{el}$. This leads to the following real and imaginary impedance terms, Z_{re} and Z_{im}, respectively

$$Z_{re} = R_{el} + \frac{R_{ct}}{\omega^2 R_{ct}^2 C_{dl}^2} \tag{4.42}$$

$$Z_{im} = -\frac{\omega R_{ct}^2 C_{dl}}{1 + \omega^2 R_{ct}^2 C_{dl}^2} \tag{4.43}$$

Two plots are used to represent the results of impedance measurements of electrochemical systems. In the Nyquist plot the negative imaginary part (y-axis) is plotted versus the real part (x-axis). The Nyquist plot of the equivalent circuits in Figures 4.9a and 4.10a is shown in Figures 4.9b and 4.10b.

In the Bode plot log |Z| and |ϕ| (y-axis) are plotted against the logarithm of frequency ω (x-axis). Figure 4.12 shows the Bode plot of the equivalent circuit of the double layer with charge transfer (Figure 4.10a).

The standard method to measure impedance values as a function of the potential is the Wheatstone bridge (Figure 4.13). The electrochemical cell is placed in one branch of the bridge. In the second branch a variable resistance and a variable capacitance are placed. An alternating current is applied to a potentiostat. In the reference circuit, capacitance and resistance are varied until the bridge current, measured with an AC instrument or an oscilloscope, is zero.

For the measurement of the impedance, a modern "impedance spectrometer" provides automatic data registration over at least five decades of frequency. With this equipment it

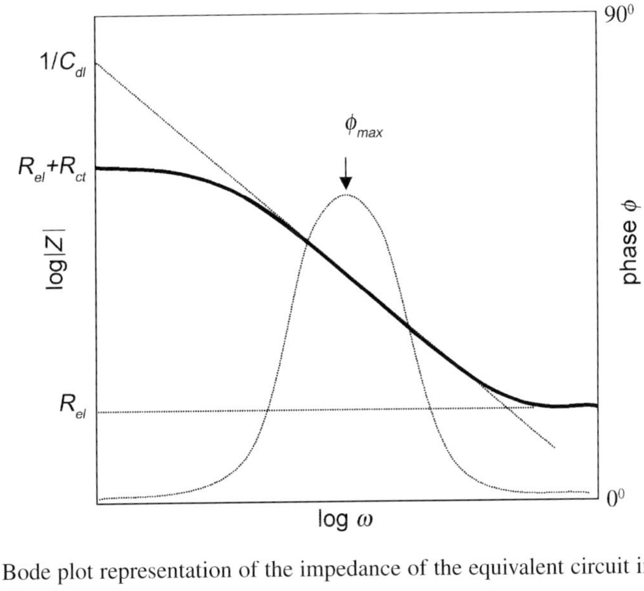

Figure 4.12 Bode plot representation of the impedance of the equivalent circuit in Figure 4.10a.

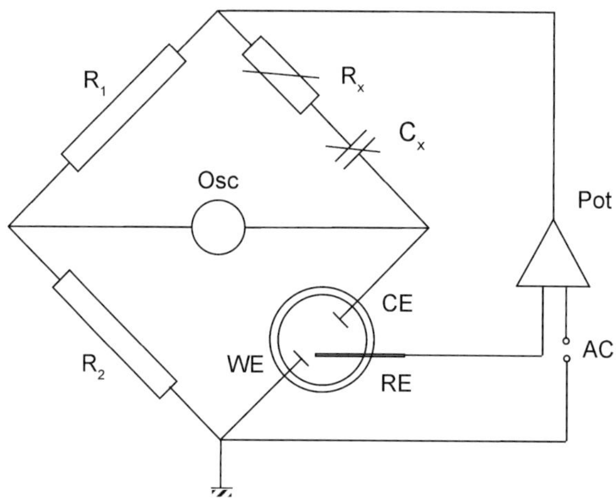

Figure 4.13 Wheatstone bridge for capacitance measurements. Electrochemical cell, with working electrode WE, counter electrode CE, and reference electrode RE. Variable resistance R_x and capacitance C_x for balancing the electrochemical cell. R_3, R_4 are fixed resistances, Pot potentiostat, and AC applied alternating potential. Osc oscilloscope for balancing the bridge (zero current).

is possible to analyze the electrical response of very complex surface structures. Detailed descriptions of the method exist in the literature.[38–41]

4.2.2 Cyclic voltammetry and chronoamperometry

The principal method for investigating the current potential dependence of an electrode–electrolyte system is cyclic voltammetry. A potentiostat is used to scan the potential with a constant scan rate v (given in V s^{-1} or mV s^{-1}) from potential E_1 to potential E_2. Then the scan direction is turned around and the associated current is measured (Figure 4.14).

At potentials where an electrochemical process can occur (a platinum electrode in sulfuric acid is shown in Figure 4.14) characteristic peaks are observed. Peak potential, peak height, and peak width are characteristic parameters.

A big advantage of cyclic voltammetry is the detection of surface processes like adsorption, oxide layer formation, etc. In the anodic scan in Figure 4.14 the oxidation of weakly and strongly bound hydrogen (peaks a and b) is followed by hydroxide adsorption (peak c) and oxide layer formation (d). In the cathodic scan the reduction of the oxide (peak e) is followed by hydrogen adsorption strongly and weakly bound to the platinum atoms (peaks f and g). Further examples will be shown in Chapters 4 (Section 4.4) and 9. In these applications cyclic voltammetry is very similar to thermodesorption spectroscopy in surface science. Cyclic voltammetry can also be used to study diffusion and kinetically controlled processes. This will be discussed in more detail in Chapters 5 and 6.

An early application of this method was reported by Knorr, Breiter, and Will.[42] Nicholson and Shain[43,44] developed one of the first theoretical analyses of a diffusion controlled redox process. The dynamic character of the method enabled kinetic investigations,

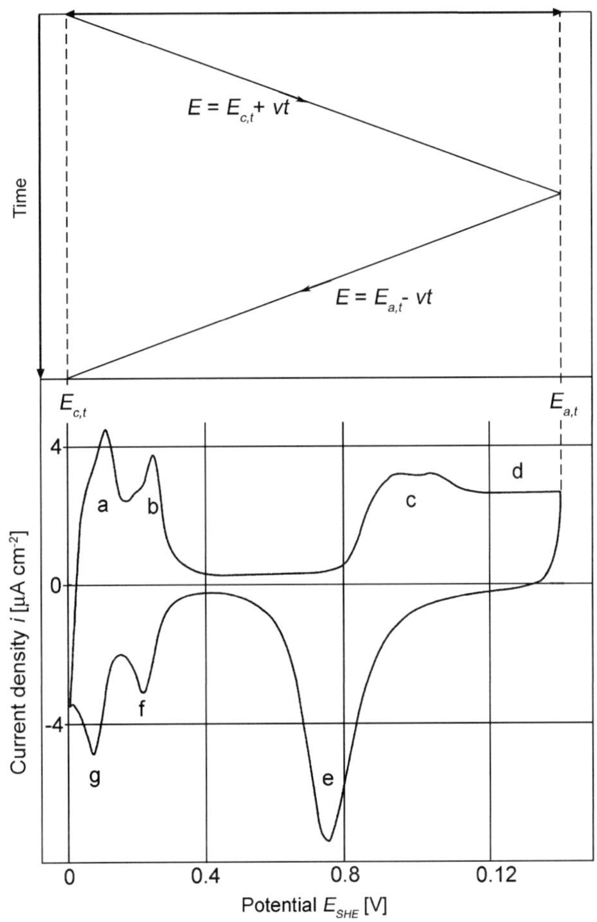

Figure 4.14 Potential variation in cyclic voltammetry and associated current for platinum in 0.5 mol·dm^{-3} H_2SO_4.

especially in organic electrochemistry. Representative for this application is, for instance, the work of Saveant, Andrieux *et al.*[45] and Amatore and Lefrou.[46]

In addition to the current potential dependence, the charge connected with the electrochemical processes can be obtained by integration of the potential peak over the time scale associated with the potential scale. This application is called coulometry. Examples will be shown in Section 4.4.

4.2.3 Determination of the adsorbed mass

Several methods are available for the determination of the adsorbed mass. One is the radiotracer method pioneered by Horanyi[47] and Kazarinov and Andreev.[48] Experimental details are found in the literature.

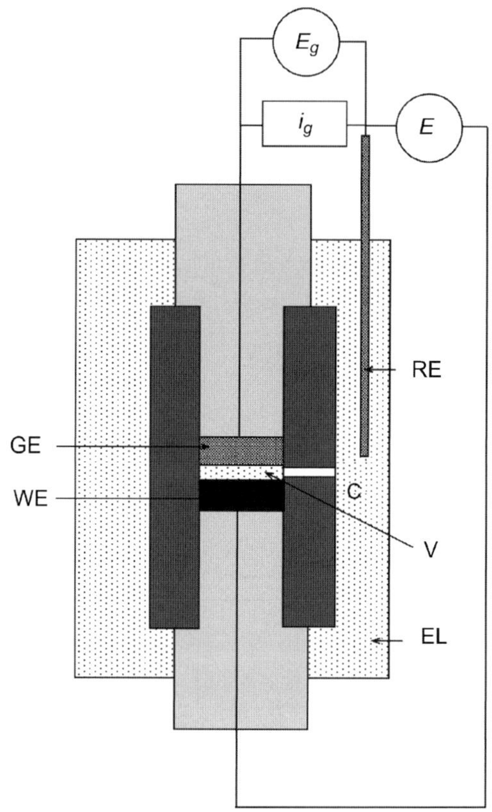

Figure 4.15 Principal construction of a thin-layer cell for determination of adsorption isotherms of metal ions. WE working electrode, GE generator electrode, RE reference electrode (not reversible for the investigated metal ions), V thin-layer compartment, C channel to keep electrical contact between the thin-layer compartment and outer electrolyte, E potential between the working and reference electrodes, E_g potential between the generator and reference electrodes (to be kept constant), and i_g current through the generator electrode after a potential jump in the potential of the working electrode.

Schmidt and Siegenthaler developed thin-layer cells to reduce the ratio between the volume of the bulk electrolyte and the volume of the interphase.[49] This method was especially useful for investigation of metal ion adsorption. The two-electrode thin-layer cell will be described in more detail (Figure 4.15). Two electrodes terminate the thin-layer compartment, the working electrode, and a generator electrode made from the metal adsorbed on the working electrode. A reference electrode is used to control the potential of the working electrode E and of the generator electrode E_g. The potential of the generator electrode (metal Me) is controlled by the concentration of the metal ions (Me^{z+}). The potential of the generator electrode is kept constant against the reference electrode. If the potential of the working electrode is changed so that metal ions are adsorbed on the working electrode, the concentration of the metal ions in the thin-layer compartment is reduced. The generator electrode, kept at its old equilibrium potential, generates metal ions until the equilibrium concentration is reached again. The time integral of the current flux through the generator electrode represents the amount of ions adsorbed on the working electrode.

An example for the determination of an adsorption isotherm is presented in Figure 4.30.

The most effective method to measure the adsorbed mass is the quartz crystal microbalance (QCMB). This method goes back to the work of Sauerbrey.[50] The application of this method is based on the following equation relating the shift Δf of the resonance frequency f_0 of a quartz crystal to the change of the mass of the crystal Δm divided by A (area)

$$\Delta f = -\frac{2f_0^2}{\sqrt{\rho_q \mu_q}} \frac{\Delta m}{A} = -C_{SB} \frac{\Delta m}{A} \tag{4.44}$$

ρ_q and μ_q are the density and the shear modulus of quartz. C_{SB} is the Sauerbrey constant. A special cut of a disc from the quartz crystal is necessary to provide a useful quartz plate. The value of the Sauerbrey constant depends on the cut. For an AT-cut C_{SB} is 226.01 $Hz \cdot cm^2 \cdot \mu g^{-1}$. The plate is then coated with the metal on which the adsorption is investigated (Figure 4.16). To provide a good adhesion of the metal film to the quartz plate, a very thin chromium layer is evaporated. The limitation to evaporated metal films weakens this method for investigating structure sensitive surface processes. Otherwise, the method is of extreme value for the investigation of surface films like conducting polymers (Chapter 11).

The application of this method is not as simple as previously assumed.[51,52] Secondary effects accompany the change of the mass, e.g., increase of roughness or changing viscosity

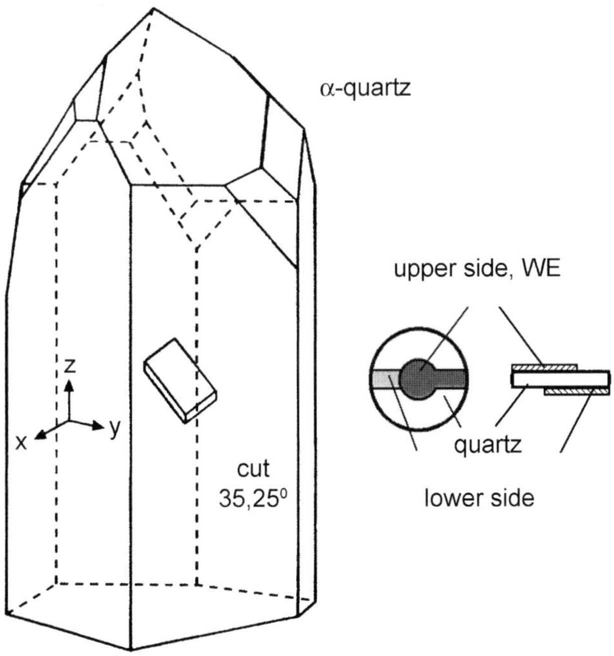

Figure 4.16 Quartz crystal and quartz crystal plate used as working electrode in QCMB.

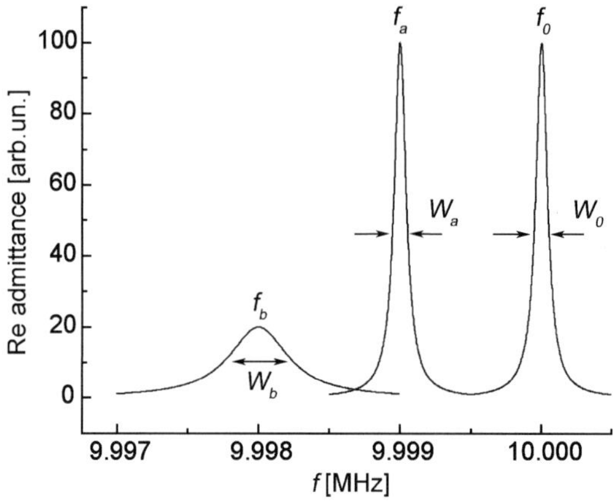

Figure 4.17 Shift and change of the resonance frequency of a quartz crystal microbalance, real part of the admittance versus frequency. f_0, w_0, resonance frequency and full width at half maximum (FWHM) of the initial gold electrode, f_a, w_a, resonance frequency and FWHM of a gold electrode after formation of a rigid and smooth surface film (no damping), f_b, w_b, resonance frequency and FWHM of a gold electrode after formation of a viscoelestic and/or rough surface film (strong damping).

of the surface film. Therefore, in addition to the frequency shift damping and broadening of the resonance frequency occurs. This is shown in Figure 4.17.

To determine frequency shift and damping a frequency analyzer is used. The additional damping can be represented by a complex frequency shift, determined by the change of the full width at half maximum (FWHM) $\Delta\omega$.

$$\Delta f^* = \Delta f + i\frac{\Delta\omega}{2} \tag{4.45}$$

The principal equipment for QCMB is shown in Figure 4.18.

4.2.4 Scanning tunneling microscopy and related methods

In situ structural investigation of the interface of solid electrodes has become possible by the development of scanning tunneling microscopy (STM).[54] This method is based on the tunneling current between a metallic tip and an electrode. The tunneling current is used to control the distance between tip and electrode surface on an atomic scale. The tunneling current starts if the tip is so close to the surface that the electronic orbitals begin to overlap. Therefore this method is principally able to scan the surface in atomic dimensions (Figure 4.19).

The tunneling current I_t depends exponentially on the distance d between nearest tip atom and surface atom

$$I_t \propto \frac{U_t}{d} e^{-\mathrm{const}\cdot d\sqrt{\Phi}} \tag{4.46}$$

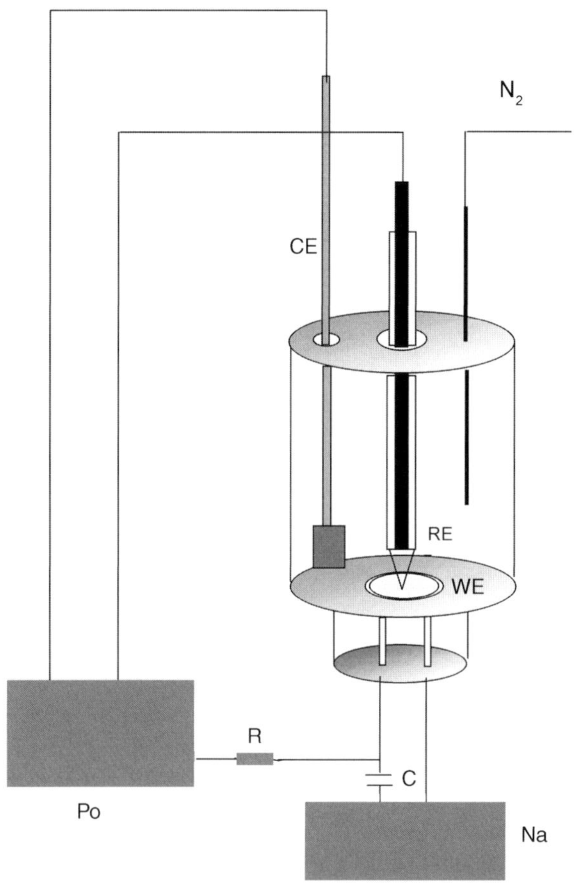

Figure 4.18 Principal construction of a quartz crystal microbalance with network analysis.[53] WE working electrode, CE counter electrode, RE reference electrode, Po potentiostat, Na network analyzer, R and C for resistance and capacitance for impedance adaptation, and N_2 inert gas purging.

U_t is the applied voltage. The function

$$\Phi = \frac{1}{2}(\Phi_t + \Phi_m) \tag{4.47}$$

is the average of the work functions of tip and metal sample.

The motion of a tip in atomic dimensions became possible by the invention of a mechanical scanning device shown in Figure 4.20 by Binning and Rohrer.[55] Three piezoelectric crystals enable a very precise motion, which in this case is with atomic resolution.

Other types of piezoelectric STM scanners were developed such as the tube scanner shown in Figure 4.21.

The invention of STM stimulated the development of several other devices to image surfaces with atomic scale resolution. In electrochemistry, atomic force microscopy is the

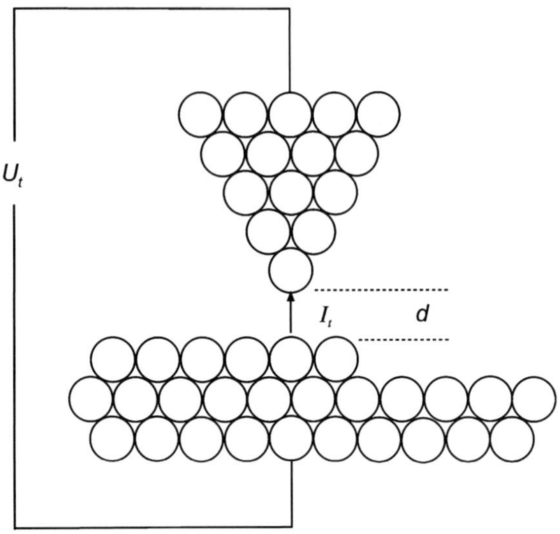

Figure 4.19 Principal function of a scanning tunneling microscope.

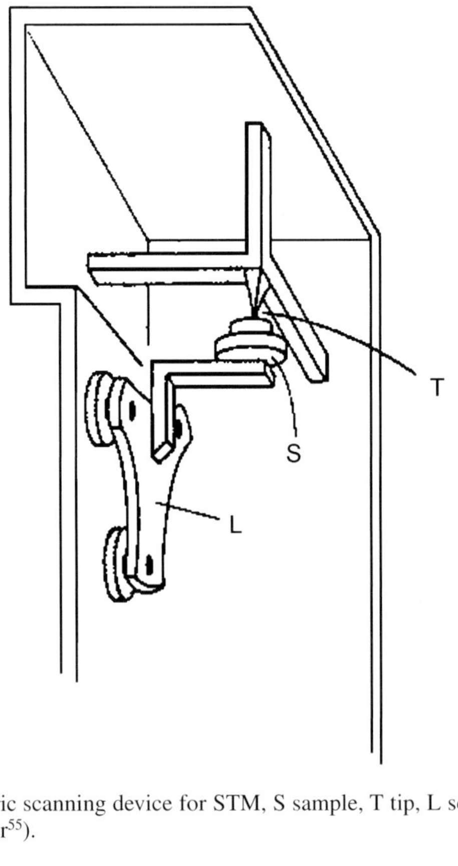

Figure 4.20 Piezoelectric scanning device for STM, S sample, T tip, L scanner ("louse") (according to Binning and Rohrer[55]).

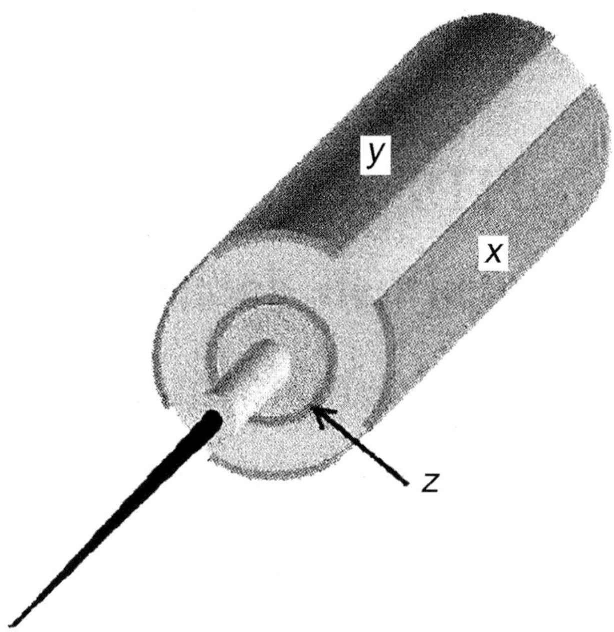

Figure 4.21 Tube scanner for STM.

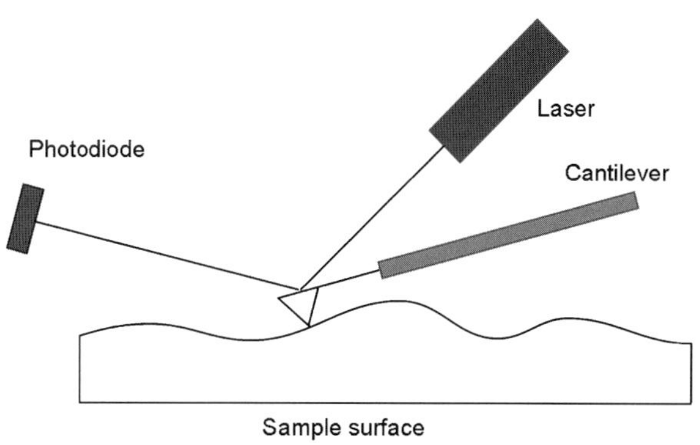

Figure 4.22 Principal function of an atomic force microscope, the tip of the cantilever is scanned across the surface and the change of the cantilever tip position by the attraction of the surface atoms is followed by an optical detection device.

alternative to STM especially when non-metallic samples must be investigated. In this case the atomic attraction force between a tip and a surface is used to control the distance between tip and surface. The deflection of the cantilever can be detected by a laser device as shown in Figure 4.22.

A different development but with similar importance for electrochemistry was the development of the electrochemical scanning microscope.[56] In this case an ultra-microelectrode

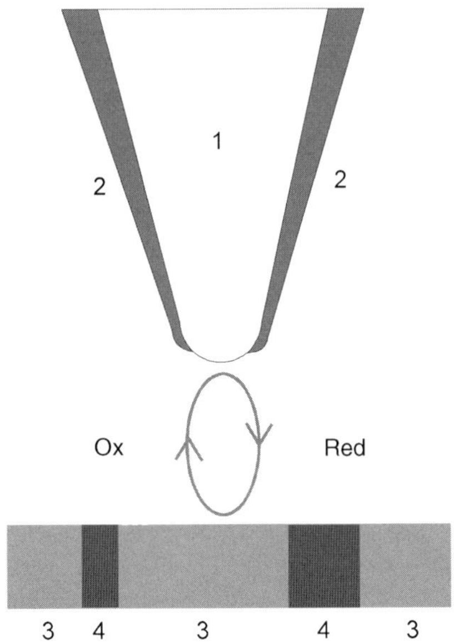

Figure 4.23 Electrochemical scanning microscopy, an isolated microelectrode is scanned at constant distance across an electrode and the local electrochemical current is monitored, (1) microelectrode, (2) isolation, (3) active electrode areas, and (4) inactive electrode areas.

is scanned across the surface and electrochemical currents are measured. The electrode diameter is in the 10-micrometer range as shown in Figure 4.23.

The advantage of this method is the local imaging of the reactivity of an electrode surface and its dependence on the surface topography, especially, if combined with simultaneous imaging by STM.[57]

4.3 AD-ATOMS

The main intermediate in the deposition or dissolution process of metals and alloys is the ad-atom, a metal ion adsorbed on the surface of the metallic phase. The ad-atom can be partially discharged and the solvation shell or the shell of complex molecules is different from the metal ion in the electrolyte. The adsorption equilibrium is described by the equation

$$Me^{z+}\nu_m H_2O + \Delta z_{ad} e^- \leftrightarrow Me^{z_{ad}}\nu_n H_2O + (\nu_m - \nu_n)H_2O \qquad (4.48)$$

with $\Delta z_{ad} = z - z_{ad}$. In the following it will be assumed that the complex molecules are the solvent molecules with constant concentration and will no longer be taken into account. The change of the charge is an example of partial charge and partial charge transfer. Using the partial charge transfer coefficient λ the charge of the ad-atom can be expressed as $z_{ad} = \lambda z$ and the partial charge transfer as $\Delta z_{ad} = (1 - \lambda)z$.

A metal atom on its own substrate is hard to detect by analytical methods. Only a few methods can be used for determination of ad-atom surface concentration. One method is impedance spectroscopy. Other methods are galvanostatic transient methods but only indirect information from these methods about the ad-atom concentration is available.

4.3.1 Adsorption and desorption of ad-atoms

The transfer rates for adsorption r_{ads} of a metal ion and desorption r_{des} of the ad-atom according to Eq. (4.48) can be described by equations that take partial discharge into account. The equations are similar to equations derived for intermediate adsorption in a redox process.[58] The energy diagram is shown in Figure 4.24. The electrical energy is shown in the lower part of Figure 4.24.

Curve 1 shows the situation for an uncharged surface and no potential difference between electrode and electrolyte, $\Delta\varphi = \varphi_m - \varphi_{el} = 0$. φ is the local Galvani potential. The situation with $\Delta\varphi < 0$ but without partial discharge is shown in curve 3. An electrical energy $zF(\varphi - \varphi_{el})$ has to be subtracted from the chemical energy. In the transition state (the maximum of the energy curve) the electrical energy is the fraction $\alpha_c\, zF\Delta\varphi$. α_c is

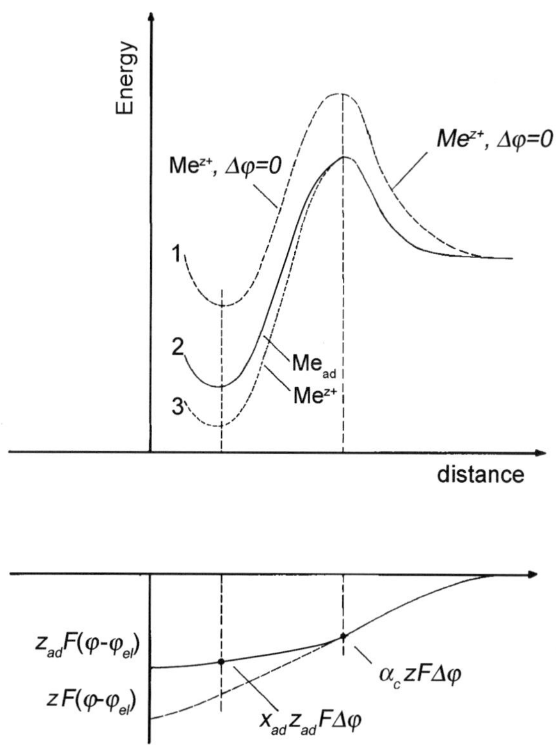

Figure 4.24 Energy diagram of adsorption and desorption of an ad-atom. The diagram shows the activation energy of adsorption of a metal ion ($z = 2$) discharged in the electrochemical reaction to a partial charge $z_{ad} = \lambda z$ and the activation energy of desorption of the ad-atom, (1) no electric field ($\Delta\varphi = 0$), (2) with the contribution of Coulomb energies, potential $\Delta\varphi = \varphi_m - \varphi_{el} < 0$, and (3) contribution of Coulomb energies with no partial discharge. The potential at the position of the ad-atom is given by $x_{ad}\Delta\varphi$, the potential in the transition state $\alpha_c\Delta\varphi$.

called charge transfer coefficient of a cathodic process, positive charge is transferred from the electrolyte to the electrode (Chapter 6). In the ad-atom position the potential is the fraction $x_{ad}\, zF\Delta\varphi$ with $x_{ad} = (\varphi_{ad} - \varphi_{el}) / (\varphi_m - \varphi_{el})$.

With partial discharge, the situation is shown in curve 2. In the transition state the electrical energy is still $\alpha_c\, zF\Delta\varphi$. In the ad-atom position the potential is the fraction $x_{ad}\, z_{ad}\, F\Delta\varphi$, with $x_{ad} = 0$ at the pre-electrode position and $x_{ad} = 1$ at the metal surface.

With the activation energies the following rate equations can be derived. For adsorption (a cathodic process with the charge transfer coefficient α_c)

$$r_{ads} = k_{ads} c_{Me^{z+}} (N_{max} - N_{ad}) \exp\left[-\frac{\alpha_c zF}{RT} E \right] \tag{4.49}$$

For desorption (an anodic process with the charge transfer coefficient α_a)

$$r_{des} = k_{des} N_{ad} \exp\left[+\frac{\alpha_a zF}{RT} E \right] \tag{4.50}$$

For the anodic charge transfer coefficient one can derive $\alpha_a = 1 - \lambda + x_{ad}\lambda - \alpha_c$. The rate constants k_{ads} and k_{des} are independent on the potential, $c_{Me^{z+}}$ is the metal ion concentration in the electrolyte, N_{ad} is the ad-atom surface density (e.g., ad-atoms cm^{-2}), and $N_{max} - N_{ad}$ is the density of free surface atoms. The exponential terms represent the potential dependence of the transfer of a metal cation through the energy barrier and, in Eq. (4.50), the additional energy connected with the rearrangement of the electron distribution in the double layer. For describing this distribution by the partial charge model, one has to consider the potential acting on the ad-atom $x_{ad}\,\Delta\varphi$, the partial charge of the ad-atom $z_{ad} = \lambda z$, and the charge transferred to the metal $(1 - \lambda)z$. The ad-atom deposition and dissolution as part of rate controlling charge transfer will be treated in Chapter 6.

4.3.2 Equilibrium ad-atom concentration

At the Nernst potential E_0 equilibrium is established for the ad-atom and metal ion concentration and one can define exchange rates of adsorption and desorption.

$$r_{ads}(E_0) = r_{des}(E_0) = r_{0,ad} \tag{4.51}$$

From this equation one can derive an equilibrium surface coverage $\Theta_{ad} = N_{ad}/N_{max}$ of the ad-atoms.

$$\frac{\Theta_{ad}}{1 - \Theta_{ad}} = \frac{k_{ads}}{k_{des}} c_{Me^{z+}} \exp\left[-\frac{(x_{ad}\lambda + 1 - \lambda) zF}{RT} E_0 \right] \tag{4.52}$$

The potential E_0 can be substituted by the Nernst equation

$$\frac{\Theta_{ad}}{1 - \Theta_{ad}} = \frac{k_{ads}}{k_{des}} c_{Me^{z+}} \exp\left[-\frac{(x_{ad}\lambda + 1 - \lambda) zF}{RT} \left(E_0^{\ominus} + \frac{RT}{zF} \ln a_{Me^{z+}} \right) \right] \tag{4.53}$$

The standard potential of the metal $E_0^{\ominus}$ is a constant (not dependent on the concentration).

The constant expressions can be combined to an equilibrium constant of adsorption K_{ad}. Neglecting differences between concentration and activity one obtains the Langmuir type adsorption isotherm of the ad-atoms

$$\Theta_{ad} = \frac{K_{ad} c_{Me^{z+}}^{\lambda(1-x_{ad})}}{K_{ad} c_{Me^{z+}}^{\lambda(1-x_{ad})} + 1} \tag{4.54}$$

with $\lambda = z_{ad}/z$ describing the partial charge and $x_{ad} = (\varphi_{ad} - \varphi_{el})/(\varphi_m - \varphi_{el})$ describing the potential at the ad-atom position. For small coverage $K_{ad} c_{Me^{z+}}^{\lambda(1-x_{ad})} << 1$ it turns into

$$\Theta_{ad} = K_{ad} c_{Me^{z+}}^{\lambda(1-x_{ad})} \tag{4.55}$$

The ad-atom concentration depends on the electrolyte concentration. The ad-atom concentration only becomes independent of the bulk concentration of the metal ions if there is a complete discharging of the ad-atom with $\lambda = 0$.

The ad-atoms are an intermediate in the mechanism of metal deposition. On one hand, they are in exchange with the metal ions in the electrolyte and, on the other, with atoms in step and kink positions. The density of step atoms and kink sites on a step is also called the step roughness.

4.3.3 Surface diffusion of ad-atoms

Each ad-atom has a certain lifetime on the substrate surface called the residence time of the ad-atom. The equation for the residence time of the ad-atoms is obtained by the reciprocal value of the rate constant of dissolution of the ad-atoms.

$$\tau_{ad} = \frac{1}{k_{des} \exp\left[+\frac{\alpha_a zF}{RT} E\right]} \tag{4.56}$$

During the residence time ad-atoms are subject to a random walk on the surface with an average displacement δ_{ad}, which is the square root of the product of the residence time and the surface diffusion coefficient

$$\delta_{ad} = \sqrt{D_{ad} \tau_{ad}} \tag{4.57}$$

D_{ad} is the surface diffusion coefficient of ad-atoms. In case of a cathodic or anodic overpotential the ad-atom density on the terraces is changed. This leads to surface diffusion to or from steps or other types of surface roughness. The model of the diffusion of ad-atoms on surface terraces is shown in Figure 4.25.

This diffusion process is described in the literature[59] and will therefore not be discussed here in greater detail. The equivalent circuit used for the simulation of the deposition process on a stepped surface is shown in Figure 4.26 and gives a rough impression of the complex experimental problems in the investigation of ad-atoms and ad-atom diffusion.

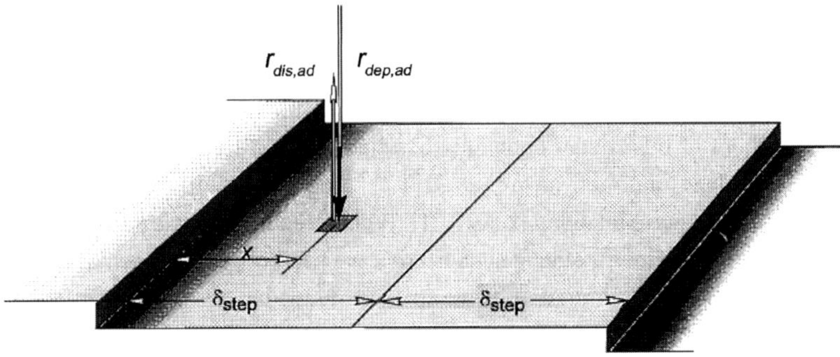

Figure 4.25 Surface diffusion of ad-atoms on step terraces.[59] The ad-atoms are deposited at a distance x from the step and carry out a random walk. During the time, the ad-atoms stay on the surface (residence time τ_{ad}) they walk a distance $\delta_{ad} = \sqrt{D_{ad}\tau_{ad}}$. (Reproduced with permission from Ref. [59], © 1996, Wiley-VCH.)

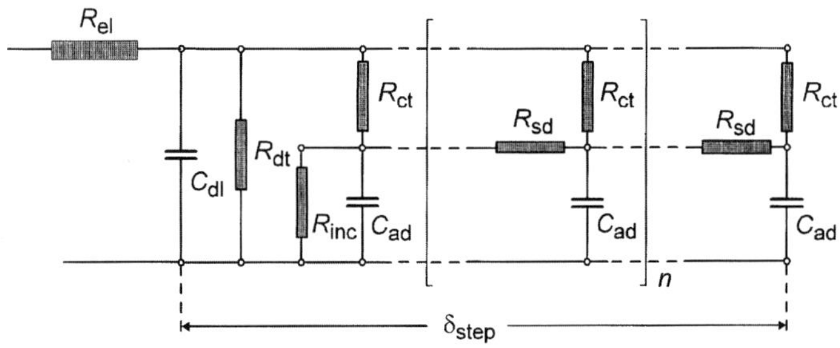

Figure 4.26 Equivalent circuit used for the simulation of a metal deposition process on a stepped surface (Ref.[59], p. 35). R_{el} electrolyte resistance, C_{dl} double layer capacitance, R_{ct} charge transfer resistance, R_{adct} ad-atom charge transfer resistance, R_{sd} ad-atom surface diffusion resistance, R_{inc} ad-atom incorporation resistance, C_{ad} ad-atom adsorption capacitance, and δ_{st} step half-width. (Reproduced with permission from Ref. [59], © 1996, Wiley-VCH.)

4.4 UNDERPOTENTIAL DEPOSITION

It was described above how the metal ion is adsorbed on its own metal. Metal ions, however, can also be adsorbed on other metal substrates. This deposition is observed before bulk deposition and sometimes it looks as if a metal deposition occurs against the rules of thermodynamics. Therefore, this phenomenon is called underpotential deposition (UPD). Budevski, Staikov, and Lorenz gave a comprehensive treatment of UPD.[59] In the following only a short description of the typical problems in this research field and its influence on surface treatment and phase formation will be described.

4.4.1 Lead on silver

UPD can be considered to be adsorption of metal ions B on the surface of a substrate metal A. In this chapter the UPD of lead on Ag(111) and Ag (100) will be described.[60]

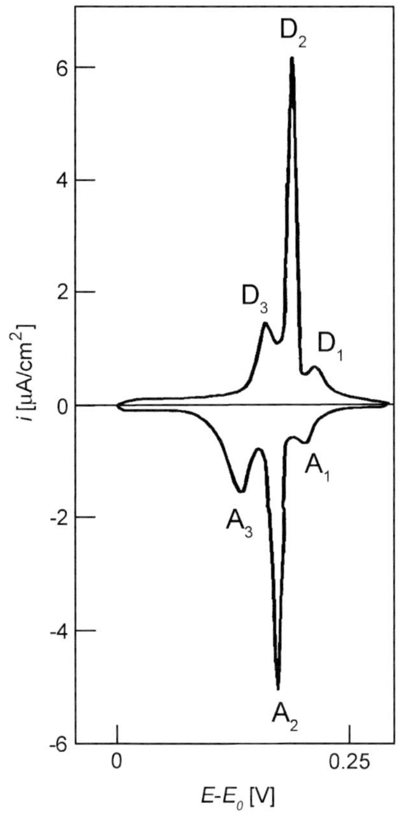

Figure 4.27 Cyclic voltammogram of UPD of lead on a Ag(111) surface.[60] The reference potential E_0 is the Nernst potential of a lead/lead ion electrode in the same solution. 5×10^{-4} mol·dm^{-3} $Pb(ClO_4)_2 + 5 \times 10^{-1}$ mol·dm^{-3} $NaClO_4 + 5 \times 10^{-3}$ mol·dm^{-3} $HClO_4$, 25 °C. Scan rate 0.42 mV·s^{-1}. (Reproduced with permission from Ref. [60], © 1978, Elsevier.)

Figure 4.27 shows a typical cyclic voltammogram of an UPD of lead on the crystallographic surfaces of a silver (111) single crystal. The cyclic voltammogram of lead UPD on silver (100) is shown in Figure 4.28. The current is plotted versus $E - E_0$, the potential referred to the Nernst potential of the lead electrode in the same solution. At $E - E_0 < 0$ the deposition of bulk lead would start. The diagram shows that a limited amount of lead is deposited at potentials that are more positive than the Nernst potential. The reason for this is an adsorption of the metal ions on the substrate metal.

The diagram on the (111) surface is characteristic for the process. Three characteristic peaks are observed prior to the bulk deposition of Pb and explained in the following manner. In peak A_1 adsorption of lead on steps and on similar irregularities of the surface is observed because these positions provide a stronger bonding than positions on the flat surface. In A_2 adsorption of lead on the terraces of the stepped crystal surface is observed to cover most of the silver surface. It was found that at this potential no adsorption is

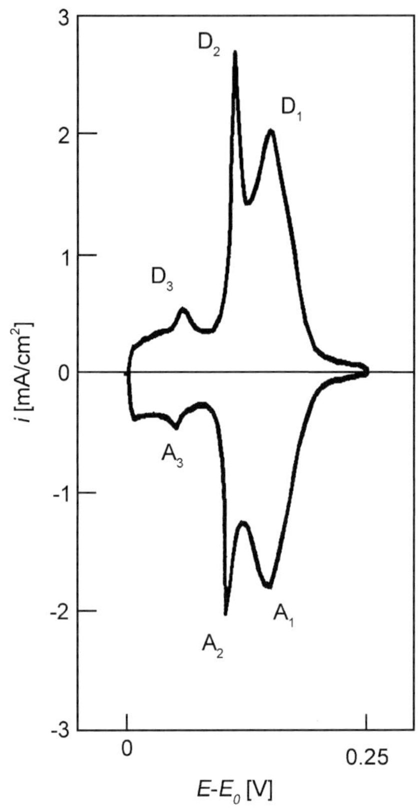

Figure 4.28 Cyclic voltammogram of UPD of lead on a Ag(100) surface.[60] The reference potential E_0 is the Nernst potential of a lead/lead ion electrode in the same solution. 5×10^{-4} mol·dm^{-3} $Pb(ClO_4)_2$ + 5×10^{-1} mol·dm^{-3} $NaClO_4$ + 5×10^{-3} mol·dm^{-3} $HClO_4$, 25 °C. Scan rate 0.42 mV·s^{-1}. (Reproduced with permission from Ref. [60], © 1978, Elsevier.)

observed at atoms on the upper side of a step. These empty places are filled with lead atoms in peak A_3 thus forming a monomolecular film of adsorbed Pb on the silver surface. This information was confirmed by experiments with ideal flat (111) surfaces prepared by a special experimental technique with no or a minimum of surface defects (Figure 4.29).[61] Only the peaks A_2/D_2 were observed for such surfaces. The peaks on the (100) surface are explained in a similar manner by adsorption on energetically different adsorption sites.

The cyclic voltammograms Figures 4.27 and 4.28 are also called chronoamperograms because the potential axis is also a time axis in the scanning process. The charge connected with the adsorption process can be determined by integration. Parallel to the integration the mass connected with the UPD was measured with the twin electrode thin-layer technique described in Section 4.2.3. The result is shown in Figure 4.30.

The electrosorption valency for lead adsorbed on silver is determined by plotting the adsorbed charge versus the adsorbed mass. This is shown for Pb on Ag(111) in Figure 4.31.

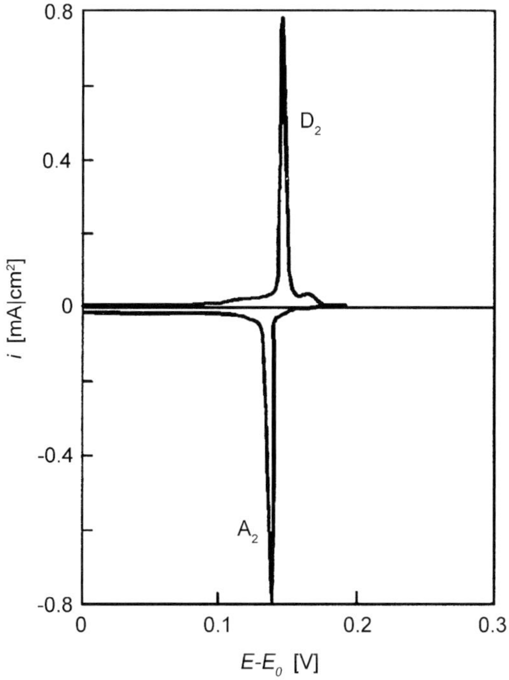

Figure 4.29 Cyclic voltammogram of Pb on a quasi perfect Ag(111) face grown by the capillary method. (Reproduced with permission from Ref. [61], © 1983, Elsevier.)

The value of the electrosorption valency determined is $z\gamma = 2$. The same value was found for Pb on Ag(100).

Strictly thermodynamically, the electrosorption valency only shows that no other co-adsorption of other ions occurs. However, there may be a compact adsorption layer. STM gives images for the two surfaces (Figure 4.32).[62] For Pb on Ag(111) the image was interpreted to be a so-called filled honeycomb 3(2×2) structure and a compressed hcp layer. For Pb on Ag(100) a c(2×2) structure was observed. Otherwise the adsorbed mass was similar to the Ag(111) substrate which lead to the model of a bi-layer.

The structure of lead UPD films on silver (hkl) faces is a consequence of the significant lattice misfit. The atomic radius of Ag is 0.144 nm, the radius of lead 0.174. This prevents a complete epitaxial structure. One can expect that the forced layer structure has a high internal strain. The elastic strain in the film can be estimated from the deviation of the distance of the atoms in the film d from the distance d_0 in the bulk crystal. The elastic strain ε_e is given by the equation

$$\varepsilon_e \approx N_A \frac{\nu}{\kappa_{2d}} \left(\frac{d_0 - d}{d_0} \right)^2 \tag{4.58}$$

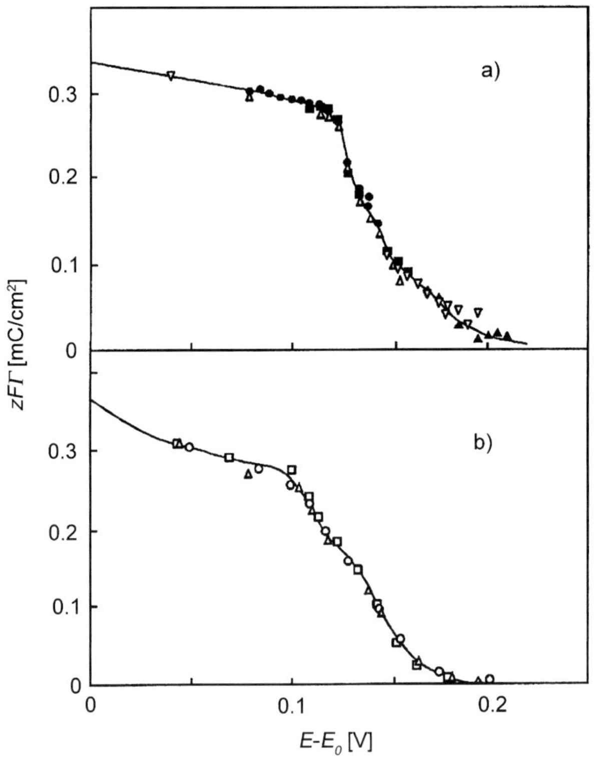

Figure 4.30 Determination of the adsorbed mass of lead with the twin electrode thin-layer method described in Figure 4.15. Pb UPD on (a) Ag(111) and (b) Ag(100). Concentrations as given in Figures 4.27 and 4.28. (Reproduced with permission from Ref. [60], © 1978, Elsevier.)

In this equation v is the atomic volume of the bulk metal (of the order of 3×10^{-29} m^3), κ_{2d} the two-dimensional compressibility (of the order of 10^{-11} m^2 N^{-1}), and N_A Avogadro's number.

4.4.2 Copper on Au

A very intensively investigated system is the UPD of copper on gold. This example will be described according to results of Kolb *et al.*[63] The voltammogram of Cu on Au(111) is shown on Figure 4.33. In this case two different deposition potentials were observed.

The STM image of the unreconstructed Au surface is shown in Figure 4.33a. A hexagonal, closely packed surface structure is observed. In the first peak a copper layer of lower density is formed, a $2(\sqrt{3}\times\sqrt{3})$ R30° Cu structure, which is a "honeycomb" structure. The associated STM image is shown in Figure 4.33b. Finally, in the second UPD peak the $2(\sqrt{3}\times\sqrt{3})$ R30° is changed into a closely packed 1×1 structure. The STM image in Figure 4.33c is similar to the STM image in Figure 4.32a. From other experiments, preferentially with *ex situ* methods, it is known that anions are incorporated in the copper

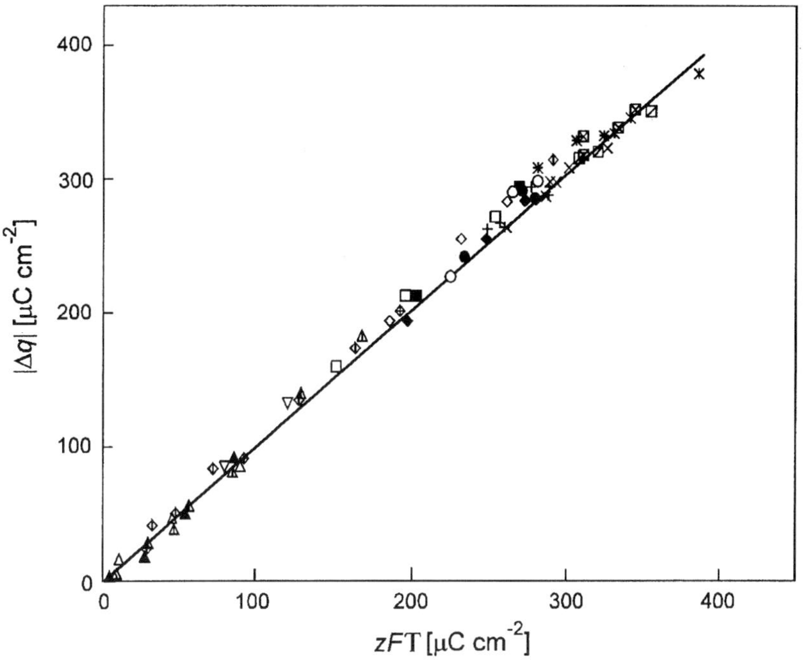

Figure 4.31 Determination of the electrosorption valency for Pb UPD on Ag(111) by adsorbed charge (determined by integration of the cyclic voltammogram in Figure 4.27) versus adsorbed mass shown in Figure 4.30. The electrosorption valency is $z\gamma = 2$. (Reproduced with permission from Ref. [60], © 1978, Elsevier.)

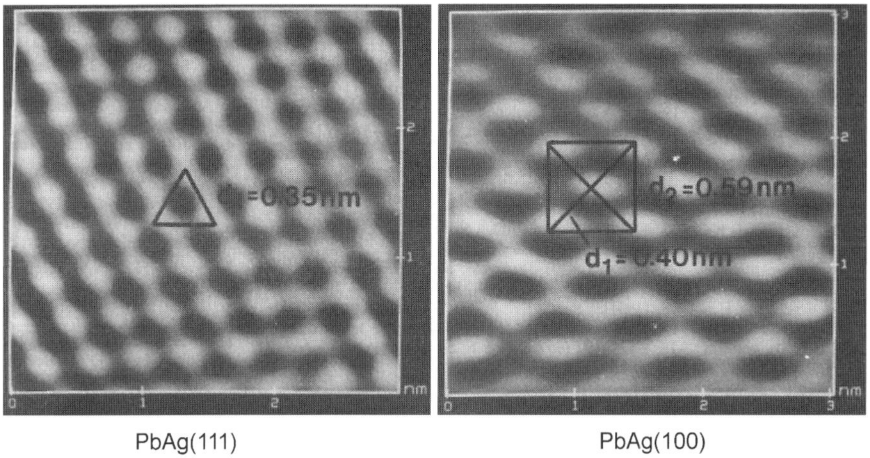

Figure 4.32 STM images of Pb UPD layers deposited on Ag(111) and Ag(100), $E - E_0 = 45$ mV. (Reproduced with permission from Ref. [62], © 1992, Elsevier.)

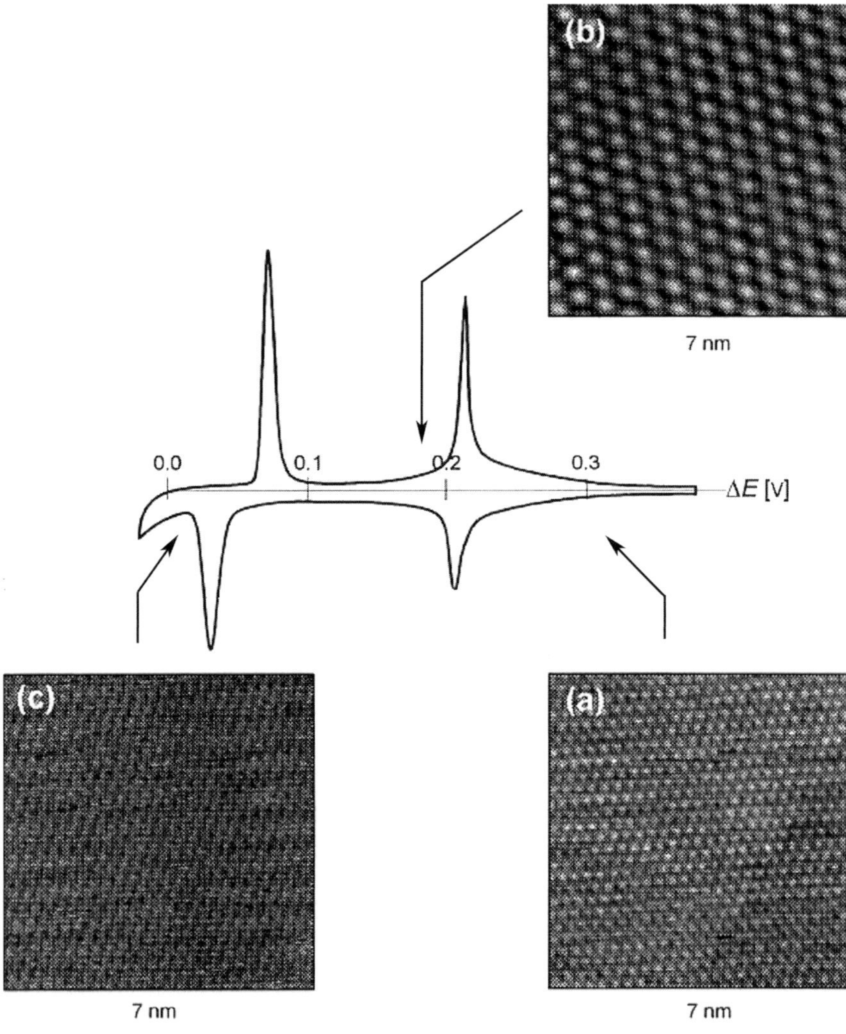

Figure 4.33 Cyclic voltammogram of UPD of copper on a Au(111) surface.[63] STM images at different potentials of the CV. 10^{-3} mol · dm^{-3} $CuSO_4 + 5 \times 10^{-2}$ mol · dm^{-3} H_2SO_4, 25 °C. (Reproduced with permission from Ref. [63], © 1995, Kluwer Academic Publishers.)

layer. The anion copper structure was investigated by *ex situ* surface techniques, e.g., by Zei *et al.*[64] The electrosorption valency experiments gave a value $z\gamma = 1$. The thermodynamic interpretation is that anions are co-adsorbed with a stoichiometry of one anion charge per one copper charge. The anions can form a $(\sqrt{3} \times \sqrt{3})$ R30° structure. The stoichiometry of the copper layer in the medium potential region can then be described as $Cu_2(SO_4)$ with a partial charge transfer of $\lambda = 0.5$ or $z_{ad} = z\lambda = 1$.

In the closely packed (1 × 1) Cu structure (Figure 4.33c) the copper atoms are obviously discharged and no co-adsorption of anions can be detected. The strain of an epitaxial layer in this example is much smaller than in the last example. The lattice misfit is small. The atomic radius of Au is 0.144 nm and that of Cu 0.127 nm.

UPD is also observed on polycrystalline metal surfaces but only a broad peak is observed with no details about the different surface positions. The reason for this is the inhomogeneous surface structure of polycrystalline electrodes.

4.4.3 Underpotential deposition as two-dimensional phase formation

The thermodynamic description of UPD is based on a deposition of metal ions with total or partial discharge of the ions as was formulated in Eq. 4.48. In the broad range of publications on UPD the interpretation of the effect concentrated primarily on an adsorption process of ions B on substrate A.

An alternative view on the UPD process provides the concept of substrate supported "low-dimensional phase formation" from Staikov, Budevski and Lorenz.[65,66] This concept explains unexpected stabilities of underpotential deposits. In this concept the UPD layer is considered to be a two-dimensional phase stabilized by the bond between substrate and UPD metal. A "linear phase" is the adsorption of atoms on steps and a phase of "zero dimension" is a group of atoms around a kink site position or around a surface dislocation. This concept is used to explain the positively shifted potentials in the cyclic voltammogram of UPD. Higher stability of the low-dimensional phase is described by introducing activities of atoms in the substrate-supported phase of lower dimension $a_{\mathrm{Me,0d}}$, $a_{\mathrm{Me,1d}}$, and $a_{\mathrm{Me,2d}}$. The deposition potential is given by the equation

$$E_{id} = E_0 + \frac{RT}{\Delta zF} \ln \frac{a_{\mathrm{Me}^{z+}}}{a_{\mathrm{Me},id}} \tag{4.59}$$

The index i either is 2, 1, or 0. The activity of metal atoms in the lower dimensional phase is smaller than the activity in a higher dimensional phase and in the bulk phase, where $a_{\mathrm{Me,3d}} = 1$. The partial discharge connected with each step is described by Δz.

Based on this concept, thermodynamic functions of the UPD phase can be determined.[67,68] The formation reaction of the UPD modification of metal B on a substrate metal A is obtained by the following reaction scheme

$$\mathrm{Me_{B,bulk}} \rightarrow \mathrm{Me}^{z+} + z\mathrm{e}^- \tag{4.60}$$

$$\mathrm{Me}^{z+} + z\mathrm{e}^- \rightarrow \mathrm{Me_{B,UPD}} + \Delta_{\mathrm{dl}}(\mathrm{UPD} - \mathrm{A}) \tag{4.61}$$

The total cell reaction is

$$\mathrm{Me_{B,bulk}} \rightarrow \mathrm{Me_{B,UPD}} + \Delta_{\mathrm{dl}}(\mathrm{UPD} - \mathrm{A}) \tag{4.62}$$

A double-layer term $\Delta_{\mathrm{dl}}(\mathrm{UPD} - \mathrm{A})$ is included in the formation reaction which takes into account changes of the double−layer structure when the substrate/electrolyte interface (A) is substituted by the UPD metal/electrolyte interface. If the chemisorption from the electrolyte components on metal A and the UPD layer B is similar, the term is very small and can be neglected.

The formation reaction (Eq. (4.62)) can be used to calculate thermodynamic functions of the UPD phase. From the cell voltage $\Delta E = E - E_0$ one can calculate the partial molar Gibbs energy $\Delta \bar{G}_{\mathrm{UPD}}$ of the UPD modification of the metal B

$$\Delta \bar{G}_{\mathrm{UPD}} = -zF\Delta E \tag{4.63}$$

The activity of the lead atoms on the silver surface is given by

$$\log a_i = -\frac{zF}{2.303RT}\Delta E \tag{4.64}$$

From the peak potential the Gibbs energy of formation $\Delta_{\mathrm{for}} G_{\mathrm{UPD}}$ of the UPD phase is obtained.

$$\Delta_{\mathrm{for}} G_{\mathrm{UPD}} = -zF\Delta E_p \tag{4.65}$$

The Gibbs energy of formation $\Delta_{\mathrm{for}} G_{\mathrm{UPD}}$ is defined by the following equation

$$\Delta_{\mathrm{for}} G_{\mathrm{UPD}} = \Delta_{\mathrm{for}} H_{\mathrm{UPD}} - T\Delta_{\mathrm{for}} S_{\mathrm{UPD}} \tag{4.66}$$

The entropy term can be determined by measuring the temperature dependence of the UPD deposition potential

$$\Delta_{\mathrm{for}} S_{\mathrm{UPD}} = zF\left(\frac{\partial \Delta E_p}{\partial T}\right)_p \tag{4.67}$$

But the relative significance of the entropy term is often very small. An approximate value of $\Delta_{\mathrm{for}} S_{\mathrm{UPD}}$ can be calculated with the equation

$$S_{\mathrm{UPD}} - S_{\mathrm{bulk}} = R\ln\frac{V_{\mathrm{UPD}}}{V_{\mathrm{bulk}}} \tag{4.68}$$

V_{bulk} is the molar volume of the unit cell of the bulk and V_{UPD} that of the UPD modification. When the bulk metal and the UPD modification have the same closely packed structure, the package density in the UPD layer is similar to the bulk, which leaves similar values of molar volumes. Deviations from complete epitaxy, tensions, misfits, etc., influence the energy of interaction between substrate A and UPD layer but have less influence on the entropy.

If the entropy difference can be neglected, $\Delta_{\mathrm{for}} G_{\mathrm{UPD}}$ is approximately equal to the formation enthalpy of the UPD modification of the metal B.

$$\Delta_{\mathrm{for}} G_{\mathrm{UPD}} \cong \Delta_{\mathrm{for}} H_{\mathrm{UPD}} \tag{4.69}$$

The formation enthalpy can be used to calculate other thermodynamic data of the UPD modification of metal B. With the Born–Haber cycle shown in Figure 4.34, the sublimation

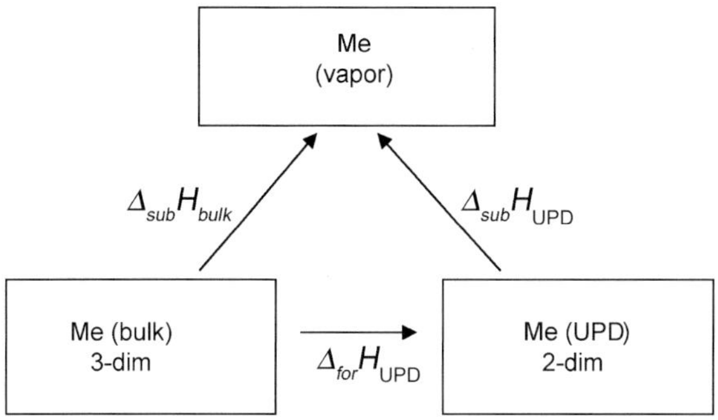

Figure 4.34 Born–Haber cycle for calculation of sublimation enthalpy of UPD phases.

Table 4.1

Formation enthalpies of some UPD layers

Substrate (S)	UPD Metal (Me)	ΔE_{UPD} [V]	$\Delta_{for}H_{UPD}$ [kJ mol^{-1}]	$\Delta_{sub}H_{UPD}$ [kJ mol^{-1}]	$\Delta_{sub}H_{UPD}$ [kJ mol^{-1}]
Ag(111)	Pb	0.15	−28.9	184.3	213.2
Ag(111)	Tl	0.25	−24.1	166.6	190.7
Au(111)	Pb	0.20	−38.6	184.3	222.9
Au(111)	Cu (β-phase)	0.23 0.06	−33.5	317.0	350.5

enthalpy of the UPD modification

$$\Delta_{sub} H_{UPD} \cong \Delta_{sub} H_{bulk} - \Delta_{for} H_{UPD} \tag{4.70}$$

can be calculated using the sublimation enthalpy of the bulk metal.

Some examples were analyzed; results are shown in Table 4.1.

4.4.4 Multiple steps of UPD film formation

The system Cu–Au(111) is an example of the formation of a compact and completely discharged UPD film in two steps. The first peak potential represents a phase structure α, which, at a second potential, is transformed into the phase structure β. The relevant free energy from the formation of the compact layer of metal B is then obtained by the sum of the free energy of formation of phase α and the free energy of transformation of phase α into phase β.

The summation must take into account the phase stoichiometry. The formation of each phase refers to the transportation of one mole of metal B into the new structure. The final value of the Gibbs energy of formation is defined for one mole UPD metal in a two-dimensional form.

The Gibbs energy of formation of the α-phase in the high voltage region is obtained from the first peak potential. This corresponds to the formation reaction

$$2Cu + H_2SO_4 \rightarrow Au(111)Cu_2SO_4 + H_2 \tag{4.71}$$

The Gibbs energy of formation is calculated for one mole Cu transferred in the neutral cell reaction from the bulk into the UPD phase (Eq. (4.66)).

From the potential of the second peak, the Gibbs energy of transformation of the α-phase into the β-phase can be calculated. The transformation of the α-phase to the β-phase is described by the reaction

$$Cu + Au(111)Cu_2SO_4 + H_2 \rightarrow Au(111)Cu_3 + H_2SO_4 \tag{4.72}$$

With Eq. (4.66) the Gibbs energy of transformation is again calculated for one mole Cu transferred in the neutral cell reaction from the bulk into the UPD phase. The Gibbs energy of formation of the β-phase, a closely packed Cu layer, is obtained by the sum of two-thirds of the Gibbs energy of formation of the α-phase plus one-third of the Gibbs energy of transformation of the α-phase to the β-phase.

$$\Delta_{\text{for}} G_{\text{UPD},\beta} = \frac{2}{3}\,\Delta_{\text{for}} G_{\text{UPD},\alpha} + \frac{1}{3}\,\Delta_{\text{tr}} G_{\alpha,\beta} \tag{4.73}$$

Eq. (4.73) can easily be generalized for a multi-step process. The values calculated for copper UPD on Au(111) are given in Table 4.1.

END NOTE

*In some other textbooks, γ is used for the free surface energy and σ for the surface charge density. Otherwise, for solid electrodes γ is used later for surface tension. It is also used for electrosorption valency. Therefore the use of σ for free surface energy and the use of q_m for surface charge are recommended.

REFERENCES

1. P. Delahay, Double Layer and Electrode Kinetics, Interscience, New York, 1965.
2. G. Gouy, J. Physique, *9*, 457 (1910); D.L. Chapman, Phil. Mag., *25*, 475 (1913).
3. J. Koryta, J. Dvorak, L. Kavan, Principles of Electrochemistry, Wiley, Chichester, 1993, p. 235.
4. A.J. Bard, R. Memming, B. Miller, Pure Appl. Chem., *63*, 569 (1991).
5. S. Trasatti, R. Parsons, Pure Appl. Chem., *58*, 437 (1986).
6. J.W. Gibbs, Collected Works, Vol. 1, Longmans, Green and Comp., London, 1928, p. 336.
7. K.J. Vetter, Elektrochemische Kinetik, Springer-Verlag, Berlin, 1961, p. 76.
8. K. Schwabe, Physikalische Chemie, Part 2, Elektrochemie, Akademie-Verlag, Berlin, 1974, p. 342.
9. D.C. Grahame, J. Am. Chem. Soc., *76*, 4819 (1954).
10. A.N. Frumkin, Z. Phys. Chem., *116*, 466 (1925).
11. M.I. Temkin, Zh. Fys. Khim., *14*, 1153 (1940); *15*, 296 (1941).

12. W.J. Plieth, K.J. Vetter, Ber. Bunsenges. Phys. Chem. (Z. f. Electrochem.), *72*, 673 (1968).
13. W.J. Plieth, J. Electroanal. Chem., *23*, 305 (1969).
14. W. Lorenz (Leipzig), G. Salié, Z. Phys. Chem., *218*, 259 (1961); Z. Phys. Chem. NF, *29*, 390 (1961).
15. For a summary see W. Lorenz (Leipzig), J. Phys. Chem., *95*, 10566 (1991).
16. W.J. Plieth, K.J. Vetter, Z. Phys. Chem. NF, *61*, 282 (1968); Coll. Czech. Chem. Commun., *36*, 816 (1970).
17. W. Lorenz (Leipzig), G. Salié, J. Electroanal. Chem., *80*, 1, 21 (1988).
18. G. Salié, J. Electroanal. Chem., *245*, 1 (1977).
19. K. Bartels, G, Salié, Z. Phys. Chem. *271*, 739 (1999).
20. E. Schmidt, Helv. Chim. Acta, *52*, 1763 (1969).
21. K.J. Vetter, J.W. Schultze, Ber. Bunsenges. Phys. Chem., *76*, 920 (1972).
22. J. Lipkowski, Z. Shi, A. Chen, B. Pettinger, C. Bilger, Electrochimica Acta, *43*, 2875 (1998).
23. D.C. Grahame, J. Am. Chem. Soc., *80*, 4201 (1958).
24. D.C. Grahame, R. Parsons, J. Am. Chem. Soc., *83*, 1291 (1961).
25. K. Bange, R. Parsons, J.K. Sass, B. Straehler, J. Electroanal. Chem., *229*, 87 (1987).
26. W. Schmickler, R. Guidelli, J. Electroanal. Chem., *235*, 387 (1985).
27. K.J. Vetter, W.J. Plieth, Z. Phys. Chem. NF, *65*, 181 (1969).
28. W. Plieth, Z. Phys. Chem. NF, *67*, 178 (1969).
29. W.J. Plieth, K.J. Vetter, Ber. Bunsenges. Phys. Chem., *73*, 79 (1969).
30. R.G. Linford, Chem. Rev., *78*, 81 (1978).
31. D.H. Everett, P.R. Couchman, J. Electroanal. Chem., *67*, 382 (1976).
32. R. Shuttleworth, Proc. Phys. Soc. London, Ser. A, *63*, 444 (1950).
33. A.Y. Gokhshtein, Russ. Chem. Rev., *44*, 921 (1975).
34. B.M. Grafov, J. Electroanal. Chem., *471*, 105 (1999); Russ. J. Electrochem., *35*, 1029 (1999).
35. G. Lang, K.E. Heusler, J. Electroanal. Chem., *377*, 1 (1994); J. Electroanal. Chem., *472*, 168 (1999).
36. R. Guidelli, J. Electroanal. Chem., *472*, 174 (1999).
37. W. Haiss, R.J. Nichols, J.K. Sass, K.P. Charlé, J. Electroanal. Chem., *452*, 199 (1998).
38. M. Sluyters-Rehbach, J.H. Sluyters, A.C. Techniques, in Comprehensive Treatise of Electrochemistry, E. Yeager, J. O'M. Bockris, B.E. Conway, S. Sarangapani (Eds.), Plenum Press, New York and London, 1984, Vol. 9, p. 177.
39. D.D. Macdonald, Transient Techniques in Electrochemistry, Plenum Press, New York, 1977, p. 229.
40. J.R. Macdonald (Ed.), Impedance Spectroscopy, Wiley, New York, 1987.
41. I. Epelboin, C. Gabrieli, M. Keddam, H. Takenouti, Non-steady state techniques, in Comprehensive Treatise of Electrochemistry, E. Yeager, J. O'M. Bockris, B.E. Conway, S. Sarangapani (Eds.), Plenum Press, New York and London, 1984, Vol. 9, p. 61.
42. F.G. Will, C.A. Knorr, M. Breiter, Z. Elektrochem. Ber. Bunsenges. Phys. Chem., *64*, 258 (1960).
43. R.S. Nicholson, S. Shain, Anal. Chem., *36*, 706 (1964).
44. R.S. Nicholson, Anal. Chem., *37*, 1351 (1965).
45. C.P. Andrieux, P. Hapiot, J.M. Saveant, J. Electroanal. Chem., *296*, 335 (1990).
46. C. Amatore, C. Lefrou., J. Electroanal. Chem., *324*, 33 (1992).
47. G. Horanyi, Electrochim. Acta, *25*, 43 (1980).
48. V.E. Kazarinov, V.N. Andreev, Tracer methods in electrochemical studies, in Comprehensive Treatise of Electrochemistry, E. Yeager, J. O'M. Bockris, B.E. Conway, S. Sarangapani (Eds.), Plenum Press, New York and London, 1984, Vol. 9, p. 393.
49. E. Schmidt, H. Siegenthaler, Helv. Chim. Acta, *52*, 2245 (1969).
50. G. Sauerbrey, Z. Physik, *155*, 206 (1955).
51. A. Glidle, A.R. Hillmann, S. Bruckenstein, J. Electroanal. Chem., *314*, 33 (1992).

52. A. Bund, G. Schwitzgebel, Electrochim. Acta, *45*, 3703 (2000).
53. S. Neudeck, Diplomarbeit, TU Dresden, 2005.
54. R.J. Behm, N. Garcia, H. Rohrer, Scanning Tunneling Microscopy and related techniques, Kluwer Academic, Dordrecht, 1990.
55. G. Binning, H. Rohrer, C, Gerber, E. Weibel, Phys. Rev. Lett., *49*, 57 (1982).
56. A. J. Bard, G. Denuault, C. Lee, D. Mandler, D. O. Wipf, Acc. Chem. Res., *23*, 357 (1990).
57. T.H. Treutler, G. Wittstock, Electrochim. Act, *48*, 2923 (2003).
58. K.J. Vetter, W. Plieth, Z. Phys Chem. NF, *65*, 181 (1969).
59. E. Budevski, G. Staikov, J.W. Lorenz, Electrochemical Phase Formation and Growth, VCH, Weinheim, 1996, p. 31.
60. H. Bort, K. Jüttner, W.J. Lorenz, E. Schmidt, J. Electroanal. Chem., *90*, 413 (1978).
61. G. Staikov, K. Jüttner, W.J. Lorenz, E. Budevski, Electrochim.Acta, *23*, 319–324 (1978).
62. W.J. Lorenz, L.M. Gassa, U. Schmidt, W. Obretenov, G. Staikov, V. Bostanov, E. Budevski, Electrochim. Acta, *37*, 2173 (1992).
63. T. Will, M. Dietterle, D.M. Kolb, The initial stages of electrolytic copper deposition: An atomistic view, in Nanoscale Probes of the Solid-Liquid Interface, A.A. Gewirth, H. Siegenthaler (Eds.), NATO ASI Series E: Applied Sciences, Vol. 288, Kluwer Academic Publ., Dordrecht 1995, p. 137.
64. M.S. Zei, G. Quiao, G. Lehmpfuhl, D.M. Kolb, Ber. Bunsenges. Phys. Chem., *91*, 349 (1987).
65. G. Staikov, W.J. Lorenz, E. Budevski, Low-dimensional metal phases and nanostructuring of solid surfaces, in Imaging of Surfaces and Interfaces – Frontiers in Electrochemistry, Vol. 5, Eds. P. Ross, J. Lipkowski, Wiley-VCH, New York, Weinheim, 1999, p. 1.
66. E. Budevski, G. Staikov, W.J. Lorenz, Electrochim Acta, *45*, 2559 (2000).
67. W. Plieth, Surf. Coat. Technol., *169–170*, 96 (2003).
68. W. Plieth, W.J. Lorenz, G. Staikov, J. Solid State Electrochem., *8*, 941 (2004).

– 5 –

Mass Transport

An important step of the reaction mechanism in a heterogeneous process is the transport of the reacting substances and reaction products between the place of reaction and the bulk phases. If the driving force for the transport is a concentration gradient, the transport process is diffusion (Figure 5.1). In the general redox system depicted in Figure 5.1a, the oxidized or reduced components diffuse between the bulk electrolyte and electrode surface. In electrodeposition (Figure 5.1b) solvated metal ions diffuse to the surface and lose the solvation shell in the process of depositing. These reaction products diffuse back into the electrolyte.

In electrochemistry, in addition to the concentration gradient, a gradient of the electric field can be the driving force. This process is called electro-migration and is often important in solid or polymer electrolytes, particularly in mass transport. Mass transport in polymer or solid phases is much slower than in aqueous electrolytes. The process is nevertheless important for many applications like charging and discharging of batteries. Control and enhancement of the charge transfer by diffusion or migration is a big problem in electrochemical kinetics.

5.1 STATIONARY DIFFUSION

Figure 5.2 shows concentration of an ion deposited on the surface of an electrode in a stirred electrochemical cell as function of distance. There is a continuous decrease of agitation in the layer formed here from the bulk of the electrolyte to the electrode surface. If an electrode process occurs on the electrode surface, in this layer of reduced influence from stirring, a concentration gradient builds up between the surface (concentration c_s) and the electrolyte bulk (concentration c_0). The concentration gradient defines a layer first described by Nernst[1] and by Brunner[2] and is called diffusion layer δ_d. The typical order of magnitude in aqueous electrolytes is 10^{-3} cm.

The reason for this layer formation in a stirred electrolyte is viscosity. While the concentration outside this layer is always kept equal to the bulk concentration by the flow of the electrolyte, the flow rate of the electrolyte in the layer is continuously decreasing and therefore approaches zero at the electrode surface.

The flow rate decrease is linear for laminar flow and nonlinear for turbulent flow. The Reynolds number[3] describes the border between laminar and turbulent flow. The Reynolds

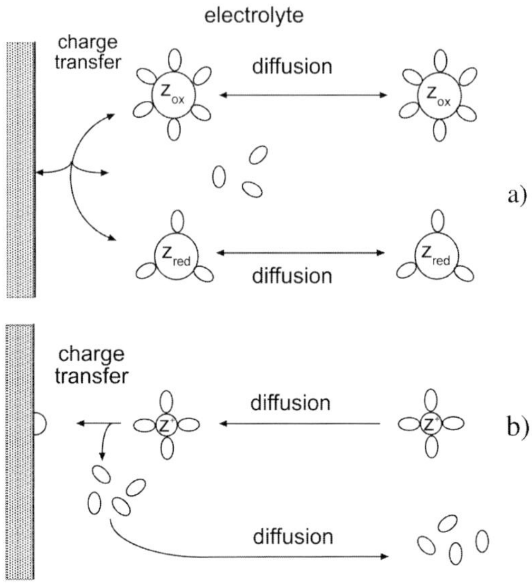

Figure 5.1 Diffusion of reacting particles and reaction products between bulk electrolyte and electrode surface. (a) Redox process in the electrolyte and (b) deposition reaction of a solvated metal ion.

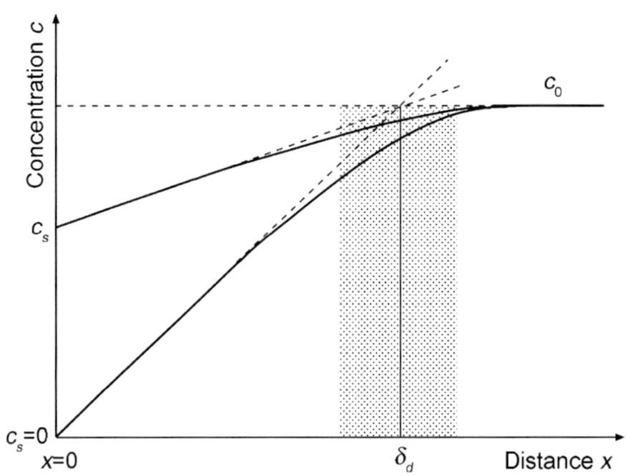

Figure 5.2 Model of stationary diffusion in a stirred electrolyte. The Nernst–Brunner diffusion layer[1,2] forms on an electrode surface of thickness δ_d because the viscosity of the electrolyte creates a gradient of convection between the bulk electrolyte and the electrode surface.

number for the flow in a pipe is given by the equation

$$Re = \frac{v \cdot l \cdot \rho}{\eta} \tag{5.1}$$

where v is the flow rate (SI unit, $m\,s^{-1}$), l the characteristic length (which is the diameter in the case of a pipe and the radius of the disc for a rotating disc), ρ the density

(SI unit, g m^{-3}), and η the viscosity (SI unit, N s m^{-2}). The ratio $\eta/\rho = \nu$ is called kinematic viscosity. With the introduction of this quantity Eq. (5.1) becomes

$$Re = \frac{\mathrm{v} \cdot l}{\nu} \tag{5.2}$$

The Reynolds number has no dimension. The critical value for the transition from laminar to turbulent flow is of the order 10^5 to 10^6.

Fick's first law can be used to describe the diffusion to the electrode surface through the diffusion layer

$$\frac{1}{A}\frac{\mathrm{d}n}{\mathrm{d}t} = D\frac{c_0 - c_\mathrm{s}}{\delta_\mathrm{d}} \tag{5.3}$$

The number of moles per time interval dn/dt and per surface area A is proportional to the concentration gradient $(c_0 - c_\mathrm{s})/\delta_\mathrm{d}$. The constant D is the diffusion coefficient and is usually measured in cm$^2 \cdot$ s^{-1}.

A cathodic reduction process is described by the equation

$$S_\mathrm{ox}^{z_\mathrm{ox}} + ne^- \rightarrow S_\mathrm{red}^{z_\mathrm{red}} \tag{5.4a}$$

In the case of a metal ion reduction, but without the molecules of the solvation shell, the process is described by the equation

$$\mathrm{Me}^{z+} + ze^- \rightarrow \mathrm{Me} \tag{5.4b}$$

The difference in both equations is the symbol used for the number of electrons. While n is usually used in a redox process, z is used in metal ion reduction because the number of electrons in the latter case corresponds with the charge of the metal ion. In this chapter n will be used as the stoichiometric number of the electrons.

The transport of $S_\mathrm{ox}^{z_\mathrm{ox}}$ (or of metal ions) can be substituted by the cathodic (negative) current density

$$i = -nFD\frac{c_0 - c_\mathrm{s}}{\delta_\mathrm{d}} \tag{5.5}$$

A limiting current is reached as the concentration on the surface, c_s, approaches zero

$$i_\mathrm{lim} = -nFD\frac{c_0}{\delta_\mathrm{d}} \tag{5.6}$$

The change in the surface concentration leads to a shift in the Nernst potential

$$E - E_0 = \frac{RT}{nF}\ln\frac{c_\mathrm{s}}{c_0} \tag{5.7}$$

The difference $E - E_0$ is the diffusion overpotential $\eta_\mathrm{d} = E - E_0$. The concentration ratio can be substituted by the current densities

$$\frac{c_{\mathrm{s}}}{c_0} = \frac{i_{\lim} - i}{i_{\lim}} \tag{5.8}$$

As a result, the diffusion overpotential is given by the equation

$$\eta_{\mathrm{d}} = \frac{RT}{nF} \ln\left(1 - \frac{i}{i_{\lim}}\right) \tag{5.9}$$

In this approach only the diffusion of $S_{\mathrm{ox}}^{z_{\mathrm{ox}}}$ (or of metal ions) is taken into account. If diffusion of more than one substance must be taken into account an additional term must be added, e.g., in a redox process an additional term for $S_{\mathrm{red}}^{z_{\mathrm{red}}}$ has to be added in Eq. (5.9). In a general approach all components with effective concentration differences between electrolyte and electrode surface must be taken into account in Eq. (5.9).

Eq. (5.9) expresses an exponential dependence of the current on the diffusion overpotential

$$\frac{i}{i_{\lim}} = 1 - \mathrm{e}^{\frac{nF}{RT}\eta_{\mathrm{d}}} \tag{5.10}$$

The dependence is shown in Figure 5.3 for different diffusion layer thicknesses.

For a small overpotential the exponential function can be substituted by $\mathrm{e}^x \approx 1 + x$ whereby the dependence is linear

$$\frac{i}{i_{\lim}} = -\frac{nF}{RT}\eta_{\mathrm{d}} \tag{5.11}$$

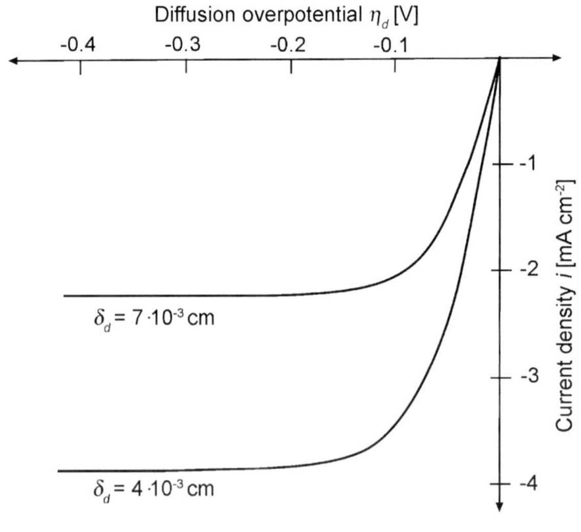

Figure 5.3 Current–potential dependence of a diffusion limited reduction process (deposition of Ag^+, $c = 10^{-2}\,\mathrm{mol \cdot dm^{-3}}$, $D = 1.6 \times 10^{-5}\,\mathrm{cm^2 \cdot s^{-1}}$), diffusion limited currents on a rotating-disc electrode, corresponding diffusion layer thicknesses δ_{d}.

The diffusion resistance can be derived from this equation

$$R_{\mathrm{d}} = \lim_{i \to 0} \frac{i}{\eta_{\mathrm{d}}} = \frac{RT}{nF} \frac{1}{\left|i_{\mathrm{lim}}\right|} \tag{5.12}$$

For a large overpotential the current becomes independent from the potential (limiting current).

Similar equations are obtained for oxidation of $S_{\mathrm{red}}^{z_{\mathrm{red}}}$ in Eq. (5.4) or surface concentrations of other diffusion limited substances.

5.2 NON-STATIONARY DIFFUSION

Concentration gradients can be time dependent. Fick's second law describes the time dependence of diffusion

$$\left(\frac{\partial c(x,t)}{\partial t}\right)_x = D\left(\frac{\partial^2 c(x,t)}{\partial x^2}\right)_t \tag{5.13}$$

Such a general equation can only be solved if the boundary conditions are known. By the selection of the boundary conditions one can adapt Eq. (5.13) for selected conditions. The principal mathematical method to solve the equation is Laplace transformation.[4] In the following some simple results will be described.

5.2.1 Chronopotentiometry

Switching on a current larger than the limiting current, the diffusion profile is described by the following boundary conditions

$$c(x,t=0) = c_0 \tag{5.14a}$$

$$\left(\frac{\partial c}{\partial x}\right)_{x=0} = -\frac{i}{nFD} \tag{5.14b}$$

$$c(x=\infty,t) = c_0 \tag{5.14c}$$

The meaning of the last condition is that the diffusion can freely expand into the electrolyte (semi-infinite condition). The solution of Eq. (5.13) gives the Sand equation.[5] If formulated for a decrease of concentration on the surface the equation is

$$c(0,t) = c_0 - \frac{|i|\sqrt{t}}{0.5nF\sqrt{\pi D}} \tag{5.15}$$

The concentration on the electrode surface decreases with the square root of time. For metal deposition, the corresponding potential is given by the equation

$$E - E_0 = \frac{RT}{nF} \ln \frac{c(0,t)}{c_0} \tag{5.16}$$

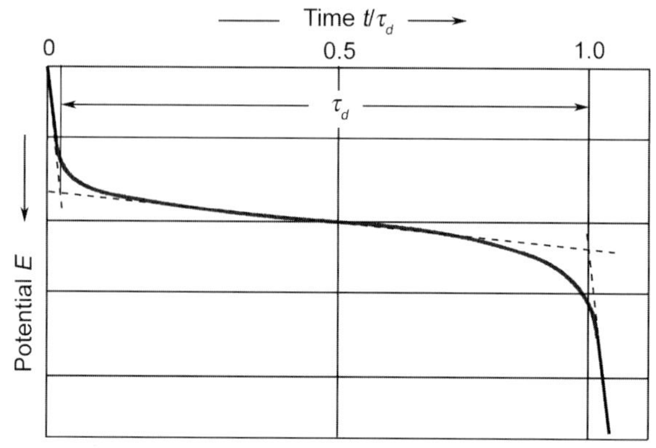

Figure 5.4 Chronopotentiometry, concentration depletion in a cathodic reduction process, potential plotted as function of time, and determination of the transition time τ_d.

At $c(0,t) = 0$ a jump of the electrode potential is observed (Figure 5.4). The associated time is called transition time τ_d. The transition time is given by the equation

$$i\sqrt{\tau_d} = 0.5nF\sqrt{\pi D} \cdot c_0 \tag{5.17}$$

The product $i\sqrt{\tau_d}$ linearly depends on the bulk concentration c_0. Figure 5.4 shows a potential versus time plot for a depletion of the oxidized component in a reduction process (e.g., metal deposition).

5.2.2 Chronoamperometry, chronocoulometry

If for potentiostatic conditions one applies a potential jump ΔE, starting at the equilibrium potential (e.g., in the case of a metal electrode from the Nernst potential to a deposition potential), the following boundary conditions exist

$$c(x,t=0) = c_0 \tag{5.18a}$$

$$c(x=0,t) = 0 \tag{5.18b}$$

$$c(x=\infty,t=0) = c_0 \tag{5.18c}$$

The last condition redefines a semi-infinite diffusion. Eq. (5.13) for these conditions describes the development of a concentration profile with time as shown in Figure 5.5.

A time-dependent thickness of a diffusion layer describes the growth of the concentration profile into the electrolyte given by the equation

$$x = \sqrt{\pi Dt} \tag{5.19}$$

The dependence of a selected concentration value on the square root of time is typical for diffusion-controlled processes (Tamman's law).

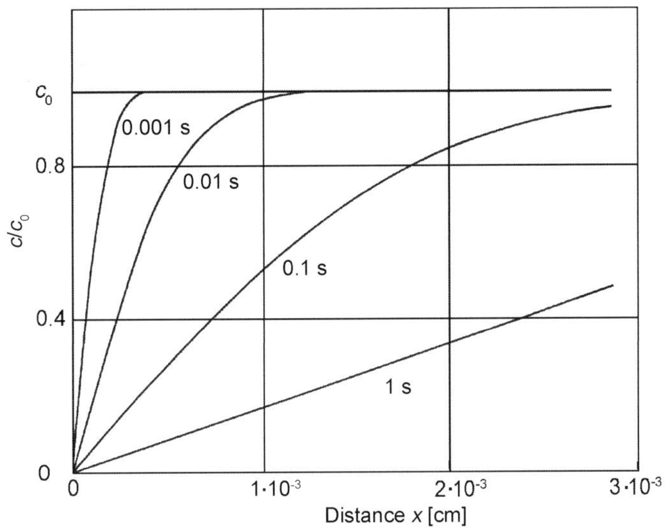

Figure 5.5 Chronoamperometry, depletion of concentration at the electrode surface, and growth with time of concentration profiles into the electrolyte.

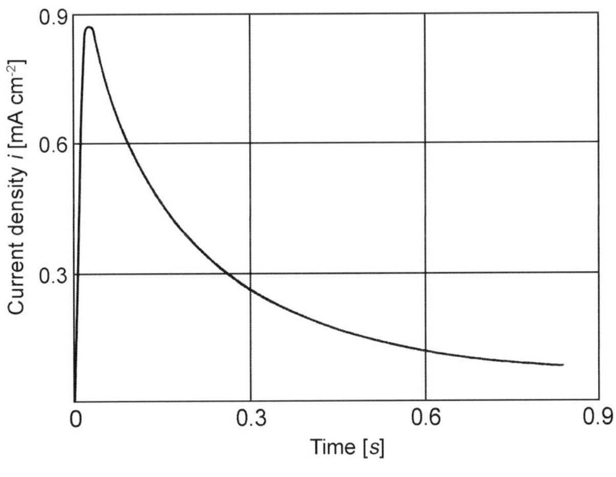

Figure 5.6 Chronoamperometry, semi-infinite condition, and current–time dependence which follows the Cottrell equation.

The equation for the dependence of the current upon time derived from the concentration–time dependence is called the Cottrell equation[6]

$$i(t) = nF\sqrt{\frac{D}{\pi}}\frac{c_0}{\sqrt{t}} \tag{5.20}$$

The dependence of the current on time is shown in Figure 5.6.

In the current–time diagram a peak usually occurs at the beginning. The initial rise of the current with time is limited by the charging of the double layer capacitance to build up the selected deposition potential.

Integration of the current gives the development of charge with time

$$Q(t) = 2nF\sqrt{\frac{D}{\pi}}c_0\sqrt{t} \tag{5.21}$$

The charge increases proportional to the square root of time. The measurement of charge as a function of time is called chronocoulometry.

5.2.3 Warburg impedance

In Section 4.2.1 the method of impedance spectroscopy was introduced. The polarization of an electrode with an alternating potential of small amplitude is also influenced by restricted mass transport. The Warburg impedance for semi-infinite diffusion describes the diffusion process. The Warburg impedance is a complex quantity with real and imaginary parts of equal magnitude.[7] The impedance is given by the equation

$$Z_W = \frac{W}{\sqrt{j\omega}} \tag{5.22}$$

W is the Warburg parameter

$$W = \frac{RT}{n^2F^2}\frac{1}{c_0\sqrt{D}} \tag{5.23}$$

Details are given in the special literature, e.g., by J.R. MacDonald[8] or Sluyters-Rehbach and Sluyters.[9]

A line of 45° versus the coordinate axis represents the Warburg impedance in the complex plain presentation (Nyquist plot, Figure 5.7a). The representation in the Bode diagram is shown in Figure 5.7b. The phase shift has a constant value of 45°, whereby the modulus of the impedance, |Z| is linearly decreasing with increasing frequency.

The so-called Randles equivalent circuit[10] describes diffusion-controlled charge transfer processes (Figure 5.8).

The impedance is represented in the Nyquist and Bode diagrams for two different situations in Figures 5.9 and 5.10.

Figure 5.9 shows a situation in which the charge transfer resistance and Warburg impedance are of equal size. In the Nyquist diagram (Figure 5.9a) the characteristic image is the semicircle with the transition into the 45° Warburg line at smaller frequencies. In the Bode diagram (Figure 5.9b) the charge transfer process dominates and the characteristic phase value of 45° is not reached in the represented frequency range, not even at the lowest frequency values (0.01 Hz).

Figure 10a shows the Nyquist plot for a situation when diffusion impedance is much larger compared to the charge transfer resistance. The 45° Warburg line dominates the

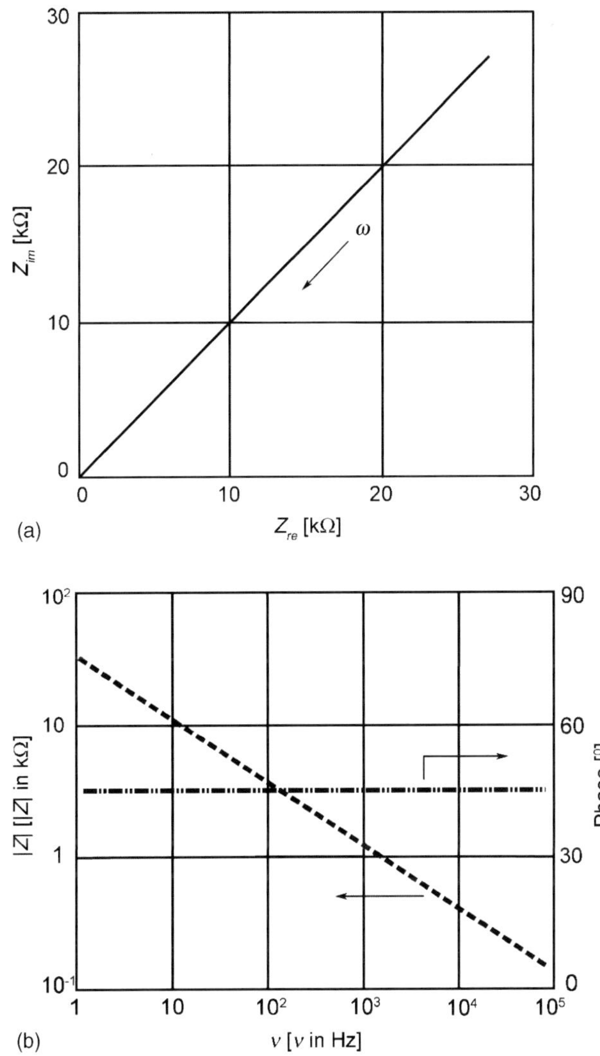

Figure 5.7 Warburg impedance for semi-infinite diffusion. (a) Nyquist plot and (b) Bode plot.

Nyquist plot. A shift of the 45° line and a small deviation from the linear behavior at highest frequencies indicate that an impure diffusion impedance is represented in the diagram. A characteristic double layer behavior dominates the high-frequency part of the Bode diagram (Figure 5.10b). At low frequencies the diagram approaches the characteristic form of diffusion.

A comparison of these cases shows that the two representations of impedance measurements provide different information. While low-frequency components dominate in the Nyquist plot, high-frequency components dominate the character of the Bode diagram.

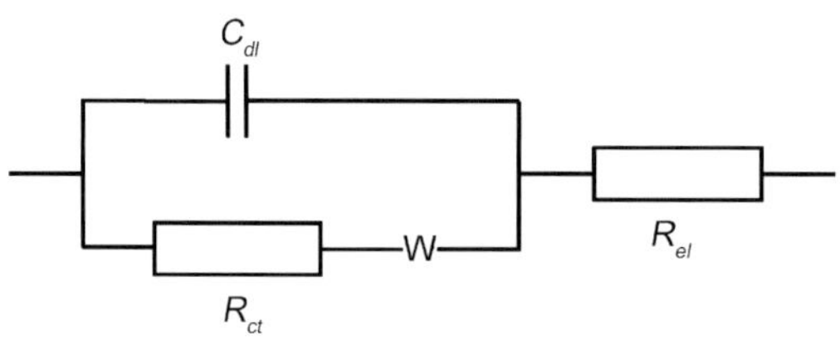

Figure 5.8 Randles equivalent circuit for a diffusion-controlled charge transfer process, C_{dl} double layer capacitance, R_{ct} charge transfer resistance, R_{el} electrolyte resistance, and W Warburg impedance.

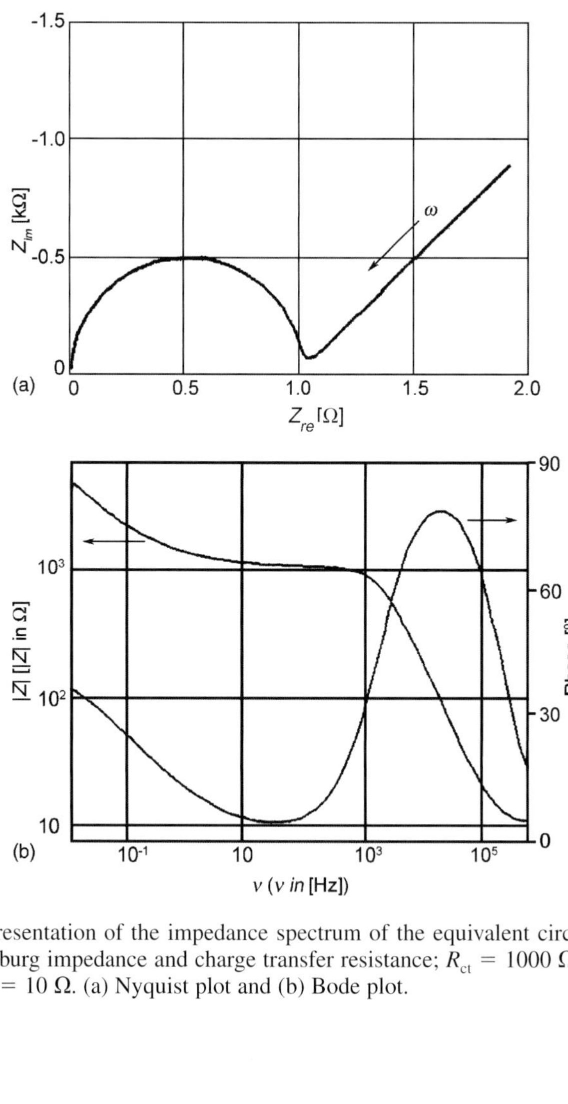

Figure 5.9 Representation of the impedance spectrum of the equivalent circuit in Figure 5.8 for equally sized Warburg impedance and charge transfer resistance; R_{ct} = 1000 Ω, $|Z_d|$ = 1000 Ω s^{-1}, C_{dl} = 100 nF, R_{el} = 10 Ω. (a) Nyquist plot and (b) Bode plot.

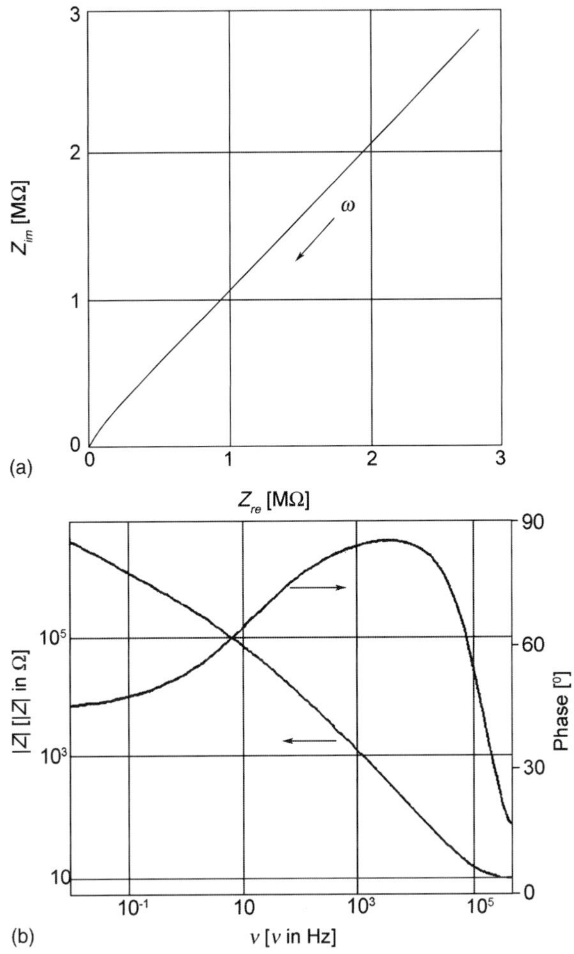

Figure 5.10 Representation of the impedance spectrum of the equivalent circuit in Figure 5.8 for when Warburg impedance is much larger than the charge transfer resistance; R_{ct} = 1000 MΩ, $|Z_d|$ = 1 MΩ s^{-1}, C_{dl} = 100 nF, R_{el} = 10 Ω. (a) Nyquist plot and (b) Bode plot.

So far unlimited diffusion has been assumed. When the diffusion layer has finite dimension, with a boundary of constant concentration of the diffusing species and permeable for these species, however, a different expression for the diffusion impedance is derived which is the so-called Nernst impedance. This impedance is given by the equation

$$Z_N = \frac{RT}{n^2F^2}\frac{1}{c_0\sqrt{D}}\frac{1}{\sqrt{j\omega}}\tanh\sqrt{\frac{j\omega}{D/\delta_d^2}} \quad (5.24)$$

The Nernst impedance is shown in the Nyquist and Bode diagrams in Figure 5.11. The ratio D/δ_d^2 (D diffusion coefficient, δ_d diffusion layer thickness) has the dimension of a

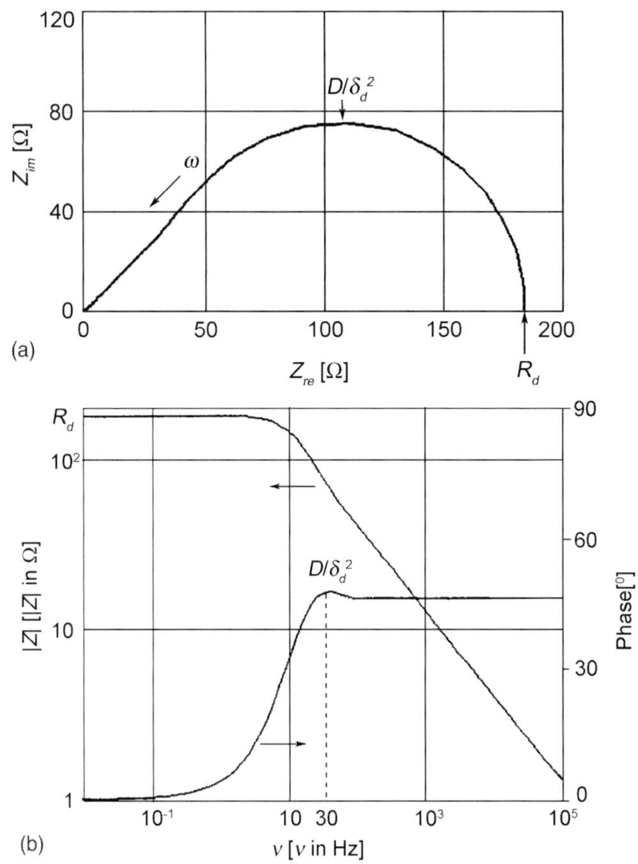

Figure 5.11 Representation of diffusion impedance spectra of diffusion processes limited by a Nernst diffusion layer, R_d 182 Ω, $D/\delta_d^2 = 30\ s^{-1}$. (a) Nyquist plot and (b) Bode plot.

frequency. In the case of finite diffusion length the diffusion resistance R_d is given by the dc limit for frequency $\nu \rightarrow 0$ (angular frequency $\omega \rightarrow 0$). D/δ_d^2 can be determined together with R_d from the representations in Figure 5.11 as indicated.

5.2.4 Cyclic voltammetry

In Section 4.2.2 cyclic voltammetry was introduced. The influence of mass transport on cyclic voltammetry for a reversible redox system is shown in Figure 5.12.

With the scanning of the potential, concentration waves extend into the electrolyte. The concentration gradients at the electrode surface go through maxima with characteristic current peaks in the current potential diagram. The Randles–Sevcik equation[11] describes the dependence of peak currents on concentration, c_0 and scan rate, v

$$i_p = 2.6 \times 10^2 n^{3/2} \sqrt{D} \cdot c_0 \sqrt{v} \tag{5.25}$$

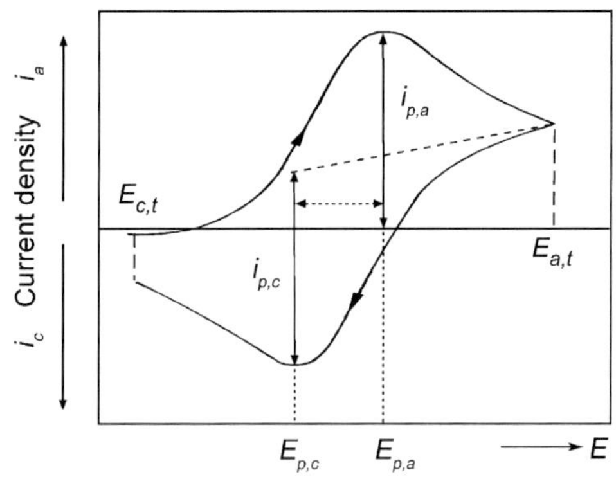

Figure 5.12 Cyclic voltammogram of a reversible redox system. For reversible systems the difference of the peak potentials is 59 mV (25 °C) with an equal height of the peak currents $i_{p,a} = i_{p,c}$.

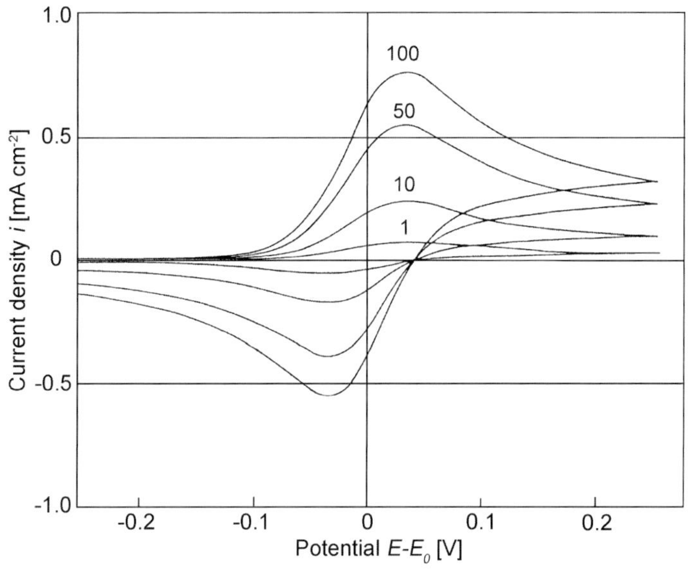

Figure 5.13 Cyclic voltammogram of a reversible redox system. Dependence of peak heights on scan rate (scan rates in $mV \cdot s^{-1}$).

D is measured in $cm^2 \cdot s^{-1}$, v in $V \cdot cm^{-1}$, and the concentration in $mol \cdot dm^{-3}$. For a reversible redox system the ratio of the peak currents is one

$$i_{p,a} = i_{p,c} \tag{5.26}$$

and the separation of the peak potentials for 25 °C

$$E_{p,a} - E_{p,c} = 0.059\,V \tag{5.27}$$

According to Eq. (5.25) the peak currents depend on the square root of the scan rate. This is another proof of reversibility. An example is shown in Figure 5.13.

5.2.5 Microelectrodes

Microelectrodes, also called ultra-microelectrodes, have some very special diffusion properties. The electrode can be prepared by melting an ultra-thin metal wire into a glass rod. The small radius of the wire with only a few micrometers leads to a situation in which the radial resistance of the electrolyte in front of the electrode substitutes the diffusion limitation by the diffusion layer. A microelectrode and the current distribution in front of the microelectrode are shown in Figure 5.14.

The current–time dependence of a cyclic voltammogram at a microelectrode for diffusion-controlled conditions is given by the equation

$$i = nFDc_0\left(\frac{1}{\sqrt{\pi Dt}} + \frac{1}{r}\right) \tag{5.28}$$

After the build-up of the diffusion layer at short times, for longer times ($t \to \infty$) the first term in Eq. (5.28) vanishes and the current density i becomes proportionally inverse to the radius r of the microelectrode and is therefore potential and time independent. This behavior leads to different forms of cyclic voltammograms. Instead of the diffusion peaks shown in Figure 5.12 for a reversible redox system, constant currents are obtained on a microelectrode.

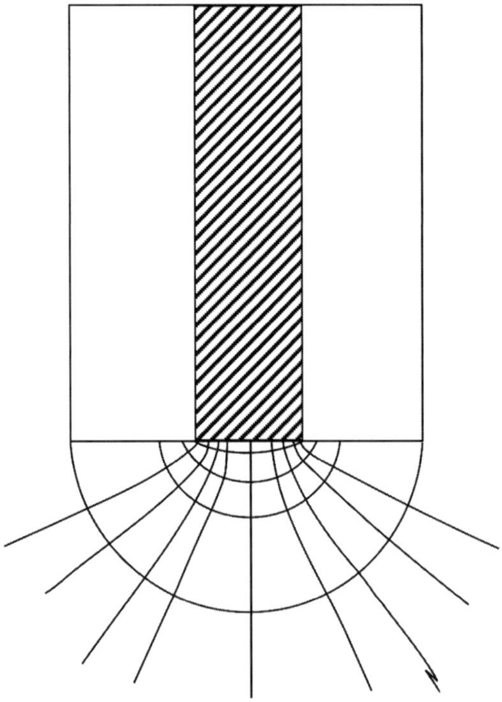

Figure 5.14 Microelectrode and radial current distribution in front of the electrode.

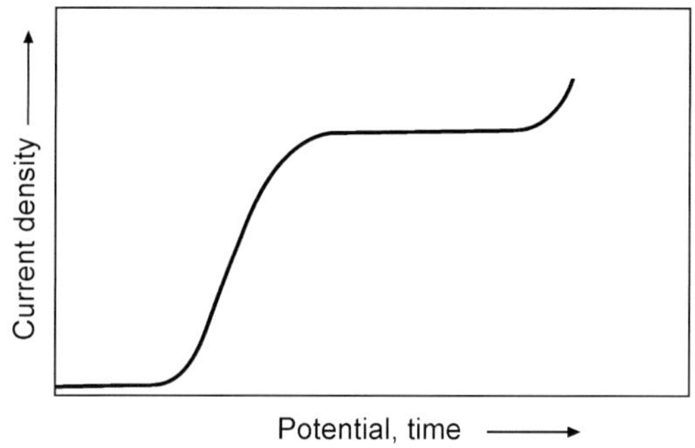

Figure 5.15 Microelectrode, cyclic voltammogram, and an x-axis that simultaneously represents potential and time. The observed constant current is not a potential-independent current but rather a time-independent current.

The reason for the potential-independent current in this case is not diffusion limitation but the control of the current density by the electrolyte resistance in front of the microelectrode. This is shown in Figure 5.15.

5.3 DIFFUSION IN SOLID PHASES

Fick's laws also describe diffusion in solid phases. In solids transport properties can be considerably different than in liquid phases. Only one component can mobile diffuse in the matrix of the second component. At higher temperatures the diffusion coefficient can be more similar in size than in liquid phases, but the diffusion coefficient at room temperature can be orders of magnitudes smaller, e.g., $D < 10^{-10}$ cm^2 s^{-1}. To overcome the time limitation one must make the diffusion length l_d smaller. Ultra-thin layers or nanoparticles provide such small dimensions. Under such conditions the diffusion is not semi-infinite but has a restricted extension. This has to be considered in the boundary conditions.

For the study of diffusion phenomena in solids it is also possible to work with potential pulses (potentiostatic pulses) or with constant current pulses (galvanostatic pulses). Examples described in the following paragraphs are based on the coulometric titration method described in Chapter 3. Weppner and Huggins reviewed these methods.[12] In a continuous series of pulses the concentration of lithium in a sheet of aluminum is increased. The diffusion in each pulse is followed by either potential or current measurements.

5.3.1 Potentiostatic method

An aluminum wire was used in a molten potassium chloride/lithium electrolyte and lithium was deposited into the aluminum by potential or current pulses. The electrode with the radius r is shown in Figure 5.16. The diffusion is no longer semi-infinite but is limited by the radius of the wire, where the radial diffusion fronts meet each other. The same conditions hold for a thin sheet of metal.

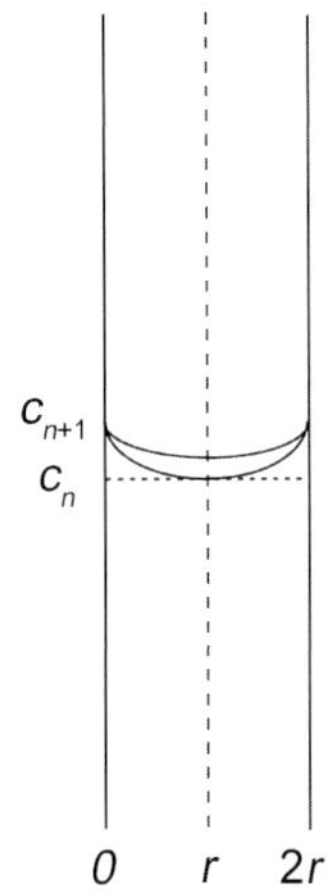

Figure 5.16 Diffusion of a metal in an alloy electrode, metal wire of radius r, concentration profiles after potential or current pulses n and $n+1$.

The experiment was made under potentiostatic conditions. This is called potential intermittent titration technique (PITT).

The concentration of lithium in the alloy at a potential E_n is c_n. After application of a potential pulse dE and increasing of the potential from E_n to E_{n+1}, the lithium concentration at the electrolyte–electrode interface is raised to c_{n+1}. A concentration profile expands into the alloy phase. At the middle of the aluminum electrode the two diffusion waves meet. Boundary conditions for this situation are

$$c(x,t = 0) = c_n \tag{5.29a}$$

$$c(x = 0,t) = c_{n+1} \tag{5.29b}$$

$$\left(\frac{\partial c}{\partial x}\right)_{x = r,t} = 0 \tag{5.29c}$$

The solution of Eq. (5.13) for these boundary conditions has two limiting solutions. For times $t << r^2/D$ the Cottrell equation is obtained. For the current density one gets

$$i(t) = F(c_{n+1} - c_n)\sqrt{\frac{D}{\pi \cdot t}} \tag{5.30}$$

The current density depends on the reciprocal square root of time. This is shown for lithium diffusion into aluminum in Figure 5.17.

The line representing the linear dependence goes through the origin of the diagram. The diffusion coefficient of lithium in aluminum can be determined from the slope of the line if one substitutes the concentration difference $c_{n+1} - c_n$, which is related to the charge passed in the pulse $Q/F = n_{\mathrm{Li}}$ (n_{Li} is the mole of Li accumulated in the metal wire).

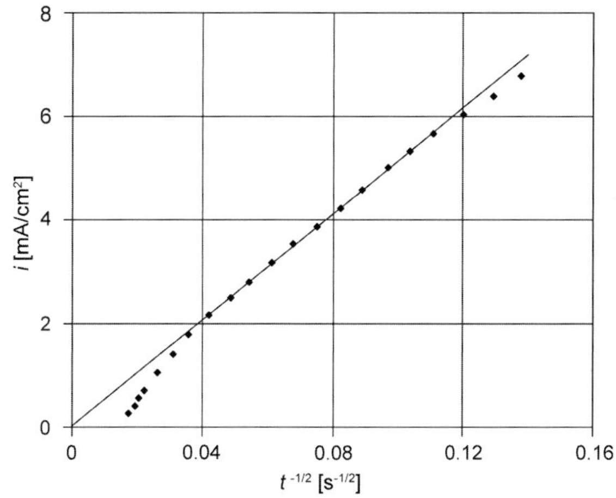

Figure 5.17 Potentiostatic pulses of lithium deposition into an aluminum alloy electrode; $i \approx t^{-1/2}$ (limiting law for $t << r^2/D$, Cottrell equation, after John Wen et al.), radius of the aluminum wire $r = 0.9$ mm, charge passed during pulse $Q = 7.47$ C, 415 °C. (Reproduced with permission from Ref. [13], © 1979, The Electrochemical Society.)

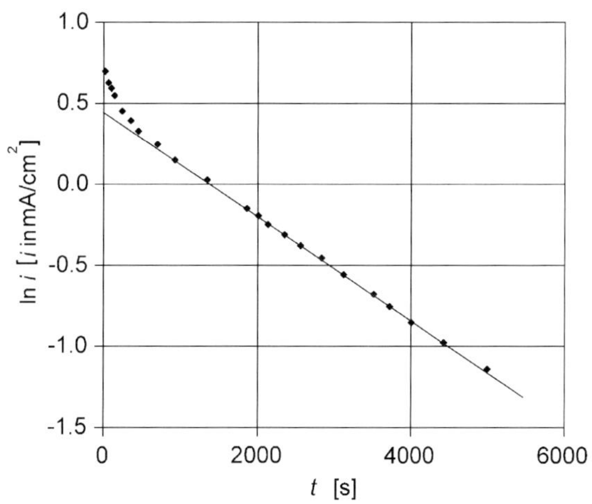

Figure 5.18 Potentiostatic pulses of lithium deposition into an aluminum alloy electrode; ln $i \approx t$ (limiting law for $t >> r^2/D$, after John Wen et al.), radius of the aluminum wire $r = 0.9$ mm, and charge passed during pulse $Q = 7.47$ C, 415 °C. (Reproduced with permission from Ref. [13], © 1979, The Electrochemical Society.)

For longer times $t >> r^2/D$ one obtains for the current density

$$i(t) = \frac{2F(c_{n+1} - c_n)D}{r} \exp\left(-\frac{\pi^2 Dt}{4r^2}\right) \tag{5.31}$$

The logarithm of the current density depends linearly on time. The example of lithium diffusion in aluminum is shown in Figure 5.18. The diffusion coefficient of lithium can be calculated from the intercept (by substituting the concentration difference by Q) or from the slope of the linear dependence (even without the knowledge of Q).

5.3.2 Galvanostatic method

The experiment was also made under galvanostatic conditions.[14] This is called the galvanostatic intermittent titration technique (GITT).

Lithium is deposited by constant current pulses. The concentration of lithium in the alloy before the application of a current pulse is c_n. The constant current creates a constant concentration gradient on the electrode surface. A concentration wave expands into the alloy electrode and raises the lithium concentration to the final value c_{n+1}.

The boundary conditions to solve the time-dependent diffusion equation (Eq. 5.13) are

$$c(x,t=0) = c_n \tag{5.32a}$$

$$\left(\frac{\partial c}{\partial x}\right)_{x=0,t>0} = -\frac{i}{nFD} \tag{5.32b}$$

$$\left(\frac{\partial c}{\partial x}\right)_{x=r,t\geq 0} = 0 \tag{5.32c}$$

Two limiting equations for the concentration–time dependence on the interface between melt and alloy electrode are obtained. For shorter times than for the characteristic diffusion time ($t << r^2/\mathrm{D}$) the Sand equation is obtained which in this case is

$$c(x=0,t) - c_n = \frac{|i|\sqrt{t}}{0.5F\sqrt{\pi D}} \tag{5.33}$$

For the potential, short times, and small concentration changes ($c_{x=0}/c_n \approx 1$), the potential changes with the square root of time according to the equation

$$E_t - E_n = -\frac{RT}{F}\frac{|i|\sqrt{t}}{0.5Fc_n\sqrt{\pi D}} \tag{5.34}$$

The change versus a LiAl reference electrode is shown in Figure 5.19. The slope can be used to determine D.

The second limiting law is obtained for times larger than the characteristic diffusion time ($t > r^2/\mathrm{D}$). The concentration change at the surface is then proportional to the time

$$c(x=0,t) - c_n = \frac{|i| \cdot t}{Fr} + \frac{|i| \cdot r}{3FD} \tag{5.35}$$

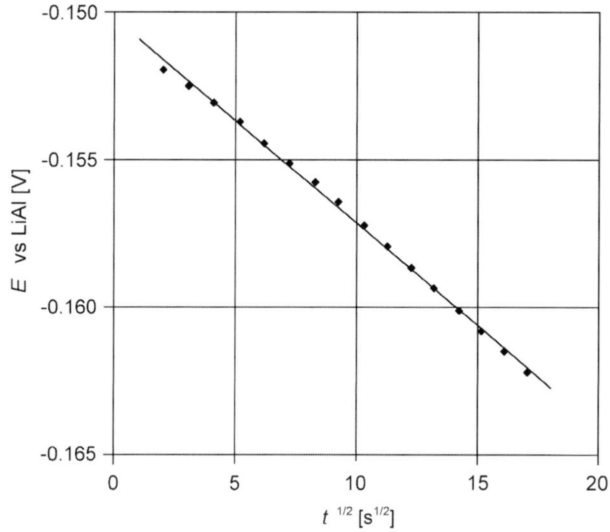

Figure 5.19 Galvanostatic pulses of lithium deposition into an aluminum alloy electrode, $E \approx \sqrt{t}$ (limiting law for $t << r^2/D$ (Sand equation), after John Wen et al.), $i = 14$ mA cm^{-2}, $r = 0.9$ mm, 415 °C. (Reproduced with permission from Ref. [13], © 1979, The Electrochemical Society.)

For the potential, times $t > r^2/D$, and small concentration changes ($c_{x=0}/c_n \approx 1$), the potential changes proportionally to time

$$E_t - E_n = -\frac{RT}{F}\left[\frac{|i| \cdot t}{Frc_n} + \frac{|i| \cdot r}{3FDc_n}\right] \tag{5.36}$$

The linear dependence of E on t is shown in Figure 5.20. The intercept with the potential axis for $t = 0$ can be used to determine D.

5.4 METHODS TO CONTROL DIFFUSION OVERPOTENTIAL

Limitation by diffusion is one of the main obstacles in an electrochemical process. Therefore one has to limit its influence and increase the limiting current density. Eq. (5.5) provides two possibilities for increasing the limiting current density, which are by an increase of concentration or by a decrease of the thickness of the diffusion layer δ_d. The principal method to decrease the thickness of the diffusion layer is convection. The simplest way to generate convection in an electrochemical cell is stirring. Another way is pumping the electrolyte parallel to the electrode or against the electrode. An example is an electrochemical plate or an electrochemical rod in an electrolyte stream. The development of a diffusion layer is shown in Figure 5.21. The thickness of the diffusion layer increases along the plate.

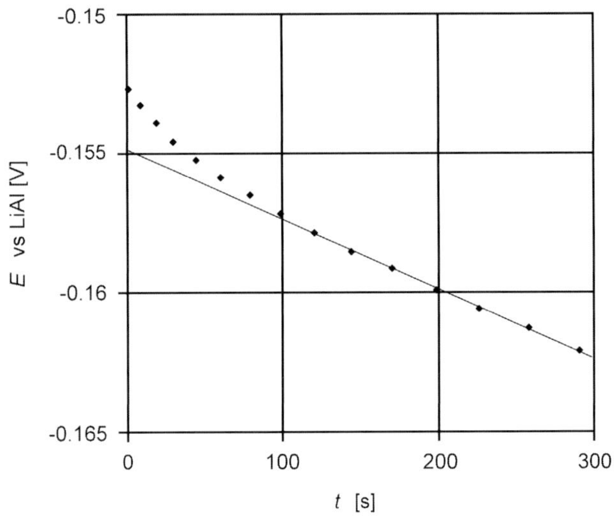

Figure 5.20 Galvanostatic pulses of lithium deposition into an aluminum alloy electrode. $E \approx t$ (limiting law for $t > r^2/D$, after John Wen et al.), $i = 14\ \text{mA cm}^{-2}$, $r = 0.9$ mm, 415 °C. (Reproduced with permission from Ref. [13], © 1979, The Electrochemical Society.)

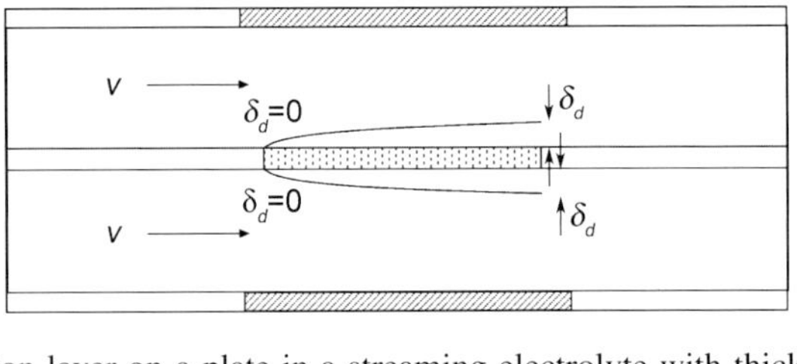

Figure 5.21 Diffusion layer on a plate in a streaming electrolyte with thickness increasing from $\delta_d = 0$ to δ_d.

5.4.1 Rotating-disc electrode

Another way to generate convection is a periodic movement of the electrode. Oscillating electrodes or rotating electrodes are examples. The most effective and reproducible way of controlling the diffusion layer thickness is the rotating-disc electrode.[15,16] The forces acting on the electrolyte are shown in Figure 5.22. The centrifugal forces of the rotation lead to a constant increase of the electrolyte velocity towards the edge of the disc. The diffusion layer thickness tends to decrease. Otherwise, the diffusion front spreads into the electrolyte with increasing time. The two effects are approximately compensated by a diffusion profile of constant thickness over the rotating disc.

The thickness of the diffusion layer on a rotating-disc electrode is given by the equation

$$\delta = 1.61 \frac{D^{1/3} \nu^{1/6}}{\omega^{1/2}} \tag{5.37}$$

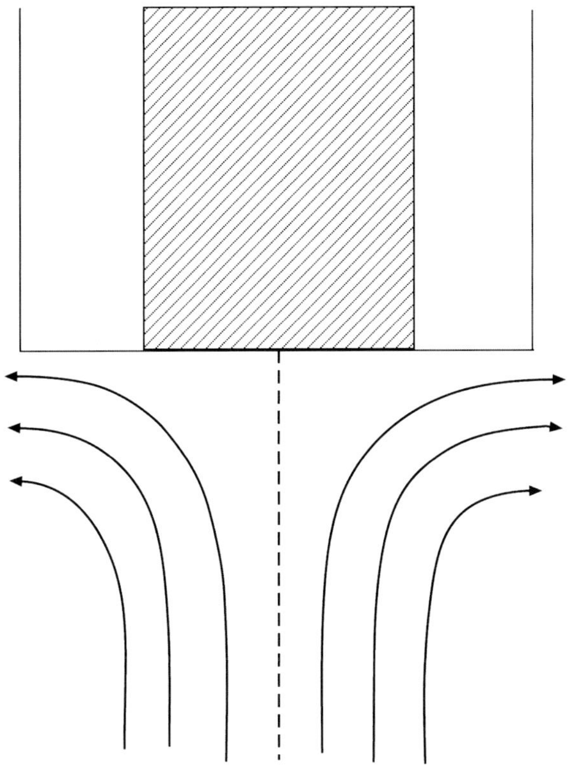

Figure 5.22 Rotating-disc electrode and radial flow in front of the disc.

D is the diffusion coefficient, $\nu = \eta/\rho$ the kinematic viscosity, the ratio between viscosity and density, and $\omega = 2\pi f$ the rotation velocity with f the rotation frequency. With typical values for the constants in Eq. (5.37) the thickness of the diffusion layer is of the order of 10^{-3} cm.

Inserting Eq. (5.37) into Eq. (5.5) gives the equation for the current density on a rotating-disc electrode

$$i = -0.62 \cdot nF \frac{D^{2/3} \omega^{1/2} (c_0 - c_s)}{\nu^{1/6}} \tag{5.38}$$

According to this equation the limiting current obtained for $c_s = 0$ is proportional to the square root of rotation rate. This is confirmed by experiments as shown in Figure 5.23.

The equations derived for the rotating-disc electrode are limited for laminar flow. The Reynolds number was introduced as a measure for the transition to turbulent flow (Section 5.1). For the rotating disc the Reynolds number is given by the equation

$$Re = \frac{\omega \cdot r}{\nu} \tag{5.39}$$

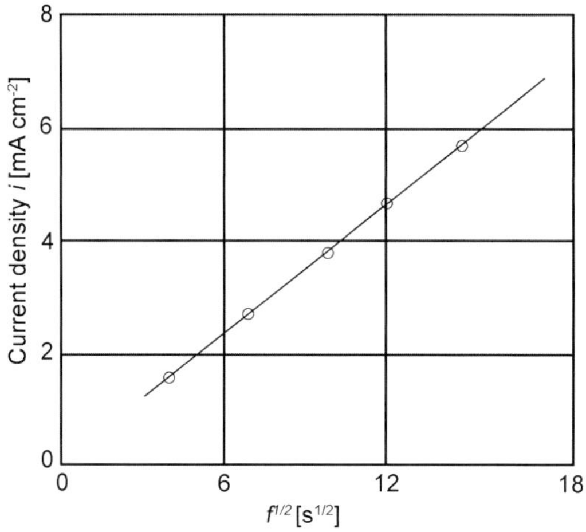

Figure 5.23 Dependence of the limiting current on the square root of the rotation frequency (rotation per s) on a rotating-disc electrode.

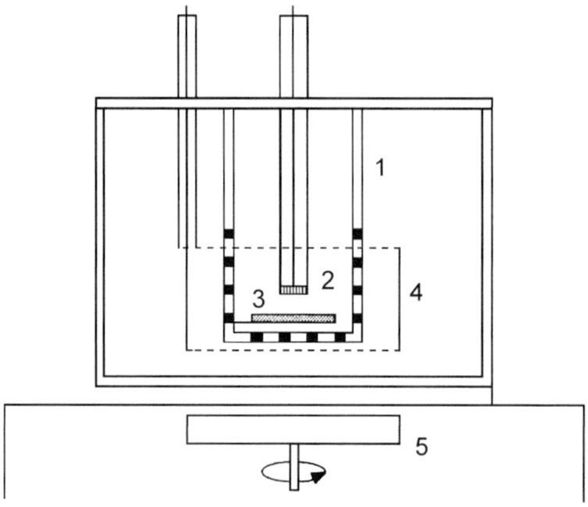

Figure 5.24 Cell for simulation of a rotating electrode by rotation of the electrolyte. A special porous cylinder is necessary for the generation of similar streaming conditions,[17] (1) cylinder for regulating the electrolyte flux, (2) disc electrode, (3) magnetic rod, (4) counter electrode, and (5) magnetic stirrer.

For a rotating-disc electrode with a radius $r = 1$ cm in an aqueous electrolyte (kinematic viscosity of water 0.01 $cm^2\ s^{-1}$ at 20 °C) the critical rotation rate is 10,000 rot min^{-1}.

There is, in principal, an alternative to the rotation of the disc, which is the rotation of the electrolyte. But just rotation of the electrolyte would not lead to similar diffusion conditions because the electrolyte, stopped at the walls of the cell, would develop a different radial flow profile. The generation of the same diffusion layer by rotation of the electrolyte was achieved by a special rotating pumping system as shown in Figure 5.24.[17]

5.4.2 Rotating ring-disc electrodes

The follow-up reactions of an electrode process can be studied with the ring-disc electrode.[18] Substances generated in an electrode process on the disc electrode are transported by the electrolyte convection in front of the rotating electrode to the ring and can be detected on the ring by an electrochemical signal. A ring-disc electrode is shown in Figure 5.25.

The potential of ring and disc is controlled independently. The generation of a substance at the disc is given, for instance, by Eq. (5.4).

The quantitative measure of the reduction is given by the disc current, i_{dsc}. The oxidation products, e.g., in the case of metal deposition molecules of the solvation shell, are transported to the ring. The potential of the ring is set to a potential at which the substance can be detected by an amperometric signal, the ring current i_{rg}. For a quantitative analysis of the ring current it is necessary to know how much disc-generated substance arrives at the ring. The transference number is a measure for this transfer

$$N_{tr} = \frac{i_{rg}}{i_{dsc}} \tag{5.40}$$

Then the ring current can be analyzed according to the rules of amperometry.

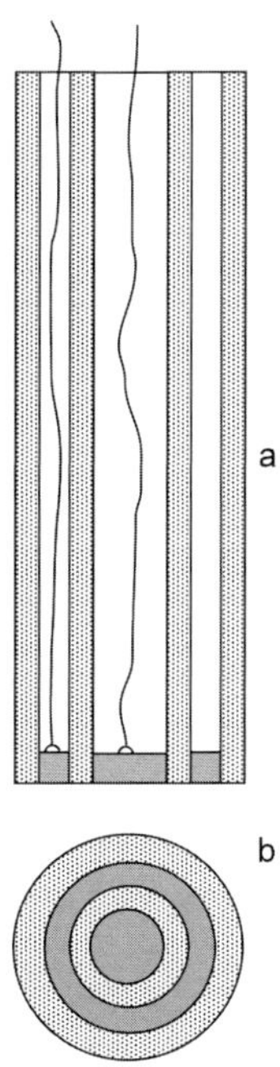

Figure 5.25 Rotating ring-disc electrode. (a) Side view and (b) top view.

5.4.3 Rotating-cylinder electrodes

A very simple method to rotate an electrode is the rotation of an electrode rod or of an electrode cylinder. The calculation of the thickness of the diffusion layer is complicated but, in a first approximation, is proportional to an exponential dependence on rotation speed. An application of this method is the cell developed by Landolt for testing plating electrolytes.[19] The standard equipment for such tests is the Hull cell (Figure 5.26).[20]

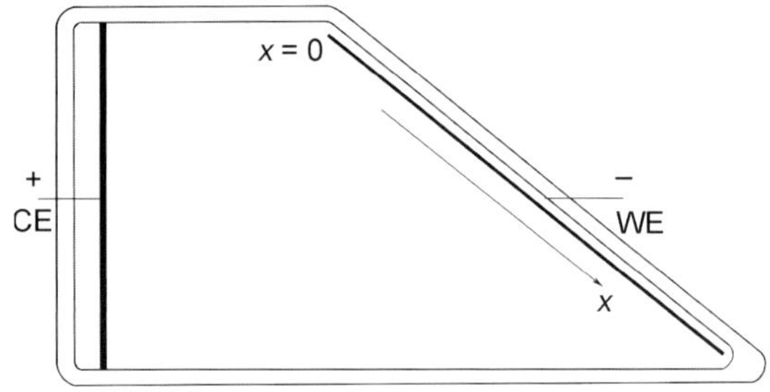

Figure 5.26 A Hull cell used for testing plating electrolytes on the influence of current density (after Kanani). The variation of distance between working and counter electrodes leads to a decrease of current along the *x*-axis in front of the working electrode, depending on the electrolyte resistance. (Reproduced with permission from Ref. [20], © 2000, Hanser Verlag.)

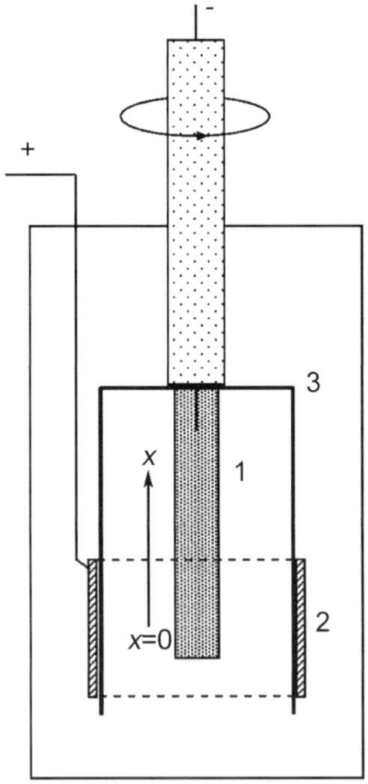

Figure 5.27 A cell for testing plating electrolytes with controlled diffusion conditions (after Landolt[16]). (1) Rotating rod (test-electrode), (2) counter electrode, and (3) shielding cylinder.

The geometry of a Hull cell enables a fast testing of the influence of current density on the plating process. What is not controlled in a Hull cell is the influence of convection of the electrolyte on the plating process. In practice paddles are used for some convection. The reproducibility of this equipment is limited.

The Landolt cell (Figure 5.27) provides controlled convection conditions by the rotating cylinder electrode. A cylindrical shield surrounds a rotating cylinder, preventing direct current flux between rotating cylinder and counter electrode. The enforced longer way of the current lines simulates the variation of current density in a conventional Hull cell.

The technical realization of such a cell is the ROTA cell described by Tannenberger.[21]

REFERENCES

1. W. Nernst, Z. Physikal. Chem., *47*, 52 (1904).
2. E. Brunner, Z. Physikal. Chem., *47*, 56 (1904).
3. O. Reynolds, Papers on mechanical and physical subjects, Cambridge, 1903.
4. G. Doetsch, Laplace Transformation, Dover, New York, 1953.
5. H.J. Sand, Phil. Mag., *1*, 45 (1900); Z. physikal. Chem., *35*, 641 (1900).
6. F.G. Cottrell, Z. physikal. Chem., *42*, 385 (1903).
7. E. Warburg, Ann. Phys. Chem., *67*, 493 (1899).
8. J.R. MacDonald (Ed.), Impedance Spectroscopy, Wiley, New York, 1987.
9. M. Sluyters-Rehbach, J. Sluyters, in A. Bard (Ed.), Electroanal. Chem., *4*, 1 (1970).
10. J.E.B. Randles, Disc. Faraday Soc., *1*, 11 (1947).
11. (a) J.E.B. Randles, Trans Faraday Soc., *44*, 327 (1948); (b) A. Sevcik, Coll. Czech. Chem. Commun., *13*, 349 (1948).
12. W. Weppner, R.A. Huggins, Ann. Rev. Met. Sci., *8*, 269 (1978).
13. C. John Wen, B.A. Boukamp, R.A. Hugguns, W. Weppner, J. Electrochem. Soc., *126*, 2258 (1979).
14. W. Weppner, R.A. Huggins, J. Solid State Chem., *22*, 297 (1977); J. Electrochem. Soc., *124*,1569 (1977).
15. V.G. Levich, Physicochemical Hydrodynamics, Prentice Hall, Englewood Cliffs, NJ, 1962.
16. Yu.V. Pleskov, V.Yu. Filinowski, The Rotating Disc Electrode, Consultants Bureau, New York, 1976.
17. W. Plieth, German Patent Application P 39 38 027.0-52, 1989.
18. W.J. Albery, M.L. Hitchmann, Ring-Disc Electrodes, Clarendon Press, Oxford, 1971.
19. C. Madore, A.C. West, M. Matlosz, D. Landolt, Electrochim. Acta, *37*, 69 (1992).
20. N. Kanani, Galvanotechnik, Carl Hanser-Verlag, München, Wien, 2000.
21. H. Tannenberger, Galvanotechnik, *93*, 646 (2002).

– 6 –

Charge Transfer

6.1 ELECTRON TRANSFER

An electrochemical reaction with electron transfer between electrode and electrolyte is called a redox reaction. The simplest case is described by the equation

$$S_{red} \cdot mH_2O \leftrightarrow S_{ox} \cdot nH_2O + e^- \tag{6.1}$$

This is a one-valent electrode reaction. The stoichiometric number of the electron is one, which is usually assumed for the elementary charge transfer step. Reduced and oxidized components have different solvation numbers. The general kinetic equations will be developed for this reaction.

In a more general description one can define an electrode reaction with n electrons, whereupon either the elementary electrode reaction occurs n times or there is a more complex situation of multi-step electrode reactions for which there is much treatment in the literature.[1]

In the usual manner of chemical kinetics the reaction rates r_{ox} of the forward (oxidation) and r_{red} backward (reduction) reactions are defined as

$$r_{ox} = k_{ox}(E) \cdot c_{red} \tag{6.2}$$

$$r_{red} = k_{red}(E) \cdot c_{ox} \tag{6.3}$$

The rate constants $k_{ox}(E)$ and $k_{red}(E)$ are functions of the potential E.

The reaction rates can be expressed as current densities, an anodic current density for oxidation

$$i_a = Fr_{ox} \tag{6.4}$$

and a cathodic current density for reduction

$$i_c = -Fr_{red} \tag{6.5}$$

The total current density is the sum of the partial current densities

$$i = i_a + i_c \tag{6.6}$$

This provides the unique possibility to investigate charge transfer reactions by current measurements.

Eq. (6.1) describes the simplest case of a redox reaction. An example could be the electron transfer between ferric/ferrous ions and an electrode

$$Fe^{2+}_{solv} \leftrightarrow Fe^{3+}_{solv} + e^- \tag{6.7}$$

The electrode reaction is identical with the elementary charge transfer step in this example. The theory of electron transfer was developed for this reaction.

6.1.1 Butler–Volmer equation

For the electrode reaction Eq. (6.1) equations for the partial current densities of oxidation and reduction can be derived by formulating the potential dependence of the rate constants. The now generally accepted formulation goes back to work of Erdey-Gruz and Volmer[2] and Butler.[3]

The equation for the anodic (oxidation, positive) current is

$$i_a = Fk_a c_{red} e^{+\frac{\alpha_a F}{RT}E} \tag{6.8}$$

The equation for the cathodic (reduction, negative) current is

$$i_c = -Fk_c c_{ox} e^{-\frac{\alpha_c F}{RT}E} \tag{6.9}$$

with

$$\alpha_a + \alpha_c = 1 \tag{6.10}$$

The constants k_a and k_c are the rate constants in Eqs. (6.2) and (6.3) at the electrode potential $E = 0$, which is related to the chosen reference electrode. The potential dependence of the partial current densities is described by a splitting of the total potential dependence in an anodic part $\alpha_a E$ and a cathodic part $\alpha_c E = (1 - \alpha_a)E$. The factors α_a and α_c are called charge transfer coefficients.

The total current density is given by the sum of the cathodic and anodic term

$$i = Fk_a c_{red} e^{+\frac{\alpha_a F}{RT}E} - Fk_c c_{ox} e^{-\frac{(1-\alpha_a)F}{RT}E} \tag{6.11}$$

The potential dependence of the partial current densities and of the total current density is shown in Figure 6.1.

One problem with the electrochemical literature is that the definition of the charge transfer coefficient α_a is not universal. Some groups prefer to use α_c for the cathodic reaction and $1 - \alpha_c$ for the anodic reaction. Other symbols (preferentially β) can be found. Therefore, it is strictly recommended to show by an index a or c which definition is applied. Moreover, the connection of anodic potential and current with the positive axis and the cathodic current with the negative axis is not universal. Therefore, a strict use of the IUPAC recommendations is recommended.[4,5]

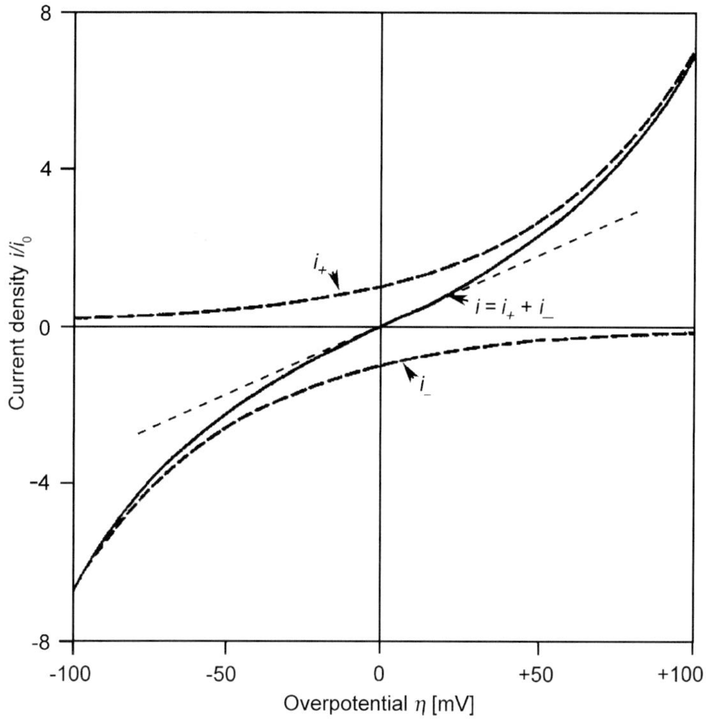

Figure 6.1 Potential dependence of the partial current densities, Eqs. (6.8) and (6.9), and of the total current density (Eq. (6.11)). The dashed, straight line represents the region of linear relation between current and potential.

Introducing the exchange current density i_0

$$\left|i_0\right| = Fk_a\bar{c}_{red}e^{+\frac{\alpha_a F}{RT}E_0} = Fk_c\bar{c}_{ox}e^{-\frac{(1-\alpha_a)F}{RT}E_0} \tag{6.12}$$

(E_0 is the Nernst equilibrium potential at zero current density, $\bar{c}_{red}$ and $\bar{c}_{ox}$ denote the bulk concentration) and the overpotential

$$\eta = E - E_0 \tag{6.13}$$

one can derive the following equation:

$$i = i_0\left[\frac{c_{red}}{\bar{c}_{red}}e^{+\frac{\alpha_a F}{RT}\eta} - \frac{c_{ox}}{\bar{c}_{ox}}e^{-\frac{(1-\alpha_a)F}{RT}\eta}\right] \tag{6.14}$$

If the surface concentrations are approximately equal to the bulk concentrations ($c \cong \bar{c}$) one obtains

$$i = i_0\left[e^{+\frac{\alpha_a F}{RT}\eta} - e^{-\frac{(1-\alpha_a)F}{RT}\eta}\right] \tag{6.15}$$

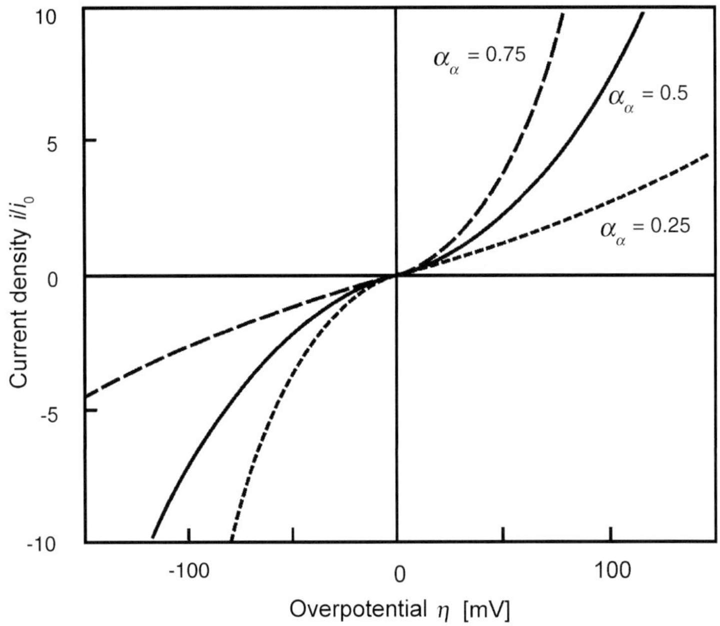

Figure 6.2 Current-potential dependence for a redox system following the Butler–Volmer equation for charge transfer coefficients $\alpha_a = 0.25$, $\alpha_a = 0.5$, $\alpha_a = 0.75$.

The dependence of the current density on the potential for different charge transfer coefficients is shown in Figure 6.2.

The exchange current density still depends on the concentration of the redox components. A constant, describing the rate of a redox reaction not dependent on concentration, is defined in Eq. (6.16) and could be recommended for comparisons and tabulation

$$k_0 = \frac{i_0}{F\bar{c}_{\mathrm{red}}^{1-\alpha_a}\bar{c}_{\mathrm{ox}}^{\alpha_a}} \tag{6.16}$$

The concentrations can be replaced by activities

$$k_0^{\ominus} = \frac{i_0}{F\bar{a}_{\mathrm{red}}^{1-\alpha_a}\bar{a}_{\mathrm{ox}}^{\alpha_a}} \tag{6.17}$$

Therefore, the constant represents the standard value of the rate of an electrochemical reaction.

6.1.2 Tafel lines

For large overpotential Tafel lines can be observed. For large anodic overpotential ($E >> E_0$) one obtains from Eq. (6.8).

$$\ln i_a = \ln Fk_a + \ln c_{red} + \frac{\alpha_a F}{RT} E \tag{6.18}$$

or after introduction of overpotential and exchange current density

$$\ln i_a = \ln i_0 + \frac{\alpha_a F}{RT} \eta \tag{6.19}$$

Similar Tafel lines can be derived for large cathodic overpotential ($E << E_0$):

$$\ln |i_c| = \ln Fk_c + \ln c_{ox} - \frac{(1-\alpha_a)F}{RT} E \tag{6.20}$$

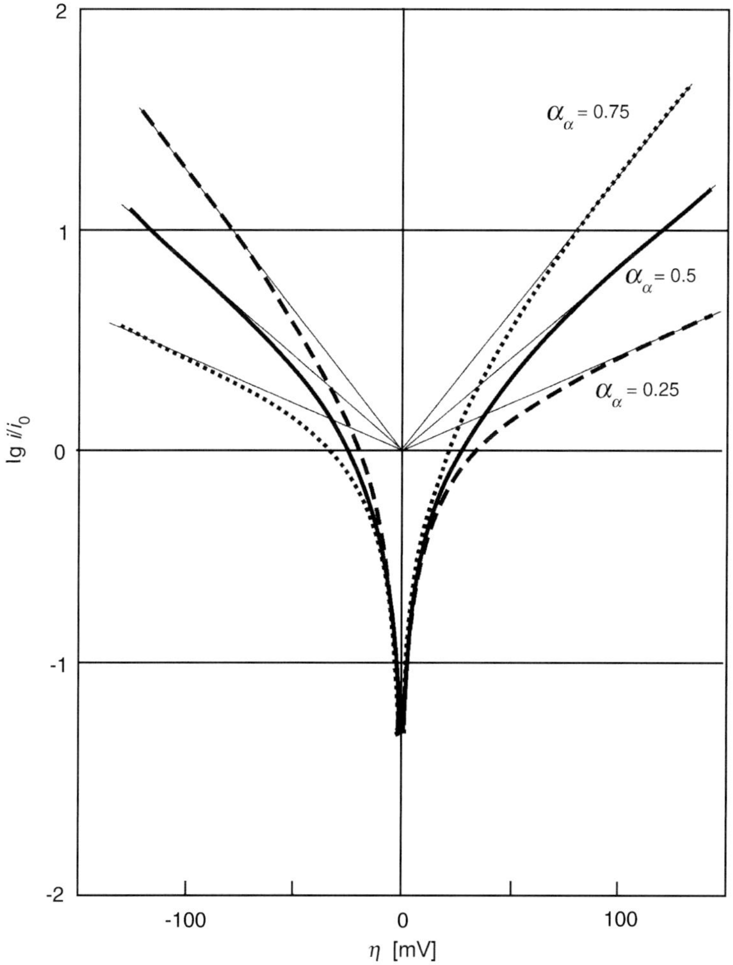

Figure 6.3 Tafel plots for three charge transfer coefficients, $\alpha_a = 0.25$, $\alpha_a = 0.50$, and $\alpha_a = 0.75$, according to Eqs. (6.19) and (6.21).

$$\ln|i_c| = \ln i_0 - \frac{(1-\alpha_a)F}{RT}\eta \tag{6.21}$$

As an example Tafel lines for $\alpha_a = 0.25$, $\alpha_a = 0.50$, and $\alpha_a = 0.75$ are plotted in Figure 6.3. One can determine the charge transfer coefficient α_a from the slope of the Tafel lines and the exchange current density i_0 from an extrapolation to the Nernst potential $E = E_0$.

6.1.3 Charge transfer resistance

For very small overpotential $E \rightarrow E_0$ a linear dependence of the current density on the potential is observed, usually for an overpotential below 10 mV. From the slope of the linear dependence the exchange current density can be determined.

$$\lim_{\eta \to 0} \frac{di}{d\eta} = \frac{F}{RT} i_0 \tag{6.22}$$

The reciprocal value is the charge transfer resistance

$$R_{ct} = \frac{RT}{F} \frac{1}{i_0} \tag{6.23}$$

The influence of the charge transfer resistance on the current-potential curves is shown in Figure 6.4.

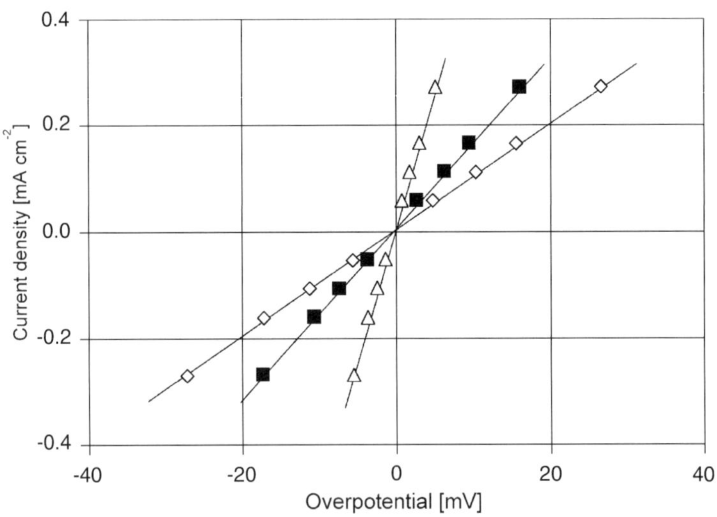

Figure 6.4 Influence of the charge transfer resistance on the current-potential curve. The current-potential curves were measured in the redox system quinon–hydroquinon on a mercury electrode after adding different inhibitors, triangles, no inhibitor, squares, methylene blue, and rhomboids, lauryl acid (0.5 mol·dm^{-3} NaF, 0.05 mol·dm^{-3} NaH_2PO_4/Na_2HPO_4, pH 6.55, and 10^{-4} mol·dm^{-3} inhibitor).[6]

6.1.4 Theories of electron transfer

How does the electron transfer occur in a redox process? One description of this process was developed by Gerischer,[7] based on the former work of Gurney[8] and Essin.[9] Another description goes back to the work of Marcus.[10] Other contributions during the development of the basic theory came from Dogonadze, Levich, Chizmadzhev, Kuznetsov, and others.[11] The model will be described for a simple redox reaction, the oxidation of a two-valent iron ion into a three-valent iron ion and vice versa.

The more phenomenological concept of Gerischer will be explained in the diagram of the electron-energy levels in the electrolyte and in the metal, introduced in Chapter 3 and shown in Figure 6.5. The transfer of an electron into the metal can occur, if the electron on an energy level E_e in the electrolyte finds an unoccupied energy level at the same energy E_e in the metal. The intensity of the electron transfer is proportional to the density of occupied electron-energy levels in the electrolyte (the density of reduced ions, D_{red}) times the density of unoccupied energy levels in the metal, N_{un}.

$$i_a(E_e) \approx D_{red}(E_e) \cdot N_{un}(E_e) \tag{6.24}$$

Otherwise, the transfer of an electron from the metal into the electrolyte can occur, if the electron on an energy level E_e in the metal finds an unoccupied energy level at the same energy E_e in the electrolyte. The intensity of electron transfer is proportional to the density of unoccupied electron-energy levels in the metal, N_{un} times the density of unoccupied energy levels in the electrolyte (the density of oxidized ions, D_{ox}).

$$i_c(E_e) \approx -N_{oc}(E_e) \cdot D_{ox}(E_e) \tag{6.25}$$

The total partial current densities are obtained by integration over the energy.

$$i_a = k_a \int_{-\infty}^{+\infty} D_{red}(E_e) \cdot N_{un}(E_e) \cdot dE_e \tag{6.26}$$

$$i_c = -k_c \int_{-\infty}^{+\infty} N_{oc}(E_e) \cdot D_{ox}(E_e) \cdot dE_e \tag{6.27}$$

The maximum overlapping of occupied and unoccupied energy levels will be found at the Fermi energy. At energies below and above the Fermi energy the density of overlapping occupied and unoccupied energy levels approaches zero. Therefore, the highest current densities are found around the Fermi energy. This is shown in Figure 6.5. In Figure 6.5a, the situation at the Nernst potential (electrode potential $E = E_0$) is shown. Anodic and cathodic current densities are equal; the total current density is zero.

In Figure 6.5b the situation is shown, if an anodic potential is applied (electrode potential $E > E_0$). The Fermi level of the metal is shifted in a direction that favorites the transfer of electrons into the metal and a total anodic current is observed.

In Figure 6.5c the situation is shown, if a cathodic potential is applied (electrode potential $E < E_0$). The Fermi level of the metal is shifted in a direction that favors the

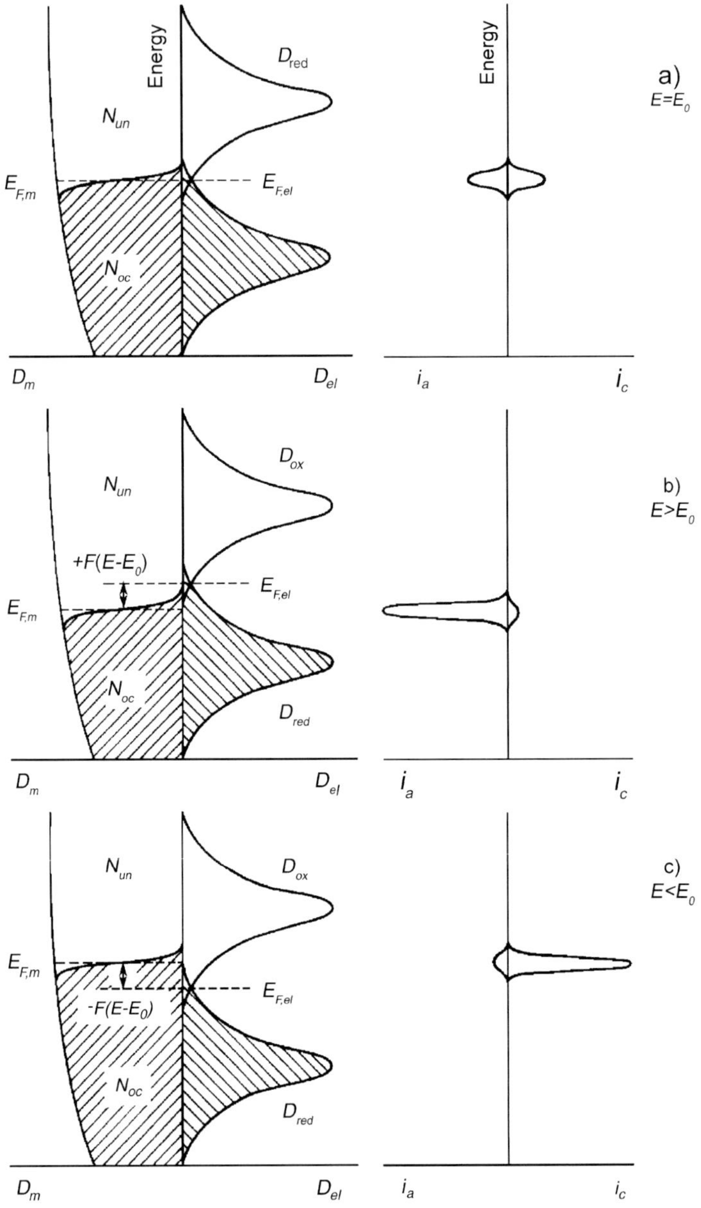

Figure 6.5 Electron transfer at equilibrium potential and for an anodic and a cathodic overpotential. (a) Equilibrium conditions (electrode potential $E = E_0$), anodic partial current equal to cathodic partial current ($i_a = i_c$), total current zero, (b) anodic polarization (electrode potential $E > E_0$), anodic partial current larger than cathodic partial current ($i_a > i_c$), total anodic current, and (c) cathodic polarization (electrode potential $E < E_0$), anodic partial current smaller than cathodic partial current ($i_a < i_c$), and total cathodic current.

transfer of electrons into the electrolyte with the result that a total cathodic current is observed.

The above equation is a simple kinetic approach. The total rate of the process is given by the rate constants k_a and k_c. These rate constants are determined by the activation energies. What is the activation energy of a process, mainly determined by electron tunneling around the Fermi energy?

The transfer mechanism is the following one. The density of electronic states in the electrolyte is represented by the distribution functions of the solvation states of the oxidized and reduced ions, D_{red} and D_{ox}, respectively. These distribution functions are of Gaussian type with maxima representing the mean electron-energy state. Because the maximum of electron transfer occurs near the Fermi energy, the solvation shell of an ion in the electrolyte must reorganize until the electron energy is equal to the Fermi energy level. Then the electron transfer can occur. The differences between Fermi energy and energy at maximum of the distribution functions are called reorganization energies, λ_{red} and λ_{ox}, respectively. After the electron transfer, the solvation shell must relax until the maximum of the new oxidation state is reached. The definition of the reorganization energies is explained in Figure 6.6

For the calculation of the activation energies, one can define the energy of the solvation complexes as functions of the mean distance between solvation shell and ion. These functions for oxidized and reduced ions are schematically shown in Figure 6.7. The distance

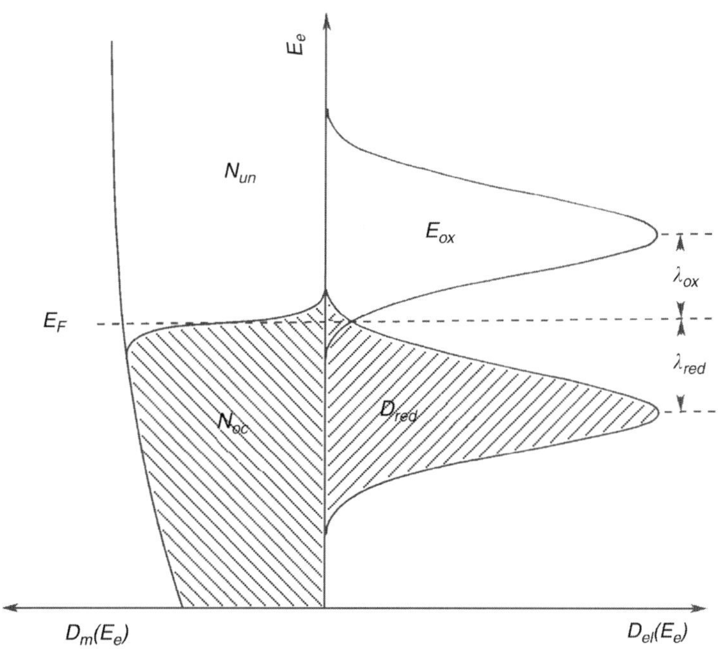

Figure 6.6 Mechanism of heterogeneous electron transfer and definition of the reorganization energy.

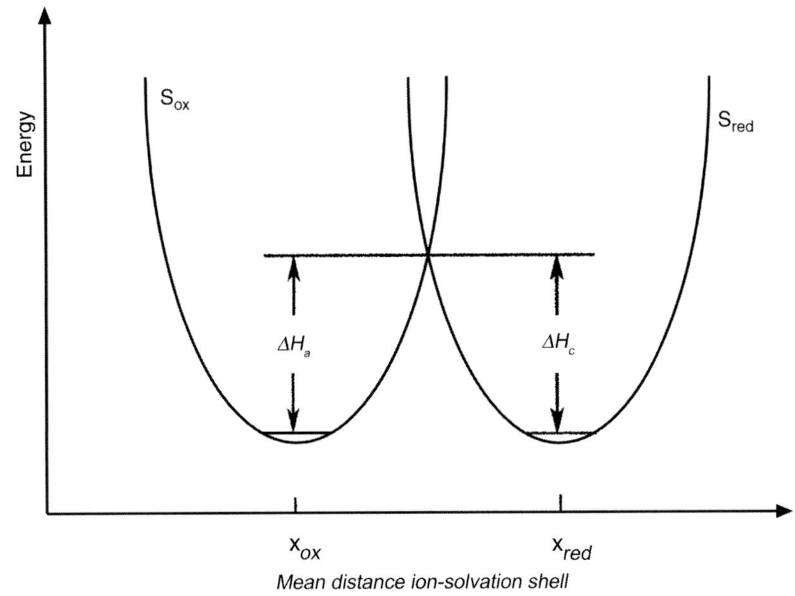

Figure 6.7 Mechanism of heterogeneous electron transfer and definition of the activation energy.

between the point where both functions cross each other and the minimum of both curves is representative for the activation energy.

Marcus developed expressions for the calculation of the activation energy and its potential dependence. From the new concept equations approximately reflecting the Butler–Volmer equation could be derived.[10]

6.2 ELECTROCHEMICAL REACTION ORDERS

Eq. (6.1) described the simplest case of a redox process. In a more general manner an electrode reaction can be formulated by the equation

$$\sum_i v_i \mathrm{S}_i \leftrightarrow \sum_j v_j \mathrm{S}_j + ne^- \tag{6.28}$$

An example of a more complex redox reaction is the oxidation of nitrous acid to nitric acid and vice versa in acidic solution. This is described by the equation

$$\mathrm{HNO_2} + \mathrm{H_2O} \leftrightarrow \mathrm{NO_3^-} + 3\mathrm{H^+} + 2e^- \tag{6.29}$$

In this case electron transfer reaction and net redox reaction will not be the same. To elaborate the mechanism of such complex reaction, it is necessary to find the elementary electron transfer reaction.

In this chapter after description of the charge transfer process, the methods of analyzing a more complex mechanism will be described. How should the more complex electrode reaction shown in Eq. (6.28) be treated? The usual model for this treatment is the separation into an elementary charge transfer reaction like reaction Eq. (6.1) and pre- and post-electron transfer reactions. Usually, the elementary charge transfer reaction is a one-electron reaction. One can write the total transfer reaction as a one-electron reaction.

$$\sum_i \frac{\nu_i}{n} S_i \leftrightarrow \sum_j \frac{\nu_j}{n} S_j + e^- \tag{6.30}$$

For equilibrium conditions, the substances S_{red} and S_{ox} in the elementary charge transfer reaction are connected with the substances S_i and S_j in the electrode reaction by pre- and post-electron transfer equilibria. The concentrations c_{red} and c_{ox} in the elementary charge transfer reaction can be substituted by the following equations:

$$c_{red} = K_{red}(E) \cdot \prod_i c_i^{n_{red,i}} \tag{6.31}$$

$$c_{ox} = K_{ox}(E) \cdot \prod_j c_j^{n_{ox,j}} \tag{6.32}$$

The exponents $n_{red,i}$ and $n_{ox,j}$ are the partial electrochemical reaction orders.

The elementary charge transfer reaction is described by the Butler–Volmer equation. The concentrations c_{red} and c_{ox} are substituted by Eqs. (6.31) and (6.32).

$$i = Fk_a^* \prod_i c_i^{n_{red,i}} e^{+\frac{\alpha_a F}{RT}E} - Fk_c^* \prod_j c_j^{n_{ox,j}} e^{-\frac{(1-\alpha_a)F}{RT}E} \tag{6.33}$$

The new rate constants k_a^* and k_c^* are the products of the former rate constants (Eqs. (6.8) and (6.9)) and the equilibrium constants in Eqs. (6.3.1) and (6.3.2).

6.2.1 Determination of electrochemical reaction orders from Tafel lines

The Tafel lines derived from this equation are

$$\ln i_a = \ln Fk_a + \sum_i n_{red,i} \ln c_i + \frac{\alpha_a F}{RT} E \tag{6.34}$$

$$\ln |i_c| = \ln Fk_c + \sum_j n_{ox,j} \ln c_j - \frac{(1-\alpha_a)F}{RT} E \tag{6.35}$$

Measurement of the Tafel lines for different concentrations, the partial electrochemical reaction orders, can be determined from Eqs. (6.34) and (6.35). This implies that the concentration at the electrode surface remains approximately equal to the bulk concentration (elimination of the diffusion overpotential). It should be pointed out that for this evaluation the logarithm of current must be plotted versus the *electrode potential* and not versus the overpotential.

As an example, results for the nitrous acid–nitric acid redox system are plotted in Figure 6.8.[12]

The electrochemical reaction orders for the anodic process are determined from the concentration dependence of lg i_a (Figure 6.9); the electrochemical reaction orders for the cathodic process are determined from the concentration dependence of lg i_c (Figure 6.10).

The charge transfer coefficients determined from the slope of the Tafel lines were $\alpha_a = 0.66$ and $\alpha_c = 1 - \alpha_a = 0.36$.

6.2.2 Determination of electrochemical reaction orders from the charge transfer resistance

The equations for the exchange current density of the complex redox reaction are

$$|i_0| = Fk_a \prod_i c_i^{n_{\mathrm{red},i}} \mathrm{e}^{+\frac{\alpha F}{RT}E_0} = Fk_c \prod_j c_j^{n_{\mathrm{ox},j}} \mathrm{e}^{-\frac{(1-\alpha)F}{RT}E_0} \tag{6.36}$$

From these equations one can derive equations for the concentration dependence of the exchange current density. The Nernst potential E_0 is substituted by concentrations according to Eq. (6.30). Setting concentrations approximately equal to activities one obtains

$$\left(\frac{\partial \lg i_0}{\partial \lg c_i}\right)_{k \neq i} = n_{\mathrm{red},\,i} + \frac{\nu_i}{n}\alpha_a \tag{6.37a}$$

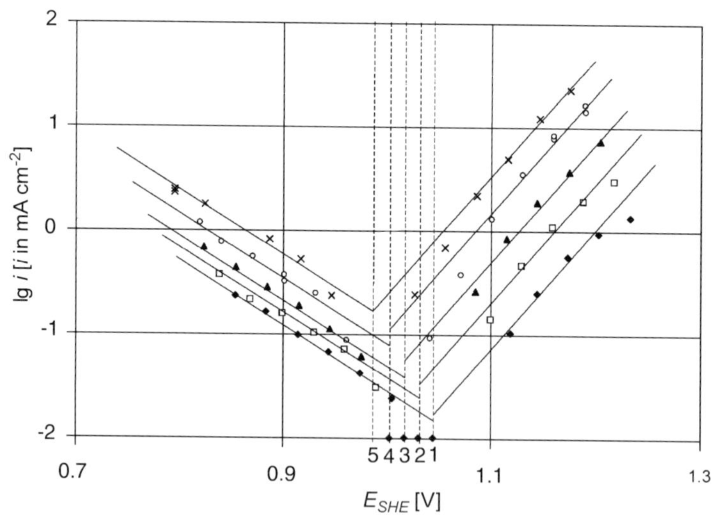

Figure 6.8 Tafel plots of lg i versus electrode potential E, nitrous acid–nitric acid redox system, 2 mol · dm^{-3} HNO_3, and dependence on HNO_2-concentration. (1) 3.05×10^{-4} mol · dm^{-3}, (2) 1.01×10^{-3} mol·dm^{-3}, (3) 2.9×10^{-3} mol · dm^{-3}, and (4) 8.8×10^{-3} mol · dm^{-3}, 2.6×10^{-2} mol · dm^{-3}, smooth Pt electrode, T = 25 °C. The diffusion overpotential was eliminated by chronoamperometric measurements (Section 6.4.2).

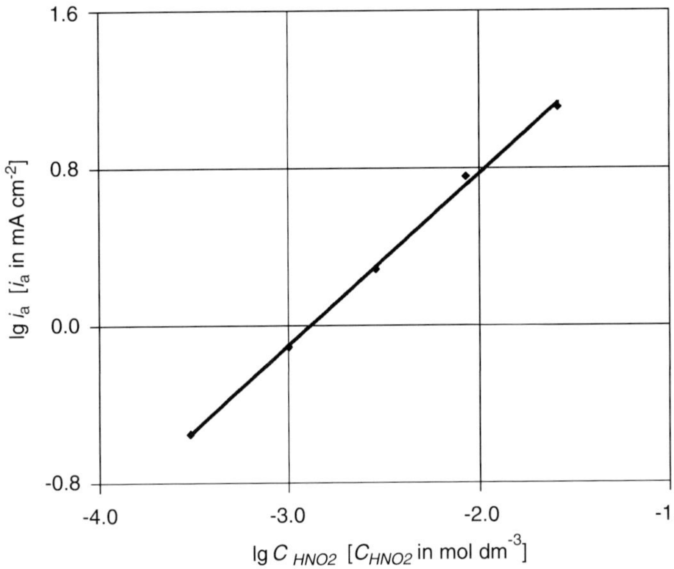

Figure 6.9 Plot of lg i_a versus lg c_{HNO_2} and determination of the electrochemical reaction order $n_{red,HNO_2} = 0.87 \approx 1$.

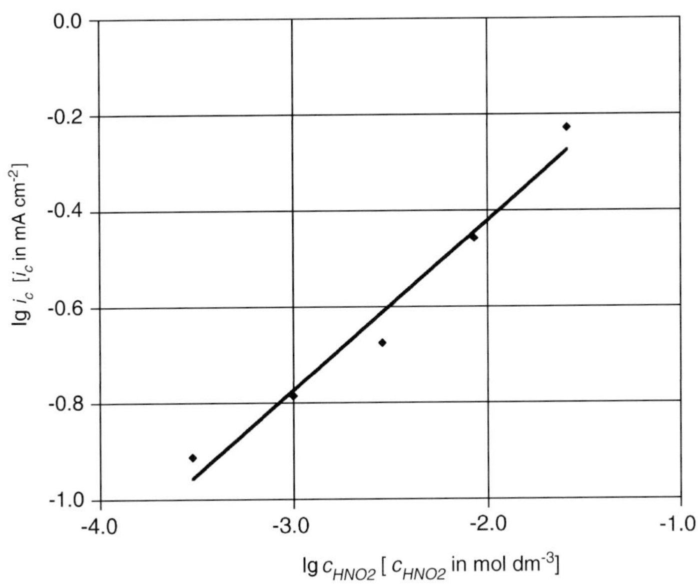

Figure 6.10 Plot of lg i_c versus lg c_{HNO_2} and determination of the electrochemical reaction order $n_{ox,HNO_2} = 0.35 \approx 1/2$.

$$\left(\frac{\partial \lg i_0}{\partial \lg c_j}\right)_{k \neq j} = n_{\text{ox},j} - \frac{\nu_j}{n}(1-\alpha_a) \tag{6.37b}$$

Applications of this equation to the example of the nitrous acid–nitric acid redox system are shown in Figures 6.11, 6.12, and 6.13.

With the charge transfer coefficient determined from the Tafel plots (Figure 6. 8) and the concentration dependence of the exchange current densities, the following partial electrochemical reaction orders were determined

$$n_{\text{red,HNO}_2} \approx +1 \qquad n_{\text{ox,HNO}_2} \approx +1/2$$

$$n_{\text{red,NO}_3^-} = 0 \qquad n_{\text{ox,NO}_3^-} \approx +1/2$$

$$n_{\text{red,H}^+} \approx -1 \qquad n_{\text{ox,H}^+} \approx +1/2$$

With these data, the following pre- and post-charge transfer equilibria could be set up

$$C_{\text{red}} = K_{\text{red}} c^{1}_{\text{HNO}_2} c^{-1}_{\text{H}^+} \tag{6.38}$$

$$C_{\text{ox}} = K_{\text{ox}} c^{1/2}_{\text{HNO}_2} c^{1/2}_{\text{H}^+} c^{1/2}_{\text{NO}_3^-} \tag{6.39}$$

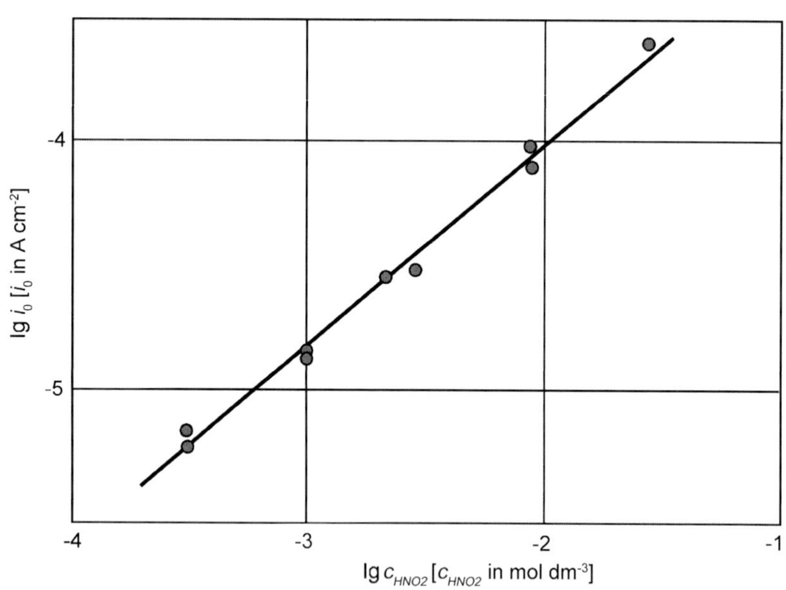

Figure 6.11 Dependence of the exchange current density on the concentration of nitrous acid, concentration of HNO_3 2 mol · dm^{-3}, smooth Pt electrode, 25 °C, $\partial \lg i_0 / \partial \lg c_{\text{HNO}_2} = 0.78$.

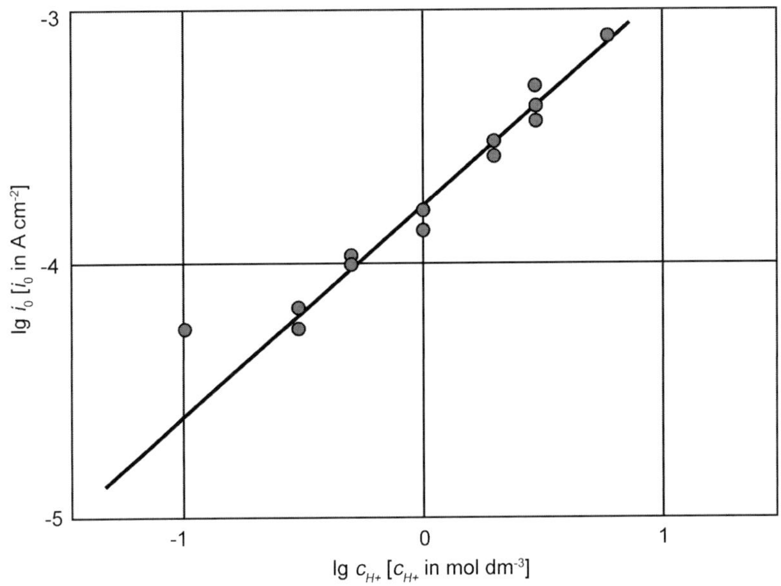

Figure 6.12 Dependence of the exchange current density on the H^+ concentration, nitrate concentration 6 mol · dm^{-3}, nitrous acid concentration 10^{-2} mol · dm^{-3}, exchange of HNO_3 by KNO_3, smooth Pt electrode, 25 °C, $\partial \lg i_0 / \partial \lg c_{H^+} = 0.82$.

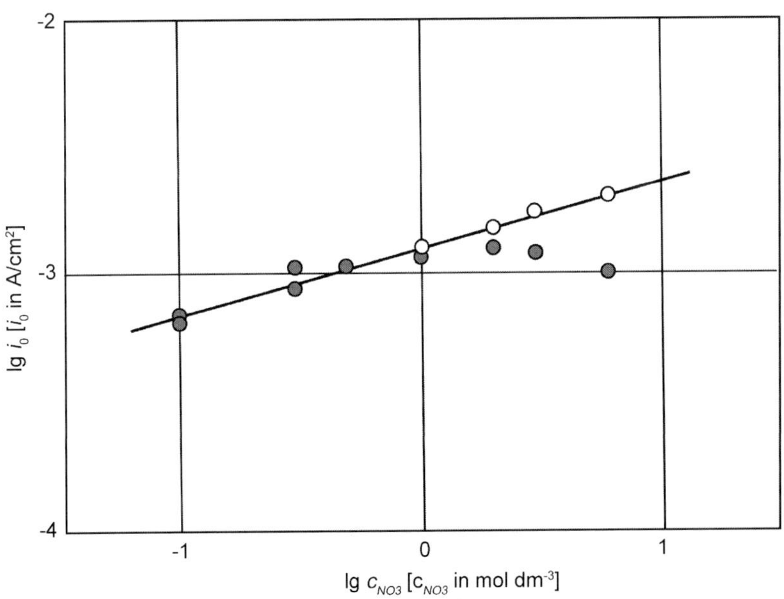

Figure 6.13 Dependence of the exchange current density on the NO_3^- concentration, H^+ concentration 6 mol · dm^{-3}, nitrous acid concentration 10^{-2} mol · dm^{-3}, open circles, H^+ concentrations corrected by mean activity coefficients, exchange of HNO_3 by $HClO_4$, smooth Pt electrode, 25 °C, $\partial \lg i_0 / \partial \lg c_{NO_3^-} = 0.27$.

This gives the charge transfer reaction

$$NO_2^- \leftrightarrow NO_2 + e^- \tag{6.40}$$

and leads to the mechanism

$$HNO_2 \leftrightarrow H^+ + NO_2^- \tag{6.41a}$$

$$NO_2^- \leftrightarrow NO_2 + e^- \tag{6.41b}$$

$$NO_2 + \frac{1}{2}H_2O \leftrightarrow \frac{1}{2}HNO_2 + \frac{1}{2}NO_3^- + \frac{1}{2}H^+ \tag{6.41c}$$

This autocatalytic mechanism explains the autocatalytic action of HNO_2 on the reduction of nitric acid.

The nitrous acid–nitric acid redox system is an example of a controversy in the literature. The above mechanism was developed by Vetter from one of the first applications of the concept of electrochemical reaction orders on a redox system.[13] G. Schmid[14–17] intuitively suggested another mechanism based on a charge transfer reaction

$$NO \leftrightarrow NO^+ + e^- \tag{6.42}$$

The presented experimental results did not confirm this charge transfer reaction for the experimental conditions of high proton activities. However, at lower proton activities a change in the mechanism was observed. Reaction Eq. (6.42) can be considered as a charge transfer reaction for lower proton concentration.[12]

6.3 ION TRANSFER

A second class of electrochemical processes in addition to electron transfer is ion transfer, e.g., metal ion transfer according to the equation

$$Me + \nu_L L \leftrightarrow M^{z+} \cdot \nu_L L + ze^- \tag{6.43}$$

This process describes the reaction of metal atoms with ligands L forming metal ions (dissolution, corrosion) or the deposition of metal atoms by reduction of metal ions. The solvent molecules are a special example of ligands that reacts with the metal ions. A partial step of the total process, ad-atom deposition and dissolution, was already discussed in Chapter 4. The rate equations are similar to Eqs. (6.2) and (6.3).

$$r_{\text{dis}} = k_{\text{dis}}(E) \cdot c_{\text{L}}^{\nu_L} \tag{6.44}$$

$$r_{\text{dep}} = k_{\text{dep}}(E) \cdot c_{\text{Me}^{z+}} \tag{6.45}$$

The current densities must now be expressed for a z-valent reaction ($i_{\text{dis}} = zFr_{\text{dis}}$ and $i_{\text{dep}} = -zFr_{\text{dep}}$).

The mechanism of the process differs considerably from the Marcus mechanism of electron transfer. The particles on the way between electrode and electrolyte must overcome an activation energy. The activation energy depends on the electric field in the double layer. The activation energy and the additional electrical term are shown in Figure 6.14.

The charge transfer coefficient has now a transparent explanation. It is the part of the potential difference across the double layer, which must be subtracted from the potential-independent part of the activation energy for the deposition reaction and must be added to the potential-independent part of the activation energy for the dissolution reaction.

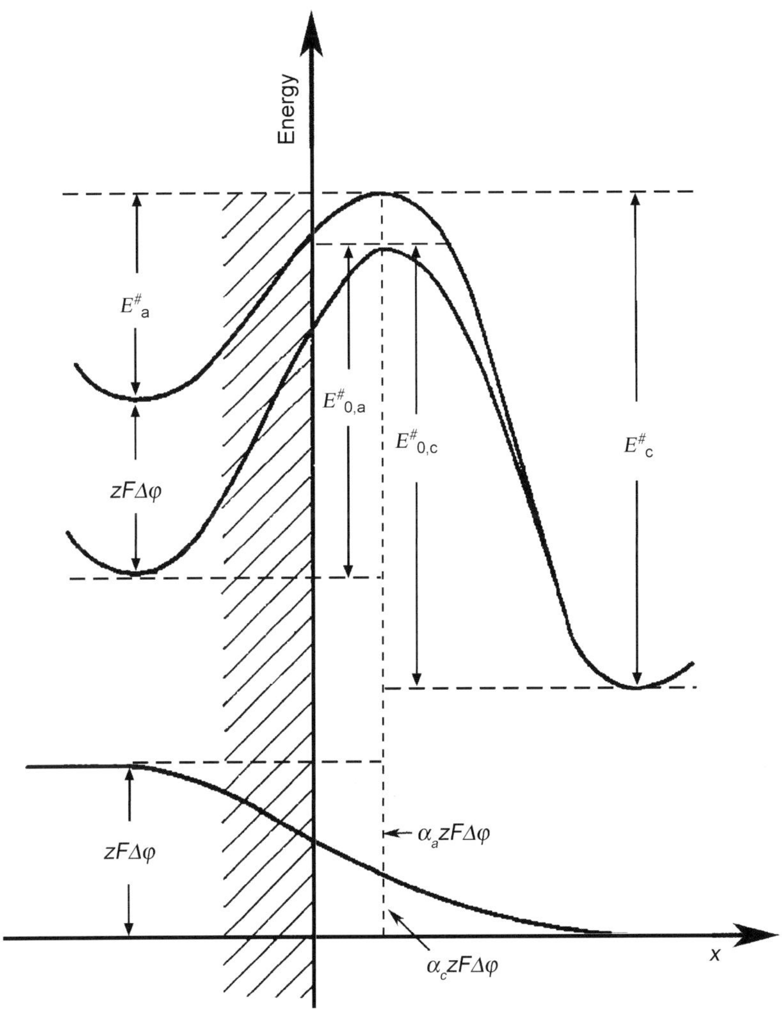

Figure 6.14 Activation energy $E^{\#}$ and electrical contributions for a metal ion crossing the electrochemical double layer.

This leads to the following activation energies:

$$E_{\mathrm{a}}^{\#} = E_{\mathrm{a},0}^{\#} - \alpha_{\mathrm{a}} zF\,\Delta\varphi \tag{6.46}$$

$$E_{\mathrm{c}}^{\#} = E_{\mathrm{c},0}^{\#} + \alpha_{\mathrm{c}} zF\,\Delta\varphi \tag{6.47}$$

Substituting the Galvani potential difference by the electrode potential versus a constant reference electrode the Butler–Volmer equation has now the form

$$i = zFk_{\mathrm{a}} c_{\mathrm{L}}^{\nu_L} \mathrm{e}^{+\frac{\alpha_{\mathrm{a}} zF}{RT} E} - zFk_{\mathrm{c}} c_{\mathrm{Me}^{z+}} \mathrm{e}^{-\frac{\alpha_{\mathrm{c}} zF}{RT} E} \tag{6.48}$$

The exchange current density is

$$i_0 = zFk_{\mathrm{a}} \bar{c}_{\mathrm{L}}^{\nu_L} \mathrm{e}^{+\frac{\alpha_{\mathrm{a}} zF}{RT} E_0} = zFk_{\mathrm{c}} \bar{c}_{\mathrm{Me}^{z+}} \mathrm{e}^{-\frac{\alpha_{\mathrm{c}} zF}{RT} E_0} \tag{6.49}$$

Assuming that the surface concentrations and bulk concentrations are approximately equal and substituting α_{c} by $1-\alpha_{\mathrm{a}}$ one gets the reduced Butler–Volmer equation

$$i = i_0 \left[\mathrm{e}^{+\frac{\alpha_{\mathrm{a}} zF}{RT}\eta} - \mathrm{e}^{-\frac{(1-\alpha_{\mathrm{a}})zF}{RT}\eta} \right] \tag{6.50}$$

From these equations the corresponding equations for Tafel lines, polarization resistance, and electrochemical reaction orders are obtained as was described in Section 6.1 for electron transfer.

The slopes of the Tafel plots now depend on the charge of the metal ion. An example is shown in Figure 6.15.

The equations of charge transfer describe the total process in a formal manner. The mechanism of the process is usually more complicated. The deposition process will be described in detail in Chapter 7, the dissolution process in Chapter 10.

6.4 CHARGE TRANSFER AND MASS TRANSPORT

The driving force for an electrochemical process is the overpotential. This is, in the case of a reversible redox electrode or a reversible metal/metal ion electrode, the deviation of the potential from the Nernst potential. The overpotential consists of several parts. One part is caused by deviation of the surface concentration from the bulk concentration. The replenishment of the surface concentration can be limited by diffusion. Then one speaks of diffusion overpotential.[*18]Another part is the overpotential necessary to drive the charge transfer.

The total overpotential is the sum

$$\eta = \eta_{\mathrm{ct}} + \eta_d \tag{6.51}$$

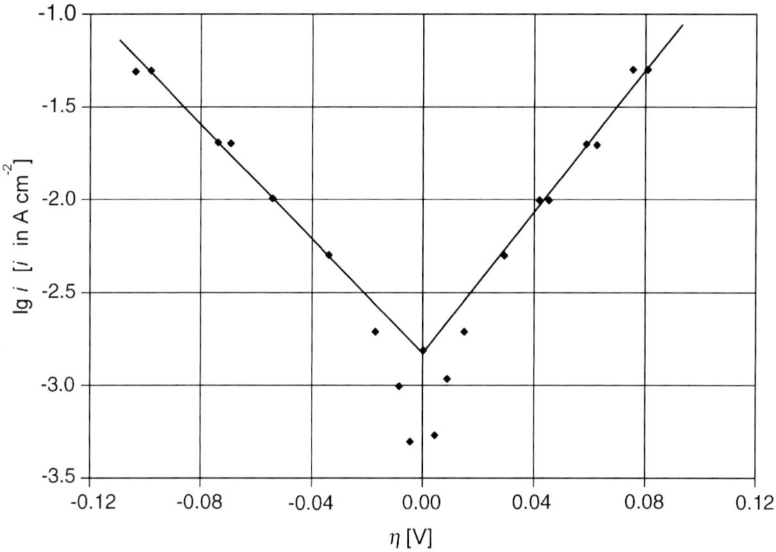

Figure 6.15 Tafel plots for a metal ion transfer reaction, dissolution and deposition of Cd/Cd^{2+}. From the slopes the charge transfer coefficient is determined, $\alpha_a z = 1.09$ and $(1 - \alpha_a)z = 0.91$. With $z = 2$ one obtains $\alpha_a = 0.55$ and $(1 - \alpha_a) = 0.45$. The exchange current density is $\lg i_0 = -2.8$.

The ratio η/i for current density approaching zero is the electrode resistance consisting of charge transfer resistance and diffusion resistance,

$$\lim_{i \to 0} \frac{\partial \eta}{\partial i} = R_{ct} + R_d \tag{6.52}$$

For analyzing the charge transfer process, it is necessary to eliminate the contribution from diffusion that means $R_d = RT/nFi_l \to 0$. Many electrochemical methods are only devoted to this goal.

The principal way to eliminate the diffusion term is reduction of the diffusion layer thickness. Under stationary condition this is achieved by enforced convection. For non-stationary conditions one can describe the increasing diffusion contribution by a time-dependent diffusion layer thickness.

$$\delta_d = \sqrt{\pi D t} \tag{6.53}$$

The diffusion layer thickness gets smaller if the overpotential is measured at short times. Investigating the time dependence of electrode processes one has as a limiting factor the charging of the double layer, because a change of the electrode potential depends on the change of the double-layer charge. The product of double-layer capacitance and electrolyte resistance $C_{dl}R_{el}$ determines the time constant of the double-layer charging. With a macroscopic size of the electrodes the limit is in the range of microseconds. A decreasing size of the electrodes shifted this limit towards nano- and femtoseconds.

6.4.1 Elimination of diffusion overpotential with a rotating disc electrode

The contribution of diffusion overpotential to the total overpotential can be achieved by an increase of convection near the electrode surface. In a very controlled manner this is possible with the rotating disc electrode. In the Koutecky—Levich[19] equation the separation of diffusion contributions and charge transfer contributions to the overpotential was achieved. A general charge transfer reaction with exchange of n electrons was chosen. The Koutecky–Levich equation is

$$\frac{1}{i} = \frac{1}{nF(k_{ox}\overline{c}_{red} - k_{red}\overline{c}_{ox})} + \frac{1.61D^{-2/3}\nu^{1/6}(k_{ox} + k_{red})}{nF(k_{ox}\overline{c}_{red} - k_{red}\overline{c}_{ox})}\frac{1}{\sqrt{\omega}} \tag{6.54}$$

D is the mean diffusion coefficient of S_{red} and S_{ox}, ν kinematic viscosity (see Section 5.3.1), and $\omega = 2\pi f$ the angular velocity with f the rotation frequency. In the anodic Tafel region ($k_{ox} >> k_{red}$) the following equation is obtained:

$$\frac{1}{i} = \frac{1}{i_a} + \frac{1.61D_{red}^{-2/3}\nu^{1/6}}{nF\overline{c}_{red}}\frac{1}{\sqrt{\omega}} \tag{6.55}$$

In the cathodic Tafel region ($k_{red} >> k_{ox}$) one gets

$$\frac{1}{|i|} = \frac{1}{|i_c|} + \frac{1.61D_{ox}^{-2/3}\nu^{1/6}}{nF\overline{c}_{ox}}\frac{1}{\sqrt{\omega}} \tag{6.56}$$

One has to measure the current density as a function of the rotation speed in the anodic and cathodic Tafel regions. Then one plots reciprocal current densities $1/i$ as a function of the square root of reciprocal rotation speed $1/\sqrt{\omega}$. The extrapolation to $1/\sqrt{\omega} \to 0$ gives charge transfer controlled current densities. Plotting these current densities versus potential, one obtains in the Tafel region Tafel lines (above ±60 mV overpotential). Charge transfer coefficients and exchange current densities can then be determined. For small overpotentials, in the region of linear dependence between current and potential the charge transfer resistance is obtained.

An example is shown in Figure 6.16. The reciprocal current is plotted versus the reciprocal value of the rotation frequency of a rotating disc electrode. The currents taken from the extrapolation to $1/\sqrt{f} = 0$ (rotation frequency $f = \infty$) are represented versus the potential in Figure 6.17. The current-potential plot shows a current-potential curve in the sub-Tafel region. An approximate current-potential line is shown in Figure 6.17. An approximate value of the charge transfer resistance and of the exchange current density

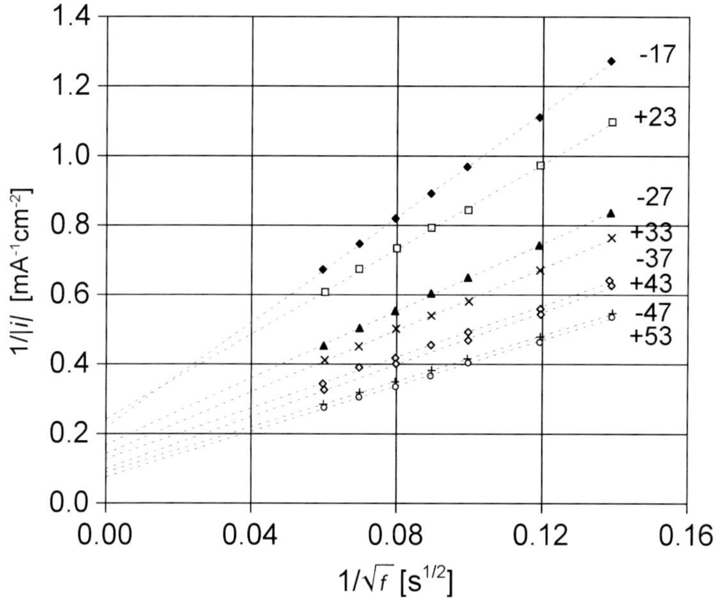

Figure 6.16 Elimination of diffusion contributions to the overpotential on the rotating disc electrode by a Koutecky–Levich plot, extrapolation to $1\sqrt{f} = 0$ (rotation frequency $f = \infty$), 10^{-2} mol · dm^{-3} $Fe(CN)_6^{4-}$/10^{-2} mol · dm^{-3} $Fe(CN)_6^{3-}$ redox system, 0.5 ml · dm^{-3} KCl, $E_0 = 0.224$ versus SCE, and Au electrode.[20]

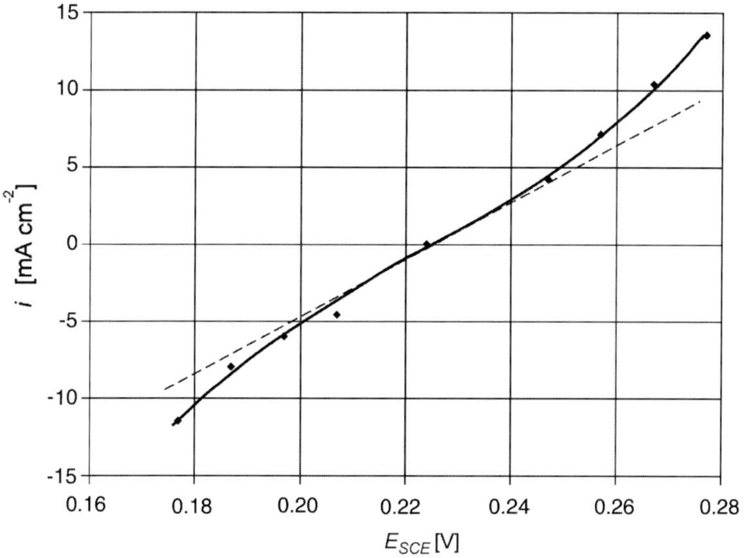

Figure 6.17 Current-potential plot of the extrapolated current data from Figure 6.15, $Fe(CN)_6^{4-}$–$Fe(CN)_6^{3-}$ redox system, charge transfer resistance $R_{ct} = 5.6\ \Omega \cdot cm^{-2}$, and exchange current density $i_0 = 5.2$ mA · cm^{-2}.

could be calculated from the slope at zero overpotential. A value of $R_{ct} = 5.6\,\Omega \cdot cm^2$ equal to $i_0 = 5.2\ mA \cdot cm^{-2}$ was obtained.

6.4.2 Elimination of diffusion contribution to the overpotential in chronoamperometry and chronopotentiometry

Elimination of the diffusion contribution to the overpotential is also possible by time-dependent measurements.

In chronoamperometry, after switching on an overpotential, the time dependence of the current is monitored. For purely diffusion-controlled processes, the current depends on time according to the Cottrell equation in Chapter 5, Eq. (5.20). The current decreases proportional to $1/\sqrt{t}$. The combination with charge transfer control leads to the following

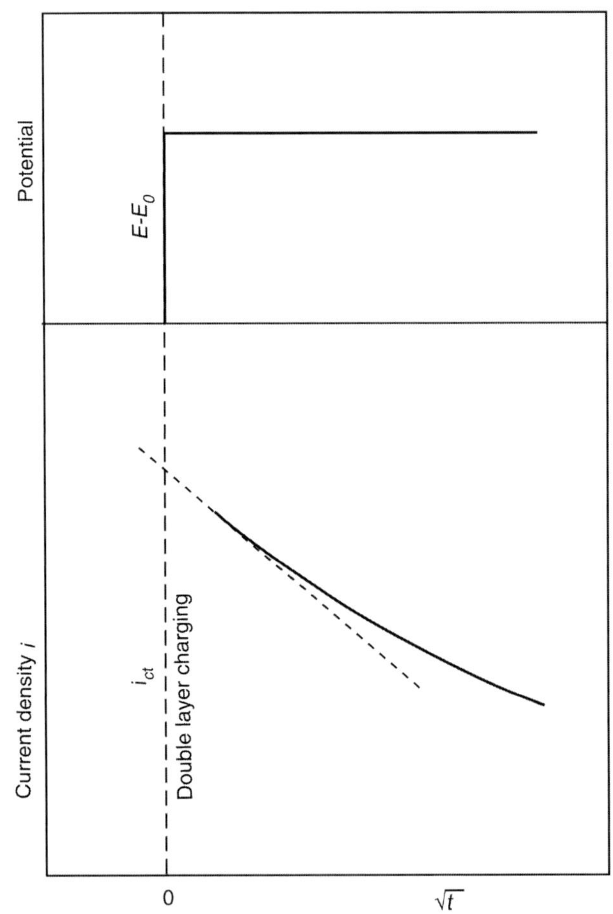

Figure 6.18 Potential and current density–time dependence in chronoamperometry. The double-layer charge peak (not shown) can overshadow the charge transfer controlled time region.

equations. Switching to potentials in the anodic Tafel region the current time dependence for short times is given by

$$i = i_a - \frac{2nFk_{ox}^2}{\sqrt{\pi D_{red}}}\sqrt{t} \tag{6.57}$$

Switching to potentials in the cathodic Tafel region one obtains

$$|i| = |i_c| - \frac{2nFk_{red}^2}{\sqrt{\pi D_{ox}}}\sqrt{t} \tag{6.58}$$

For short times (below milliseconds or even microseconds) the current decreases with $\sqrt{t}$. Potential- and current-time dependence are shown schematically in Figure 6.18.

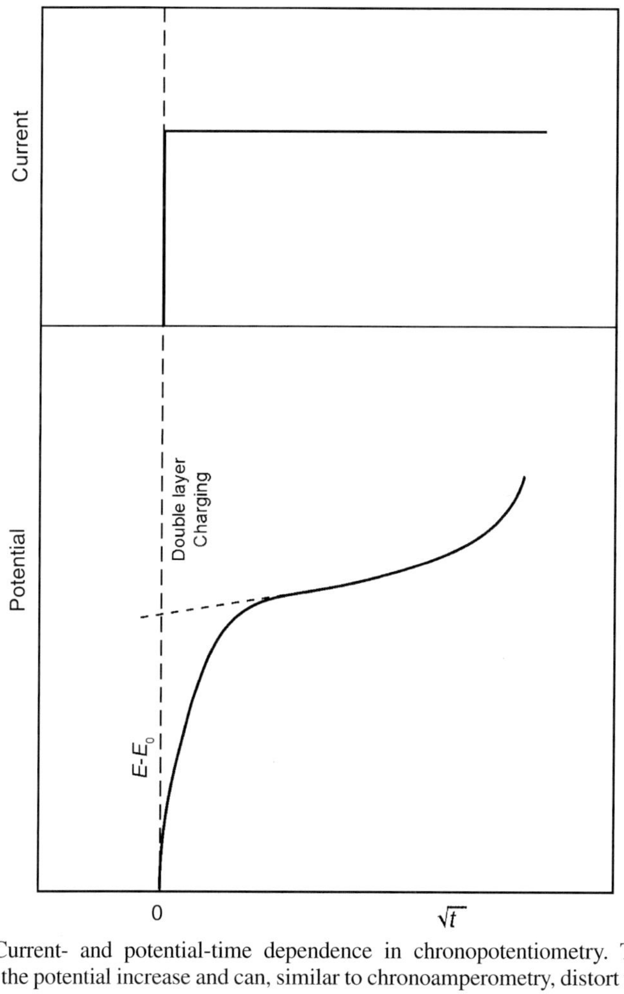

Figure 6.19 Current- and potential-time dependence in chronopotentiometry. The double-layer charging retards the potential increase and can, similar to chronoamperometry, distort the extrapolation.

An example was the determination of charge transfer controlled current densities in the HNO_2–HNO_3 redox system (Section 6.2.1).

In such measurements the limitation by the double-layer charging is a serious problem. The charging peak can dominate the current-time dependence at short times.

In chronopotentiometry, after switching on a current, the time dependence of the potential is monitored. Potential- and current-time dependence are shown schematically in Figure 6.19. Similar to chronoamperometry, the charging of the double layer can overshadow the region of charge-controlled behavior.

Currents flowing through the electrochemical cell can cause a potential drop in the electrolyte and the measured potential must be corrected for this potential drop. The potential drop is called IR-drop.

In a three-electrode cell the measured IR-drop occurs between working electrode and reference electrode (Figure 6.20). Therefore, the reference electrode should be brought as close to the working electrode as possible. Experimental procedures were suggested to eliminate the IR-drop. Some potentiostats have the possibility to correct a known electrolyte resistance. Another method is to interrupt the current for a short time and to measure the potential at the interruption.

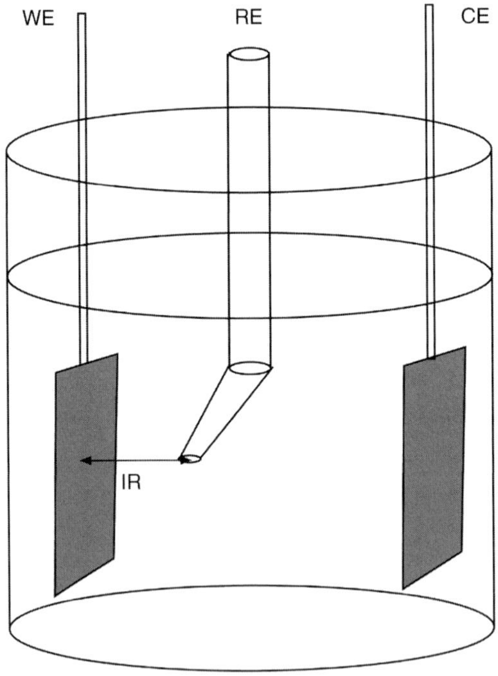

Figure 6.20 IR-drop in an electrochemical cell. The IR-drop occurs between working electrode WE and reference electrode RE (a Haber–Luggin capillary is shown), CE counter electrode.

6.4.3 Elimination of diffusion contributions to the overpotential by impedance spectroscopy

Impedance spectroscopy in the high-frequency region is another way of eliminating diffusion overpotential. The equations were given in Chapter 5. The combination of the double-layer equivalent circuit with the diffusion impedance was described in Section 5.23 and examples for the determination of the charge transfer resistance at high frequencies were given.

END NOTE

*The replenishment can also be limited by a slow reaction in the diffusion layer or on the electrode surface, in which case one speaks of reaction overpotential. This will not be treated and the reader is referred to the special literature.

REFERENCES

1. C.P. Andrieux, Pure Appl. Chem., *66*, 2445 (1994).
2. T. Erdey-Gruz, M. Volmer, Z. physikal. Chem., *150A*, 203 (1930).
3. J.A.V. Butler, Trans Faraday Soc., *28*, 379 (1932).
4. IUPAC, Quantities, Units and Symbols in Physical Chemistry (so-called "Green Book"), 3rd Ed., Blackwell Science, Oxford.
5. R. Parsons, Pure Appl. Chem., *52*, 233 (1979).
6. W. Plieth, I. Stellmacher, B. Quast, Electrochim. Acta, *20*, 335 (1975).
7. H. Gerischer, Z. Physikal. Chem. NF, *26*, 223, 325 (1960).
8. R.W. Gurney, Proc. Roy. Soc. London, A, *134*, 137 (1931).
9. O. Essin, Acta Physicochim. USSR, *13*, 123 (1940).
10. R. Marcus, J. Chem. Phys. *24*, 966 (1956); *26*, 867 (1957); Ann. Rev. Phys. Chem., *15*, 155 (1954).
11. V.G. Levich, Present status of the theory of oxydation-reduction in solution (bulk and electrode reactions) in P. Delahay (Ed.), Advances in Electrochemistry and Electrochemical Engineering, Vol. 4, Interscience Publ., New York, 1966, p. 249.
12. W. Plieth, Nitrogen, in A. Bard (Ed.), Electrochemistry of the Elements, Vol. 8, Marcel Dekker, New York, 1978, p. 321.
13. K.J. Vetter, Z. Physical. Chem., *194*, 109, 284 (1950).
14. G. Schmid, Z. Elektrochem., *63*, 1183 (1959); *65*, 531 (1961).
15. G. Schmid, J. Delfs, Z. Elektrochem., *63*, 1192 (1959).
16. G. Schmid, G. Krichel, Ber. Bunsenges. Phys. Chem. (formerly Z. Elektrochem.), *68*, 677 (1964).
17. G. Schmid, M.A. Lobeck, Ber. Bunsenges. Phys. Chem. (formerly Z. Elektrochem.), *70*, 1165 (1966); *73*, 189 (1969).
18. K.J. Vetter, Elektrochemische Kinetik, Springer, Berlin, 1961.
19. J. Koutecky, V.G. Levich, Dokl. Akad. Nauk. SSSR, *117*, 441 (1957); Zhur. Fiz. Khim., *32*, 1565 (1958).
20. Southampton electrochemistry group: R. Greef *et al.* (Eds.), Instrumental Methods in Electrochemistry, Ellis Horwood, Chichester, 1990, p. 130.

– 7 –

Nucleation and Growth of Metals

The electrochemical deposition of a metal is one of the important applications of electrochemistry. In this chapter some aspects of this process will be discussed. However, only pure metal deposition will be treated.

From a look at the periodic table it can be seen that only a selected number of metals can be deposited from an aqueous electrolyte due to the limited potential window of aqueous electrolytes. An even smaller group, i.e., copper, zinc, nickel chromium, tin, and lead, is of greater technological importance.

Organic solvents provide suitable potential windows for some important metals such as aluminum and lithium that cannot be deposited from aqueous electrolyte. Some metals are deposited from molten salts. The so-called room temperature molten salts or ionic liquids are a further group of solvents used to deposit metals like aluminum again that cannot be deposited from aqueous electrolytes.

7.1 NUCLEATION

The growth of a new phase starts with the formation of a nucleus. Erdey-Gruz and Volmer gave the first description of nucleation.[1] In this chapter, the basic concepts will only be shortly described because Budevski, Staikov, and Lorenz have comprehensively treated this subject.[2]

7.1.1 Three-dimensional nucleation

In electrochemical phase formation nucleation must be on a surface of a foreign or native substrate. If the nucleus formed on a surface is a cluster that grows in all three dimensions, one speaks of a three-dimensional nucleation. The nucleus is a cluster of only a few atoms and can be depicted as a semi-sphere (Figure 7.1).

The Gibbs energy of the nucleus that has formed as a cluster and as a function of the size with the semi-sphere approximating the nucleus radius is the sum of bulk free energy and surface free energy

$$\Delta g_{\mathrm{nucl}} = \Delta g_{\mathrm{bulk}} + \Delta g_{\mathrm{surf}} \tag{7.1}$$

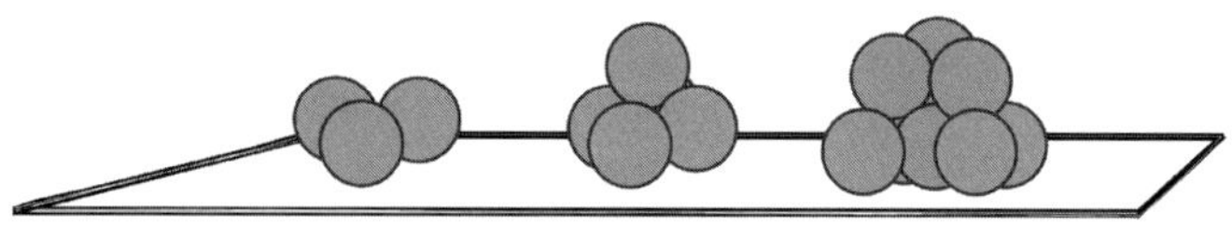

Figure 7.1 Different nucleus models on a surface (a) a cluster of three atoms, (b) a cluster of four atoms, and (c) a cluster of ten atoms in more or less a semi-sphere.

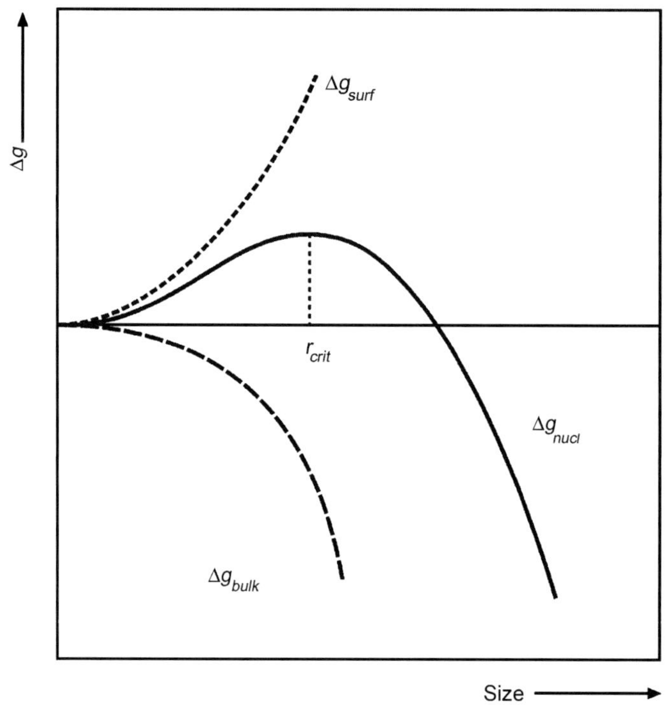

Figure 7.2 Gibbs energy of nucleation as a function of size. r_{crit} is the critical nucleus or the size where the nucleus becomes stable.

$$\Delta g_{\mathrm{nucl}} = -\frac{2\pi}{3} r^3 \frac{\Delta G_{\mathrm{ss}}}{V_{\mathrm{m}}} + 2\pi r^2 \sigma \tag{7.2}$$

where ΔG_{ss} is the molar free energy of super saturation, V_m the volume of one mole of metal, and σ the free surface energy.

The Gibbs energy of nucleation as a function of the size of the nuclei is shown in Figure 7.2. Very small clusters are instable. The formation in this size region is a statistical process. A maximum marks the critical size of the nucleus and the critical Gibbs energy of nucleation between instability and stable growths.

For a semi-spherical nucleus, the radius of the critical nucleus can be derived from the equation $d\Delta g_{\mathrm{nucl}}/\mathrm{d}r = 0$. One obtains

$$r_{\mathrm{crit}} = \frac{2\sigma}{\Delta G_{\mathrm{ss}}} V_{\mathrm{m}} \tag{7.3}$$

The Gibbs energy of super saturation is connected to the overpotential by the equation

$$\Delta G_{\mathrm{ss}} = -zF|\eta| \tag{7.4}$$

By inserting Eq. (7.3) into Eq. (7.2) and substituting $\Delta G_{\mathrm{ss}} = -zF|\eta|$ one obtains

$$\Delta g_{\mathrm{crit}} = \frac{8\pi}{3} \frac{V_{\mathrm{m}}^2 \sigma^3}{z^2 F^2} \frac{1}{|\eta|^2} \tag{7.5}$$

With the volume of the critical nucleus $V_{\mathrm{crit}} = (4\pi/3)r_{\mathrm{crit}}^3$ one can also calculate the number of atoms in the critical nucleus by dividing the volume of N_{A} critical nuclei (N_{A} Avogadro constant) through the molar volume of the metal V_{m}. The result is

$$N_{\mathrm{crit}} = \frac{16\pi}{3} \frac{N_{\mathrm{A}} V_{\mathrm{m}}^2 \sigma^3}{z^3 F^3} \frac{1}{|\eta|^3} \tag{7.6}$$

7.1.2 Two-dimensional nucleation

Sometimes a two-dimensional nucleus forms on a surface (Figure 7.3), which is therefore a monomolecular layer growth because of the strong interaction between the substrate and deposited metal as in under potential deposition. This cluster of N atoms almost has a circular shape.

Similar equations can be derived for two-dimensional nucleation as were given for three-dimensional nucleation.[2]

Figure 7.3 A two-dimensional nucleation with the nucleus in the approximate form of a circle.

7.1.3 Rate of nucleation

Volmer and Weber derived an equation for the nucleation rate.[3]

$$J = A\exp\left(-\frac{\Delta g_{\text{crit}}}{kT}\right) \tag{7.7}$$

Substituting the value for Δg_{crit} by Eq. (7.5) and forming the logarithm one obtains

$$\ln J = \ln A - \frac{B}{|\eta|^2} \tag{7.8}$$

where B stands for the different constants in Eq. (7.5). This equation was confirmed by most experiments, despite a continuous discussion of the validity of this classical approach. An example is shown in Figure 7.4.

The following equation was also given in the literature:[6,7]

$$\frac{\partial \ln J}{\partial |\eta|} = -\frac{zF}{RT} N_{\text{crit}} \tag{7.9}$$

The logarithm of the nucleation rates of Figure 7.4 is plotted as a linear function of the overpotential in Figure 7.5. As N_{crit} is a function of η, no linear relationship is expected.

According to Eq. (7.9), the differentiation of the plot of ln J versus η, shown in Figure 7.5, should give N_{crit}. Otherwise, N_{crit} should be proportional to $1/\eta^3$ (Eq. (7.6)). In Figure 7.6

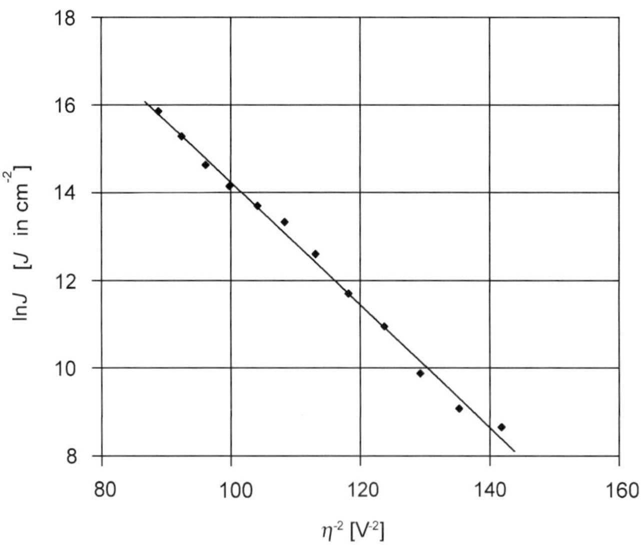

Figure 7.4 Logarithm of the rate of formation of mercury nuclei J on the (111) faces of a spherical platinum single crystal versus $|\eta|^{-2}$.[4,5]

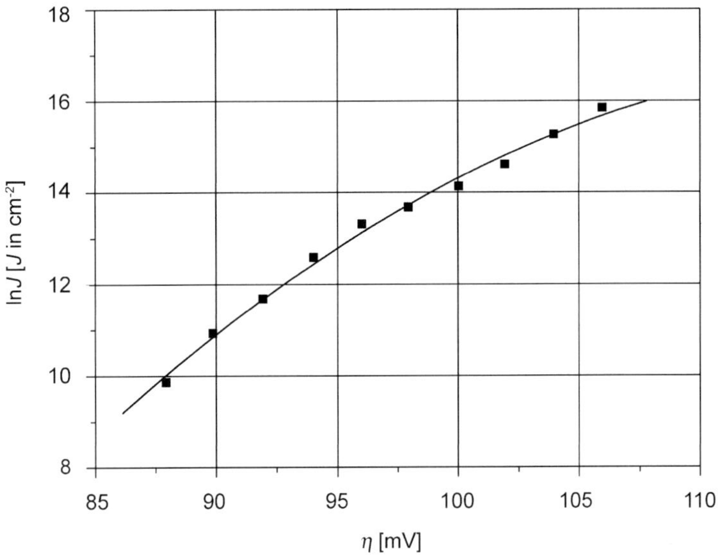

Figure 7.5 Logarithm of the rate of nucleation J versus overpotential η (data from Figure 7.4).

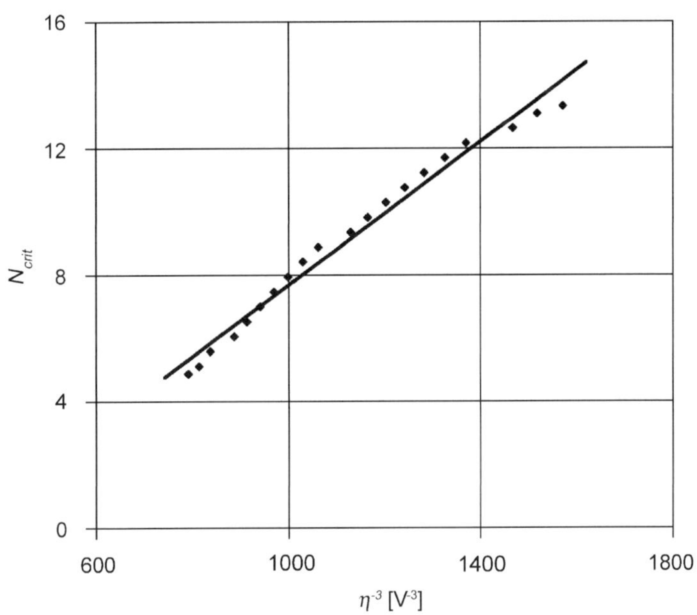

Figure 7.6 N_{crit}, determined by differentiation of a polynomial fit of the data in Figure 7.5, shown as function of the overpotential $1/\eta^3$.

N_{crit} is plotted versus $1/\eta^3$. The approximately linear dependence indicates that Eq. (7.9) may be used as a basis for obtaining information about the size of the critical nucleus.

7.1.4 Instantaneous and progressive nucleation

The many defects on the surface of a real crystal form nucleation centers. After switching to a potential in the deposition region, there will already be a fixed density of defects N_0 where local growth can start.

$$N = N_0 \tag{7.10}$$

This is called instantaneous nucleation.

As an example, diffusion-controlled hemispherical growths will be described. Hills, Schiffrin, and Thomson calculated the current of N_0 individual nuclei that grow independently of each other.[8]

$$i = zF\pi(2Dc)^{3/2}\left(\frac{M}{\rho}\right)^{1/2} N_0 t^{1/2} \tag{7.11}$$

Hills, Schiffrin, and Thomson also calculated an equation for a progressive increase of the number of nucleation centers that occur after switching to a potential in the deposition region. This is called progressive nucleation. The density of nuclei as a linear function of time is described by the equation

$$N(t) = N_0 A_n t \tag{7. 12}$$

where A_n is the nucleation rate. The current density is given by the equation

$$i = \frac{2}{3} zF\pi(2Dc)^{3/2}\left(\frac{M}{\rho}\right)^{1/2} N_0 A_n t^{3/2} \tag{7. 13}$$

Diffusion zones develop during the growths of the nuclei. The diffusion zones first have a semi-spherical shape and then overlap with time (Figure 7.7).

The area of the diffusion zone projected on the surface plane can be described by the equation

$$S = \pi kDt \tag{7.14}$$

The overlapping of the growing diffusion areas is shown in Figure 7.8.

An equation given by Avrami[9] can be used to relate the real surface coverage with diffusion zones Θ to the surface coverage, if the growth of the diffusion zones is unlimited, $\Theta_{ex} = N\pi kDt$.

$$\Theta = 1 - \exp(-N\pi kDt) \tag{7.15}$$

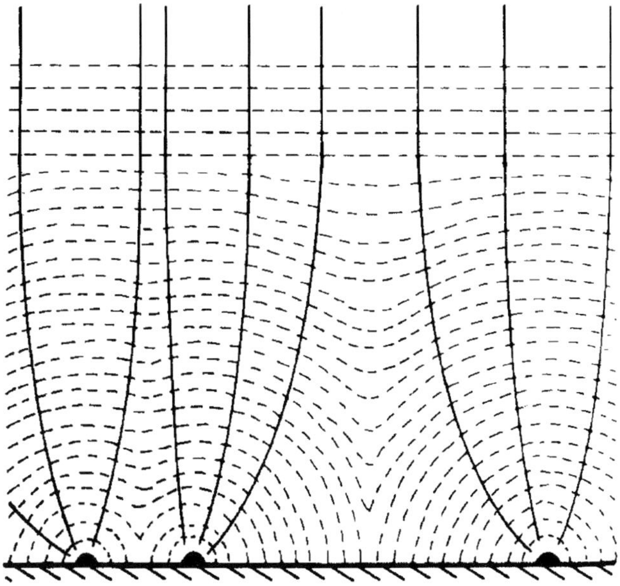

Figure 7.7 Development of diffusion zones around hemispherical nuclei on an electrode surface.

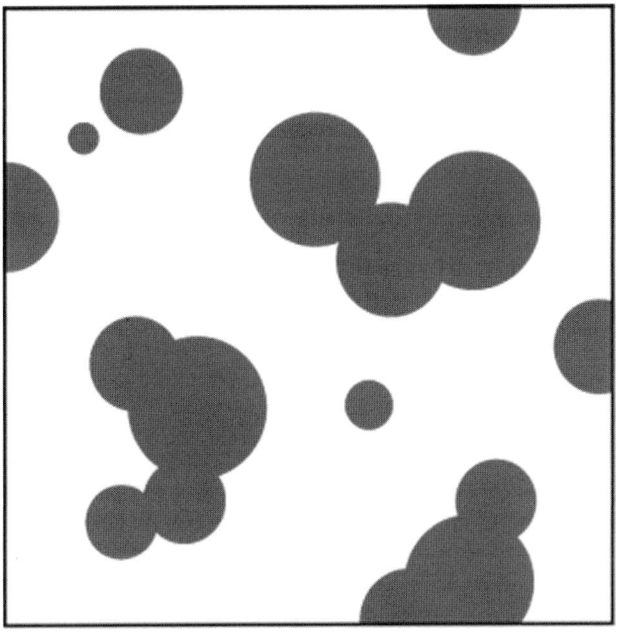

Figure 7.8 Two-dimensional projection of overlapping diffusion zones around hemispherical nuclei on an electrode surface.

On the basis of these considerations Scharifker and Hills[10] derived equations that describe the current transients. The equation for the current density as a function of time for instantaneous nucleation is

$$i = zFc\left(\frac{D}{\pi t}\right)^{1/2}\left[1 - \exp\left(N\pi kDt\right)\right] \tag{7.16}$$

Comparison of Eq. (7.16) for $t \to 0$ with Eq. (7.11) gives for k

$$k = \left(\frac{8\pi cM}{\rho}\right)^{1/2} \tag{7.17}$$

The equation for the current density as a function of time for progressive nucleation is

$$i = zFc\left(\frac{D}{\pi t}\right)^{1/2}\left[1 - \exp\left(-\frac{A_n}{2}\pi k' Dt^2\right)\right] \tag{7.18}$$

Comparison of Eq. (7.18) for $t \to 0$ with Eq. (7.13) gives for k'

$$k' = \frac{4}{3}\left(\frac{8\pi cM}{\rho}\right)^{1/2} \tag{7.19}$$

In Figure 7.9, current transients for instantaneous and progressive nucleation are shown in a reduced presentation, plotting i/i_{max} versus t/t_{max}.

Work on nucleation and current transients is continuously being developed. Examples are the papers of Scharifker and Mostany[11] or of Heermann and Tarallo.[12] Liu and Penner

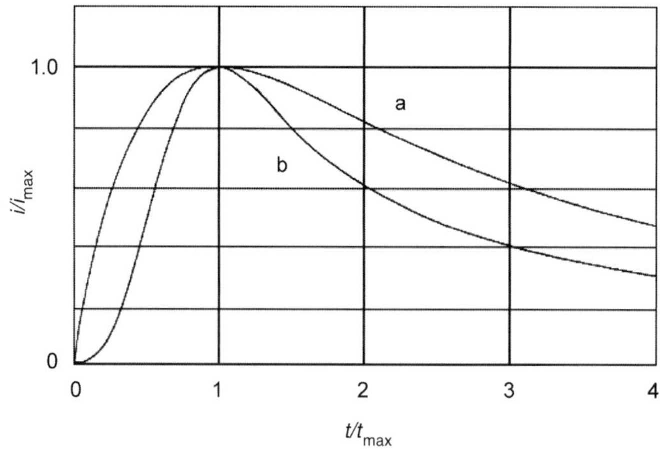

Figure 7.9 Instantaneous and progressive nucleation, according to Eqs. (7.16) and (7.18), reduced presentation: i/i_{max} versus t/t_{max}; (a) instantaneous nucleation, (b) progressive nucleation.

took into account minimization of inter-particle coupling.[13] A double-pulse technique was used for controlled preparation of metallic nanoparticles.[14]

7.2 INTERMEDIATE STATES OF ELECTRODEPOSITION

In the process of deposition of metal ions into a crystalline phase an atom has to pass several intermediate positions. This is shown in Figure 7.10.

In the model the main three surface processes are shown, which are (1) the charge transfer process, (2) surface diffusion, and (3) the transfer from the ad-atom position into a step or kink position. Not shown are the diffusion processes in the electrolyte. Each step can be rate determining, such as the charge transfer between electrolyte and metal surface (Chapter 6), surface diffusion (Chapter 4), or the transfer into a step or kink position, the final crystallization process.

In the following treatment, the density of kink site positions is considered to be in equilibrium with the ad-atom concentration and with the ion concentration in the electrolyte.[15] This is shown in Figure 7.11.

In this diagram the reaction of an ad-atom with the kink site position is reversible. But in the kink site position an atom may return to its previous position by separation from the kink site position, or it may become fixed in the crystal lattice by the next atom deposited in the kink site position. The built-in process is then completed. In this model, the kink site position has a similar function as the transition state in the theory of chemical reactions.

7.2.1 Crystallization overpotential

The first connection between the surface processes in electrocrystallization and the overpotential were discussed in papers of Lorenz[16] and Vermilyea.[17] In these papers an overpotential was defined by the difference between the ad-atom concentration in the

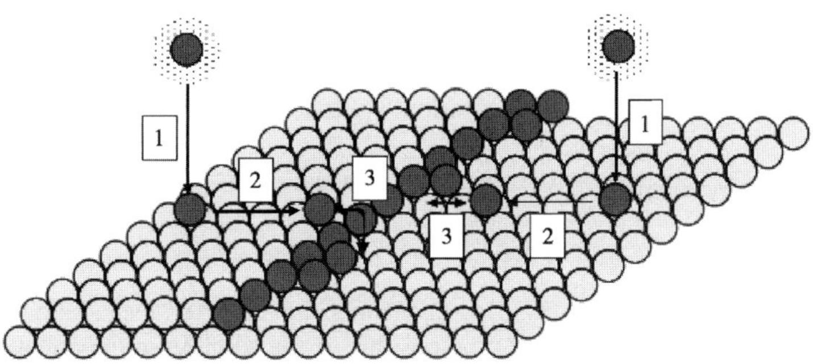

Figure 7.10 Model of an electrochemical deposition process: (1) charge transfer of an ion in the electrolyte to an ad-atom position, (2) surface diffusion of ad-atoms, and (3) transfer of an ad-atom into a step or kink position.

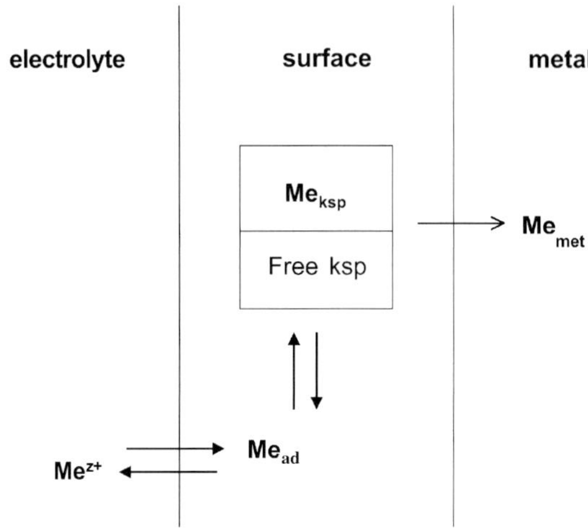

Figure 7.11 Transition state model of an electrochemical deposition; charge transfer and surface diffusion are in equilibrium.[18]

midst of a terrace $c_{\mathrm{ad,m}}$ and at the edges of a terrace $c_{\mathrm{ad,e}}$. This overpotential was called crystallization overpotential η_{k}

$$\eta_k = \frac{RT}{zF} \ln \frac{c_{\mathrm{ad,m}}}{c_{\mathrm{ad,e}}} \tag{7.20}$$

Gerischer made an experimental attempt to measure the crystallization overvoltage of Ag deposition in a chronopotentiometric experiment.[18] In this experiment two results were obtained. From the slope at $t \to 0$ the capacitance was determined. This capacitance was much larger than the double-layer capacitance and was interpreted as adsorption capacitance C_{ad}. The ad-atom concentration c_{ad} was calculated from the adsorption capacitance C_{ad}.

$$c_{\mathrm{ad}} = \frac{RT}{z^2 F^2} C_{\mathrm{ad}} \tag{7.21}$$

Figure 7.12 shows the ad-atom concentration c_{ad} as a function of the electrolyte concentration $c_{\mathrm{Ag^+}}$.

The following adsorption isotherm was determined:

$$\lg c_{\mathrm{ad}} = \mathrm{const} + \beta \cdot \lg c_{\mathrm{Ag+}} \tag{7.22}$$

This is a confirmation of Eq. (4.55) derived in Chapter 4 for the ad-atom adsorption isotherm with $\beta = \lambda(1 - x_{\mathrm{ad}}) = 0.2$. The position of the ad-atom in the potential gradient of the double layer can be guessed to be 0.2. The partial charge coefficient would then be $\lambda = 0.25$. The remaining overpotential was interpreted as crystallization overpotential. Experimental evidence was presented that the charge transfer overpotential can be neglected.

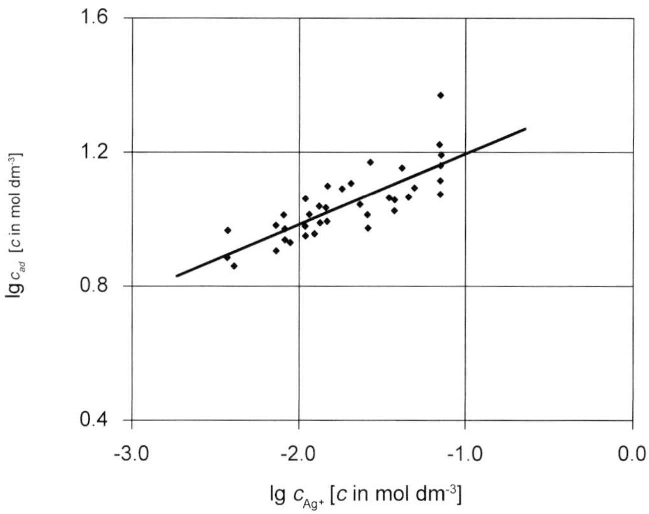

Figure 7.12 Ad-atom isotherm, showing $\lg c_{\mathrm{ad}}$ versus $\lg c_{\mathrm{Ag^+}}$.[18]

Finally, the crystallization exchange current density was calculated and interpreted as the rate of deposition and separation of ad-atoms into kink site positions.

7.3 SURFACE DYNAMICS

In the following section, a statistical description of the surface processes will be presented. This is based on the definition of a residence time of an atom in a kink site position.

7.3.1 Residence time in kink site positions

Figure 7.13 shows the (100) surface of a metal crystallizing in the face-centered cubic lattice. Three planes of atoms and a step running in the [110] direction are shown. On the step a kink site position is shown.

Ad-atoms can be deposited into a kink site position either directly or via a step position. Otherwise, the atom in a kink site position can be separated from the kink site position to a step position or to an ad-atom position as shown in Figure 7.13.

There is, in principle, the direct transfer between kink site position and electrolyte. This process, which must be taken into account at larger overpotentials, will not be discussed further.

The transfer of an ad-atom into a kink site position can be described by a reaction of a metal ion with a kink site position

$$r_{\mathrm{dep}} = k_{\mathrm{dep}} c_{\mathrm{Me}^{z+}} [ksp] \tag{7.23}$$

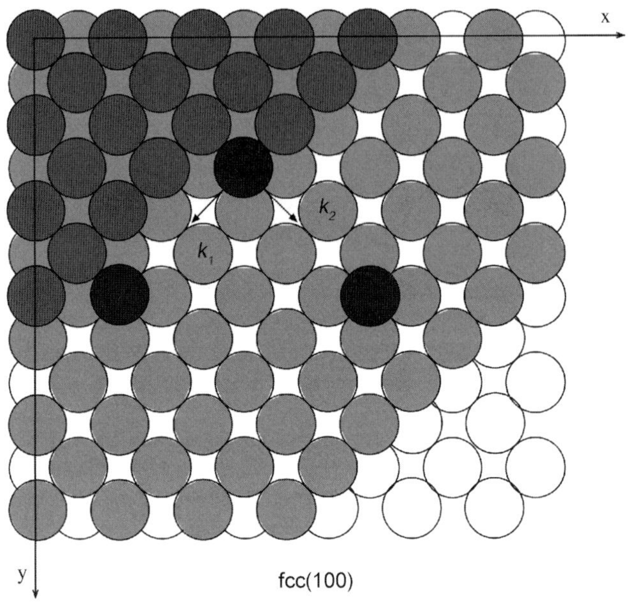

Figure 7.13 View on the (100) surface of a face-centered cubic lattice in which three planes of atoms and a step running in the [110] direction are shown. An atom in a kink site position (black) can move into a step position or into an ad-atom position. Taking into account all the neighbors of the separating atom, a transition complex is defined. Therefore, calculation of the complex energy is a method to determine activation energies.

The separation from a kink site position is a reaction of first order. The reaction rate is given by the rate equation

$$r_{sep} = k_{sep}[ksp] \tag{7.24}$$

In principle, the potential dependence of the rate constants must be taken into account. The potential dependence can be described by a Butler–Volmer exponential term $\exp(+\alpha_{a,i}\Delta z_i\, FE/RT)$. The coefficient $\alpha_{a,i}$ represents the part of the potential, which acts on the separation step and Δz_i is the charge transfer connected with the separation step.

What happens to an atom in the kink site position? This may be determined by the time interval in which the atom stays in the kink site position. This time is called residence time in the kink site position (in German "Verweilzeit").[19] The residence time is the statistical average of the time between the arrival of the atom in the kink site position and its separation from the kink site position. During the residence time the process of separation from the kink site position competes with the arrival of new atoms. The residence time τ in a kink site position is given by the reciprocal value of the rate constant of separation.

$$\tau = \frac{1}{k_{sep}} \tag{7.25}$$

7.3.2 Calculation of the residence time

The rate constant of separation consists of two terms, step-atom formation and direct ad-atom transfer

$$k_{sep} = k_{sep,1} + k_{sep,2} \tag{7.26}$$

The rate constant is given by the equation

$$k_{sep} = k_0 \exp\left(-\frac{E_{a,1} + E_{a,2}}{kT}\right) \tag{7.27}$$

The pre-exponential factor k_0 is approximately the oscillation frequency of the atom in the kink site position. An estimated value of $10^{12}\ s^{-1}$ can be used.

The activation energies in Eq. (7.27) are the energy barriers for an atom to diffuse from the kink site position to the step position or to the ad-atom position (Figure 7.13). The energy barrier is the difference between the energy of the diffusing atom in the transition complex and the energy in the kink site position. The transition complex depends on the surface structure and differs for different crystallographic lattices and crystallographic faces. An example for the transition complex of step diffusion in Figure 7.13 is shown in Figure 7.14. The energy of such a cluster can be calculated. The difference between the energy of the transition complex and the energy of the atom in the kink site position gives the activation energy.

At the interface to a liquid phase the transition complex is not only described by the atoms of the solid phase. The moving atom can form complexes with molecules or ions of

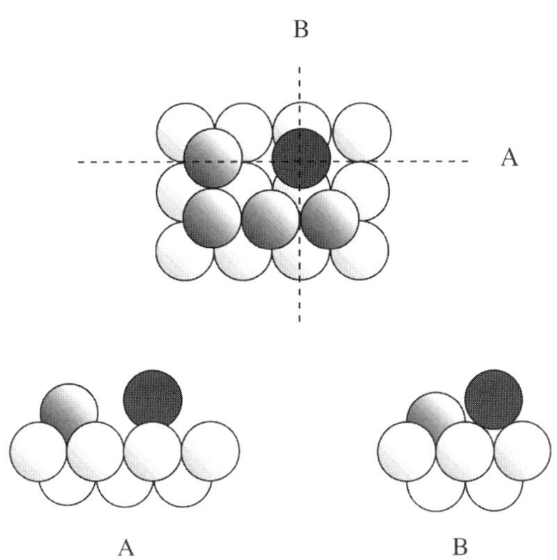

Figure 7.14 Calculation of the activation energy for step diffusion on the (100) surface of a cubic close-packed (ccp) lattice structure. The neighbors of the separating atom in the transition state form a transition complex. A top-view of the transition complex is shown as well as cross-sectional views marked by A and B.

the liquid phase. The interaction of the surface atoms with atoms from the electrolyte will considerably reduce the magnitude of the activation energy. But a calculation demands special models of the transition complex. Moreover, the partial charge of the surface atoms must be known to calculate the dipole interaction with the water molecules.

For comparison of different metals the following rough approximation is possible. The interaction of metals with solvent molecules in the electrolyte will be of similar intensity. Then the difference of the activation energy for separation from a kink site position of different metals will be proportional to the cleavage of bonds between the separating atom and the metal lattice. *A possible assumption is that one bond is broken when the atom separates from the kink site position into a step position.* Separation to an ad-atom position requires more energy and a possible assumption is that this activation energy would be equal to the cleavage of two bonds. Because of the exponential dependence on the activation energy only the first term is of significance. Based on these assumptions values k_{sep} can be calculated using the following equation:

$$k_{sep} \approx k_0 \exp\left(-\frac{\varphi_{1/2}}{6kT}\right) \tag{7.28}$$

where $\varphi_{1/2}$ is the energy in the kink site position. By adding a Butler–Volmer term $\exp(+\alpha_{a,i}\,\Delta z_i FE/RT)$ a potential dependence can be taken into account. The coefficient $\alpha_{a,i}$ represents the part of the potential, which acts on the separation step and Δz_i is the charge transfer connected with the separation step. The residence time is given by the equation

$$\tau \approx \frac{1}{k_0 \exp(-(\varphi_{1/2}/6kT))} \tag{7.29}$$

Rate constants of separation and residence times calculated for some ccp and hcp metals are shown in Table 7.1.

Table 7.1

Comparison of the dynamic of some metals with ccp and hcp structures; rate constants of separation from kink site positions k_{sep} (calculated with the assumption that the activation energy is proportional to the cleavage of one bond φ_{Me-Me}) and residence time τ

Metal	Structure	$\phi_{1/2}$ (eV)	$\phi_{1/2}/6$ (eV)	k_{sep} (s^{-1})	τ(s)
Ag	ccp	2.95	0.492	5×10^{3}	2×10^{-4}
Au	ccp	3.79	0.632	2×10^{1}	5×10^{-2}
Cd	hcp	1.16	0.193	5×10^{8}	2×10^{-9}
Co	hcp	4.40	0.733	4×10^{-1}	3×10^{0}
Cu	ccp	3.50	0.583	1×10^{2}	7×10^{-3}
Ni	ccp	4.45	0.742	3×10^{-1}	3×100
Pb	ccp	2.02	0.337	2×10^{6}	5×10^{-7}
Pd	ccp	3.92	0.653	9×10^{0}	1×10^{-1}
Pt	ccp	5.86	0.977	3×10^{-5}	3×10^{4}
Rh	ccp	5.77	0.962	6×10^{-5}	2×10^{4}
Ru	hcp	6.66	1.110	1×10^{-7}	7×10^{6}
Zn	hcp	1.35	0.225	2×10^{8}	6×10^{-9}

The residence times presented in Table 7.1 are relative values, but they provide a feeling for the large differences in the surface dynamics of these metals under deposition conditions. While the differences in the bond energies are inconspicuous, the differences in the residence times cover several orders of magnitude.

7.4 DENSITY OF KINK SITE POSITIONS

7.4.1 Equilibrium conditions

In a first approximation it can be assumed that the density of kink site positions is in equilibrium with the ad-atom concentration. More generally, all partial reactions are in equilibrium at the Nernst potential. The following equation describes equilibrium between ions in the electrolyte and atoms in kink site positions:

$$A^* + A^{z+} + ze^- \underset{k_{sep}}{\overset{k_{dep}}{\leftrightarrow}} AA^* \tag{7.30}$$

This reaction expresses the repeating nature of the reaction of an ion with a kink site position A^* that forms a new kink site position AA^*. The equilibrium constant K_{AA} is related to the rate constants k_{dep} and k_{sep} by the following equation:

$$K_{AA} = \frac{k_{dep}}{k_{sep}} \tag{7.31}$$

The rate of the deposition reaction corresponds to the cathodic current density

$$i_- = -zFk_{dep} \cdot c_{Me^{z+}} \cdot [ksp]/N_A \tag{7.32}$$

The rate of the separation reaction is proportional to the anodic current density

$$i_+ = zFk_{sep} \cdot [ksp]/N_A \tag{7.33}$$

The density of kink site positions A^* and AA^* at equilibrium potential is $[ksp]_0$. For equilibrium conditions one gets, $i_+ = i_0$. Thus it follows with Eq. (7.25):

$$[ksp]_0 = \frac{i_0 \cdot \tau \cdot N_A}{zF} \tag{7.34}$$

With N_A Avogadro's number and F the Faraday constant. The density of kink site positions at the equilibrium potential $[ksp]_0$ is given by the product of exchange current density and residence time. This equation will be discussed in the next section.

7.4.2 Deposition conditions

The residence time provides a measure of how long an atom can stay in a kink site position. The next atom accumulated in the kink site position determines its future. Therefore one has to compare residence time and rate of deposition.

The rate of deposition is measured by the deposition current. From the current, the number of atoms discharged per second and per unit of surface area can be calculated. Multiplying this value with the residence time, the number of atoms N_τ arriving during the residence time per unit of surface area is obtained (Eq. (7.35))

$$N_\tau = \frac{i \cdot \tau \cdot N_A}{zF} \tag{7.35}$$

where N_A is the Avogadro number, F the Faraday constant, and z the charge of the metal ions in the electrolyte.

Dividing Eq. (7.35) through the density of kink site positions [*ksp*], the number of atoms arriving during the residence time per kink site position is obtained

$$N_{\tau,ksp} = \frac{i \cdot \tau \cdot N_A}{zF[ksp]} \tag{7.36}$$

If $N_{\tau,ksp}$ is much smaller than [*ksp*], the number of kink site positions decrease, because the renewal of the kink site positions by accumulation of metal ions is so slow that more and more kink site positions disappear because of the natural surface reconstruction. If $N_{\tau,ksp}$ is much larger than [*ksp*], however, the number of kink positions will increase because new surface structures can develop. Thus one can say for stationary conditions that $N_{\tau,ksp}$ should approach the value $N_{\tau,ksp} = 1$. The density of kink site positions [*ksp*] finally therefore approach the value

$$[ksp] \cong \frac{i \cdot \tau \cdot N_A}{zF} \tag{7.37}$$

For example, the density of kink site positions for stationary conditions is 10^{13} cm^{-2} for a current density of $i = 10$ mA cm^{-2}, $z = 2$, and a residence time of $\tau = 10^{-3}$ s.

A special situation that can cause stationary conditions is at the electrochemical equilibrium potential (Nernst potential). The current density at the Nernst potential is the exchange current density i_0. The density of kink site positions $[ksp]_0$ is therefore proportional to the product of exchange current density i_0 and residence time, as was already shown in Eq. (7.34).

Exchange current densities have been determined for many metals. Using i_0 values from the literature and the values for the residence times presented in Table 7.1 a rough estimation of the order of magnitude of the density of kink site positions at the Nernst potential is possible. Examples are presented in Table 7.2.

The calculation of $[ksp]_0$ takes into account the nature of the metal (by τ) and the electrochemical conditions (by i_0). The value of i_0 reflects the different experimental parameters like electrolyte composition, and additives, etc. Therefore, the values of the density of kink site positions provide an image of the surface dynamic at equilibrium conditions. While the absolute values might be questionable, the comparative nature of the described procedures allows at least a comparison of the dynamics of experimental systems. The values for Ag reflect the density of kink site positions for three experimental conditions. The largest value of 5×10^{15} cm^{-2} corresponds to one of the highest experimental exchange current densities observed. In this case, $[ksp]_0$ approaches the surface density of atoms on an ideally smooth surface, but, similar to the residence times in Table 7.1, the $[ksp]_0$ values in Table 7.2 are relative values for the purpose of comparison.

The density of kink positions under deposition conditions $[ksp]$ is a factor that determines the structure of the growing metal phase. At small current densities and small residence times, the crystal growth occurs at the natural growth centers of the substrate for the special electrochemical conditions. The basic reproduction type (BR, in the notation of Fischer[20]) is developed. Otherwise, high current densities and large residence times induce large values

Table 7.2

Density of kink site positions at the Nernst potential calculated with Eq. (7.34) for some metals, residence times taken from Table 7.1, and exchange current densities from the literature[22]

Metal	Electrolyte	τ(s)	i_0(A·cm^{-2})	[ksp] (cm^2)
Ag	0.028 M $[Ag(CN)_3]^{2-}$; 1 M CN^-	2×10^{-4}	3×10^{-3}	4×10^{12}
	0.1M $AgClO_4$; 1M $HClO_4$		4×10^{0}	5×10^{15}
	0.001M $AgClO_4$; 1M $HClO_4$		2×10^{-1}	2×10^{14}
Cd	0.02 M $CdSO_4$; 1.6 M K_2SO_4	2×10^{-9}	2×10^{-3}	1×10^{07}
Cu	1.0 M $CuSO_4$; 1M H_2SO_4	7×10^{-3}	1×10^{-10}	2×10^{06}
	0.1 M $CuSO_4$; 1M H_2SO_4		1×10^{-11}	2×10^{05}
Ni	0.5M $NiSO_4$; pH 2.3 or 4.4	3×10^{0}	1×10^{-6}	9×10^{12}
Zn	1.0 M $ZnSO_4$; 0.01M H_2SO_4	6×10^{-9}	2×10^{-5}	4×10^{05}

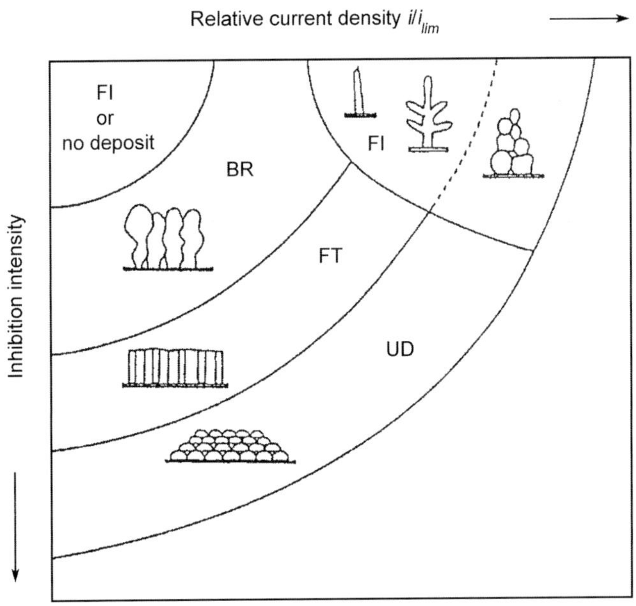

Figure 7.15 Winand diagram. Depending on current density and inhibition activity, Fischer classified four structural types of preferential observation: a base-oriented reproduction type BR, a field-oriented texture type FT, an un-oriented dispersion type UD, and the field-oriented isolation type FI, which is near the origin of deposition or formation of isolated crystals and is not often observed.

of [*ksp*] and irregular crystallographic structure; the field-oriented texture type (FT) or the un-oriented dispersion type (UD) will be formed. Winand has developed an empirical diagram for the prediction of the developing structure by plotting a ratio of the deposition current through the diffusion-limiting current versus the inhibition activity (Figure 7.15).[21]

7.5 EXPERIMENTAL INVESTIGATIONS OF ELECTRODEPOSITION

7.5.1 Electrodeposition on amalgam electrodes

Vetter summarized the main results obtained up to 1967.[22] The principal difficulty in the experimental investigation of metal electrodes in the past was the poorly reproducible preparation of the electrode surface. This problem could be avoided for some electrodes by dissolving metals in liquid mercury. Therefore, many results about the mechanism were measured with amalgam electrodes. On such an electrode no crystallization overvoltage is expected. When it was possible to suppress diffusion depletion, the charge transfer process could be investigated. The charge transfer mechanism could then be determined from the electrochemical reaction orders of the complex molecules or ions.

7.5.2 Investigations on solid electrodes

On solid electrodes reliable former results suffered from the problem of reproducible surface preparation. A big step was made with the introduction of single crystalline electrodes and the reproducible preparation of single crystalline surfaces. The special problems connected with electrochemistry on single crystalline electrodes are described in a number of publications, e.g., by Clavilier *et al.*[23]

Nevertheless, even on polycrystalline electrodes the main steps of the deposition mechanism can be determined. These investigations are based on the evaluation of the structure of the transferred ion. In the following, some classical examples of evaluation of the deposition mechanism on polycrystalline electrodes will be described.

Cadmium

Cadmium plating is still an important plating process in industry. The mechanism of electrodeposition from a sulfate electrolyte was one of the first to be investigated in more detail.[24] The electrode reaction is

$$\mathrm{Cd} \leftrightarrow \mathrm{Cd}^{2+} + 2e^{-} \tag{7.38}$$

The current potential diagram in the Tafel region was already shown in Figure 6.15 with a charge transfer valence $z = 2$. The dependence of the exchange current density on the cadmium ion concentration is shown in Figure 7.16.

The equation for the dependence of the logarithm of the exchange current density on the logarithm of the cadmium ion concentration is

$$\frac{\partial \log i_0}{\partial \log c_{\mathrm{Cd}^{2+}}} = n_{\mathrm{ox,Cd}^{2+}} - \nu_{\mathrm{Cd}^{2+}} \frac{z}{n}(1 - \alpha_{\mathrm{a}}) \tag{7.39}$$

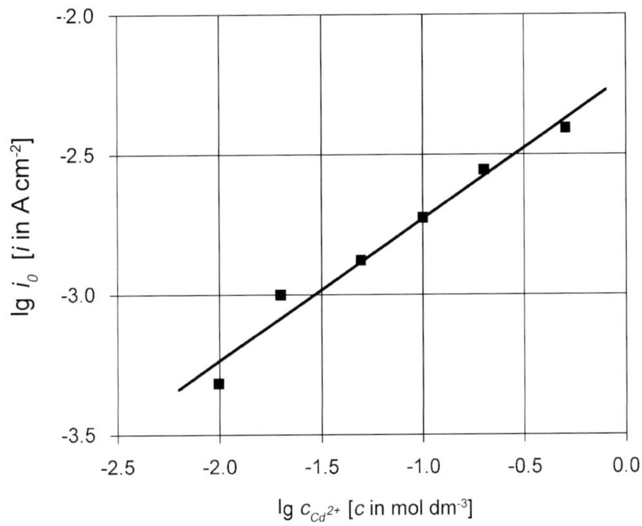

Figure 7.16 Determination of electrochemical reaction orders for the cadmium electrode. $\lg i_0$ versus $\lg c_{Cd^{2+}}$, slope of the line 0.51, and the electrochemical reaction order $n_{ox,Cd^{2+}} = 1$.

The stoichiometric number of cadmium ions in the electrode reaction is $\nu_{Cd^{2+}} = 1$, the stoichiometric number of the electrons in the electrode reaction is $n = 2$, and the valence of the elementary charge transfer reaction is $z = 2$. The charge transfer coefficient determined in Figure 6.15, was $\alpha_a = 0.55$. With the slope of 0.51 in Figure 7.16, the electrochemical rate constant of the cadmium ions is

$$n_{ox,Cd^{2+}} = 0.51 + (1 - \alpha_a) = 0.96 \approx 1.$$

The charge transfer process is identical with the electrode reaction.

Silver

An example for a deposition of a metal from a metal ion complex is the silver electrode.[25] The electrolyte varies between 3×10^{-2} and 1 mol · dm^{-3} KCN. The total ion concentration is kept at a concentration of 1 mol · dm^{-3} by partial substitution of the KCN by KCl. The Ag^+ concentration is in the range of 10^{-2} mol·dm^{-3}.

The electrode process under these conditions is

$$Ag + 3CN^- \leftrightarrow Ag(CN)_3^{2-} + e^- \tag{7.40}$$

The dependence of the logarithm of the exchange current density on the logarithm of the cyanide concentration has been studied and the results are shown in Figure 7.17.

The change of the slope is connected with a change of the mechanism.

$$\frac{\partial \lg i_0}{\partial \lg c_{CN^-}} = -0.29 \quad (c_{Cd^{2+}} < 0.1 \text{mol} \cdot \text{dm}^{-3})$$

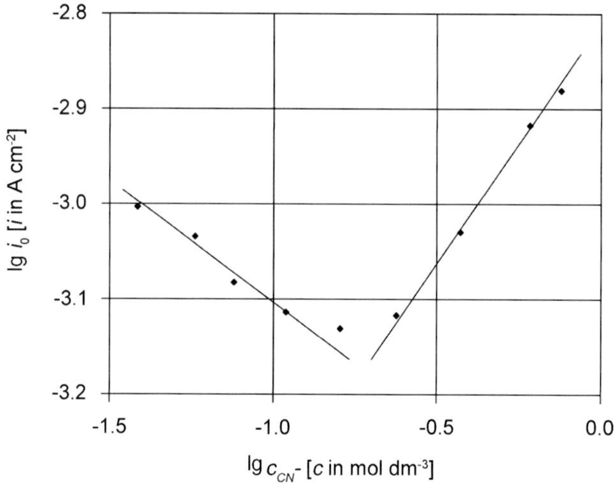

Figure 7.17 Dependence of the logarithm of exchange current density on the logarithm of cyanide concentration for a silver/silver cyanide electrode. The total ion concentration is kept at a concentration of 1 mol·dm^{-3} by partial substitution of the KCN by KCl, and the Ag^+ concentration was 5×10^{-3} mol·dm^{-3}.

$$\frac{\partial \lg i_0}{\partial \lg c_{\mathrm{CN^-}}} = +0.49 \quad (c_{\mathrm{Cd^{2+}}} > 0.2\,\mathrm{mol \cdot dm^{-3}})$$

The charge transfer coefficient (determined by chronoamperometric elimination of diffusion overvoltage) is

$$\alpha_{\mathrm{a}} = 0.44 \quad (c_{\mathrm{Cd^{2+}}} < 0.1\,\mathrm{mol \cdot dm^{-3}})$$

$$\alpha_{\mathrm{a}} = 0.50 \quad (c_{\mathrm{Cd^{2+}}} > 0.2\,\mathrm{mol \cdot dm^{-3}})$$

The valence of the elementary charge transfer reaction was $z = 1$ in both cases. The equation for the electrochemical reaction orders of the complex ions is

$$\frac{\partial \lg i_0}{\partial \lg c_{\mathrm{CN^-}}} = n_{\mathrm{red,CN^-}} + v_{\mathrm{CN^-}} \frac{z}{n} \alpha_{\mathrm{a}} \tag{7.41a}$$

$$\frac{\partial \lg i_0}{\partial \lg c_{\mathrm{CN^-}}} = n_{\mathrm{ox,CN^-}} - v_{\mathrm{CN^-}} \frac{z}{n} (1 - \alpha_{\mathrm{a}}) \tag{7.41b}$$

Then, with $z = 1$, $n = 1$, and $v_{\mathrm{CN^-}} = -3$, the following electrochemical reaction orders are determined for the silver electrode:

$$n_{\mathrm{red,CN^-}} = -0.29 - (-3)\frac{1}{1}0.44 = 1.03$$

$$n_{\mathrm{ox,CN^-}} = -0.29 + (-3)\frac{1}{1}0.56 = -1.97 \quad (c_{\mathrm{Cd^{2+}}} < 0.1\,\mathrm{mol \cdot dm^{-3}})$$

$$n_{\text{red,CN}^-} = +0.44 - (-3)\frac{1}{1}0.50 = 1.99$$

$$n_{\text{ox,CN}^-} = +0.44 + (-3)\frac{1}{1}0.50 = -1.06 \quad (c_{\text{Cd}^{2+}} > 0.2\,\text{mol} \cdot \text{dm}^{-3})$$

The electrochemical reaction orders lead to the following pre- and post-electron transfer equilibrium:
For $c_{\text{Cd}^{2+}} < 0.1$ mol · dm^{-3}

$$S_{\text{red}} = K_{\text{red}}[\text{Ag}]\left[(\text{CN})^-\right]^{+1} \tag{7.42a}$$

$$S_{\text{ox}} = K_{\text{ox}}\left[\text{Ag(CN)}_3^{2-}\right]\left[\text{CN}^-\right]^{-2} \tag{7.42b}$$

the following mechanism is derived:

$$\text{Ag} + \text{CN}^- \leftrightarrow \text{Ag(CN)} + e^- \tag{7.43a}$$

$$\text{AgCN} + 2\,\text{CN}^- \leftrightarrow \text{Ag(CN)}_3^{2-} \tag{7.43b}$$

For $c_{\text{Cd}^{2+}} > 0.2$ mol·dm^{-3}

$$S_{\text{red}} = K_{\text{red}}[\text{Ag}]\left[(\text{CN})^-\right]^{+2} \tag{7.44a}$$

$$S_{\text{ox}} = K_{\text{ox}}\left[\text{Ag(CN)}_3^{2-}\right]\left[\text{CN}^-\right]^{-1} \tag{7.44b}$$

the mechanism is

$$\text{Ag} + 2\text{CN}^- \leftrightarrow \text{Ag(CN)}_2^- + e^- \tag{7.45a}$$

$$\text{Ag(CN)}_2^- + \text{CN}^- \leftrightarrow \text{Ag(CN)}_3^{2-} \tag{7.45b}$$

Spectroscopic methods in general and Raman spectroscopy in particular provide further details for silver deposition from cyanide complexes. This will be described in Section 7.8.

7.5.3 Applications of electrodeposition from aqueous solvents

Many metals are made by electrodeposition. Several metals are prepared by electrolysis in molten salts (Section 7.6). A variety of plating processes have been developed for the metals deposited by electroplating. Quite different, usually aqueous, electrolytes are used. Each electrolyte has characteristic features and produces deposits with quite different properties. As an example, some electrolytes for the plating of copper will be described. Other important plating processes, e.g., zinc, nickel, tin, and lead, are described in detail in the literature.[26]

The deposition mechanism can be quite complex, e.g., chromium can be deposited from hexavalent chromic acid. In this case, a layer of chromium oxides of lower valence is formed on the metal surface.

Electrodeposition of copper is one of the most important deposition processes in industry. Copper is already being refined by electrolysis. Copper plating is the main deposition process for the production of printed circuit boards. Plating processes for many applications are known with varying results for the characteristic properties of the deposits. Examples are

- Copper deposition from sulfuric acid electrolytes
  A standard electrolyte has the following composition:
  160–240 g·dm^{-3} $CuSO_4 \cdot 5H_2O$,
  40–100 g·dm^{-3} H_2SO_4
  30–150 mg·dm^{-3} chloride.
  Typical additives are polyethylene glycol, an organic sulfur compound, e.g., thiophene and a brightener, e.g., cresyl violet.
  The temperature is 20–40 °C and the current density is usually 2–8 A·dm^{-2} but up to 50 °C and up to 100 A·dm^{-2} are also known.
- Copper for printed circuit boards
  Less copper sulfate (20–80 g·dm^{-3} $CuSO_4 \cdot 5H_2O$) and more sulfuric acid are typical for this electrolyte, while the other conditions are similar.
- Cyanide copper
  A typical bath composition is
  25 g·dm^{-3} CuCN
  35 g·dm^{-3} NaCN
  20 g·dm^{-3} Na_2CO_3.
  The temperature is 20–30 °C; the current density is usually 0.2–0.5 A·dm^{-2}.
- An alternative electrolyte for higher current densities contains sodium–potassium tartrate (Rochelle salt) and is therefore called the Rochelle electrolyte. A typical composition is
  70 g·dm^{-3} CuCN
  80 g·dm^{-3} NaCN
  20 g·dm^{-3} Na_2CO_3
  30 g·dm^{-3} $NaKC_4H_4O_6 \cdot 4H_2O$.
  The temperature is 50–70 °C and the current density is usually 1–5 A·dm^{-2}. The electrolyte can be used without additives or with additives, like thiosulfate, selene compounds, and alcanediols.

The mechanism of deposition from these electrolytes is different. While in the sulfuric acid electrolyte copper is dissolved as a two-valent ion, in the cyanide copper electrolyte the copper forms cyanide complexes, $Cu(CN)_2^-$, $Cu(CN)_3^{2-}$, and $Cu(CN)_4^{3-}$.

The mechanism of deposition from sulfuric acid electrolytes proceeds via two consecutive steps with Cu^+_{ad} as intermediate (Bockris-Mattson mechanism). But the dependence of the intermediate of the experimental parameters is not in agreement with all expectations.[27]

The mechanism of deposition of copper from copper cyanide electrolytes is more complex. The mechanism of deposition could be similar to the described mechanism of Ag from silver cyanide electrolytes. Some mechanisms were proposed in the literature.[28]

7.5.4 Parallel reactions

A principal feature of metals if deposited cathodic from the hydrogen electrode is the co-development of hydrogen. This reaction reduces the current efficiency of the deposition process and can have a negative influence on the plated metal layer. The bubbles are responsible for defects in the film surface. Incorporation leads to poor mechanical properties. The hydrogen can penetrate into the plated parts. For example, if depositing zinc on steel it can reduce the mechanical properties of the steel parts.

Another process is reduction of salts of the phosphorous acid. The phosphorous acid is reduced to phosphor, which is then incorporated into the metal in a larger amount. This can improve the properties. An example is nickel–phosphor with increased hardness.

7.6 DEPOSITION FROM NON-AQUEOUS SOLVENTS

So far deposition from an aqueous electrolyte has been discussed. The number of metals that can be deposited from aqueous electrolytes is limited. Metal aluminum is used to demonstrate traditional and new ways of deposition from non-aqueous electrolytes.

7.6.1 Aluminum deposition from a molten salt

The classical process for winning aluminum is one developed by Hall and by Herault in 1886 and 1888.[29] Aluminum oxide (Al_2O_3) of high purity is mixed with cryolite (Na_3AlF_6) in the ratio of 1:10. The mixture melts at approximately 950 °C. The preparation of the pure aluminum oxide, which is usually from bauxite, is the first step in winning aluminum. This process was developed by K.J. Bayer in 1892. The bauxite is treated with concentrated sodium hydroxide and reacted with the aluminum minerals to get sodium aluminate. After aluminum hydroxide precipitates from the aluminate solution, it is then calcinated at 1200–1300 °C.

The modern electrolysis cell is part of a long development. Schematic view of a cell is shown in Figure 7.18.

The electrolysis is made at 4–5 V with 100,000 to 180,000 A. To produce 1000 kg of aluminum one needs 13.5 MW h electrical energy.

In the melt part of the cryolite dissociates[30]

$$Na_3AlF_6 \leftrightarrow 3Na^+ + AlF_6^{3-} \tag{7.46}$$

The aluminum oxide and the ions in the cryolite melt form oxy-fluoride complexes, e.g.,

$$Al_2OF_6^{2-},\ Al_2OF_8^{4-},\ Al_2OF_{10}^{6-},\ \text{or}\ Al_2O_2F_4^{2-}.$$

From kinetic experiments, the following mechanism of aluminum deposition at the liquid aluminum electrode was derived:[31]

$$AlF_6^{3-} \leftrightarrow AlF_4^- + 2F^- \tag{7.47a}$$

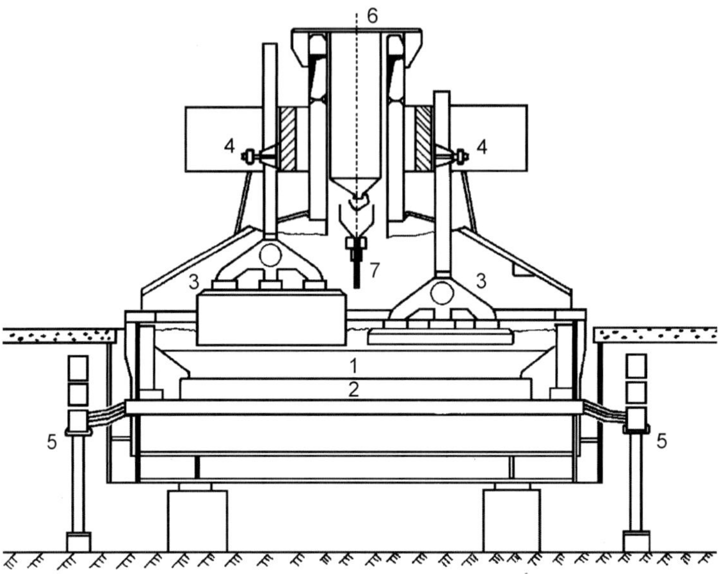

Figure 7.18 View of an aluminum electrolysis cell, in which (1) the aluminum collects as liquid metal at the bottom of the cell (2) on the carbon cathode, (3) the height-adaptable carbon anodes react with the aluminum oxide to form CO/CO_2, (4 and 5) the contact rails are for currents up to 180,000 A, (6) the charging is made by the construction in the middle of the cell, and (7) the liquid aluminum is regularly sucked off.

$$AlF_4^- + 2e^- \leftrightarrow AlF_2^- + 2F^- \tag{7.47b}$$

$$AlF_2^- + e^- \leftrightarrow Al + 2F^- \tag{7.47c}$$

In a technical deposition process one also has to look for the counter-electrode, which in aluminum electrolysis is a graphite electrode. The processes at the carbon electrode are described by the equations

$$Al_2OF_{10}^{6-} \rightarrow Al_2OF_6^{2-} + 4F^- \tag{7.48a}$$

$$Al_2OF_6^{2-} + C \rightarrow 2AlF_3 + CO_{ad} + 2e^- \tag{7.48b}$$

$$Al_2O_2F_4^{2-} + CO_{ad} \rightarrow Al_2OF_4 + CO_2 + 2e^- \tag{7.48c}$$

The electrolysis needs a large amount of energy. Increasing energy prices may open the way to alternative processes, e.g., to win aluminum from chloride melts.

7.6.2 Aluminum deposition from an organic electrolyte

Several technical processes were developed to deposit aluminum from an organic electrolyte. One example is the Sigal process using alkaline halide–alkyl complexes of aluminum in toluene.[32] Another example is the Real process that works with an electrolyte of aluminum chloride and aluminum hydride in tetrahydrofuran. A third example is the Natal process which deposits aluminum from an electrolyte of aluminum chloride and aluminum hydride in toluene.

In the following, the Sigal process will be described in more detail. The typical complex in this process is shown in Figure 7.19.[33] The deposition of aluminum from these complexes is a rather slow process. A mechanism was discussed by Kautek and Birkle.[34] Typical process parameters are 5–30 V, 0.2–2 A·dm^{-2} DC, and 90 °C.

7.6.3 Aluminum deposition from ionic liquids

The electrodeposition of aluminum from ionic liquids has been intensively investigated. The principal rule for ionic liquids is the special design of the anion–cation combination for the metal to be deposited. In Figure 7.20 the usual ionic liquid for aluminum deposition is shown.

In this electrolyte, after an *in situ* electrochemical etching of the steel surface to remove oxides, a film with very good adhesion is formed (Figure 7.21).[35]

A quite reversible cyclic voltammogram has been observed (Figure 7.22).

It has also been observed that the particle size is a sensitive function of the cation. The deposition of nanostructured aluminum is a further option of ionic liquids.[36]

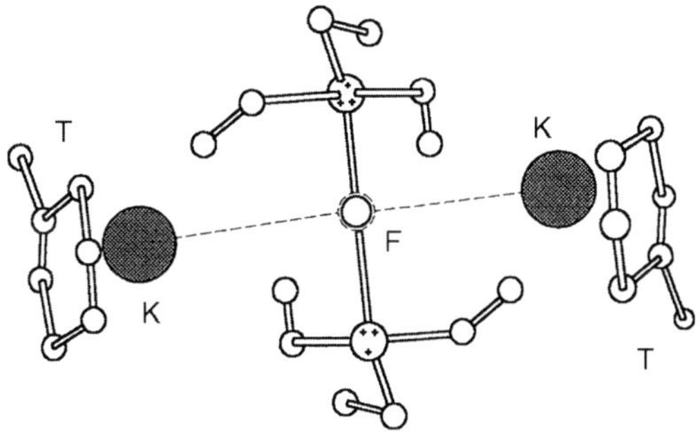

Figure 7.19 KF-ethyl-aluminum complex or $K_2Al_2C_3H_5F^+$ in toluene, K for the potassium ions, F for fluoride ion, and T for the toluene molecules.

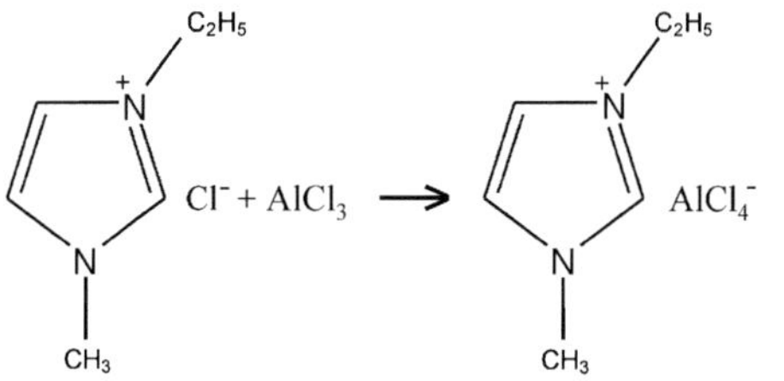

Figure 7.20 A typical anion–cation combination in an ionic liquid used for aluminum deposition on steel, ethyl-methyl-imidazolium cation, and tetra-aluminum anion.

Figure 7.21 Side view of an aluminum film on steel deposited from an ionic melt that has been formed of ethyl-methyl-imidazolium-chloride and aluminum-tri-chloride. (Reproduced with permission from Ref. [35], © 2006, Elsevier.)

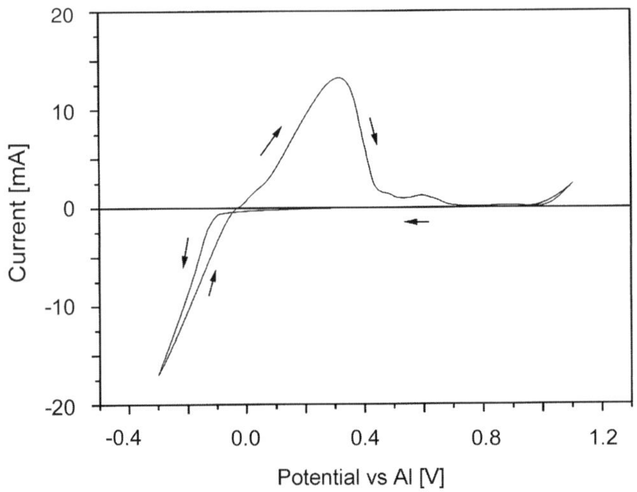

Figure 7.22 A cyclic voltammogram of an aluminum film on steel deposited from an ionic melt formed of ethyl-methyl-imidazolium-chloride and aluminum-tri-chloride. (Reproduced with permission from Ref. [35], © 2006, Elsevier.)

7.7 ADDITIVES

So far an ideal interface between a metal and an electrolyte has been considered. Neither the solvent nor any other component of the electrolyte (except the metal ions) can adsorb on the metal interface. The advantage of electrochemical deposition is the possibility to modify this ideal surface by additives, which adsorb on the metal surface. The additives have different effects, which can be used as a characterizing feature like brightening or leveling the deposit.

The general description of the action of additives is very complicated and difficult to predict in a general manner. But it is possible to extract rules from the complex nature of the absorption process, which can guide us through the chaos. The description of adsorption by adsorption isotherms has already been described in Chapter 4. The Gibbs free energy is the characteristic quantity that describes how strongly a complex molecule will interact with metal surfaces.

7.7.1 Adsorption, the hard–soft concept

The adsorption isotherms provide a quantitative description of the surface in the presence of an additive. But there is only enough data to calculate the isotherms for a few additives. In order to develop new electrolytes for electrochemical deposition, more general rules for obtaining a feeling for the chemical nature of the additives are needed. Such a guide is the Pearson concept of hard and soft acids and bases. It was originally developed to predict the tendency of molecules to form complexes with metal ions.[37] Later on, it was applied to the adsorption of molecules or ions on metal surfaces,[38] and to adsorption on charged faces of different crystallographic structures.[39]

Pearson looked into the stability between Lewis acids (electron acceptors) and Lewis bases (electron donors) and found that a stable complex forms between a hard acid and a hard base or between a soft acid and a soft base.

For the central ion of a complex the character of the electronic shell is connected with the nature of the atomic orbital. The s, p, d, and f electrons are characterized by an increasing distance to the kernel and therefore can be seen as increasingly polarizable electrons. In the language of Pearson the character of these electrons varies from extremely hard (s-electrons) to extremely soft (f-electrons). The hardness increases with ionization and for different valance states with an increased oxidation state.

One way to measure softness and hardness of molecules is the polarizability of the electron shell. A first criterion is the number of σ- and π-electrons. With an increasing number of π-electrons, the polarizability and the soft character usually increase.

Another characteristic property related to hardness and softness is electronegativity. Low electronegativity is connected with a soft character and high electronegativity with a hard character. This is illustrated in Figure 7.23. For the interaction of the molecules with a metal atom or metal ion only part of a molecule might be responsible. Therefore, it is possible to apply the hard/soft concept to special parts of the molecule.

For the elements in the periodic table one can use electronegativity values as a measure of hardness (Chapter 2, Figure 2.4). The hardness increases from Li to F and decreases from Li to Cs. For the transition metals the hardness would increase from Sc to Zn and decrease from Cu to Au.

The Pearson concept can be transferred to metal surfaces. In this case, in addition to the character of the bonding electrons, the crystallographic structure of the metal surface and the surface topography must be taken into account. For cubic close-packed structures the most prominent crystallographic surfaces are the (111), (100), and (110) interfaces. The (111) face is the closest packed structure and has the hardest surface topography followed by the (100) and (110) interfaces (the softest one). The hard–soft character of a polycrystalline

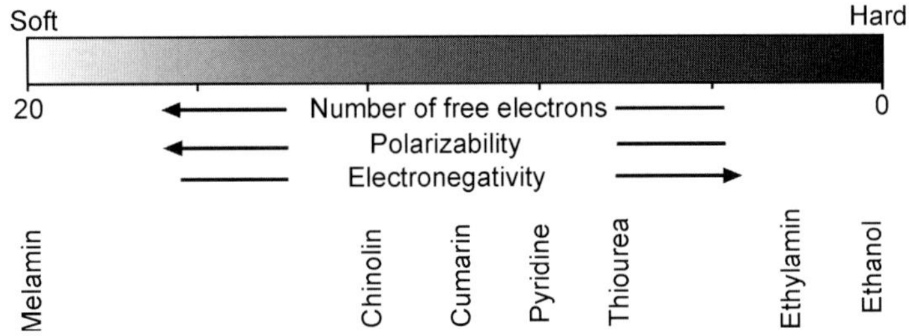

Figure 7.23 Dependence of the hard–soft character of a molecule or molecule group on polarizability and electronegativity.

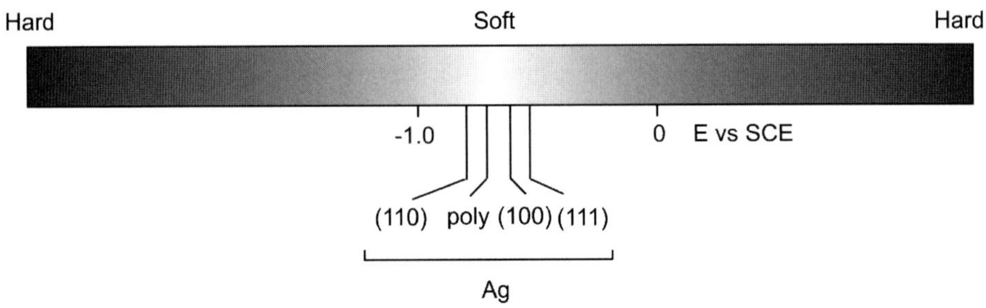

Figure 7.24 Hard–soft characterization of different crystallographic faces.

surface is represented by an average value, usually between the (100) and (110) faces. This is illustrated in Figure 7.24.

On an electrode surface the hard–soft character becomes potential dependent. The metal will have a soft character at the potential of zero charge. It is obvious that at high positive potentials (low electron density) the metal surface has a hard character. Pearson suggested the same hard character for high negative potentials (high electron density) as is shown in Figure 7.24.

With increasing step density the softness increases. Furthermore, the softness increases with the surface density of kink site positions, or, quite generally, with increasing surface roughness. On polycrystalline surfaces, the same rules apply and the softness increases with an increasing amount of surface roughness.

On the basis of these general rules, a first characterization of the interaction of additives with electrode surfaces is provided. A hard molecule or ion will primarily adsorb on hard surfaces and hard surface areas. Soft molecules will preferentially adsorb on soft surfaces and soft surface structures.

7.7.2 Influence of additives on deposition at different crystallographic faces

The rate of the crystallization process of a pure metal differs for different crystallographic faces. For comparison, the net deposition current density can be split into partial current densities i_{hkl}.

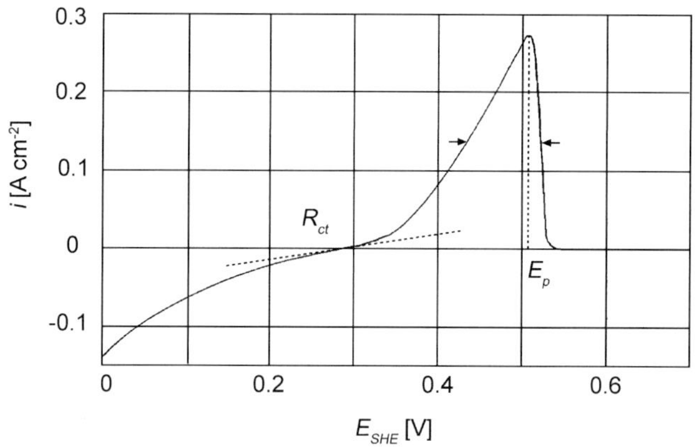

Figure 7.25 Principles of anodic voltammetric stripping to study the influence of additives on electrodeposition.

The growth rate, characterized by the partial current density i_{hkl},. is the factor determining the equilibrium shape of a crystal. The faces with the smallest partial current densities i_{hkl}. form the equilibrium form of the crystal.

Additives can control the growth rate by blocking surface processes, e.g., surface diffusion or the transfer in and out of kink site positions. Therefore, interaction of additives with the different crystallographic structures is a dominating factor in the development of crystallographic shape and structure.

7.7.3 Anodic stripping to study additive behavior

How should the adsorption of additives be measured? The most straightforward way is the determination of surface concentration. Some methods were already discussed in Chapter 4. These methods are complemented by a very effective method, anodic stripping. The application of this method to study the influence of additives in electrocrystallization was first described by Ogden and Tench.[40] The most important application was achieved in the process control of copper deposition. This is explained in Figure 7.25.

On the cathodic potential sweep copper is deposited in the cathodic region with a rate depending on the experimental parameters. In the anodic scan, after passing the equilibrium potential, the copper is dissolved in the pronounced anodic peak. The charge obtained by integration over this peak is representative for the copper deposited in the cathodic region. The influence of additives on the charge is a measure for the action of additives on the copper deposition process. An example is the influence of dithiadecyl-disodium-sulfonate (Figure 7.26).[41]

7.8 OPTICAL SPECTROSCOPY TO STUDY METAL DEPOSITION

Reflection spectroscopy, Raman spectroscopy, and ellipsometry complement the various electrochemical methods to study metal deposition. The optical methods can be used for a direct monitoring of the deposition process. The great advantage of optical spectroscopy

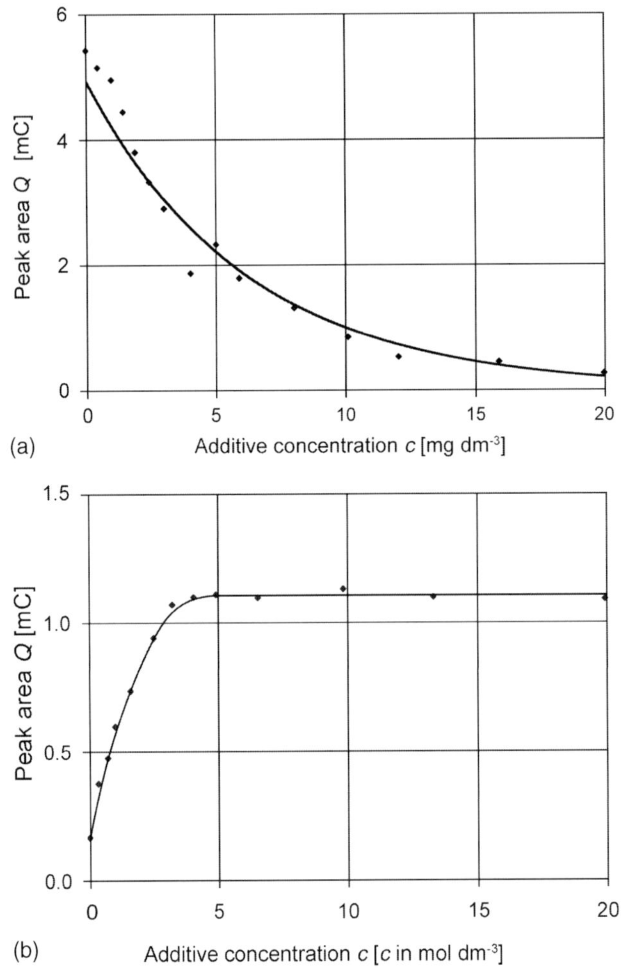

Figure 7.26 Influence of concentration of dithiadecyl-disodium-sulfonate on the amount of deposited copper, determined by anodic stripping voltammetry. (a) Pure sulfuric acid electrolyte (2 mol·dm^{-3} H_2SO_4, 0.1 mol·dm^{-3} $CuSO_4$, and (b) sulfuric acid electrolyte (10^{-4} mol·dm^{-3} polyethylene glycol).

is the detailed information about structure, orientation, and other properties of adsorbed molecules or ions. Besides collections of summarizing papers,[42] the IUPAC commission on electrochemistry has published a short compilation of the different methods.[43] Aspects of spectroscopic investigations of organic monolayers have been discussed in a further review.[44]

7.8.1 Raman spectroscopy on silver in cyanide electrolytes

Silver deposition is an example of a metal deposition, which can be monitored by Raman spectroscopy. A Raman spectrometer is now usually a Raman microscope using confocal

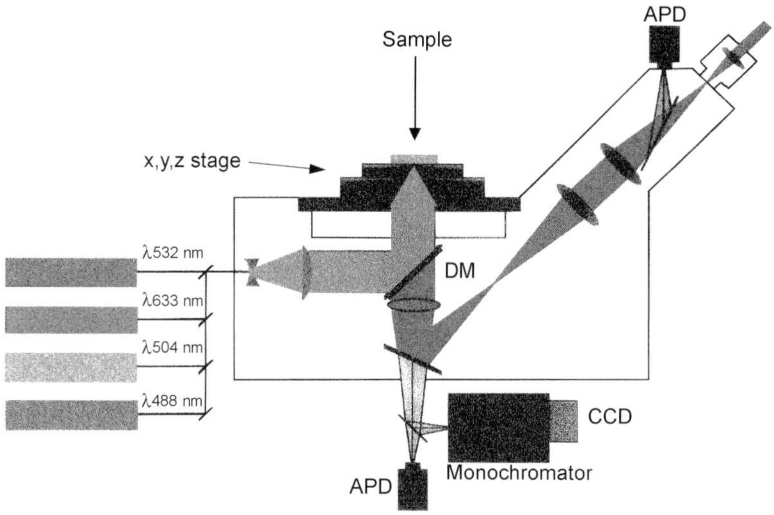

Figure 7.27 Confocal Raman microscope spectrometer. The sample is illuminated by a laser light reflected on a dielectric mirror (DM) combined with a highly selective Notch filter. The Raman signals scattered from the sample are detected by highly selective avalanche photo diodes (APD) or via a monochromator on a CCD detector. The sample is scanned in two or three dimensions and the image is built-up point by point in a computer.[45]

imaging of the surface and registration of the spectrum. A schematic description of a spectrometer is shown in Figure 7.27.[45] The sample is illuminated by a laser beam, which causes the scattering of Raman signals. A so-called Notch filter is only transparent for the Raman frequencies, which can then be detected by high sensitive avalanche photodiodes or, after passing a monochromator, on a CCD detector.

Silver is an example of a metal that shows the surface-enhanced Raman effect. After a special surface treatment, the signal of a molecular group on the surface of the metal is enhanced by several orders of magnitude. One successful surface treatment is deposition of silver. So, after starting silver deposition from a cyanide electrolyte on a platinum electrode, a Raman signal of the CN-stretch vibration develops and reaches a limiting value (Figure 7.28).[46]

The different complexes in a silver cyanide electrolyte show different Raman signals of the CN-stretch vibration. The peak observed in Figure 7.28 is usually explained as the $Ag(CN)_3^{2-}$ peak.[47] The peak around 2110 cm^{-1} (2080–2140 cm^{-1}) was observed at –0.7 V_{SCE}. Switching to 0 V_{SCE} a shift to 2140 cm^{-1} (2110–2175 cm^{-1}) was observed, which is the frequency of $Ag(CN)_2^-$. Compared to the deposition mechanism (Section 7.5.2) one has to conclude that the $Ag(CN)_3^{2-}$ complex dominates even on the surface.

There is evidence that the extreme enhancement of the Raman signal on a silver deposit comes from colloidal silver particles. Considering the special situation of a silver nucleus with adsorbed CN^- ions, other explanations of the Raman results are possible, e.g., an adsorption of CN^- on clusters of different sizes and charges.[48]

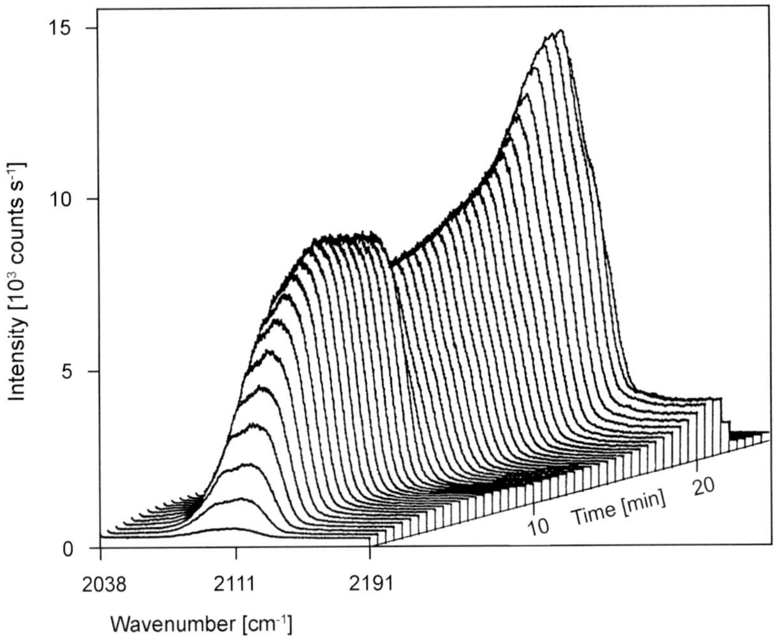

Figure 7.28 Development of the Raman spectrum of the CN-stretch vibration in the $Ag(CN)_3^{2-}$ complex after switching to a deposition potential of silver, 0.4 mol dm^{-3} KNO_3, 0.028 mol dm^{-3} KCN, 0.005 mol dm^{-3} $AgNO_3$, λ = 514.5 cm^{-1}, 30 mW cm^{-2}, and deposition on a Pt electrode. After switching to an anodic potential of dissolution, the Raman signal is observed (even increases) until the silver deposit has dissolved.

7.8.2 Raman spectroscopy of organic additives

Surface-enhanced Raman spectroscopy was intensively used to study organic additives on silver, copper, and gold.[49,50] An example is an investigation of the orientation of nicotinic acid on silver.[51,28] The Raman spectra were measured at different pH values. This is shown in Figure 7.29a. Preferential adsorption of the nicotinic acid anion was observed. The dissociation reactions are shown in Figure 7.29b.

Another measurement was made as a function of the potential (Figure 7.30). The final conclusion is that the nicotinic acid anion was adsorbed flat on the surface for pH < 9 and was in an upright position for pH > 9. The nicotinic acid anion was bound to the silver surface by the nitrogen atom (Figure 7.31).[31]

The origin of surface-enhanced Raman spectroscopy is still being discussed. In addition to the average enhancement of 10^3–10^5, the so-called "hot spots" have been observed with enhancement factors up to 10^{10}.[52–54] Further research of application of this technique to electrodeposition is promising.

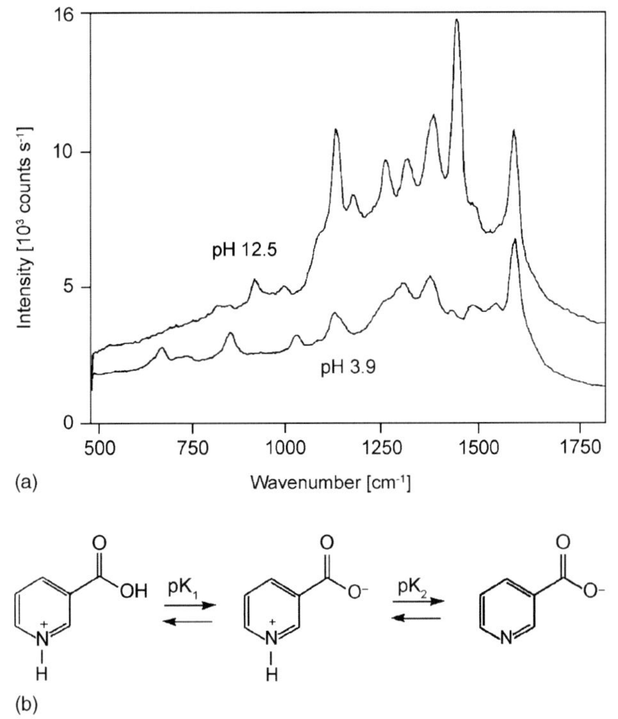

Figure 7.29 (a) Raman spectra of Ag in an electrolyte-containing nicotinic acid at pH 3.9 and 12.5, the spectra can be attributed to the spectrum of the anion. (b) Dissociation equilibria of nicotinic acid.[51]

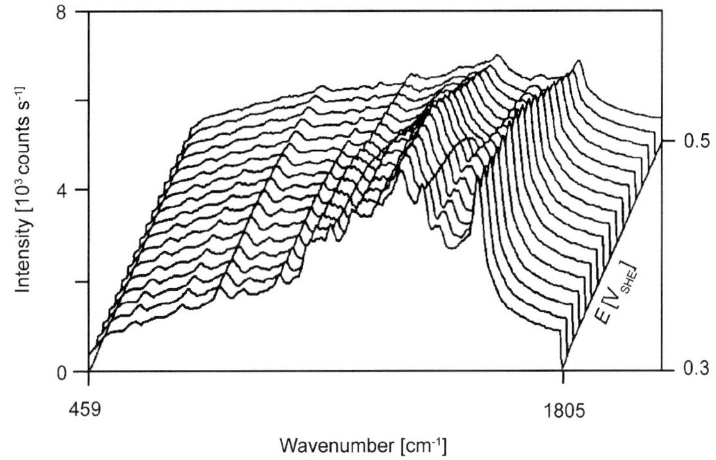

Figure 7.30 Raman spectra of Ag in an electrolyte containing nicotinic acid as a function of the potential.[51]

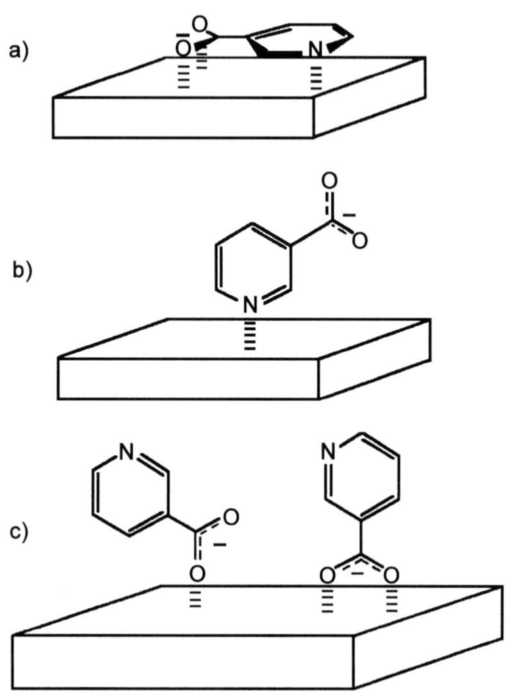

Figure 7.31 Orientation and structure of nicotinic acid on silver determined by surface-enhanced Raman spectroscopy. (a) Flat orientation, pH < 9, (b) perpendicular orientation, pH > 9, and (c) alternative perpendicular orientation (not observed).[51]

REFERENCES

1. T. Erdey-Gruz, M. Volmer, Z. Phys. Chem., *157*, 165 (1931).
2. E. Budevski, G. Staikov, J.W. Lorenz, Electrochemical Phase Formation and Growth, VCH, Weinheim, 1996.
3. M. Volmer, A. Weber, Z. Phys. Chem., *119*, 277 (1926).
4. A. Milchev, S. Stoyanov, J. Electroanal. Chem., *72*, 33 (1976).
5. S. Toshev, I. Markov, Ber. Bunsenges. Phys. Chem., *73*, 184 (1969).
6. D. Kaschchiev, J. Chem. Phys., *76*, 5098 (1982).
7. D.W. Oxtoby, D. Kaschchiev, J. Chem. Phys., *100*, 7665 (1994).
8. G.J. Hills, D.J. Schiffrin, J. Thomson, Electrochim. Acta, *19*, 657 (1974).
9. M. Avrami, J. Chem. Phys., *7*, 1103 (1939); *8*, 212 (1940); *9*, 107 (1941).
10. B. Scharifker, G. Hills, Electrochim. Acta, *28*, 879 (1983).
11. B. Scharifker, J. Mostany, J. Electroanal. Chem., *177*, 13 (1984).
12. L. Heerman, A. Tarallo, J. Electroanal. Chem., *470*, 70 (1999).
13. H. Liu, R.M. Penner, J. Phys. Chem. B, *104*, 9131 (2000).
14. M. Ueda, H. Dietz, A. Anders, H. Kneppe, A. Meixner, W. Plieth, Electrochim. Acta, *48*, 377 (2002).
15. W.K. Burton, N. Cabrera, F.C. Frank, Phil. Trans. Roy. Soc. A, *243*, 299 (1951).
16. W. Lorenz (Leipzig), Z. Elektrochem., *57*, 382 (1953).
17. D.A. Vermilyea, J. Chem. Phys., *25*, 1254 (1956).
18. H. Gerischer, Z. Elektrochem., *62*, 256 (1958).
19. W. Plieth, Z. Phys. Chem., *217*, 383 (2003).

20. H. Fischer, Elektrolytische Abscheidung von Metallen, Springer, Berlin, 1954.
21. M. Winand, J. Appl. Electrochem., *21*, 377 (1991).
22. K.J. Vetter, Electrochemical Kinetics, Academic Press, New York, 1967.
23. J. Clavilier, R. Faure, G. Guinet, R. Durand, J. Electroanal. Chem., *107*, 205 (1980).
24. W. Lorenz (Leipzig), Z. Elektrochem., *58*, 912 (1954).
25. W. Vielstich, H. Gerischer, Z. Phys. Chem. NF, *4*, 10 (1955).
26. M. Schlesinger, M. Paunovic (Eds.), Modern Electroplating, 4th Ed., Wiley, New York, 2000.
27. A. De Agostino, E. Schmidt, W.J. Lorenz, Electrochim. Acta, *34*, 1243 (1989).
28. D. Chu, P.S. Fedkiw, J. Electroanal. Chem., *345*, 107 (1993).
29. W.E. Haupin, Electrometallurgy of Aluminum, in B.E. Conway, E. Yeager, R.E. White (Eds.), Comprehensive Treatment of Electrochemistry, Plenum Press, New York, 1982.
30. K. Grjotheim, C. Krohn, M. Malinovsky, K. Maliasovsky, J. Thronstad, Aluminium Electrolysis, 2nd Ed., Aluminium-Verlag, Düsseldorf, 1992.
31. A. Kisza, J. Kazmierczak, J. Ihonstad, T. Eidet, J. Hives, The Kinetics and Mechanism of the Electrode Reactions in Aluminium Electrolysis, in H. Wendt (Ed.), Molten Salt Forum 5, Trans Tech Publications LTD, New York, 1998, p. 263.
32. R. Dötzer, Chem. Ing. Tech., *35*, 616 (1965); *45*, 653 (1973).
33. W. Kautek, J. Fischer, Galvano-Aluminium, Alcotec Beschichtungsanlagen GmbH, Overath-Vilkerath, 1994.
34. W. Kautek, S. Birkle, Electrochim. Acta, *34*, 1213 (1989).
35. Q.X. Liu, S. Zein El Abedin, F. Endres, Surf. Coating Tech., *201*, 1352 (2006).
36. S. Zein El Abedin, E.M. Moustafa, R. Hempelmann, H. Natter, F. Endres, Electrochem. Commun., *7*, 1116 (2005).
37. R.G. Pearson, Hard and Soft Acids and Bases, Dowden, Hutchinson and Ross, Stroudsberg, PA, 1973.
38. R.G. Pearson, J. Am. Chem. Soc., *90*, 319 (1968).
39. W. Plieth, Electrochim. Acta, *37*, 2115 (1992).
40. C. Ogden, D. Tench, J. Electrochem. Soc., *125*, 194, 1218 (1978).
41. W. Plieth, Metalloberfläche, *44*, 67 (1990).
42. C. Gutierrez de la Fe, C. Melendres (Eds.), Spectroscopic and Diffraction techniques in Interfacial Electrochemistry, Kluwer Academic Publisher, Dordrecht, 1990.
43. W. Plieth, G.S. Wilson, C. Gutierrez de la Fe, Pure Appl. Chem., *70*, 1395, 2409 (1998).
44. W. Plieth, W. Kozlowski, T. Twomey, Reflectance Spectroscopy and Ellipsometry of Organic Monolayers, in J. Lipkowski, P.N. Ross (Eds.), Frontiers of Electrochemistry, Adsorption of Molecules at Metal Electrodes, VCH, Weinheim, Cambridge 1992, Chapter 5, p. 239.
45. W. Plieth, H. Dietz, G. Sandmann, A. Meixner, D. Weber, P. Moyer, J. Schmidt, Electrochim. Acta, *44*, 2025 (1999).
46. G. Lacconi, B. Reents, W. Plieth, J. Electroanal. Chem., *325*, 207 (1992).
47. A. Otto, J. Billmann, J. Eickmaus, U. Ertürk, C. Pettenkofer, Surf. Sci., *138*, 319 (1084).
48. B. Reents, W. Plieth, Growth Morphology and Molecular Orientation of Additives in Electrocrystallization Studied by Surface Enhanced Raman Spectroscopy, in W.J. Lorenz, W. Plieth (Eds.), Electrochemical Nanotechnology, Wiley-VCH, Weinheim, 1998, p. 277.
49. B. Bossini, Trans. IMF, *80*, 25 (2002).
50. Z.Q. Tian, Y.Z. Lian, M. Fleischmann, Electrochim. Acta, *35*, 879 (1990).
51. J. Barthelmes, G. Pofahl, W. Plieth, J. Electroanal. Chem., *329*, 329 (1992).
52. S. Nie, S.R. Emeroy, Science, *275*, 1102 (1997).
53. K. Kneip, Y. Wang, H. Kneip, I. Itzkan, R.R. Dasari, M.S. Feld, Phys. Rev. Lett., *78*, 1667 (1996).
54. W. Plieth, H. Dietz, G. Sandmann, A.J. Meixner, H. Kneppe, M. Weber, in D. Landolt, D. Matlosz, Y. Sato (Eds.), Fundamentals in Electrochemical Deposition and Dissolution, The Electrochemical Society, Pennington, Vol. 99–33, 1999.

– 8 –

Deposition of Alloys

8.1 DEPOSITION POTENTIAL AND EQUILIBRIUM POTENTIAL

The bulk deposition of a pure metal A occurs if the applied potential is more negative than the Nernst equilibrium potential $E < E_{0,A}$. If a second metal B having a more negative Nernst potential, $E_{0,B} < E_{0,A}$ is deposited together with metal A, the deposition of B can begin at a potential E in the interval $E_{0,B} \leq E < E_{0,A}$ and in any case for $E < E_{0,B}$.

The electrode reaction is described by the equation

$$A_xB_{1-x} \leftrightarrow xA^{m+} + (1-x)B^{n+} + [mx + (1-x)n]e^- \tag{8.1}$$

one can define an equilibrium potential of an alloy.[1]

$$E_{0,A_xB_{1-x}} = E^{\ominus}_{0,A_xB_{1-x}} + \frac{RT}{\left[xm + (1-x)n\right]F} \ln a_{A^{m+}} a_{B^{n+}} \tag{8.2}$$

with

$$E^{\ominus}_{0,A_xB_{1-x}} = -\frac{\Delta^{\ominus} G_{A_xB_{1-x}}}{[xm + (1-x)n]F} \tag{8.3}$$

Equilibrium potential, standard potential, and the standard Gibbs energy depend on the alloy composition.

$$\Delta G_{A_xB_{1-x}} = \Delta_{for} G_{A_xB_{1-x}} - x\mu_{A^{m+}} - (1-x)\mu_{B^{n+}} \tag{8.4}$$

The alloy Gibbs energy is the difference between the Gibbs energy of formation and the chemical potentials of the ions in the electrolyte.

The limiting value of the standard Gibbs energy for $1 - x \rightarrow 0$ can be called first standard Gibbs energy, a notation used for a diluted solution of B in A. The first deposition can occur for potentials negative of the potential derived from this first standard Gibbs energy.

8.2 ALLOY NUCLEATION AND GROWTH: THE PARTIAL CURRENT CONCEPT

The first steps of alloy deposition differ from the nucleation of a pure metal. If one metal interacts stronger with the substrate, an upd layer formation or even a formation of several layers of a pure phase can precede the formation of alloy nuclei. Special models have been developed for the nucleation process. Milchev and Lacmann[2] and Milchev[3] have discussed specific questions of alloy nucleation.

The growth follows the same principles as were discussed for pure metals. The rate of deposition of each component can be described by a partial deposition current density i. The total deposition rate is then the sum of all partial current densities (Wagner and Traud[4]), for a binary alloy

$$i = i_A + i_B$$

If the deposition of the two components is independent of each other, the total current potential diagram is the sum of the current potential diagrams of the components. An example is the deposition of copper–nickel shown in Figure 8.1. In the voltammogram one observes at more positive potentials the deposition of pure copper, which is diffusion controlled and follows the rules of mass transport (Chapter 5) with the limiting current of Cu^{2+} diffusion. At the potential, where Ni deposition begins, the Ni deposition is first charge transfer controlled. The partial current is described by the Butler–Volmer equation (Chapter 6). If the charge transfer dominates at larger overvoltage, Tafel lines will be found. The increasing partial current of nickel deposition is added to the limiting partial current of Cu deposition. With increasing polarization in cathodic direction, the Ni ion discharge also becomes diffusion controlled and the total deposition rate is the sum of the limiting currents.

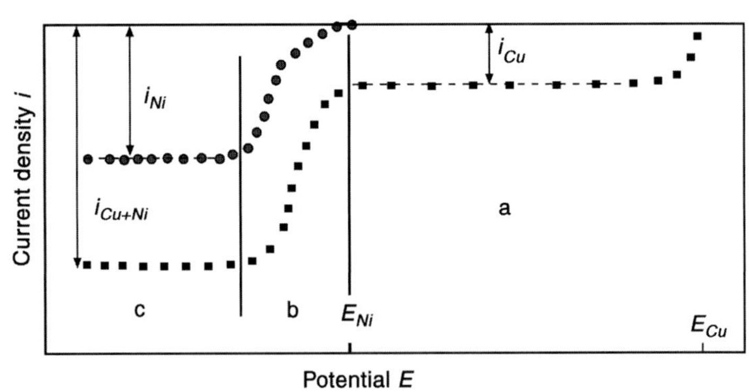

Figure 8.1 Voltammogram of CuNi alloy deposition with partial currents of Cu and Ni deposition and total deposition current; (a) copper deposition, (b) irregular co-deposition, (c) regular co-deposition. The names of the different alloy deposition regions refer to the classification of Brenner (see the following section).

8.3 BRENNER'S ALLOY CLASSIFICATION

Alloys are usually classified following a phenomenological suggestion of Brenner.[5] For his definition Brenner used a so-called composition reference line. On this basis the mole fraction of the component A in the alloy can be plotted versus the concentration ratio in the electrolyte $c_A/(c_A + c_B)$ (actually the mole fraction of ions A in the electrolyte, Eq. (8.5)):

$$x_A = \text{function}\left(\frac{c_A}{c_A + c_B}\right) \tag{8.5}$$

Brenner distinguished between a normal and abnormal alloy deposition

1. *Normal alloy deposition*
   Deposition is as expected from the equilibrium potentials of the components.
   Within this group there are three subgroups:
   (a) Equilibrium co-deposition, which could be observed for very similar values of the equilibrium potentials of the alloy components. Kinetic differences of the deposition of both components can mask the ideal behavior. (b) Regular co-deposition which is diffusion controlled, i.e., the CuNi deposition at high overvoltage (Figure 8.1). (c) Irregular co-deposition which is fully or partially charge transfer controlled, i.e., the CuNi deposition at medium overvoltage (Figure 8.1).

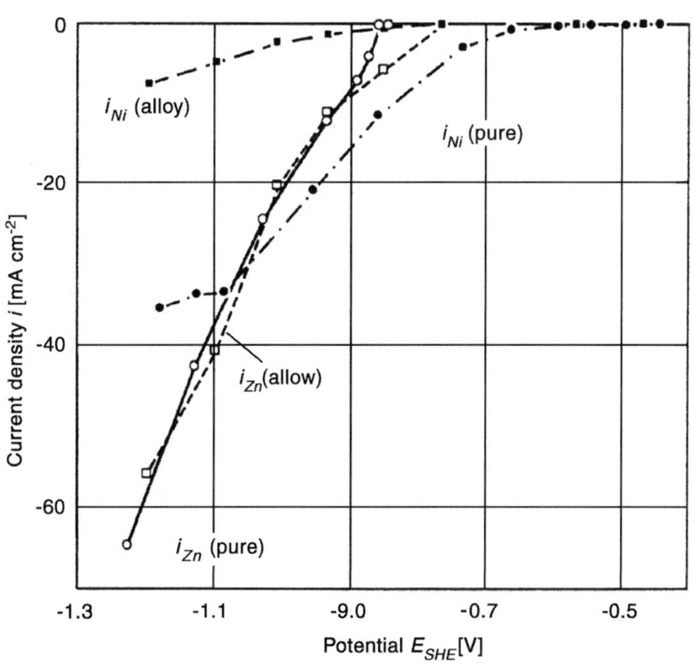

Figure 8.2 Anomalous deposition of zinc in zinc–nickel alloys. (Reproduced with permission from Ref. [6], © 1994, Elsevier.)

2. *Abnormal alloy deposition*
Deposition is unexpected because of the equilibrium potentials; in this group two subgroups were defined:
(a) Anomalous co-deposition, preferentially the less noble metal deposits. An example is the preferential deposition of zinc in nickel zinc alloys (Figure 8.2) observed at higher current densities (higher potentials).
(b) Induced co-deposition in which an inducing metal stimulates deposition of a reluctant metal. Examples are the induced deposition of tungsten or molybdenum with the metals iron, cobalt, or nickel. This will be described in more detail in Section 8.7.3.

8.4 MIXED POTENTIAL THEORY

Landolt gave a more detailed description of the Wagner–Traud concept of partial deposition currents.[6] In Figure 8.3 different situations of combinations of partial current densities are described in Evans diagrams. In the gray areas the less noble component is preferentially deposited.

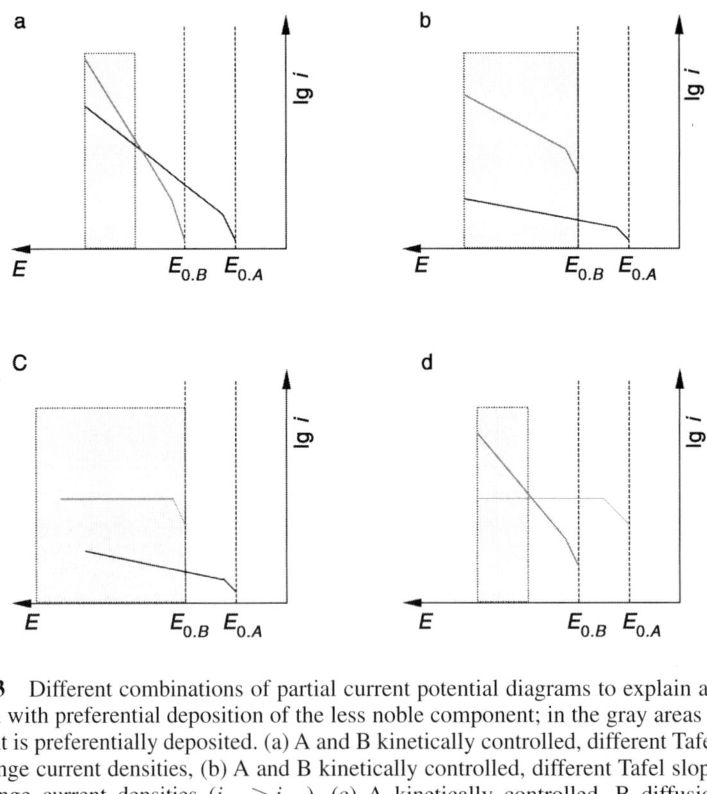

Figure 8.3 Different combinations of partial current potential diagrams to explain anomalous co-deposition with preferential deposition of the less noble component; in the gray areas the less noble component is preferentially deposited. (a) A and B kinetically controlled, different Tafel slopes, similar exchange current densities, (b) A and B kinetically controlled, different Tafel slopes and different exchange current densities ($i_{0,B} > i_{0,A}$), (c) A kinetically controlled, B diffusion controlled, different exchange current densities ($i_{0,B} > i_{0,A}$), (d) A diffusion controlled, B kinetically controlled, similar exchange current densities. (Reproduced with permission from Ref. [6], © 1994, Elsevier.)

In a classification based on the kinetic model, Landau distinguished three groups of alloys.

- Non-interactive systems such as the copper–nickel system.
- Charge transfer coupled systems in which the partial currents influence each other. One component inhibits the second one, e.g., Zn inhibits Ni (Figure 8.2). Or one component catalyzes the deposition of the second component, e.g., Ni catalyzes Mo deposition (Section 8.7.3).
- Mass transport coupled systems in which the product of another cathodic reaction catalyzes the deposition process. It is assumed that adsorbed hydrogen atoms catalyze nickel deposition. The hydrogen development is coupled with Ni reduction and is therefore called a mass transport coupled system.

8.5 SURFACE SELECTIVITY IN ALLOY DEPOSITION

8.5.1 Kink site positions of alloys

In Chapter 7, the outstanding role of a kink site position in the process of formation of a crystalline phase was shown. Kink site positions are of similar importance in alloy deposition.[7] But for alloys several kink site positions must be defined depending on the lattice neighbors of the kink site position. In the following the situation of a bi-component alloy AB of stoichiometric lattice structure will be described. In this case one must consider two different kink site positions A* and B* (Figure 8.4).

The two kink site positions can react with the two types of metal ions in the electrolyte to form four new kink site positions. Kink site position A* reacts with ion A^{z+} and forms the new kink site position AA*. Kink site position A* reacts with ion B^{z+} and forms the new kink site position AB*. Kink site position B* reacts with ion A^{z+} and forms the new kink site position BA*. Kink site position B* reacts with ion B^{z+} and forms the new kink

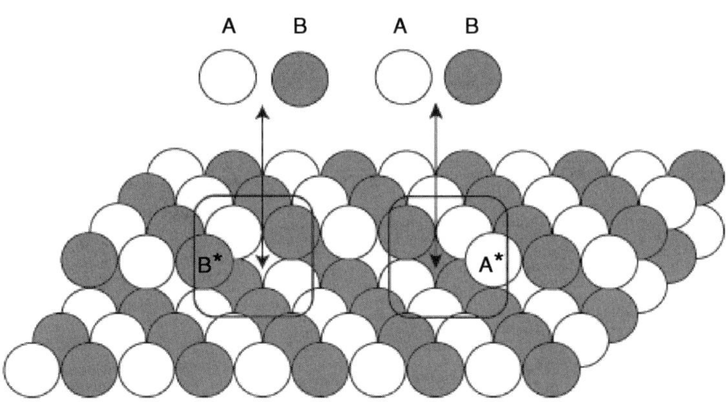

Figure 8.4 Different kink site positions A* and B* in alloy deposition (surrounded areas). Atoms in kink site positions can react with ions A^{z+} or B^{z+} of the electrolyte to form one of the four kink site positions AA*, AB*, BA*, or BB*.

site position BB* (Figure 8.4). The atoms last included in the kink site positions are denoted with an asterisk.

The four reactions of kink site positions A* and B* with ions of the alloy components in the electrolyte can be described by four rate equations

$$A^* + A^{z+} + ze^- \rightarrow AA^* \tag{8.6a}$$

$$A^* + B^{z+} + ze^- \rightarrow AB^* \tag{8.6b}$$

$$B^* + B^{z+} + ze^- \rightarrow BB^* \tag{8.6c}$$

$$B^* + A^{z+} + ze^- \rightarrow BA^* \tag{8.6d}$$

The nature of the atom in the kink site position determines the reactivity of the kink site position in the next step, when a new atom is added to the kink site thus incorporating the atom in the kink site position into the lattice and forming a new type of kink site position, or the atom last included is separated from the kink site position (transition to the previous type). These reactions either lead to a reproduction of the existing kink site position, AB* or BA* or to new kink site positions, AA* and BB*. It depends on the stability of the new kink site positions whether the strong stoichiometry continues or whether new structures will be developed, either a eutectic mixture or a solid solution.

The concept of kink site positions can be extended to alloys of variable stoichiometry and of irregular composition (solid solutions). Atoms of different chemical nature A or B surround each atom in a kink site position. Restricting the discussion to first neighbors in the case of a cubic close packed or hexagonal close packed lattice, the following different situations are possible:

A_6, A_5B, A_4B_2, A_3B_3, A_2B_4, AB_5, B_6.

A similar variety of kink site position is expected for a body centered cubic lattice where the second nearest neighbors must be included into the discussion.

In the case of the primitive lattice structure in Figure 8.4 (NaCl structure), the possible kink site positions are

A_3, A_2B, AB_2, B_3.

The bond energy of atoms A or B in the different kink site position differs according to the different nature of the surrounding atoms. This can be separated into the bonds to a crystallographic background and the bond to the last added atom. With this approximation, the reactivity of a kink site position can be described and reduced to the nature of the last added atom.

8.5.2 Rate of separation and residence times

The process of separation is described by similar rate equations as for the pure metals (Eqs. (7.24) and (7.28)). Knowing the number of bonds cleaved and the bond energies between neighboring atoms, $\varphi_{AB} = \varphi_{BA}$, φ_{AA}, φ_{BB}, respectively, the rate constants $k_{sep,AB} = k_{sep,BA}$, $k_{sep,AA}$, and $k_{sep,BB}$ can be approximately calculated. One obtains

$$k_{sep,AB} = k_o \{e^{-E_{AB}/kT}\} \tag{8.7a}$$

$$k_{\text{sep,AA}} = k_o \,\{\text{e}^{-E_{\text{AA}}/kT}\} \quad (8.7\text{b})$$

$$k_{\text{sep,BB}} = k_o \,\{\text{e}^{-E_{\text{BB}}/kT}\} \quad (8.7\text{c})$$

Electrochemical conditions can again be taken into account by an additional Butler–Volmer factor $\exp(+\alpha_{a,i}\Delta z_i FE/RT)$ to the exponential terms of Eq. (8.7). The coefficient $\alpha_{a,i}$ represents the part of the potential, which acts on the separation step and Δz_i is the charge transfer connected with the separation step.

A residence time can be defined for each atom newly added to a kink site position.

$$\tau_{\text{AB}} = \frac{1}{k_{\text{sep,AB}}} \quad (8.8\text{a})$$

$$\tau_{\text{AA}} = \frac{1}{k_{\text{sep,AA}}} \quad (8.8\text{b})$$

$$\tau_{\text{BB}} = \frac{1}{k_{\text{sep,BB}}} \quad (8.8\text{c})$$

8.5.3 Residence time and structure of alloys

By comparing the residence times an image of the selectivity of the surface for the electrolyte components can be developed. The deposition of an alloy of two similarly sized components A and B has been chosen as an example. By a suitable selection of the experimental parameters (concentration of metal ions, complexing agents, potential, etc.), the deposition rates of A and B can be made equal and equivalent to the equal partial current densities $i_{\text{A}} \cong i_{\text{B}}$. Therefore one can distinguish between three cases.

(i) $\tau_{\text{AB}} \cong \tau_{\text{AA}} \cong \tau_{\text{BB}}$

If a metal atom A or B is deposited into a kink site position, regardless of the character of the kink site position, in each case it will stay for a similar time in this position as all other atoms deposited. All kink site positions are of similar stability. If during this time a second atom arrives (B or A) the position of the firstly deposited atom in the crystal lattice is fixed. The deposition rates are similar; an attachment of atoms A or B is of similar probability. The situation leads to a homogeneous mixture between A and B. This is called a solid solution.

(ii) $\tau_{\text{AB}} << \tau_{\text{AA}} \cong \tau_{\text{BB}}$

If the kink site position AB* is much less stable then the position AA* or BB* and has equal rates of deposition, the formation of pure crystallites of atoms A and B is preferred. This is the typical situation when a eutectic mixture is formed.

(iii) $\tau_{\text{AB}} >> \tau_{\text{AA}}, \tau_{\text{BB}}$

Under these conditions the kink site position AB is much more stable than for AA or BB. If kink site positions AA or BB are formed, one can expect that either atom A or B will

separate from the kink site position before the next attached atom stabilizes the position. Only if the kink site position AB forms, can one expect the atom in this position to stay there until the next built-in atom stabilizes its position.
Under these conditions a stoichiometric alloy or an intermetallic compound is formed.

8.6 MARKOV CHAIN THEORY; DEFINITION OF THE PROBABILITY MATRIX

The follow-up of reactions A^{z+} or B^{z+} with A*, respectively, A^{z+} or B^{z+} with B* is a chain process known as a finite Markov chain.[8] This is in accordance with the kink site property that the last atom added to the kink site position determines the nature and activity of the kink site. Applications to deposition processes were published.[9–11]

A probability matrix can be set up to describe the probability of the reactions

$$P = \begin{array}{c|cc} & A^{z+} & B^{z+} \\ \hline A^{*} & p_{AA} & p_{AB} \\ B^{*} & p_{BA} & p_{BB} \end{array} \tag{8.9}$$

For the definition of the probabilities p_{AA}, p_{AB}, p_{BA}, and p_{BB} one has two limiting situations.

8.6.1 Equilibrium of the crystallization process

In the case of equilibrium of the crystallization process, equilibrium equations of reactions (8.6a–c) can be used to define the probabilities. Equilibrium means equilibrium of the surface processes. The equilibria are, if the following is restricted to ions of equal charge

$$A^{*} + A^{z+} + ze^{-} \underset{k_{sep,AA}}{\overset{k_{dep,AA}}{\leftrightarrow}} AA^{*} \qquad K_{AA} = \frac{k_{dep,AA}}{k_{sep,AA}} \tag{8.10a}$$

$$A^{*} + B^{z+} + ze^{-} \underset{k_{sep,AB}}{\overset{k_{dep,AB}}{\leftrightarrow}} AB^{*} \qquad K_{AB} = \frac{k_{dep,AB}}{k_{sep,AB}} \tag{8.10b}$$

$$B^{*} + B^{z+} + ze^{-} \underset{k_{sep,BB}}{\overset{k_{dep,BB}}{\leftrightarrow}} BB^{*} \qquad K_{BB} = \frac{k_{dep,BB}}{k_{sep,BB}} \tag{8.10c}$$

$$B^{*} + A^{z+} + ze^{-} \underset{k_{sep,BA}}{\overset{k_{dep,BA}}{\leftrightarrow}} BA^{*} \qquad K_{BA} = \frac{k_{dep,BA}}{k_{sep,BA}} \tag{8.10d}$$

These equilibrium reactions can be compared to the equilibrium reaction between reactants and transition state in the kinetic theory of homogeneous reactions. The kink site position in this model is comparable to the transition state of the reactants.

The concentrations of the newly formed kink site positions can be deduced from Eqs. (8.10a)–(8.10d).

$$\left[AA^*\right] = K_{AA}\left[A^*\right]c_{A^{z+}} \tag{8.11a}$$

$$\left[AB^*\right] = K_{AB}\left[A^*\right]c_{B^{z+}} \tag{8.11b}$$

$$\left[BB^*\right] = K_{BB}\left[B^*\right]c_{B^{z+}} \tag{8.11c}$$

$$\left[BA^*\right] = K_{BA}\left[B^*\right]c_{A^{z+}} \tag{8.11d}$$

$c_{A^{z+}}$, $c_{B^{z+}}$ are the ion concentrations in the electrolyte. It is then possible to express the probabilities p_{ij} (i, j for A, B, respectively) in the transition matrix by comparison of the equilibrium surface densities of the kink site positions:

$$p_{AA} = \frac{\left[AA^*\right]}{\left[AA^*\right]+\left[AB^*\right]} \tag{8.12a}$$

$$p_{AB} = \frac{\left[AB^*\right]}{\left[AA^*\right]+\left[AB^*\right]} = 1 - p_{AA} \tag{8.12b}$$

$$p_{BB} = \frac{\left[BB^*\right]}{\left[BB^*\right]+\left[BA^*\right]} \tag{8.12c}$$

$$p_{BA} = \frac{\left[BA^*\right]}{\left[BB^*\right]+\left[BA^*\right]} = 1 - p_{BB} \tag{8.12d}$$

Substituting the surface densities of kink site positions by equilibrium constants and ion concentrations in the electrolyte bath one obtains

$$p_{AA} = \frac{K_{AA} \cdot c_{A^{z+}}}{K_{AA} \cdot c_{A^{z+}} + K_{AB} \cdot c_{B^{z+}}} \tag{8.13a}$$

$$p_{AB} = \frac{K_{AB} \cdot c_{B^{z+}}}{K_{AA} \cdot c_{A^{z+}} + K_{AB} \cdot c_{B^{z+}}} \tag{8.13b}$$

$$p_{BB} = \frac{K_{BB} \cdot c_{B^{z+}}}{K_{BB} \cdot c_{B^{z+}} + K_{BA} \cdot c_{A^{z+}}} \tag{8.13c}$$

$$p_{BA} = \frac{K_{BA} \cdot c_{A^{z+}}}{K_{BB} \cdot c_{B^{z+}} + K_{BA} \cdot c_{A^{z+}}} \tag{8.13d}$$

From the finite Markov chain theory, it follows that the atomic fractions (x_A and x_B) of atoms A and B in the alloy can be calculated with the probabilities and the following equations

$$x_A = \frac{p_{BA}}{p_{AB} + p_{BA}} \tag{8.14a}$$

$$x_B = \frac{p_{AB}}{p_{AB} + p_{BA}} = 1 - x_A \tag{8.14b}$$

For the equilibrium approach one obtains by substitution of the probabilities

$$\frac{x_A}{x_B} = \frac{c_{A^{z+}}}{c_{B^{z+}}} \cdot \frac{(K_{AA}/K_{AB}) \cdot c_{A^{z+}} + c_{B^{z+}}}{(K_{BB}/K_{BA}) \cdot c_{B^{z+}} + \cdot c_{A^{z+}}} \tag{8.15}$$

One can introduce in Eq. (8.15) the following substitutions:

$$g_{AB} = \frac{K_{AB}}{K_{AA}} \tag{8.16}$$

$$g_{BA} = \frac{K_{BA}}{K_{BB}} \tag{8.17}$$

The values of g_{AB} and g_{BA} are the ratios of the corresponding equilibrium constants of the competing elementary reactions. The g_{AB} and g_{BA} values represent the selectivity of the kink site positions for the accumulation of the electrolyte ions. Therefore, these values are called selectivity constants.

8.6.2 Rate controlled processes

In the case of rate-controlled processes, e.g., diffusion controlled or charge transfer controlled, the probabilities in this matrix are given by the reaction rates[12]

$$A^* + A^{z+} + ze^- \xrightarrow{k_{dep,AA}} AA^* \tag{8.18a}$$

$$A^* + B^{z+} + ze^- \xrightarrow{k_{dep,AB}} AB^* \tag{8.18b}$$

$$B^* + A^{z+} + ze^- \xrightarrow{k_{dep,BA}} BA^* \tag{8.18c}$$

$$B^* + B^{z+} + ze^- \xrightarrow{k_{dep,BB}} BB^* \tag{8.18d}$$

Then the probabilities have the form

$$p_{AA} = \frac{k_{dep,AA} \cdot c_{A^{z+}}}{k_{dep,AA} \cdot c_{A^{z+}} + k_{dep,AB} \cdot c_{B^{z+}}} \tag{8.19a}$$

$$p_{AB} = \frac{k_{dep,AB} \cdot c_{B^{z+}}}{k_{dep,AA} \cdot c_{A^{z+}} + k_{dep,AB} \cdot c_{B^{z+}}} = 1 - p_{AA} \tag{8.19b}$$

$$p_{\mathrm{BB}} = \frac{k_{\mathrm{dep,BB}} \cdot c_{\mathrm{B}^{z+}}}{k_{\mathrm{dep,BA}} \cdot c_{\mathrm{A}^{z+}} + k_{\mathrm{dep,BB}} \cdot c_{\mathrm{B}^{z+}}} \tag{8.19c}$$

$$p_{\mathrm{BA}} = \frac{k_{\mathrm{dep,BA}} \cdot c_{\mathrm{A}^{z+}}}{k_{\mathrm{dep,BA}} \cdot c_{\mathrm{A}^{z+}} + k_{\mathrm{dep,BB}} \cdot c_{\mathrm{B}^{z+}}} = 1 - p_{\mathrm{BB}} \tag{8.19d}$$

For the kinetic approach one obtains from Eqs. (8.13) and (8.14)

$$\frac{x_{\mathrm{A}}}{x_{\mathrm{B}}} = \frac{c_{\mathrm{A}^{z+}}}{c_{\mathrm{B}^{z+}}} \cdot \frac{\frac{k_{\mathrm{dep,AA}}}{k_{\mathrm{dep,AB}}} \cdot c_{\mathrm{A}^{z+}} + c_{\mathrm{B}^{z+}}}{\frac{k_{\mathrm{dep,BB}}}{k_{\mathrm{dep,BA}}} \cdot c_{\mathrm{B}^{z+}} + \cdot c_{\mathrm{A}^{z+}}} \tag{8.20}$$

In Eq. (8.20) the values g_{AB} and g_{BA} are the ratios of the corresponding rate constants of the partial reactions (partial currents) of deposition of A or B.

$$g_{\mathrm{AB}} = \frac{k_{\mathrm{dep,AB}}}{k_{\mathrm{dep,AA}}} \tag{8.21}$$

$$g_{\mathrm{BA}} = \frac{k_{\mathrm{dep,BA}}}{k_{\mathrm{dep,BB}}} \tag{8.22}$$

This represents the kinetic approach of the mixed potential theory.

From the selectivity constants g_{AB} and g_{BA}, the determination will be described in the following section. No distinction can be made between whether the composition is determined by the surface selectivity or by the experimental forced deposition rates, because there is a continuous transition from the equilibrium approach to the kinetic approach with an increasing current density. But this is possible by comparison with the current potential dependence. One possibility to eliminate the kinetic contributions is the extrapolation to zero current density.

8.6.3 Determination of selectivity constants

The linearization of Eqs. (8.15) and (8.20) provides a way to determine g_{AB} and g_{BA} from the intercepts and slopes of the experimental straight lines. Making the substitutions

$$v = \frac{x_{\mathrm{A}}}{x_{\mathrm{B}}} \tag{8.23}$$

$$w = \frac{c_{\mathrm{A}^{z+}}}{c_{\mathrm{B}^{z+}}} \tag{8.24}$$

two linear equations, Eqs. (8.25) and (8.26), can be derived

$$\frac{v-1}{w} = g_{\mathrm{AB}}^{-1} - \frac{v}{w^2} g_{\mathrm{BA}}^{-1} \tag{8.25}$$

$$w\left(1 - \frac{1}{v}\right) = -g_{\mathrm{BA}}^{-1} + \frac{w^2}{v} g_{\mathrm{AB}}^{-1} \tag{8.26}$$

The values of g_{AB} and g_{BA} can be determined by plotting the data according to Eqs. (8.25) and (8.26), if the dependence of the alloy composition on the ion concentration in the electrolyte is known.

8.6.4 Alloy characterization by selectivity constants

The dependence of the alloy composition (mole fraction x_{A}) on the electrolyte concentration ratio $c_{\mathrm{A}}/(c_{\mathrm{A}} + c_{\mathrm{B}})$ for different combinations of g_{AB} and g_{BA} values is shown in Figure 8.5.

The classification of Brenner is now represented by different values of the selectivity constants. The composition reference line introduced by Brenner is given by $g_{\mathrm{AB}} = g_{\mathrm{BA}} = 1$. If the component A is the less noble component, curve 5 would represent normal and curve 4 anomalous co-deposition. Curve 3 represents segregation and curve 2 intermetallic compound formation AB.

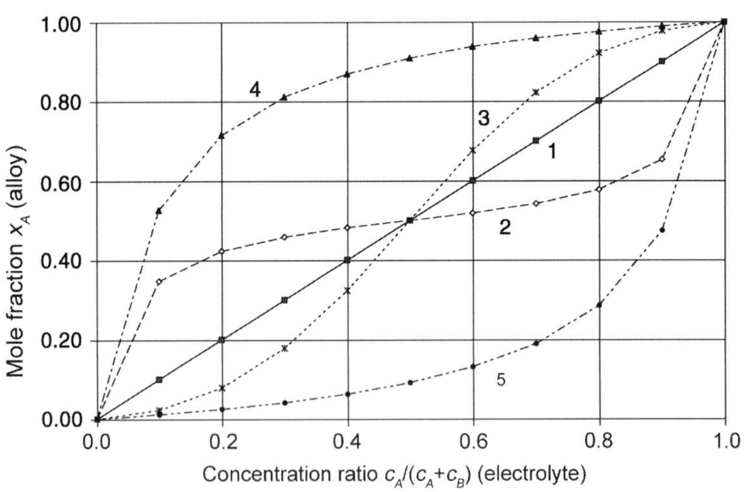

Figure 8.5 Dependences of the mole fraction of component A in the deposited alloy (x_{A}) on the concentration ratio (the mole fraction) in the electrolyte $c_{\mathrm{A}}/(c_{\mathrm{A}} + c_{\mathrm{B}})$ are depicted here at different combinations of g_{AB} and g_{BA} values. Curve 1: $g_{\mathrm{AB}} = g_{\mathrm{BA}} = 1$, curve 2: $g_{\mathrm{AB}} = g_{\mathrm{BA}} = 10$, curve 3: $g_{\mathrm{AB}} = g_{\mathrm{BA}} = 0.1$, curve 4: $g_{\mathrm{AB}} = 0.1$; $g_{\mathrm{BA}} = 10$, and curve 5: $g_{\mathrm{AB}} = 10$; $g_{\mathrm{BA}} = 0.1$.

8.6.5 Selectivity constants and residence times in kink site positions

In the equilibrium case, the equilibrium constants in the selectivity constants can be substituted by the ratio of the rate constants of deposition and separation

$$g_{AB} = \frac{k_{dep,AB} \cdot k_{sep,AA}}{k_{dep,AA} \cdot k_{sep,AB}} \tag{8.27}$$

$$g_{B} = \frac{k_{dep,BA} \cdot k_{sep,BB}}{k_{dep,BB} \cdot k_{sep,BA}} \tag{8.28}$$

The rate constants of deposition for ions of similar size, structure, and equal charge are very similar: $k_{dep,AA} \approx k_{dep,AB}$ and $k_{dep,BB} \approx k_{dep,BA}$. After substitution of the rate constants of separation by the residence times it follows:

$$g_{AB} \approx \frac{\tau_{AB}}{\tau_{AA}} \tag{8.29}$$

$$g_{BA} = \frac{\tau_{BA}}{\tau_{BB}} \tag{8.30}$$

These equations show the similarity between residence times and selectivity constants in the equilibrium case.

8.7 EXPERIMENTAL EXAMPLES

To apply the Markov chain theory to experimental examples, experimental data must be available between the composition of an alloy and the concentrations of the components in the electrolyte. Examples can be found in a book by Watanabe.[13] Some representative systems will be described.

8.7.1 The cobalt–iron alloy system

In the case of the CoFe alloy system, the elementary kink site transformations for alloy crystallization can be represented by the following four equilibrium reactions

$$Co^{*} + Co^{2+} + 2e^{-} \leftrightarrow CoCo^{*} \tag{8.31a}$$

$$Co^{*} + Fe^{2+} + 2e^{-} \leftrightarrow CoFe^{*} \tag{8.31b}$$

$$Fe^{*} + Co^{2+} + 2e^{-} \leftrightarrow FeCo^{*} \tag{8.31c}$$

$$Fe^{*} + Fe^{2+} + 2e^{-} \leftrightarrow FeFe^{*} \tag{8.31d}$$

The surface density of the principal four kink site positions (in square brackets) is related to the ion concentration of the two components Co and Fe in the electrolyte, described by c_{Co} and c_{Fe}, by the equations

$$\left[\mathrm{CoCo}^*\right] = K_{\mathrm{CoCo}} \exp\left(-\frac{2FE}{RT}\right)\left[\mathrm{Co}^*\right] c_{\mathrm{Co}^{2+}} \tag{8.32a}$$

$$\left[\mathrm{CoFe}^*\right] = K_{\mathrm{CoFe}} \exp\left(-\frac{2FE}{RT}\right)\left[\mathrm{Co}\,*\right] c_{\mathrm{Fe}^{2+}} \tag{8.32b}$$

$$\left[\mathrm{FeCo}^*\right] = K_{\mathrm{FeCo}} \exp\left(-\frac{2FE}{RT}\right)\left[\mathrm{Fe}^*\right] c_{\mathrm{Co}^{2+}} \tag{8.32c}$$

$$\left[\mathrm{FeFe}^*\right] = K_{\mathrm{FeFe}} \exp\left(-\frac{2FE}{RT}\right)\left[\mathrm{Fe}^*\right] c_{\mathrm{Fe}^{2+}} \tag{8.32d}$$

The exponential term describes the potential dependence of the equilibrium.

The conditional probabilities p_{Co} and p_{Fe} for both conservation and alteration of the active atom (denoted by asterisks) in each elementary crystallization step are given by the equations:

$$p_{\mathrm{CoCo}} = [\mathrm{CoCo}^*] / ([\mathrm{CoCo}^*] + [\mathrm{CoFe}^*]) \tag{8.33a}$$

$$p_{\mathrm{CoFe}} = [\mathrm{CoFe}^*] / ([\mathrm{CoCo}^*] + [\mathrm{CoFe}^*]) = 1 - p_{\mathrm{CoCo}} \tag{8.33b}$$

$$p_{\mathrm{FeCo}} = [\mathrm{FeCo}^*] / ([\mathrm{FeCo}^*] + [\mathrm{FeFe}^*]) \tag{8.33c}$$

$$p_{\mathrm{FeFe}} = [\mathrm{FeFe}^*]/([\mathrm{FeCo}^*] + [\mathrm{FeFe}^*]) = 1 - p_{\mathrm{FeCo}} \tag{8.33d}$$

The probability can be expressed by the equilibrium constants and ion concentrations in the electrolyte bath by insertion of the relations (8.32) for the active atoms in the kink site positions in Eq. (8.33a–d). The relation between the alloy composition and electrolyte concentration is derived by the procedure described in Section 8.6.1.

$$\frac{x_{\mathrm{Co}}}{x_{\mathrm{Fe}}} = \frac{p_{\mathrm{FeCo}}}{p_{\mathrm{CoFe}}} = \frac{c_{\mathrm{Co}^{2+}}}{c_{\mathrm{Fe}^{2+}}} \cdot \frac{\dfrac{K_{\mathrm{CoCo}}}{K_{\mathrm{CoFe}}} c_{\mathrm{Co}^{2+}} + c_{\mathrm{Fe}^{2+}}}{\dfrac{K_{\mathrm{FeFe}}}{K_{\mathrm{FeCo}}} c_{\mathrm{Fe}^{2+}} + c_{\mathrm{Co}^{2+}}} \tag{8.34}$$

The selectivity constants are

$$g_{\mathrm{CoFe}} = \frac{K_{\mathrm{CoFe}}}{K_{\mathrm{CoCo}}} \tag{8.35}$$

$$g_{\mathrm{FeCo}} = \frac{K_{\mathrm{FeCo}}}{K_{\mathrm{FeFe}}} \tag{8.36}$$

The selectivity constants do not depend on the potential.
Introducing the substitutions

$$\frac{x_{Co}}{x_{Fe}} = v \quad \text{and} \quad \frac{c_{Co^{2+}}}{c_{Fe^{2+}}} = w$$

the linear equations are

$$\frac{v-1}{w} = g_{CoFe}^{-1} - \frac{v}{w^2} g_{FeCo}^{-1} \tag{8.37}$$

and

$$w\left(1 - \frac{1}{v}\right) = -g_{FeCo}^{-1} + \frac{w^2}{v} g_{CoFe}^{-1} \tag{8.38}$$

Cobalt–iron alloys were deposited from a sulfate electrolyte containing a sum of 1 mol · dm^{-3} metal salt. 0.5 mol · dm^{-3} H_3BO_4 were added and subsequently the pH was adjusted to pH = 2.0. No additives were used. The deposition was carried out at three different potentials and under potentiostatic control on copper substrates. The results at 1.0 V deposition potential are presented in Figure 8.6. The dependence is approximately described by curve 1 in Figure 8.5. The system can be described as an equilibrium system. The deviation below 30% of iron ions in the electrolyte may be connected with differences in the crystallographic structure. In this region a mixture of fcc and bcc structure was

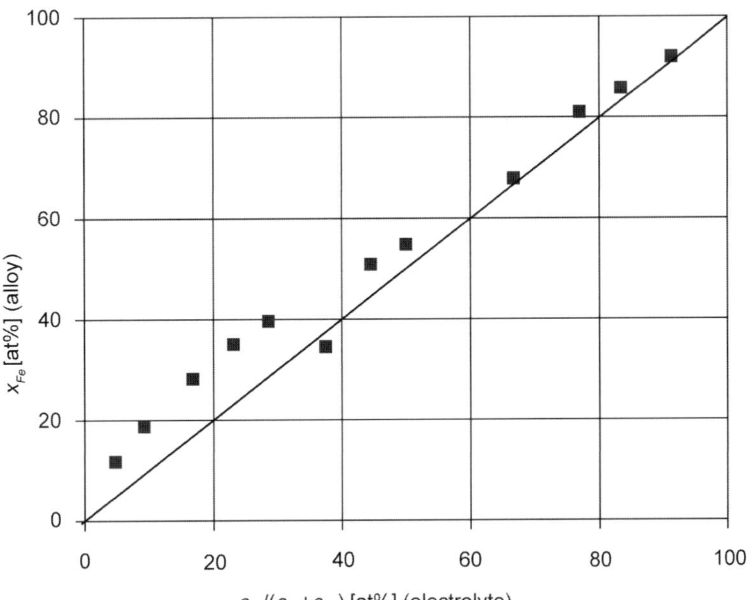

Figure 8.6 Dependence of the mole fraction of iron in the alloy on the concentration ratio (mole fraction) of iron in the electrolyte (data from Watanabe[13]), deposition potential 1.0 V.

observed while at higher contents of iron the pure bcc α-phase was found. The linearization according to Eq. (8.37) is shown in Figure 8.7, the linearization according to Eq. (8.38) in Figure 8.8. The following selectivity constants were determined: $g^{-1}_{\mathrm{FeCo}} = 0.35$ and $g^{-1}_{\mathrm{CoFe}} = 0.61$ (Figure 8.7), $g^{-1}_{\mathrm{CoFe}} = 1.06$ and $g^{-1}_{\mathrm{FeCo}} = 0.58$ (Figure 8.8). The different values obtained from the two plots reflect the different balancing of the experimental

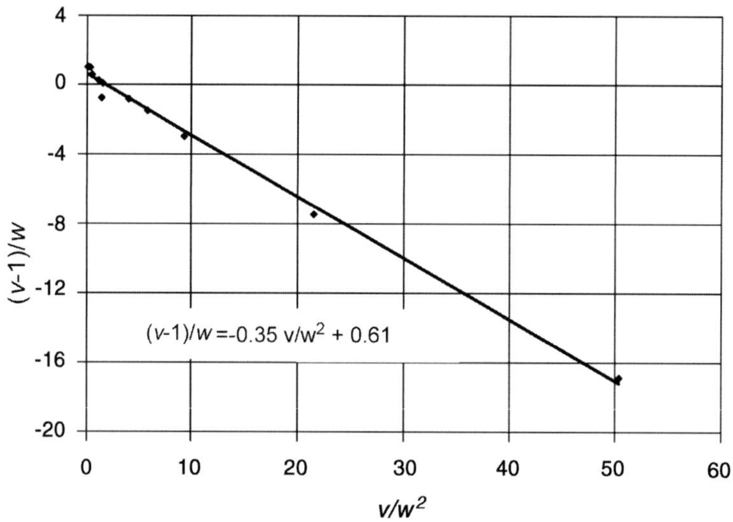

Figure 8.7 Linearization of the data of Figure 8.6 according to Eq. (8.37).

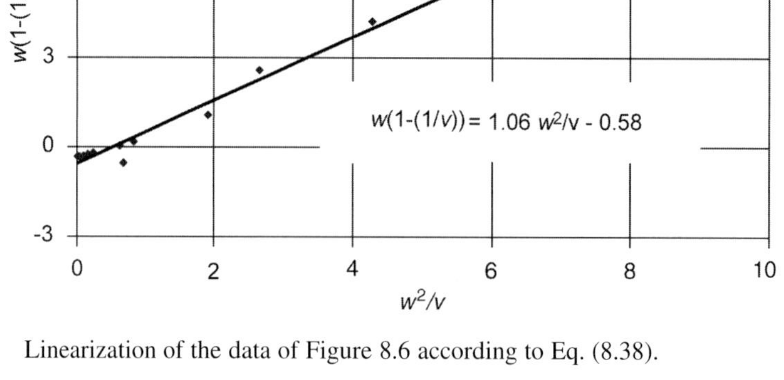

Figure 8.8 Linearization of the data of Figure 8.6 according to Eq. (8.38).

points in the two linear equations. The average values are

$$g_{\mathrm{FeCo}}^{-1} = 0.46 \pm 0.12$$

$$g_{\mathrm{CoFe}}^{-1} = 0.84 \pm 0.23$$

8.7.2 Cobalt–nickel

A system very similar to the cobalt–iron alloy system is CoNi. CoNi alloys were deposited from a sulfate electrolyte containing a sum of 1 mol · dm^{-3} metal salt. 0.5 mol · dm^{-3} H_3BO_4 were added and subsequently the pH was adjusted to pH = 2.0. No additives were used. The deposition was carried out under potentiostatic control on copper substrates. The deposition was independent on the potential between 0.8 and 1.6 V_{SCE}. The results are presented in Figure 8.9.

The dependence is approximately described by curve 4 in Figure 8.5. The less noble component (cobalt) is preferentially deposited (anomalous co-deposition).

Figures 8.10 and 8.11 show the linearization according to equations equivalent to Eqs. (8.37) and (8.38). In these equations the substitutions are

$$\frac{x_{\mathrm{Co}}}{x_{\mathrm{Ni}}} = v \quad \text{and} \quad \frac{c_{\mathrm{Co^{2+}}}}{c_{\mathrm{Ni^{2+}}}} = w$$

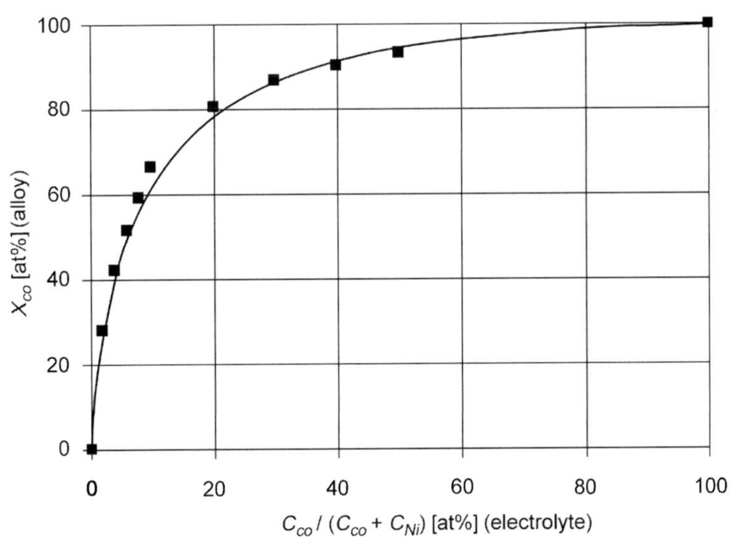

Figure 8.9 Dependence of the mole fraction of cobalt in the alloy CoNi on the concentration ratio (mole fraction) of cobalt in the electrolyte (data from Watanabe[13]), deposition independent on the potential for potentials from 0.8 to 1.6 V_{SCE}.

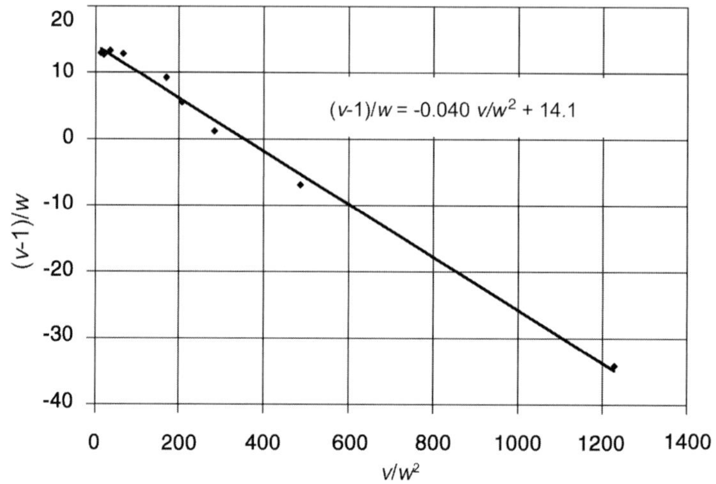

Figure 8.10 Linearization of the data of Figure 8.9, according to an equation equivalent to Eq. (8.37).

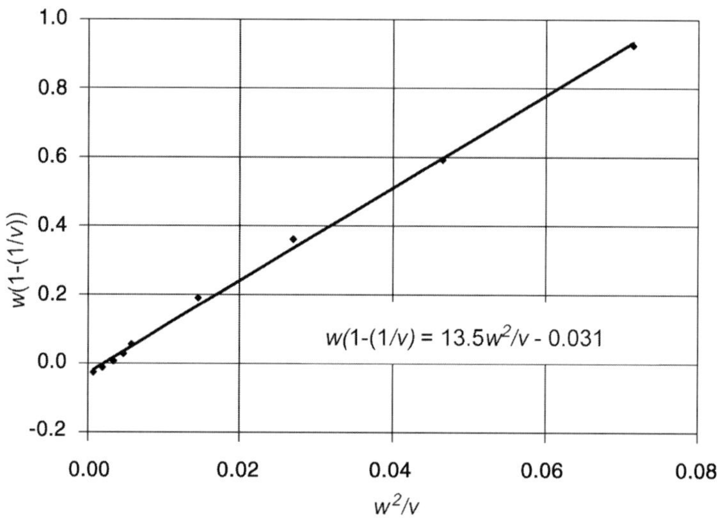

Figure 8.11 Linearization of the data of Figure 8.9, according to an equation equivalent to Eq. (8.38).

The selectivity constants are

$$g_{\mathrm{CoN}i} = \frac{K_{\mathrm{CoNi}}}{K_{\mathrm{CoCo}}} \tag{8.39}$$

$$g_{\mathrm{NiCo}} = \frac{K_{\mathrm{NiCo}}}{K_{\mathrm{NiN}i}} \tag{8.40}$$

The selectivity constants do not depend on the potential.
The following selectivity constants were determined: $g_{\mathrm{NiCo}}^{-1} = 0.040$ and $g_{\mathrm{CoNi}}^{-1} = 14.1$ (Figure 8.10), $g_{\mathrm{CoNi}}^{-1} = 13.5$ and $g_{\mathrm{NiCo}}^{-1} = 0.031$ (Figure 8.11). The average values are

$$g_{\mathrm{CoNi}}^{-1} = 13.8 \pm 0.3$$

$$g_{\mathrm{NiCo}}^{-1} = 0.036 \pm 0.004$$

8.7.3 Iron–nickel

Another system very similar to the cobalt–iron alloy system is FeNi. FeNi alloys were deposited from a sulfate electrolyte containing a sum of 1 mol · dm^{-3} metal salt. 0.5 mol·dm^{-3} H_3BO_4 were added and subsequently the pH was adjusted to pH = 2.0. The electrolyte was made with deoxidized water. No additives were used. The deposition was carried out under potentiostatic control on copper substrates. The composition was approximately independent on the potential between 1.1 and 1.3 V_{SCE}. The results are presented in Figure 8.12.

The dependence is approximately described by curve 4 in Figure 8.5. The less noble component (iron) is preferentially deposited (anomalous co-deposition).

Figures 8.13 and 8.14 show the linearization according to equations equivalent to Eqs. (8.37) and (8.38).

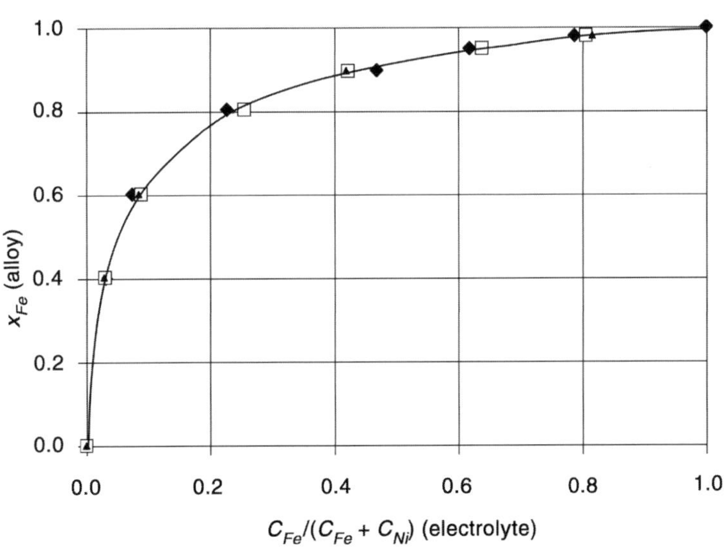

Figure 8.12 Dependence of the mole fraction of iron in the alloy FeNi on the concentration ratio (mole fraction) of iron in the electrolyte (data from Watanabe[13]). The composition was approximately independent on the potential for potentials 1.1 V_{SCE} (rhomboids), 1.2 V_{SCE} (open squares), and 1.3 V_{SCE} (triangles).

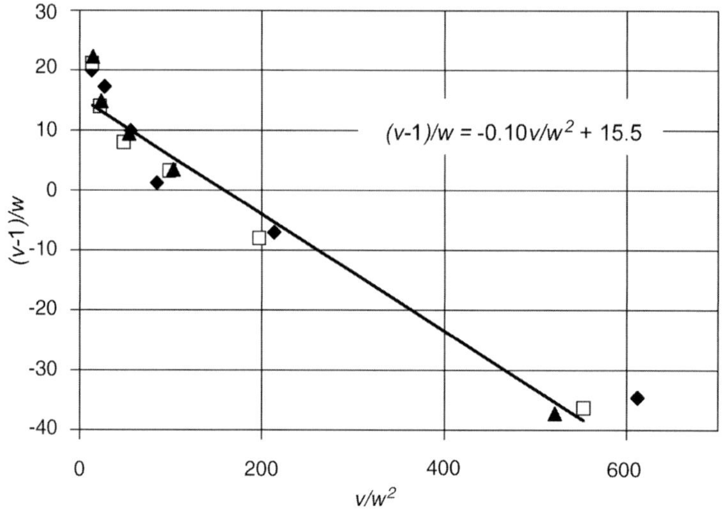

Figure 8.13 Linearization of the data of Figure 8.12, according to an equation equivalent to Eq. (8.37).

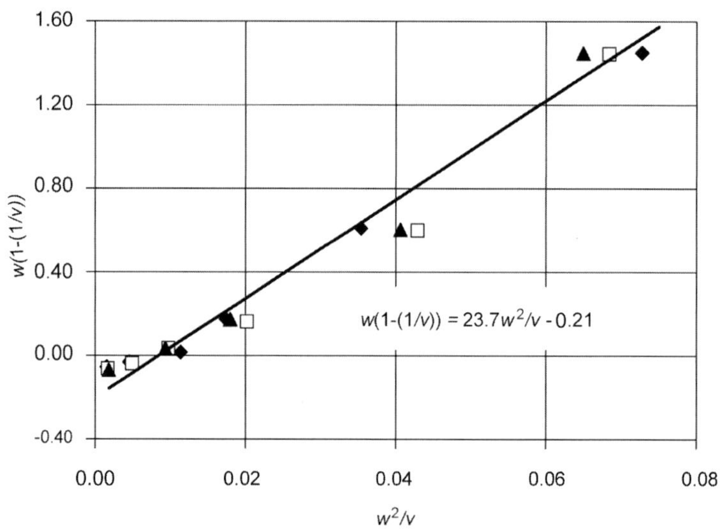

Figure 8.14 Linearization of the data of Figure 8.12, according to an equation equivalent to Eq. (8.38).

In these equations the substitutions are

$$\frac{x_{\mathrm{Fe}}}{x_{\mathrm{Ni}}} = v \qquad \text{and} \qquad \frac{c_{\mathrm{Fe}^{2+}}}{c_{\mathrm{Ni}^{2+}}} = w$$

The selectivity constants are

$$g_{\mathrm{FeNi}} = \frac{K_{\mathrm{FeNi}}}{K_{\mathrm{FeFe}}} \tag{8.41}$$

$$g_{\mathrm{NiFe}} = \frac{K_{\mathrm{NiFe}}}{K_{\mathrm{NiNi}}} \tag{8.42}$$

The selectivity constants do not depend on the potential.

The following selectivity constants were determined: $g_{\mathrm{FeNi}}^{-1} = 15.5$ and $g_{\mathrm{NiFe}}^{-1} = 0.10$ (Figure 8.13), $g_{\mathrm{FeNi}}^{-1} = 23.7$ and $g_{\mathrm{NiFe}}^{-1} = 0.21$ (Figure 8.14). The average values are

$$g_{\mathrm{FeNi}}^{-1} = 19.6 \pm 4.1$$

$$g_{\mathrm{NiFe}}^{-1} = 0.16 \pm 0.05$$

The scattering of the data in this third example is larger than in the previous examples.

8.7.4 Induced electrodeposition: the NiMo system

A special case is observed for the electrodeposition of molybdenum and tungsten from molybdate or tungstate solutions. Deposition of the metallic form is not possible from a solution of the pure molybdate or tungstate ions. Only oxide layers of pure quality are formed. But a deposition is observed in combination with metals of the iron group. These alloys, especially NiW and NiMo, have become of interest because of their mechanical properties like hardness or wear resistance. Therefore, they are discussed as substitutes for hard chromium.

Several authors have investigated the mechanism of the deposition of these alloys. Podlaha and Landolt[14] suggested a model of catalytic electrodeposition. The main observations of these authors concern the dependence of composition on the concentration ratio, rotation speed, and current density. A rotating Hull cell was used for the experiments, which has already been described in Chapter 5, Figure 5.27.

In an electrolyte with a high concentration of $NiSO_4$ ($1.0\ \mathrm{mol \cdot dm^{-3}}$) and low concentration of Na_2MoO_4 ($0.005\ \mathrm{mol \cdot dm^{-3}}$) and in the presence of $0.7\ \mathrm{mole \cdot dm^{-3}}$ sodium citrate, pH = 7.5, the deposition of molybdenum is diffusion limited. The composition as a function of the total current density is shown in Figure 8.15.

The authors found in their study of the partial current of Ni and Mo that the partial current of Ni deposition increases with the total current with increasing negative potential but has only a marginal dependence on the rotation rate. The molybdenum content of the deposit is independent of the total current density but increases with the rotation rate (diffusion limitation).

The opposite situation with a high Na_2MoO_4 concentration ($0.4\ \mathrm{mol \cdot dm^{-3}}$) and low Ni^{2+} concentration ($0.005\ \mathrm{mol \cdot dm^{-3}}$) is shown in Figure 8.16. The Ni content of the deposit only slightly increases with the current density (from 35 to 50% weight).

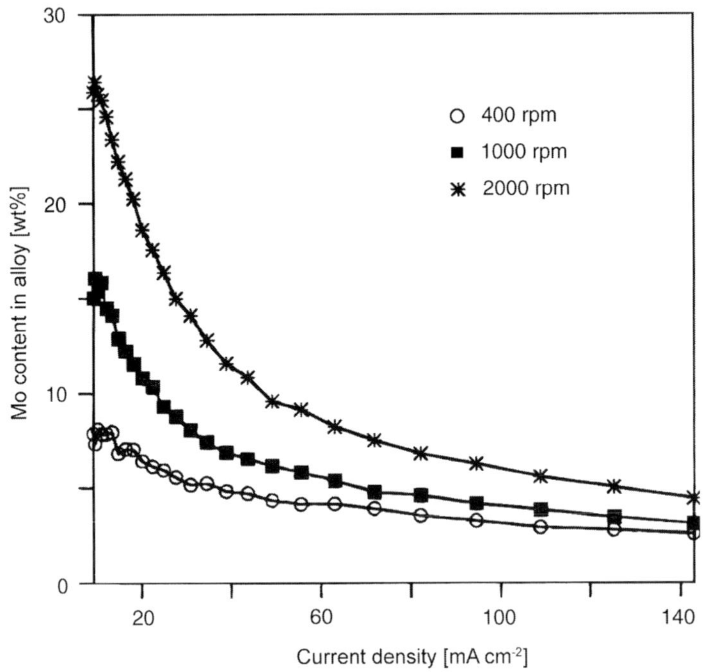

Figure 8.15 Nickel–molybdenum alloy deposition, high concentration of $NiSO_4$ (1.0 mol · dm^{-3}) and low concentration of Na_2MoO_4 (0.005 mol · dm^{-3}), 0.7 mol · dm^{-3} sodium citrate, pH = 7.5, dependence on the rotation rate for 400, 1000, and 2000 rpm (Podlaha and Landolt).[14] (Reproduced with permission from J. Electrochem. Soc. 143, 885 (1996), © 1996, The Electrochemical Society.)

The partial current densities of both components only slightly depend on the current density after some irregularities at low current densities but are both proportional to the rotation rate. The composition is, however, independent of the rotation rate.

The authors explained the experimental results by catalytic action of the citrate (Ci) with the following mechanism:

$$\mathrm{Ni(Ci)^-} + 2e^- \rightarrow \mathrm{Ni} + \mathrm{Ci} \tag{8.43}$$

$$(\mathrm{MoO_4})^{2-} + \mathrm{Ni(Ci)^-} + 2\mathrm{H_2O} + 2e^- \rightarrow [\mathrm{Ni(Ci)MoO_2}]^-_{\mathrm{ad}} + 4\mathrm{OH^-} \tag{8.44}$$

The complex on the electrode surface induces the catalytic reduction of the molybdate to the metallic state releasing the unchanged Ni-complex. The Ni-complex is catalytic active. The mechanism was used as a basis for a simulation of the deposition.

The deposition reactions according to the model of alternating kink site positions can be formulated by the following equations:

$$\mathrm{Ni^*} + \mathrm{Ni^{2+}} + 2e^- \leftrightarrow \mathrm{NiNi^*} \tag{8.45a}$$

$$\mathrm{Ni^*} + \mathrm{MoO_4^{2-}} + \mathrm{Ci} + \Delta z e^- \leftrightarrow \left[\mathrm{NiCiMoO_4^{(2+\Delta z)-}}\right]^* \tag{8.45b}$$

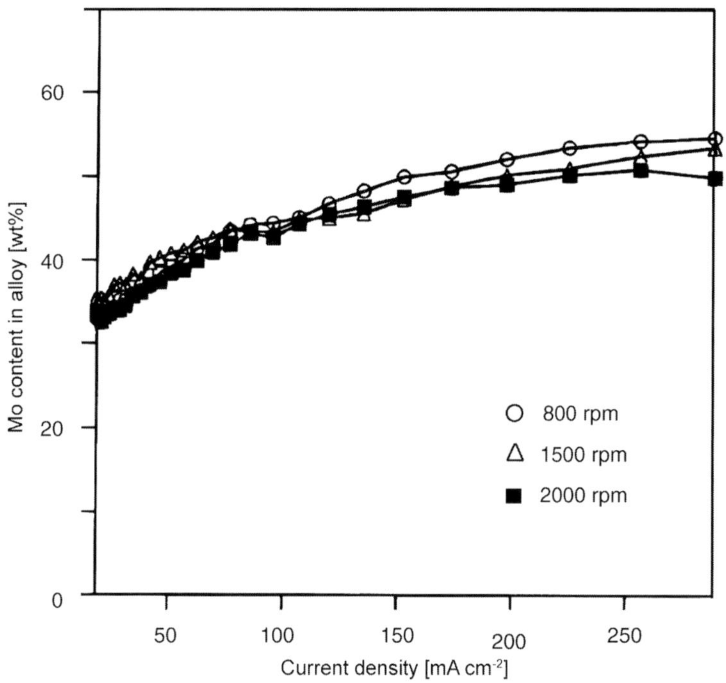

Figure 8.16 Nickel–molybdenum alloy deposition with a large Na_2MoO_4 concentration ($0.4\ mol \cdot dm^{-3}$) and low Ni^{2+} concentration ($0.005\ mol \cdot dm^{-3}$), $0.005\ mol \cdot dm^{-3}$ sodium citrate, and pH 11.0. The composition is independent of the rotation rate (Podlaha and Landolt).[14] (Reproduced with permission from J. Electrochem. Soc. 143, 885 (1996), © 1996, The Electrochemical Society.)

$$\left[NiCiMoO_4^{(2+\Delta z)-}\right]^* + Ni^{2+} + 8H^+ + (2+\Delta z)e^- \leftrightarrow MoNi^* + 4H_2O + Ci \qquad (8.45c)$$

Reaction (8.45b) takes into account the catalytic activity of the citrate and a partial discharge of the molybdenum in the NiMo-citrate complex. In the following $[NiCiMoO_4^{(2+\Delta z)-}]^*$ is substituted by NiMo*. c_{Mo} stands for the concentration of molybdate.

Experimental evidence shows that reaction of NiMo* with MoO_4^{2-} is not possible. The kink site position MoMo* does not exist.

Then the following probabilities can be formulated:

$$p_{NiNi} = [NiNi^*]/([NiNi^*] + [NiMo^*]) \qquad (8.46a)$$

$$p_{NiMo} = [NiMo^*]/([NiNi^*] + [NiMo^*]) = 1 - p_{NiNi} \qquad (8.46b)$$

$$p_{MoNi} = [MoNi^*]/([MoMo^*] + [MoNi^*]) = 1 \qquad (8.46c)$$

$$p_{MoMo} = [MoMo^*]/([MoMo^*] + [MoNi^*]) = 0 \qquad (8.46d)$$

The probability matrix has the form

$$P = \begin{matrix} \\ x^*_{\text{Ni}} \\ x^*_{\text{Mo}} \end{matrix} \begin{matrix} c_{\text{Ni}^{2+}} & c_{\text{Mo}} \\ \left[\begin{matrix} p_{\text{NiNi}} \\ 1 \end{matrix}\right. & \left.\begin{matrix} p_{\text{NiMo}} \\ 0 \end{matrix}\right] \end{matrix} \tag{8.47}$$

The equilibrium equations corresponding to Eq. (8.10) are

$$\frac{[\text{NiNi}\ *]}{[\text{Ni}\ *]c_{\text{Ni}^{2+}}} = K_{\text{NiNi}*} \exp\left(-\frac{2FE}{RT}\right) \tag{8.48a}$$

$$\frac{[\text{NiMo}\ *]}{[\text{Ni}\ *]c_{\text{Mo}}c_{\text{Ci}}} = K_{\text{NiMo}*} \exp\left(-\frac{\Delta zFE}{RT}\right) \tag{8.48b}$$

The following equation for the ratio of the mole fractions in the alloy is obtained.

$$\frac{x_{\text{Ni}}}{x_{\text{Mo}}} = \frac{1}{p_{\text{NiMo}}} = \frac{(K_{\text{NiNi}^*}/K_{\text{NiMo}^*})\exp(-((2-\Delta z)FE)/RT)c_{\text{Ni}^{2+}} + c_{\text{Mo}}c_{\text{Ci}}}{c_{\text{Mo}}c_{\text{Ci}}} \tag{8.49}$$

Similar to the previous example one can define selectivity constant for nickel kinks.

$$g^{-1}_{\text{NiMo}} = \frac{K_{\text{NiNi}^*}}{K_{\text{NiMo}^*}} \exp\left(-\frac{(2-\Delta z)F\Delta E}{RT}\right) \tag{8.50}$$

This constant is now potential dependent. The substitutions for v and w are the same as in the previous examples

$$v = \frac{x_{\text{Ni}}}{x_{\text{Mo}}} \qquad w = \frac{c_{\text{Ni}^{2+}}}{c_{\text{Mo}}}$$

Then one obtains for the v–w relationship

$$v = \frac{g^{-1}_{\text{NiMo}} \cdot w + c_{\text{Ci}}}{c_{\text{Ci}}} \tag{8.51}$$

Can the results of Podlaha and Landolt be represented by this equation?

The results of Figure 8.15 are shown in Figure 8.17. The concentrations are $c_{\text{Ni}^{2+}} = 1\ \text{mol} \cdot \text{dm}^{-3}$, $c_{\text{Mo}} = 0.005\ \text{mol} \cdot \text{dm}^{-3}$, and a citrate concentration 0.7 mol·dm^{-3}. The equation describing this situation is:

$$v = \frac{g_{\text{Ni}} \cdot w}{c_{\text{Ci}}} \tag{8.52}$$

For a constant citrate concentration ($0.7\ \text{mol} \cdot \text{dm}^{-3}$) v is proportional to w. But w depends on the current density and on the rate of rotation because the concentration of MoO_4^{2-} on the surface depends on the mass transport conditions.

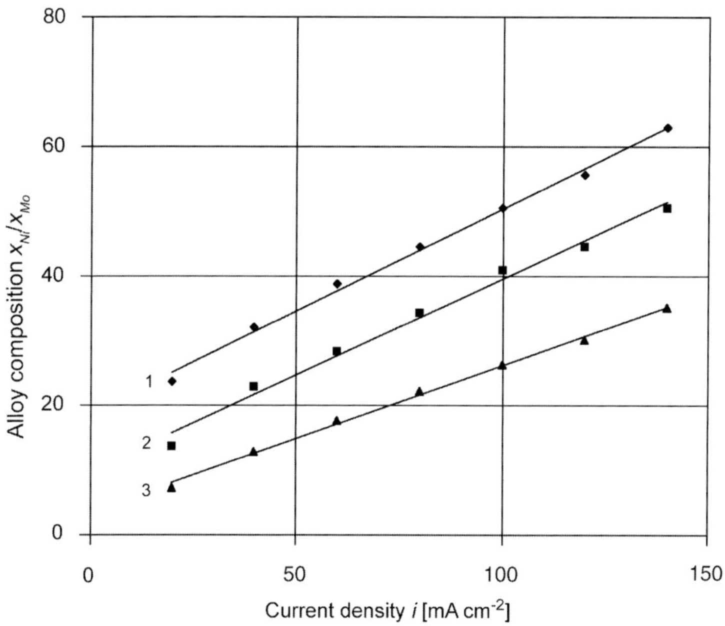

Figure 8.17 Alloy composition $v = x_{Ni}/x_{Mo}$ versus current density for three rotation rates r, (1) 400, (2) 1000, and (3) 2000 rpm, $c_{Ni^{2+}} = 1$ mol · dm^{-3} and $c_{Mo} = 0.005$ mol · dm^{-3}, citrate concentration 0.7 mol · dm^{-3}.

Mass transport caused by stationary diffusion is usually described by the equation

$$i = -k_{dif}(c - c_0) \tag{8.53}$$

The bulk concentration is c_0 and $k_{dif} = 2FD/\delta$ (representing divalent ions) contains Faraday constant, diffusion coefficient, and the thickness of the Nernst diffusion layer. One obtains for c

$$c = c_0(1 - k^* i) \tag{8.54}$$

The constant k^* stands for $k^* = 1/(c_0\, k_{dif})$.

Assuming A to be diffusion controlled and substituting in the ratio $w = c_A/c_B$, the concentration of A by Eq. (8.54) and the concentration of B by $c_B = c_{0,B}$ one obtains

$$w = w_0(1 - k^* i) \tag{8.55}$$

with $w_0 = c_{0,A}/c_{0,B}$. Then the linear Eq. (8.25) gets the form

$$\frac{v-1}{w_0} = g_{AB}^{-1}(1 - k^* i) - \frac{v}{w_0^2}\frac{g_{BA}^{-1}}{1 - k^* i} \tag{8.56}$$

For $k^*i << 1$ this equation can be approximated by

$$\frac{v-1}{w_0} = g_{AB}^{-1}(1-k^*i) - \frac{v}{w_0^2} g_{BA}^{-1}(1\ 1\ k^*i) \tag{8.57}$$

Assuming B to be diffusion controlled and substituting in the ratio $w = c_A/c_B$, the concentration of B by Eq. (8.54) and the concentration of A by $c_A = c_{0,A}$ one obtains

$$w = \frac{w_0}{1-k^*i} \tag{8.58}$$

with $w_0 = c_{0,A}/c_{0,B}$. Then the linear Eq. (8.25) gets the form

$$\frac{v-1}{w_0} = \frac{g_{AB}^{-1}}{1-k^*i} - \frac{v}{w_0^2} g_{BA}^{-1}(1-k^*i) \tag{8.59}$$

For $k^*i << 1$ this equation can be approximated by

$$\frac{v-1}{w_0} = g_A^{-1}(1+k^*i) - \frac{v}{w_0^2} g_B^{-1}(1-k^*i) \tag{8.60}$$

In the case of the NiMo system, additionally to the diffusion limitation with increasing current, the convection, induced by the rate of rotation of the working electrode, has to be taken into account. An increasing rate of rotation decreases the thickness of the diffusion layer. In a rough approximation the diffusion layer is inversely proportional to the rate of rotation r. Thus, one obtains for diffusion limited molybdenum concentration

$$c_{Mo} = c_{0,Mo}\left(1 - k^*\frac{i}{r}\right) \tag{8.61}$$

The concentration ratio is, for $k^*\ (i/r) << 1$

$$w \approx w_0\left(1 + k^*\frac{i}{r}\right) \tag{8.62}$$

One obtains for the alloy composition:

$$v \approx \frac{g_{NiMo}^{-1} \cdot w_0}{c_{Ci}}(1 + k^*\frac{i}{r}) \tag{8.63}$$

Figure 8.17 shows that $v = x_{Ni}/x_{Mo}$ is proportional to i and is inversely proportional to the rotation speed r. The dependence on the rotation speed r is also shown in Figure 8.18.

The linear plots of the composition on i in Figure. 8.17 should have the same intersection point with the v-axis. The real values lead to an average intersection point of $v = g_{NiMo}^{-1} w_0/c_{Ci} \approx 10$. With $w_0 = c_{Ni^{2+}}/c_{Mo} = 200$ and $c_{Ci} = 0.7\ \text{mol} \cdot \text{dm}^{-3}$ a value of $g_{NiMo} \approx 29$ is calculated, which is a measure of the selectivity of the alloy surface for Mo.

The results of Figure 8.16 are shown in Figure 8.19. The nickel ion concentration is $c_{Ni^{2+}} = 0.005\ \text{mol dm}^{-3}$, the concentration of MoO_4^{2-} is $c_{Mo} = 0.4\ \text{mol} \cdot \text{dm}^{-3}$, and the

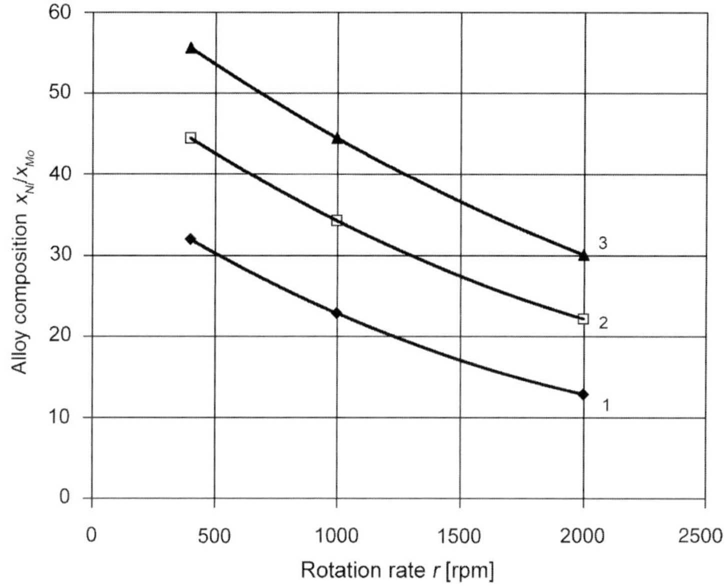

Figure 8.18 Alloy composition $v = x_{Ni}/x_{Mo}$ versus rotation rate for three different current densities. (1) 40 mA · cm^{-2}, (2) 80 mA · cm^{-2}, and (3) 120 mA · cm^{-2}, $c_{Ni^{2+}} = 1$ mol dm^{-3} and $c_{Mo} = 0.005$ mol · dm^{-3}, and citrate concentration 0.7 mol dm^{-3}, pH = 7.5.

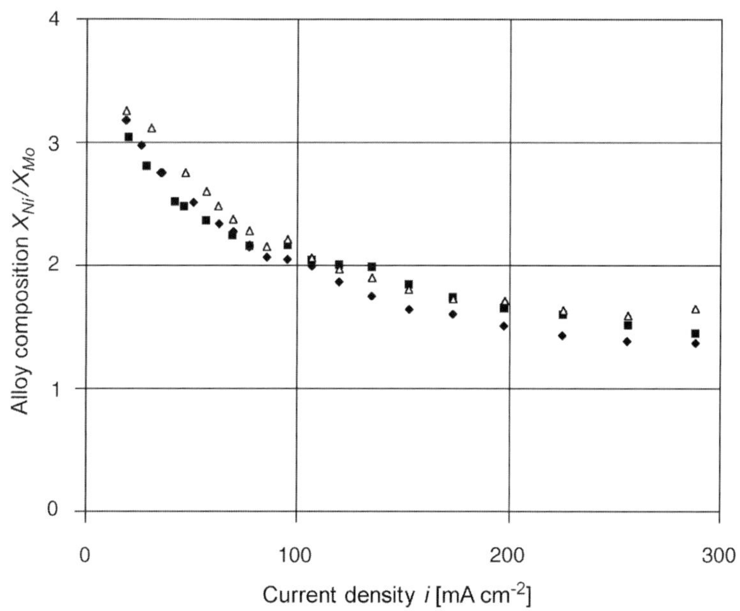

Figure 8.19 Alloy composition $v = x_{Ni}/x_{Mo}$ as a function of the current density for three rotation rates: 800 rpm (open triangles), 1500 rpm (squares), 2000 rpm (rhomboids); nickel ion concentration of $c_{Ni^{2+}} = 0.005$ mol · dm^{-3}, concentration of molybdate, $c_{Mo} = 0.4$ mol · dm^{-3}, and citrate concentration 0.01 mol · dm^{-3}.

citrate concentration is 0.01 mol·dm^{-3}. The experimental conditions correspond to w<<1 and if $g_{NiMo}^{-1} \cdot$ w<<c_{Ci}. Then Eq. (8.56) can be approximated by

$$v \equiv \frac{x_{Ni}}{x_{Mo}} \approx 1 \tag{8.64}$$

Figure 8.19 shows that $v = x_{Ni}/x_{Mo}$ is 3 for lower current densities and is 1.5 at higher current densities. The value $v = 1$ has not been reached for the experimental conditions. A possible explanation could be (a) $g_{NiMo}^{-1} \cdot w$ cannot be neglected if compared to the citrate concentration (0.01 mol·dm^{-3}) and (b) $g_{NiMo}^{-1} \cdot w$ depends on the current density or the deposition potential. Otherwise, $v = x_{Ni}/x_{Mo}$ does not depend on the convection of the electrolyte as predicted by the approximation.

8.8 TERNARY SYSTEMS

8.8.1 Kink site positions of ternary systems

The concept of kink site positions and the selectivity function of these positions can be extended to ternary systems. The variety of kink site positions on the surface grows with the number of components. This is shown for a close packed cubic structure if only taking into account first neighbors. Then the following kink site positions are possible on the surface.

$$A_6,\ A_5B,\ A_4B_2,\ A_3B_3,\ A_2B_4,\ AB_5,\ B_6.$$
$$A_5C,\ A_4CB,\ A_3CB_2,\ A_2CB_3,\ ACB_4,\ CB_5$$
$$A_4C_2,\ A_3C_2B,\ A_2C_2B_2,\ AC_2B_3,\ C_2B_4$$
$$A_3C_3,\ A_2C_3B,\ AC_3B_2,\ C_3B_4$$
$$A_2C_4,\ AC_4B,\ C_4B_2$$
$$AC_5,\ C_6B$$
$$C_6$$

The rules of selection between the different kink site positions lead either to the dominance of pure components, A_6, B_6, or C_6, or to the development of dominating mixtures.

It was approximated that in the treatment of a binary alloy, the character of a kink site position is determined by the previously deposited atom and not so much by the environment of the atom. If this is extended to a ternary alloy one can describe the processes on the surface by nine main reactions

$$A^* + A^{z+} + ze^- \rightarrow AA^* \tag{8.65a}$$

$$A^* + B^{z+} + ze^- \rightarrow AB^* \tag{8.65b}$$

$$A^* + C^{z+} + ze^- \rightarrow AC^* \tag{8.65c}$$

$$B^* + A^{z+} + ze^- \rightarrow BA^* \tag{8.66a}$$

$$B^* + B^{z+} + ze^- \rightarrow BB^* \tag{8.66b}$$

$$B^* + C^{z+} + ze^- \rightarrow BC^* \tag{8.66c}$$

$$C^* + A^{z+} + ze^- \rightarrow CA^* \tag{8.67a}$$

$$C^* + B^{z+} + ze^- \rightarrow CB^* \tag{8.67b}$$

$$C^* + C^{z+} + ze^- \rightarrow CC^* \tag{8.67c}$$

8.8.2 The Markov chain theory for ternary systems

Similar to the binary system, one can define probabilities for the formation of new kink sites. For equilibrium conditions one can use the equilibria of the reactions, Eqs. (8.65)–(8.67). For kinetic control one can formulate probabilities with the rate equations.

For equilibrium conditions, the probabilities for the reactions with the kink site atom A* are given by the equations

$$p_{AA} = \frac{c_{A^{z+}}}{c_{A^{z+}} + g_{AB} \cdot c_{B^{z+}} + g_{AC} \cdot c_{C^{z+}}} \tag{8.68a}$$

$$p_{AB} = \frac{g_{AB} \cdot c_{B^{z+}}}{c_{A^{z+}} + g_{AB} \cdot c_{B^{z+}} + g_{AC} \cdot c_{C^{z+}}} \tag{8.68b}$$

$$p_{AC} = \frac{g_{AC} \cdot c_{C^{z+}}}{c_{A^{z+}} + g_{AB} \cdot c_{B^{z+}} + g_{AC} \cdot c_{C^{z+}}} \tag{8.68c}$$

with

$$g_{AB} = \frac{K_{AB}}{K_{AA}} \qquad g_{AC} = \frac{K_{AC}}{K_{AA}} \tag{8.69}$$

The probabilities for reactions with the kink site position B* are

$$p_{BA} = \frac{g_{BA} \cdot c_{A^{z+}}}{g_{BA} \cdot c_{A^{z+}} + c_{B^{z+}} + g_{BC} \cdot c_{C^{z+}}} \tag{8.70a}$$

$$p_{BB} = \frac{c_{B^{z+}}}{g_{BA} \cdot c_{A^{z+}} + c_{B^{z+}} + g_{BC} \cdot c_{C^{z+}}} \tag{8.70b}$$

$$p_{BC} = \frac{g_{BC} \cdot c_{C^{z+}}}{g_{BA} \cdot c_{A^{z+}} + c_{B^{z+}} + g_{BC} \cdot c_{C^{z+}}} \tag{8.70c}$$

with

$$g_{BA} = \frac{K_{BA}}{K_{BB}} \qquad g_{BC} = \frac{K_{BC}}{K_{BB}} \tag{8.71}$$

The probabilities for reactions with the kink site position C* are

$$p_{CA} = \frac{g_{CA} \cdot c_A}{g_{CA} \cdot c_A + g_{CB} \cdot c_B + c_C} \tag{8.72a}$$

$$p_{CB} = \frac{g_{CB} \cdot c_B}{g_{CA} \cdot c_A + g_{CB} \cdot c_B + c_C} \tag{8.72b}$$

$$p_{CC} = \frac{c_C}{g_{CA} \cdot c_A + g_{CB} \cdot c_B + c_C} \tag{8.72c}$$

with

$$g_{CA} = \frac{K_{CA}}{K_{CC}} \qquad g_{CB} = \frac{K_{CB}}{K_{CC}} \tag{8.73}$$

The components of the Eigen vector $\boldsymbol{\alpha}$ of the (3×3) probability matrix, α_A, α_B, and α_C give the equations for the mole fractions of the components in the alloy

$$x_A = \frac{p_{CA}(1 + p_{CB} - p_{BB}) - p_{CB}(p_{CA} - p_{BA})}{(1 + p_{CA} - p_{AA})(1 + p_{CB} - p_{BB}) - (p_{CA} - p_{BA})(p_{CB} - p_{AB})} \tag{8.74a}$$

$$x_B = \frac{p_{CB}(1 + p_{CA} - p_{AA}) - p_{CA}(p_{CB} - p_{AB})}{(1 + p_{CA} - p_{AA})(1 + p_{CB} - p_{BB}) - (p_{CA} - p_{BA})(p_{CB} - p_{AB})} \tag{8.74b}$$

$$x_C = 1 - x_A - x_B \tag{8.74c}$$

The selectivity constants g_{ij} in these equations can be determined by similar procedures as derived for binary alloys.

The derived equations connect the mole fraction of a component of the alloy with the probabilities of a binary deposition process. It becomes possible to predict the composition of a ternary alloy from the information known for the binary combination. The selectivity constants required can be taken from the analysis of binary alloys. The analysis must be carried out for all combinations of the components: AB, AC, and BC. The only requirement is that there are similar deposition conditions for all combinations and for the final deposition of the ternary alloy. This was confirmed by experiments and by data taken from the literature.[15]

8.8.3 Example: prediction of the composition of CoFeNi alloys

As an example for such a calculation, data for CoFeNi are shown in this chapter. The selectivity constants for the combinations, CoFe, CoNi, and FeNi, were determined in

Sections 8.71–8.73. Then, using the equations of the last chapter, the composition of the ternary system was predicted. Experimental data of the ternary system could also be taken from the literature.[16] The comparison of the calculated data with the experimental data is shown in Figures 8.20 and 8.21.

Similar predictions were made for several other ternary alloys. It was possible to include the influence of experimental parameters like temperature or additive concentration into

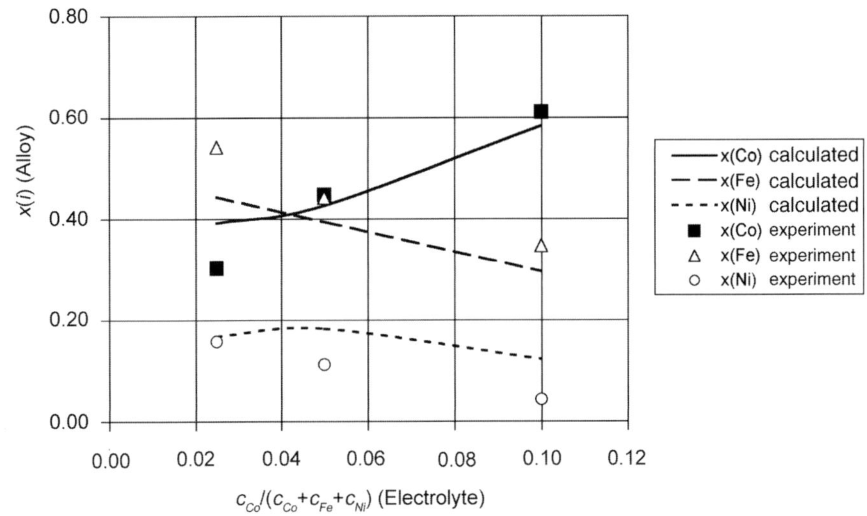

Figure 8.20 CoFeNi alloy system with dependence of the mole fraction x_{Co}, x_{Fe}, and x_{Ni} in the alloy on the concentration fraction $c_{Co}/(c_{Co} + c_{Fe} + c_{Ni})$ in the electrolyte (i stands for Co, Fe, or Ni).

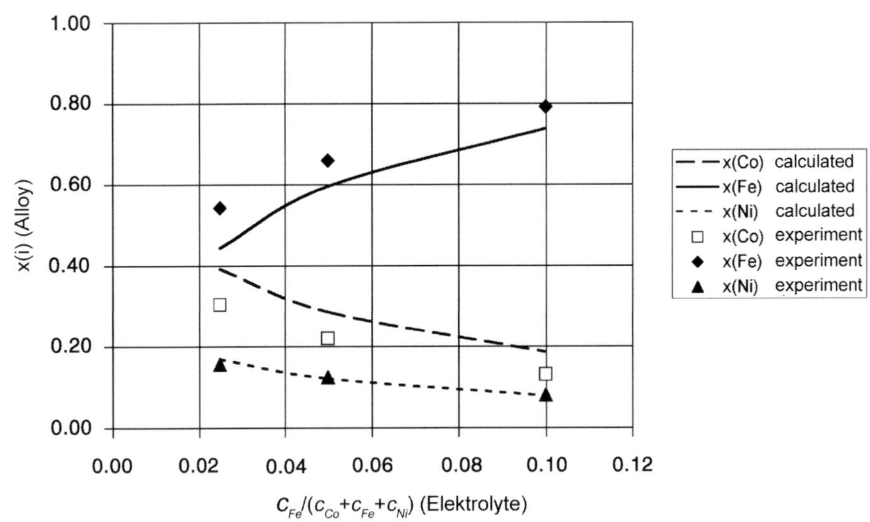

Figure 8.21 CoFeNi alloy system with dependence of the mole fraction x_{Co}, x_{Fe}, and x_{Ni} in the alloy on the concentration fraction $c_{Fe}/(c_{Fe} + c_{Fe} + c_{Ni})$ in the electrolyte (i stands for Co, Fe, or Ni).

the predictions. The principles of this method can be applied to more complex systems, e.g., quaternary alloy systems.

The equations do not compete with quantitative simulations based on detailed mechanistic models of the deposition process. But the equations provide a tool for the prediction of trends. This is especially important, if the influence of several experimental parameters has to be predicted.

REFERENCES

1. B. Despić, V. Jović, Electrochemical Deposition and Dissolution of Alloys and Metal Composites - Fundamental Aspects, in R.E. White, J.O'M. Bockris, B. Conway (Eds.), Modern Aspects of Electrochemistry, Vol. 27, Plenum Press, New York, 1995, p. 143.
2. A. Milchev, R. Lacmann, J. Cryst. Growth, *110*, 919, 925 (1991); Electrochimica Acta, *40*, 1475 (1995).
3. A Milchev, Electrocrystallization: Fundamentals of Nucleation and Growth, Kluwer Academic Publishers, Boston, 2002.
4. C. Wagner, W. Traud, Z. Elektrochem., *44*, 391 (1938).
5. A. Brenner, Electrodeposition of Alloys, Vol. 2, Academic Press, New York, 1963, p. 152.
6. D. Landolt, Electrochimica Acta, *39*, 1075 (1994).
7. W. Plieth, Z. Phys. Chem., *217*, 383 (2003); J. Solid State Electrochem., *8*, 338 (2004).
8. A.T. Bharucha-Reid, Elements of the Theory of Markov Processes and Their Applications, MacGraw-Hill, New York, 1960.
9. G.S. Georgiev, V.T. Georgieva, W. Plieth, Electrochimica Acta, *50*, 870 (2005).
10. W. Plieth, G.S. Georgiev, J. Solid State Electrochem., *9*, 859 (2005).
11. W. Plieth, G.S. Georgiev, Elektrokhimiya, *42*, 1216 (2006); Russ. J. Electrochem., *42*, 1093 (2006).
12. W. Plieth, Electrochimica Acta, *53*, 245 (2007).
13. T. Watanabe, Nanoplating, Elsevier, Oxford, 2004.
14. E.J. Podlaha, D. Landolt, J. Electrochem. Soc., *143*, 885 (1996); J. Electrochem. Soc., *144*, 1672 (1997).
15. W. Plieth, H Kühnlein, Meet. Abstr.-Electrochem. Soc. *602*, 1929 (2006).
16. V. Zhuang, E.J. Podlaha, J. Electrochem. Soc., *150*, C219 (2003).

– 9 –

Oxides and Semiconductors

In this chapter the electrochemical properties of materials with covalent bonds and stoichiometric composition will be described. Many of these materials are semiconductors. In the literature there is a classical description of their electrochemistry by Morrison.[1] Memming[2] has given a comprehensive review on semiconductor electrochemistry, Sato[3] has described the electrochemistry for oxides and semiconductor electrodes, and Trasatti[4] has edited a book on catalytic aspects of oxides.

9.1 ELECTROCHEMICAL PROPERTIES OF A SEMICONDUCTOR

9.1.1 Band model of a semiconductor

The electronic properties of solids were described in Chapter 2 using the band model. A characteristic feature of semiconductors is the separation of the electron energy levels into two bands, the valence band with occupied energy levels and the conduction band with unoccupied energy levels. Both bands are separated by an energy gap. The band gap energy E_g determines the intrinsic conductivity because electricity can only be transported through the semiconductor if some electrons are excited from the valence band to the conduction band. Then either holes in the valence band or electrons in the conduction band become mobile. The mobility of valence band holes μ_+ and conduction band electrons μ_- are characteristic values for a semiconductor. For $E_g \approx 1$ eV at room temperature if enough electrons are excited, the material is an intrinsic semiconductor, e.g., Si, $E_g = 1.1$ eV. For high band gap energies, e.g., TiO_2, $E_g = 3.2$ eV intrinsic conductivity becomes negligible.

The energy of an energy level is defined as the energy released from the material, if an electron with zero kinetic energy is brought from the vacuum into the energy level. The first energy level of a neutral material that an electron can occupy is the lower edge of the conduction band (Figure 9.1). This energy is sometimes called electron affinity. It is easier to determine the energy needed to emit an electron from the semiconductor. The lowest energy to emit an electron is the energy of the upper edge of the valence band. The lowest energy to emit an electron into the vacuum with zero kinetic energy is called work function Φ. The negative work function is equal to the energy of the upper edge of the valence band.

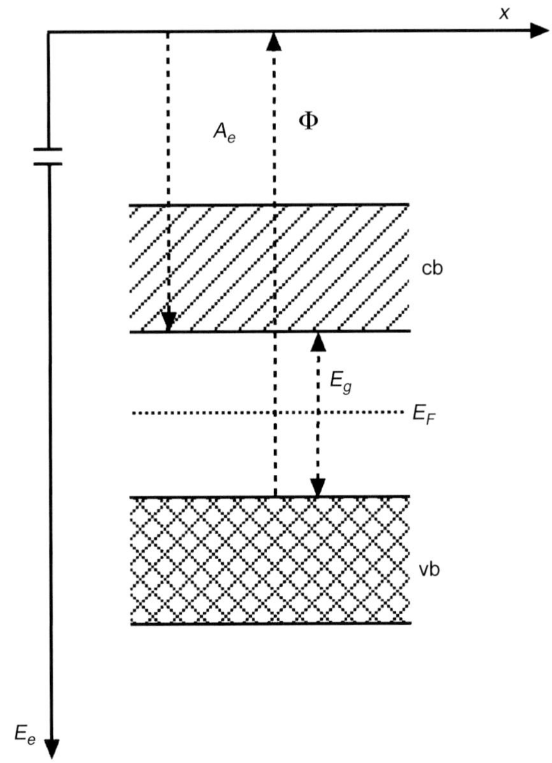

Figure 9.1 Band model of an intrinsic semiconductor and definition of energies. A_e is the electron affinity, Φ the work function, and E_g the gap energy.

In metals the Fermi energy is the energy of the highest occupied energy level (at $T = 0$ K). In an intrinsic semiconductor the Fermi energy is per definition midway between the valence and conduction bands. Therefore, the Fermi energy for an intrinsic semiconductor is

$$E_F = -\Phi + \frac{1}{2}E_g \tag{9.1}$$

Conductivity can also be achieved by doping. By adding small amounts of substances with excess electrons, these electrons can be inserted into the conduction band (n-doping, n-semiconductor). Adding substances with electron deficiency attracts the electrons from the valence band and creates holes in the valence band (p-doping, p-semiconductor). The properties of the semiconductor are then determined by the new majority careers, either by the electrons in the conduction band or the electron holes in the valence band.

Doping shifts the Fermi energy. For an n-semiconductor the Fermi energy moves to a position near to the lower edge of the conduction band. For a p-semiconductor the Fermi energy moves to a position near to the upper edge of the valence band. This was shown in Figure 2.32.

9.1.2 Semiconductor–electrolyte contact

In electrochemistry the semiconductor (phase I) is connected to an electrolyte (phase II). In equilibrium the electrochemical potential for the electrons in both phases must be equal. The electrochemical potential of the electrons in the semiconductor is equal to the Fermi energy.

$$\tilde{\mu}_e(\mathrm{I}) = \tilde{\mu}_e(\mathrm{II}) \tag{9.2}$$

$$\tilde{\mu}_e(\mathrm{I}) = E_{\mathrm{F}}(\mathrm{I}) \tag{9.3}$$

Transfer of the charge between both phases establishes the equilibrium condition. For an n-semiconductor with excess electrons, electrons move from the conduction band to the electrolyte and a positive charge is built up in the space-charge region of the semiconductor. This leads to the build up of a positive Galvani potential difference $\Delta\varphi$ between semiconductor and electrolyte (Figure 9.2A).

For a p-semiconductor with an excess of holes in the valence band, electrons accumulate in the space-charge region of the semiconductor. This leads to the build up of a negative Galvani potential difference between semiconductor and electrolyte (Figure 9.2B). The potential gradient across the space-charge layer leads to a shift of the electron energy levels. This is called band bending. Potential gradient and band bending are shown for n-semiconductors in Figure 9.2A, and for p-semiconductors in Figure 9.2B.

Application of an external potential mainly changes the potential gradient in the space-charge layer while the surface potential and the band position at the surface remain nearly constant. This is called band pinning. The band bending changes with the potential.

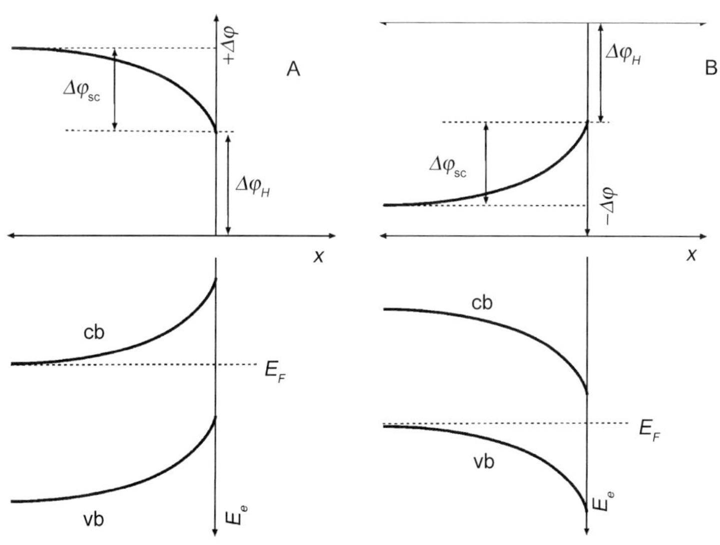

Figure 9.2 Potential gradient and band bending in the double layer of a semiconductor–electrolyte contact. (A) n-Semiconductor and (B) p-semiconductor.

Different polarization situations for the semiconductor–electrolyte interface are shown in Figure 9.3.

A characteristic potential for band bending when it is compensated is the flat band potential. In the case of an n-semiconductor, polarization of the electrode from the flat band potential in the anodic direction causes a depletion of the charge in the space-charge or depletion region. At high anodic potentials when the Fermi energy reaches and passes the upper edge of the valence band, holes accumulate in the surface region, which is called the degeneration region. Otherwise, if the potential is polarized and is therefore more cathodic than the flat band, potential electrons accumulate in the surface or accumulation region and the electrode behaves metallically.

The situation for the p-semiconductor is similar. Beginning at the flat band potential and if the electrode is polarized in the cathodic direction, the electrode is brought into the depletion region, i.e., depletion of holes. At high cathodic potentials when the degeneration region is reached, the Fermi energy passes the lower edge of the conduction band and electrons accumulate in the surface region. Otherwise, in the accumulation region, which is anodic from the flat band potential, holes accumulate in the surface region.

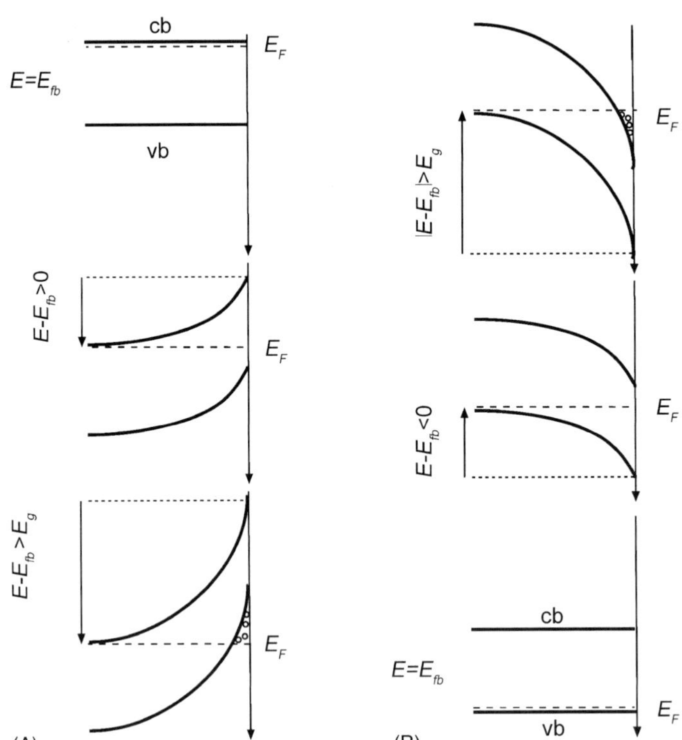

Figure 9.3 Polarization of the semiconductor–electrolyte interface. (A) n-Semiconductor and (B) p-semiconductor.

9.1.3 Gap states and surface states

Ideal crystallographic conditions are optimal. For semiconductors with a large concentration of defects or even amorphous systems, a more realistic band diagram must be considered (shown in Figure 9.4). These bands have no sharp boundaries, however. Instead diffused band tails mark the transition from the area of free mobile careers to the band gap. This band gap is now called a mobility gap. Energy states exist within the gap that can act as career traps. These traps are of special importance in photoelectrochemical processes because they are catalysts for recombination processes. Some surface states are formed by the truncation of the crystallographic structure. Other ones are created by specific adsorption. At a high density of surface states the Fermi level is fixed or pinned at the surface, which is called Fermi level pinning.

9.1.4 Current–potential curves

Application of a potential between a semiconductor electrode and a counter electrode in an electrolyte leads to a current flux. For an n-semiconductor in the accumulation region

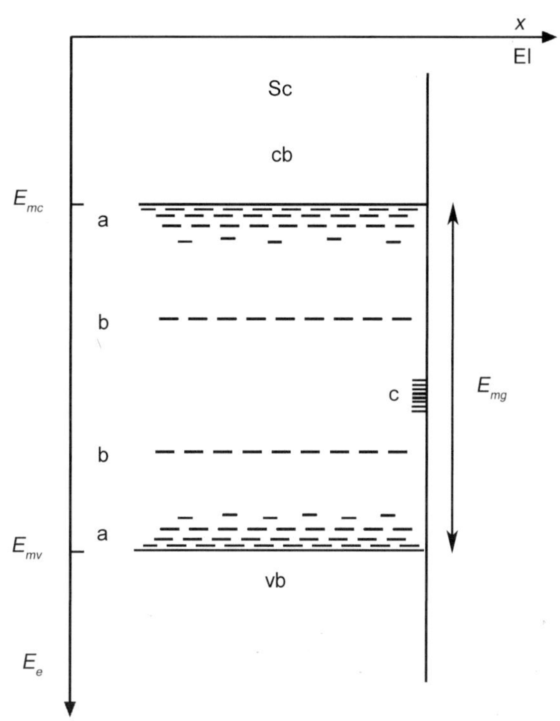

Figure 9.4 Electron energy states of semiconductors with a large concentration of defects (or amorphous semiconductors) and surface states. E_{mc} energy limit of free mobility in the conduction band, E_{mv} energy limit of free mobility in the valence band, E_{mg} mobility gap, (a) and (b) gap states, and (c) surface states.

electrons accumulate at the surface and a cathodic current can flow. In the depletion region where current cannot flow, the electrode behaves like a diode. At high anodic potentials holes accumulate in the surface region and an anodic current can flow. For a p-semiconductor in the accumulation region an anodic current is observed. In the depletion region current cannot flow, and in the degeneration region a cathodic current can flow. This is shown in the schematic current–potential curve in Figure 9.5.

The charge transfer on a metal electrode depends on the overlap between occupied and unoccupied energy levels in metal and electrolyte. This can be found around the Fermi level (Chapter 6). The contact of an electrolyte with a redox system and a semiconductor

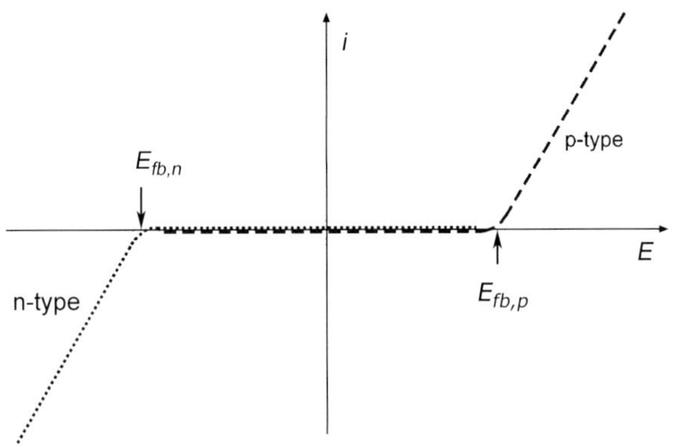

Figure 9.5 Schematic current–potential curve of an n-semiconductor and a p-semiconductor.

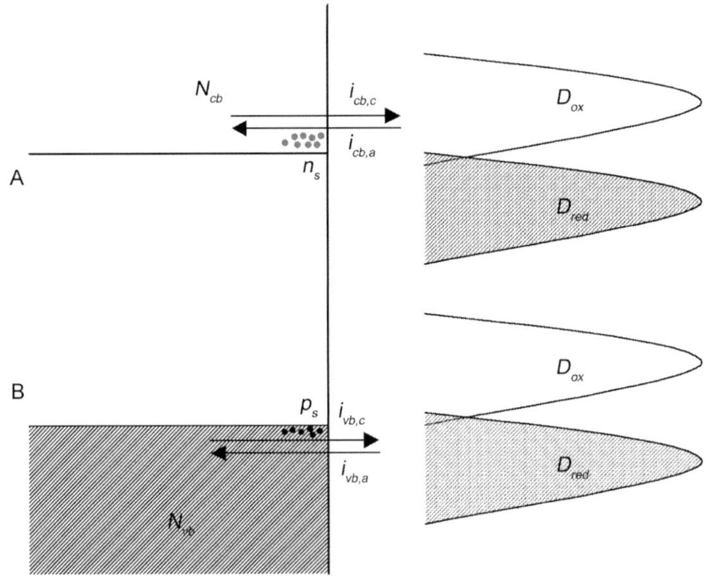

Figure 9.6 Charge transfer between a semiconductor electrode and a redox system. (A) Conduction band processes and (B) valence band processes.

electrode is shown for an n-semiconductor in Figure 9.6A and for a p-semiconductor in Figure 9.6B. As shown in this schematic representation, the redox system can overlap with both bands. Depending on the redox system and electrode potential, charge transfer is possible between the conduction band and redox system and between the valence band and redox system. In the first case, one speaks of a conduction band process, in the second, of a valence band process.

If the density of electrons in the conduction band on the surface of the semiconductor is n_s and the density of unoccupied energy levels in the electrolyte is D_{ox}, the partial current density of conduction band reduction is

$$i_{cb,c} = k_{cb,c} \int_{-\infty}^{+\infty} n_s D_{ox} dE \tag{9.4}$$

Otherwise, the anodic conduction band current is proportional to the density of occupied states in the electrolyte D_{red} times the density of states in the conduction band

$$i_{cb,a} = k_{cb,a} \int_{-\infty}^{+\infty} D_{red} N_{cb} dE \tag{9.5}$$

Similarly, one can formulate for the anodic valence band process with p_s denoting the surface density of holes in the valence band

$$i_{vb,a} = k_{vb,a} \int_{-\infty}^{+\infty} D_{red} p_s dE \tag{9.6}$$

For the cathodic valence band process one obtains with N_{vb} the density of energy levels in the valence band

$$i_{vb,c} = k_{vb,c} \int_{-\infty}^{+\infty} D_{ox} N_{vb} dE \tag{9.7}$$

In a first approximation D_{red}, D_{ox}, N_{cb}, and N_{vb} are independent of the potential. The surface density of electrons in the conduction band and the surface density of holes in the valence band, n_s and p_s, depend exponentially on the potential. The current density at the flat band potential can be used as a reference point. In this approximation one obtains

$$i_{cb,c} = i_{cb,0} \exp\left[-\frac{F}{RT}(E - E_{fb})\right] \tag{9.8}$$

$$i_{cb,a} = \text{const} \tag{9.9}$$

$$i_{vb,a} = i_{vb,0} \exp\left[+\frac{F}{RT}(E - E_{fb})\right] \tag{9.10}$$

$$i_{vb,c} = \text{const} \tag{9.11}$$

This behavior is schematically represented in Figure 9.5. The slope of the rise of the current in the accumulation region in Figure 9.5, often differs from the 59 mV expected according to Eqs. (9.8) and (9.10). Otherwise, in the accumulation region the semiconductor electrodes approach metal-like properties and the classical theory of electron transfer must be applied.

The characteristic feature of the dark current–potential curves of n- and p-semiconductors is the depletion region where the electrode behaves like a diode. The diode region should be approximately equal to the gap energy E_g. Examples of current–potential curves are shown together with photocurrents for n-GaAs in Figure 9.10A, and for p-GaAs in Figure 9.10B.

9.1.5 Space-charge capacitance

Another characteristic electrochemical property of a semiconductor/electrolyte contact is the double-layer capacitance, which is an approximation of the space-charge capacitance (Chapter 4). The space-charge capacitance can be determined by impedance measurements. If no current flows in the depletion region, the impedance is given by the reciprocal value of the space-charge capacitance. For other conditions the capacitance can be calculated from the complex impedance measurements. How to measure the impedance and to evaluate the data was described in Chapter 4 as well as the influence of diffusion processes in Chapter 5.

The capacitance C_{sc} is a function of the electrode potential E. The relation between capacitance and potential is called the Mott–Schottky equation

$$\frac{1}{C_{sc}^2} = \frac{2}{\varepsilon\varepsilon_0 e_0 N_{cc}}\left(E - E_{fb} - \frac{k_B T}{e_0}\right) \tag{9.12}$$

where N_{cc} is the charge career (donor or acceptor) density, ε and ε_0 the relative and vacuum electric permittivities, e_0 the electron charge, and k_B the Boltzmann constant. By plotting $1/C_{sc}^2$ versus the potential E, a linear dependence can be expected. The extrapolation of the linear dependence to $1/C_{sc}^2 = 0$ gives the potential $E_{fb} + k_BT/e_0$. Usually k_BT/e_0 can be neglected

$$E_{fb} + \frac{k_B T}{e_0} \approx E_{fb}$$

From the slope of the Mott–Schottky line one can determine the charge career density N_{cc} (donor or acceptor density) if the relative dielectric permittivity is known.

An example of such a plot for n- and p-GaAs is shown in Figure 9.7. The distance of the two flat band potentials is equal to the band gap energy.

Current–potential measurements, capacitance measurements, and complementary photoelectrochemical measurements have provided a detailed picture of the electrochemical data of the band structure of semiconductors. Some experimental results are presented in Figure 9.8.

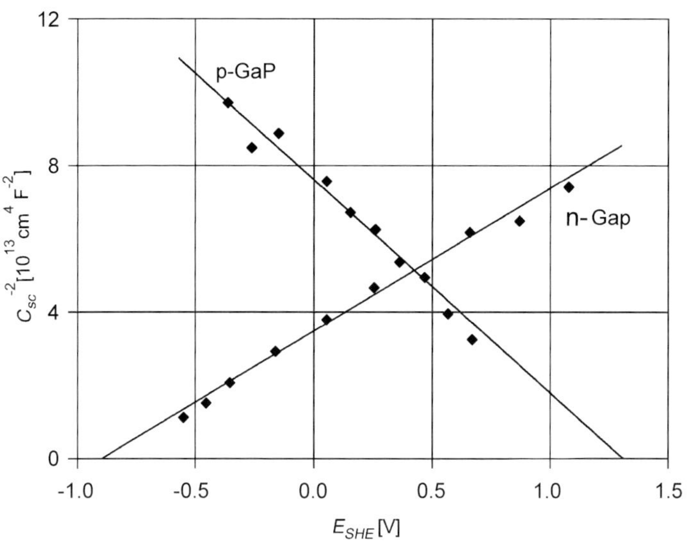

Figure 9.7 Mott–Schottky plot of n- and p-GaAs; 0.05 mol·cm^{-3} H_2SO_4 (after Memming).[5]

9.2 PHOTOELECTROCHEMISTRY OF SEMICONDUCTORS

The separation of the conduction and valence bands provides a way to excite photons from the valence band to the conduction band. This is also the origin of a variety of photoelectrochemical effects. Special monographs and reviews have been published, e.g., the book by Pleskov and Gurevich.[6]

9.2.1 Photocurrents

If the light of energy is greater than the gap energy electrons can then be excited from the valence band to the conduction band. The photoelectrons and photoholes are separated by the field gradient in the space-charge region. The result is an anodic photocurrent for n-semiconductors and a cathodic photocurrent for p-semiconductors. The principles are shown in Figure 9.9; photocurrents of n- and p-GaP are shown (in comparison to the dark currents) in Figure 9.10.

The photocurrent should begin at the flat band potential but the start potential usually shifts to more anodic potentials (n-semiconductor) or respectively to more cathodic potentials (p-semiconductor). Kinetic effects can explain the shift of the start-potential,[5] but other explanations have also been given.[8]

The general theory of photocurrents goes back to Gärtner.[9] The photocurrent is proportional to the light intensity I_0 and depends on the ratio between the width of the space-charge layer w_{sc}, the thickness of the absorption layer in the semiconductor given by the

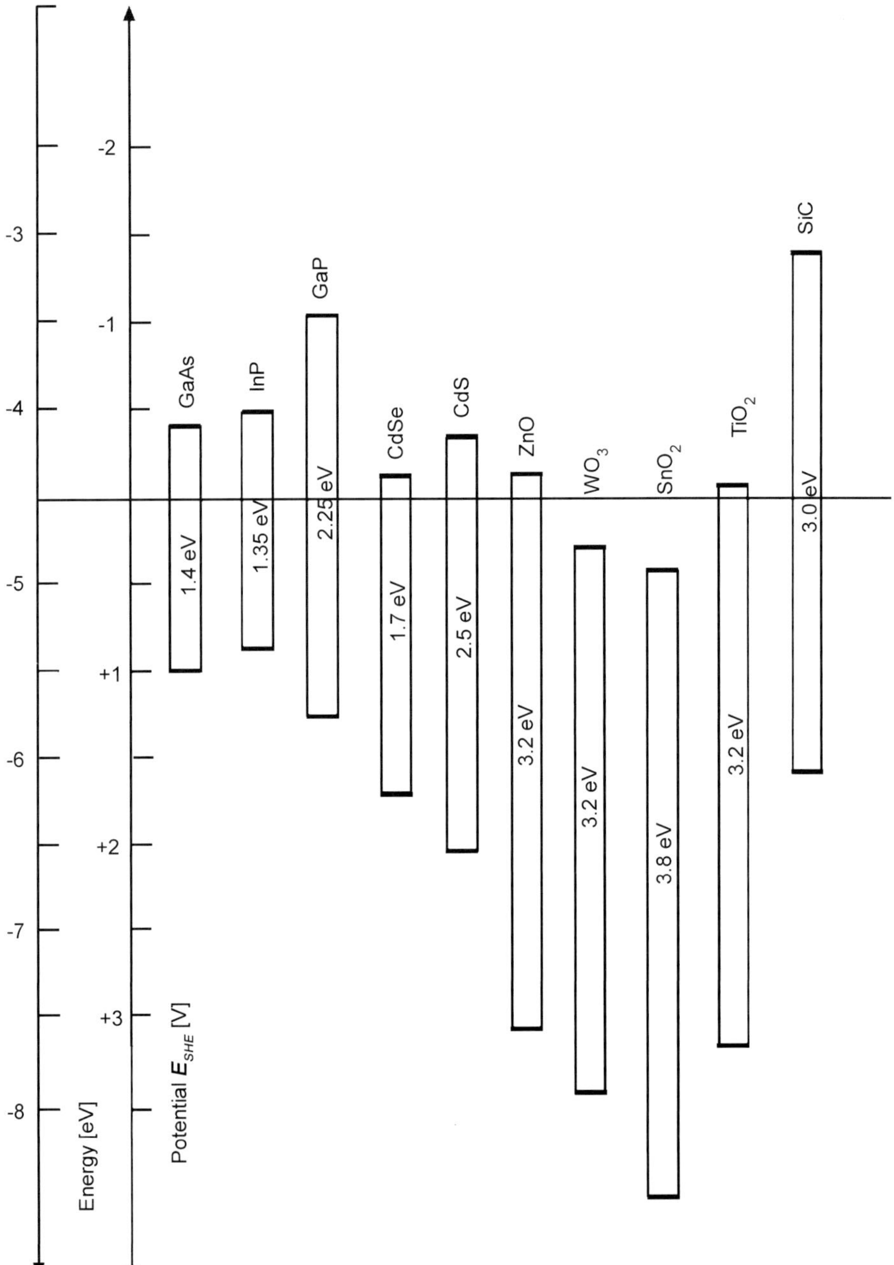

Figure 9.8 Band structure of some semiconductors on the electrochemical potential scale.

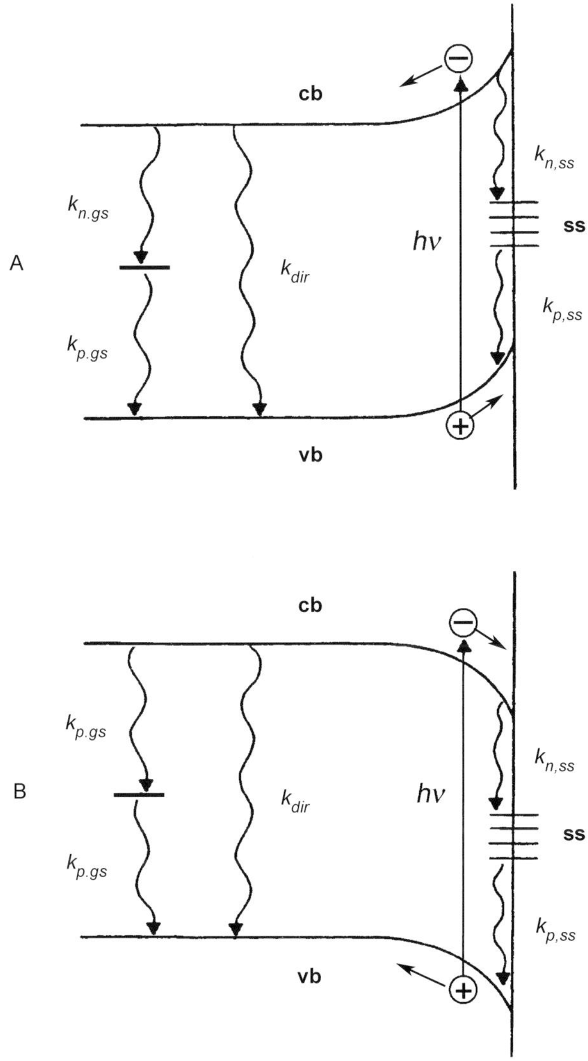

Figure 9.9 Photoexcitation at (A) an n-semiconductor and (B) a p-semiconductor (schematic representation). Direct recombination and recombination via band states and surface states.

reciprocal absorption coefficient α_{sc}, and the mean diffusion length of the charge carrier l_{cc} according to the equation

$$i_{ph} = I_0 \left[1 - \frac{\exp(-\alpha_{sc} \cdot w_{sc})}{1 + \alpha_{sc} \cdot l_{cc}} \right] \tag{9.13}$$

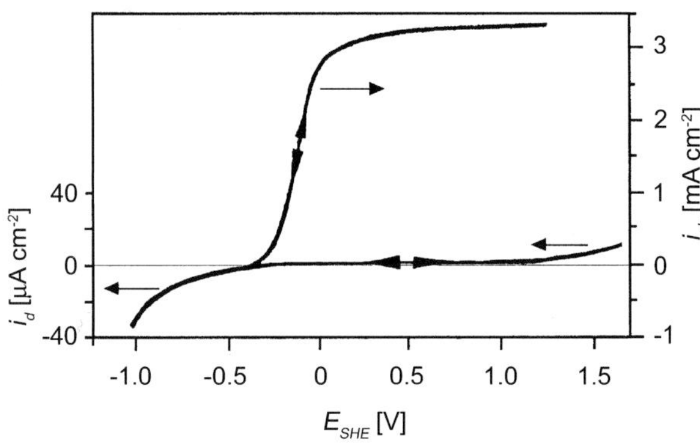

Figure 9.10a Photocurrent density i_{ph} and dark current density i_d depend on the potential. n-GaAs, 0.2 mol·cm^{-3} H_2SO_4 (according to Plieth, Pfuhl, and Matz).[7]

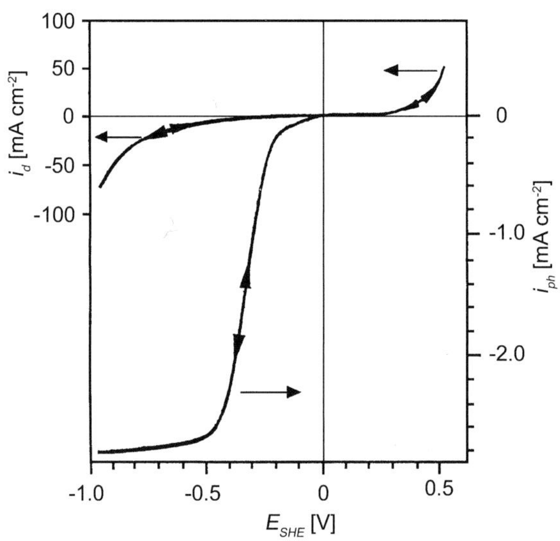

Figure 9.10b Photocurrent density i_{ph} and dark current density i_d depend on the potential. p-GaAs, 0.2 mol·cm^{-3} H_2SO_4 (according to Plieth, Pfuhl, and Matz).[7]

For small band bending, $\alpha_{sc} \cdot w_{sc} << 1$ and small diffusion length $\alpha_{sc} \cdot l_{cc} << 1$, one derives the approximation $i_{ph} = I_0 \cdot \alpha_{sc} \cdot w_{sc}$. Substituting the width of the space-charge layer w_{sc} by an expression obtained from the Mott–Schottky theory

$$w_{sc} = \sqrt{\frac{2\varepsilon\varepsilon_0}{e_0 N_{cc}}\left(E - E_{fb}\right)} \tag{9.14}$$

one obtains

$$i_{ph} = I_0 \alpha_{sc} \sqrt{\frac{2\varepsilon\varepsilon_0}{e_0 N_{cc}}\left(E - E_{fb}\right)} \tag{9.15}$$

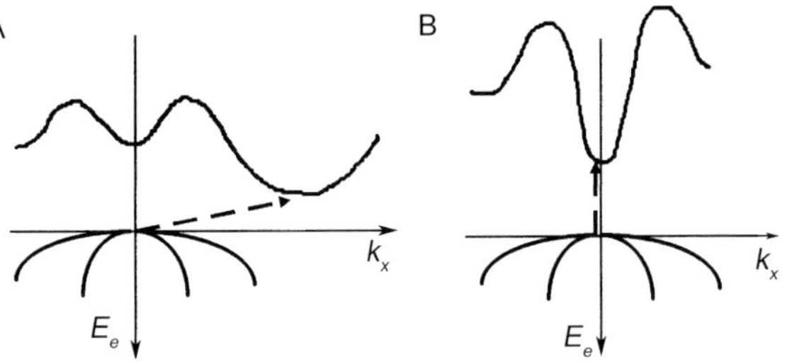

Figure 9.11 Photoexcitations in semiconductors. (A) Indirect transition, e.g., GaAs and (B) Direct transition, e.g., SiO_2. E_e is the energy and k_x the wave vector.

The absorption coefficient depends on the light energy $h\nu$ according to the equation

$$\alpha_{sc} = \text{const}\frac{\left(h\nu - E_g\right)^n}{h\nu} \tag{9.16}$$

The exponent is $n = 0.5$ for semiconductors with direct transition, and $n = 2$ for semiconductors with indirect transition. Direct and indirect transitions are explained in Figure 9.11.

By inserting Eq. (9.16) into Eq. (9.15) one gets for constant potential two types of representations of photocurrents

For direct transition $$\left(\frac{i_{ph} \cdot h\nu}{I_0}\right)^2 = \text{const} \cdot (h\nu - E_g) \tag{9.17}$$

For indirect transition $$\left(\frac{i_{ph} \cdot h\nu}{I_0}\right)^{1/2} = \text{const} \cdot (h\nu - E_g) \tag{9.18}$$

The better linearity of these plots is used to distinguish direct and indirect transitions and to determine the band gap energy from photocurrent measurements.

For the description of the effects of illuminated semiconductor electrodes the concept of the quasi Fermi level was developed. For stationary illumination Δn photoelectrons and Δp photoholes are generated in the surface region with the result that there is no longer equilibrium between the conduction and valence bands. One can define individual electrochemical potentials for the photoelectrons (quasi Fermi level of electrons) and the photoholes (quasi Fermi level of holes).[10]

9.2.2 Intensity modulated photocurrent spectroscopy (IMPS)

If the electrode is illuminated by a light of modulated intensity, the current response of the electrode is a complex quantity consisting of the photocurrent amplitude and the phase shift between light and photocurrent. The method was developed primarily by Peter and

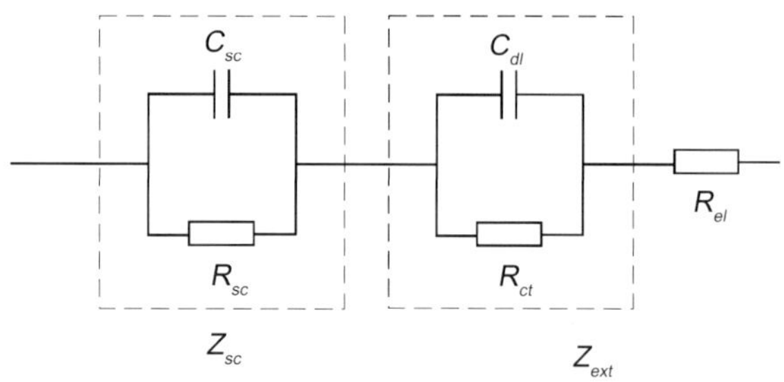

Figure 9.12 Equivalent circuit for intensity modulated photocurrent spectroscopy (IMPS).[12]

was called intensity modulated impedance spectroscopy.[11] For the explanation of the processes connected with the periodical illumination of the electrode, the equivalent circuit in Figure 9.12 was suggested.[12]

The light from separating electrons and holes periodically charges the space-charge capacitance C_{sc}. The recombination of the separated charge can occur internally via the resistance of the space-charge layer R_{sc}. Impedance Z_{sc} represents the internal recombination. The second way for the recombination, which is via the external circuit represented by impedance Z_{ext}, takes into account the charge transfer resistance R_{ct} and double layer capacitance C_{dl}. Resistance R_{el} represents the electrolyte resistance.

Evaluation of IMP spectra demands comparison with impedance spectra. Commercial equipment for measuring parallel impedance spectra and IMP spectra is available.[13]

Details can be found in the literature, e.g., Peter and Vanmaeckelbergh.[14]

9.2.3 Photopotentials and photopotential transients

Experiments with laser light pulses provide further insight into the action of the light. The separation of the charge careers by the light pulse generates a photopotential. A photopotential transient follows the decay of the photopotential.[15] The equivalent circuit in Figure 9.12 used for the interpretation of IMPS also explains the processes on the semiconductor surface after a laser pulse.

For the measurement of the photopotential and photopotential decay the usual potentiostat/galvanostat could not be used and special equipment was developed.[16,17] An example of a photopotential transient measured on n-GaAs is shown in Figure 9.13.

With a large external resistance one can force the internal recombination. This was achieved in the example seen in Figure 9.13 with an external resistance of 1 MΩ. The internal recombination followed a second-order rate equation as is shown by the linear time dependence of the reciprocal values.

Short laser pulses can also be used to study the primary processes in the build up of the photopotentials. These are extreme fast processes in the femto- and picosecond domain and special developments have been made to make measurements in these time domains, e.g., Bitterling and Willig.[18]

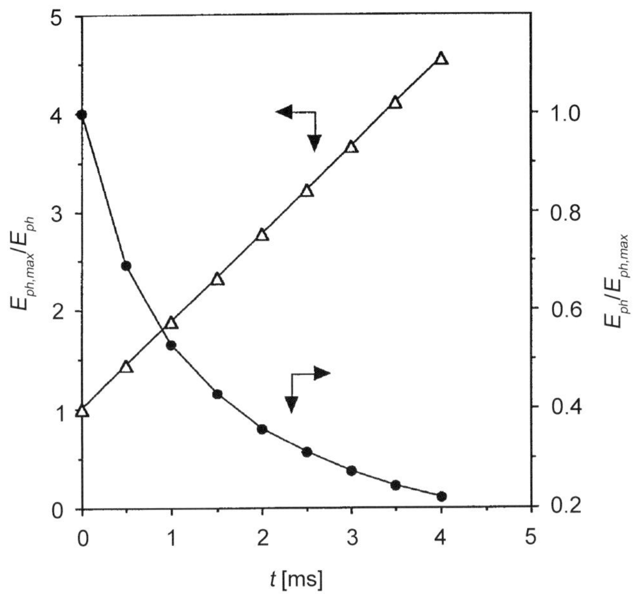

Figure 9.13 Photopotential transient of p-GaAs. Reduced presentation $E_{ph}/E_{ph,max}$ and presentation of the reciprocal values, 0.05 mol·cm^{-3} H_2SO_4; Excimer laser, λ = 308 nm; and external resistance 1 MΩ.[7]

9.3 SPECTROSCOPIC METHODS

A variety of spectroscopic methods are available to investigate properties of semiconductors and oxides with relevance to the electrochemistry of these materials. These methods are often divided into *in situ* and *ex situ* methods. Some methods are described in other chapters (Chapters 7 and 11). Some aspects that are more closely related to oxides and semiconductors will be described in this chapter. Detailed descriptions exist and summaries of reviewing papers are found in the literature.[19,20]

9.3.1 *In situ* spectroscopic methods

The application of ellipsometry for investigating optical properties and film thickness has been known for a long time. This method is based on the measurement of changes of the state of polarization during reflection from the film covered metal surface. The method has become more widely applicable through the development of computer, because the evaluation of data (thickness, refraction, and absorption index) from the polarization parameters (intensity and phase changes) is tedious and complicated. An example for the application of this method will be shown in Chapter 10.

Spectroscopic reflectance methods are UV/vis reflectance spectroscopy and infrared reflection absorption spectroscopy (IRRAS) with several variations. For the application of these methods a mirror-like electrode surface is needed. This can be avoided if the scattered

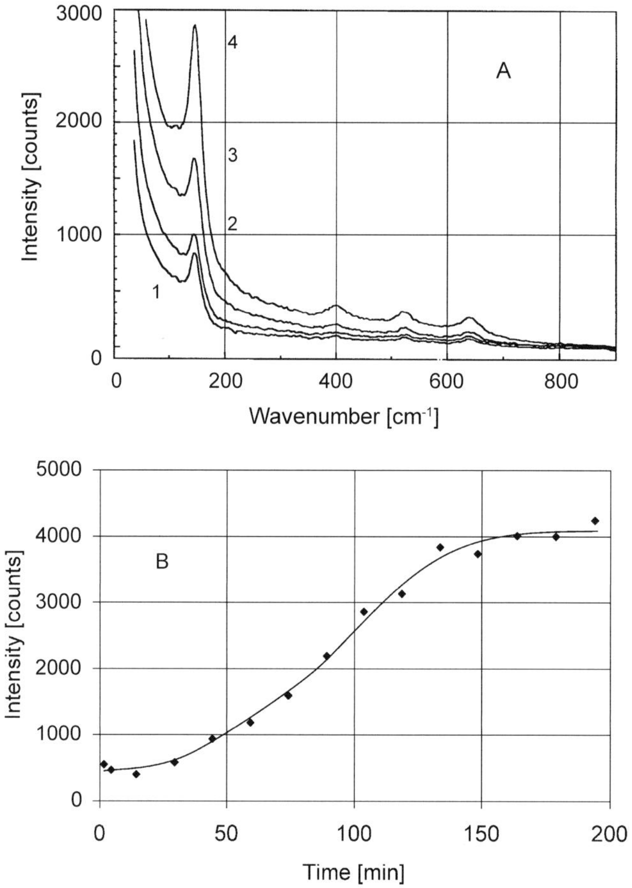

Figure 9.14 (A) Raman spectra of electrochemical formation of titanium dioxide. 0.5 mol · dm^{-3} H_2SO_4, potentiostatic control at 100 V, formation times: (1) 2 min, (2) 15 min, (3) 45 min, and (4) 90 min. (B) Time dependence of the main peak, anatas modification.[21]

light is analyzed. An example is surface Raman spectroscopy. In Chapter 7, surface-enhanced Raman spectroscopy was described as a method to study additives. Investigation of oxide films and semiconductors is not restricted to surface Raman enhancement. As an example, an investigation on titanium dioxide is shown in Figure 9.14.[21] It was possible to follow the transformation of amorphous titanium dioxide (the modification formed at lower potentials) into anatas, brookite, and rutile. In Figure 9.14 the transformation into anatas is shown.

9.3.2 *In situ* X-ray diffraction (XRD) and X-ray absorption spectroscopy (XAS)

Connected with the development of synchrotron radiation sources, X-ray diffraction and X-ray spectroscopy could be applied to *in situ* investigating surfaces. One method especially suited for studying electrode surfaces is extended X-ray absorption fine structure (EXAFS) analysis.

The principle arrangement for studying the surface area or thin films is the suppression of reflexes coming from the bulk of the electrode. To solve this problem, special cells for grazing incidence diffraction and spectroscopy were developed. Particularly helpful here are grazing incidence X-ray diffraction (GIXRD) and grazing incidence X-ray absorption spectroscopy (GIXAS). Two examples are shown in Figure 9.15.[22]

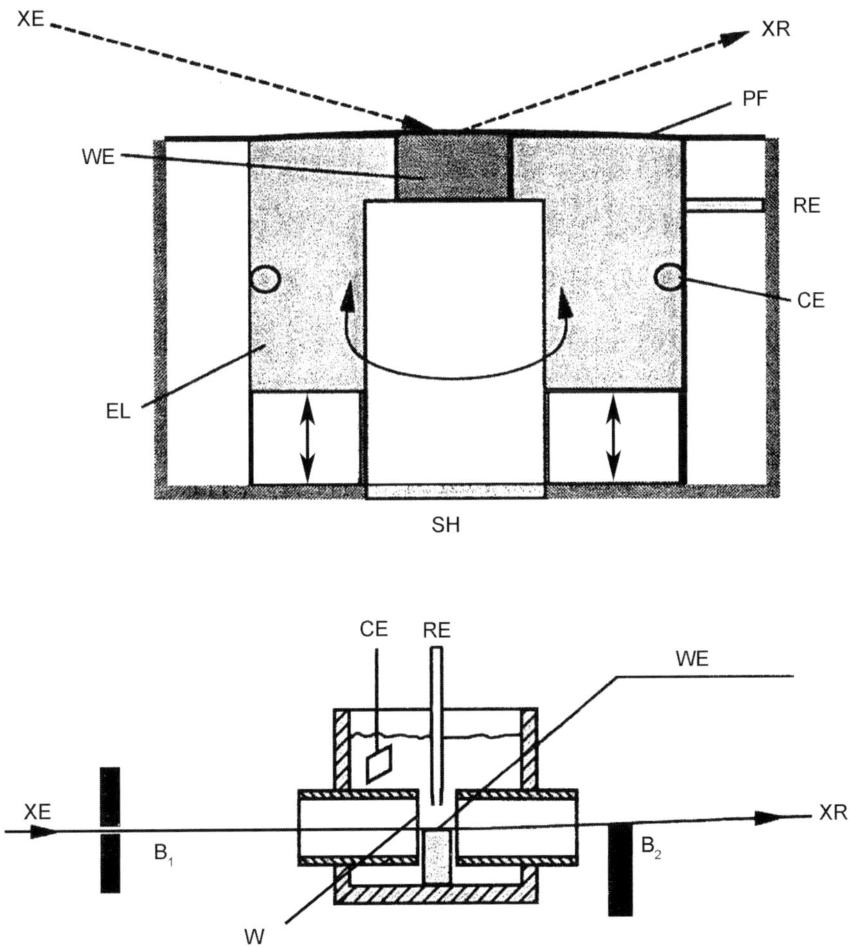

Figure 9.15 Electrochemical cells for *in situ* grazing incidence X-ray diffraction (GIXRD) and grazing incidence X-ray absorption spectroscopy (GIXAD). (A) Thin-layer cell for XAS in reflection and grazing incidence XRD; XE, entrance X-ray beam; XR, reflected X-ray beam; PF, polyethylene foil; WE, working electrode; RE, reference electrode; CE, counter electrode; EL, electrolyte; SH, sample holder and (B) cell for XAS in transmission and reflection. XE entrance X-ray beam, XR reflected X-ray beam, WE working electrode, CE counter electrode RE reference electrode, W windows, B1 beam aperture, entrance beam, B2 beam aperture, reflected beam, and blocking direct beam (according to Strehblow). (Reproduced with permission from Ref. [22], © 2006, Walter de Gruyter & Co.)

In Figure 9.15A a thin-layer cell for XAS in reflection and grazing incidence XRD is shown. A polyethylene foil covers the electrolyte. After electrochemical surface treatment, part of the electrolyte is sucked off and the foil is pressed on the surface of the electrode, usually a single crystal.

In Figure 9.15B a cell for XAS in transmission and reflection is shown. The thickness of the electrolyte, which can be accepted depends on the X-ray energy and is usually limited to a few milimeter. This limits the size of the electrode, which is placed between the entrance and exit windows. Otherwise, with this method the angle of incidence can be made smaller than the angle of total reflection (0.13° for Ag, 0.1° for Ag_2O, 0.45° for Cu, and 0.23° for Cu_2O). At these angles the penetration depths of the X-ray beam into the electrode material is 2 nm.

Special programs must be used for the data evaluation. Results are usually obtained in a feedback of calculations of crystallographic distances and the recalculation of the initial diagrams.

9.3.3 *In situ* Mössbauer spectroscopy

Some nuclei, if embedded into a solid matrix, are capable of repulsion-free γ-ray resonance absorption. Examples of elements are Fe,[57] Ru,[99] and Sn.[119] The extremely sharp resonance line can be used to measure with high resolution the influence of the chemical environment of the nucleus. For details of the experimental equipment the special literature should be consulted, e.g., Scherson.[23]

9.3.4 *Ex situ* methods

The instruments of surface analysis have become extremely important for electrochemistry. Investigations have been performed to study the double layer and corroborate results of electrochemical methods.

One type of method is an energy analysis of electrons emitted from the surface with Auger-electron-spectroscopy (AES), X-ray photoelectron spectroscopy (XPS), and ultraviolet photoelectron spectroscopy (UPS). A second way is to sputter the surface and analyze the emitted particles by mass spectroscopy. This method is called secondary ion mass spectroscopy (SIMS). A third method is based on the scattering of He ions from the surface and the analysis of the scattering parameters (Rutherford backscattering).

Diffraction of electrons, either of high energy (RHEED) or of low energy (LEED), has been used for studying the structure of surfaces.

9.4 MICROSCOPY

The general importance of microscopy for surface investigations is reflected in the *in situ* applications of optical microscopy, scanning tunneling microscopy (STM), and atomic force microscopy (AFM) to oxide and semiconductor electrodes. The methods were described in Chapter 4. Of equal importance for oxide and semiconductor electrodes is the application of *ex situ* methods like scanning electron microscopy (SEM).

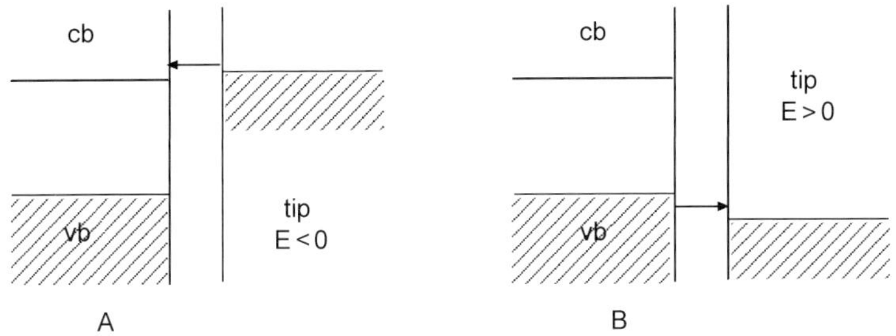

Figure 9.16 Band model of semiconductor and tip in vacuum. (A) Negatively charged tip to image conduction band states and (B) positively charged tip to image valence band states (according to Allongue).[24]

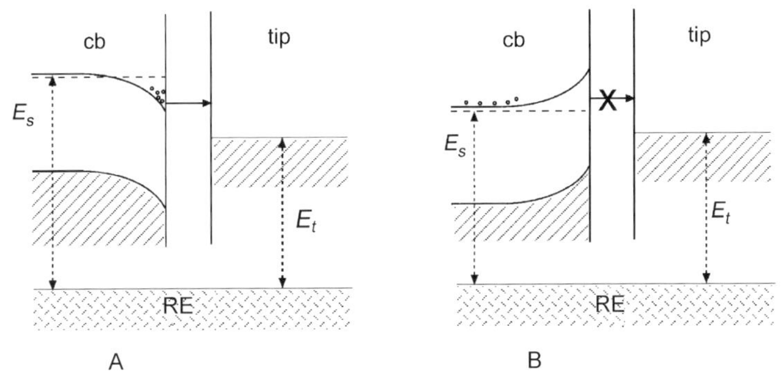

Figure 9.17 Band model of an n-semiconductor and tip in an electrochemical cell. The Fermi level of the tip is fixed in the band gap of the semiconductor. (A) The semiconductor sample is negatively polarized with respect to the flat band potential. The current between semiconductor and tip stabilize the position of the tip above the n-type electrode. (B) The semiconductor sample is positively polarized with respect to the flat band potential. The depletion blocks the current and the tip comes into contact with the surface (according to Allongue).[24]

The application of STM to oxide and semiconductor electrodes differs from STM on metal surfaces. The problems are described in the literature and in special reviews, e.g., by Allongue.[24]

The special situation of the tip in front of a semiconductor is shown in Figure 9.16 for vacuum conditions and in Figure 9.17 in an electrochemical cell.

In vacuum the tip can be adjusted to the band energy, but in an electrochemical environment the tip Fermi energy is located within the gap energy. In Figure 9.17 it is shown that an n-type semiconductor can be imaged for accumulation conditions. Corrosion prevents p-type semiconductors from being imaged under similar conditions. But n- and p-type semiconductors can be imaged in the depletion region, if the electrode is illuminated (Figure 9.18). The photoholes (n-type, Figure 9.18A) or the photoelectrons (p-type, Figure 9.18b) accumulate at the semiconductor surface and carry the tip current.

The complications of STM on semiconducting electrodes vanish, if the semiconductor is highly doped and becomes more and more metallic. A doping level of $N_{cc} = 10^{19}-10^{20}$ marks the limit.

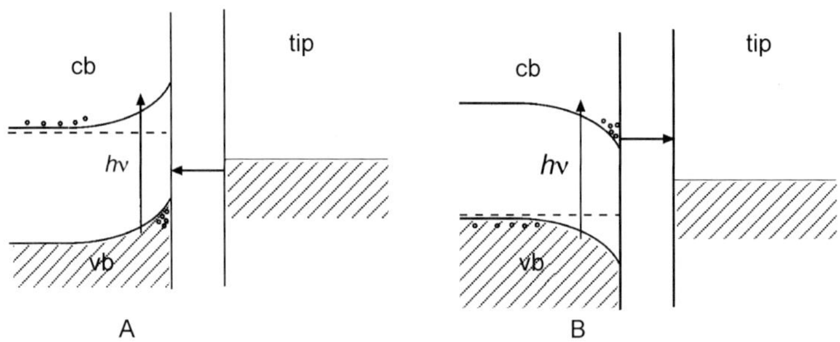

Figure 9.18 Band model of a semiconductor and tip in an electrochemical cell under illumination. (A) n-Type semiconductor and (B) p-type semiconductor (according to Allongue).[24]

9.5 OXIDE PARTICLES

9.5.1 Batteries

Throughout the long history of oxide electrochemistry oxide particles have played a prominent role. Examples are the use of oxides in primary batteries like manganese dioxide in the Leclanché cell (Figure 9.19). This cell consists of a zinc cup filled with the electrolyte ($ZnCl_2 \cdot NH_4Cl$) and an MnO_2 pocket. The reactions in the cell can be described by the following reactions:

$$Zn \rightarrow Zn^{2+} + 2e^- \tag{9.19}$$

$$2MnO_2 + 2H_2O + 2e^- \rightarrow 2MnO(OH) + 2OH^- \tag{9.20}$$

The cell voltage is 1.5 V. The modern type is the alkaline manganese battery but the classical cell is still on the market.

The classical example of an oxide powder in a secondary (rechargeable) battery is lead dioxide in the lead acid battery (Figure 9.20). The reactions in the cell are described by the equations

$$Pb + H_2SO_4 \leftrightarrow PbSO_4 + 2H^+ + 2e^- \tag{9.21}$$

$$PbO_2 + H_2SO_4 + 2H^+ + 2e^- \leftrightarrow PbSO_4 + 2H_2O \tag{9.22}$$

The cell voltage of the charged battery is 2.06 V. The electrodes are placed in 36% sulfuric acid. The electrodes consist of grids of lead alloys that are resistant against corrosion. The grids are filled with lead oxide (PbO). After filling the battery with sulfuric acid in the first charging process the lead oxide is oxidized to lead dioxide on the positive plate and to a felt-like structure of lead needles on the negative plate.

A second classical example is the use of nickel oxide in the nickel–cadmium rechargeable battery (Figure 9.21). The reactions in this battery are described by the equations

$$Cd + 2OH^- \leftrightarrow Cd(OH)_2 + 2e^- \tag{9.23}$$

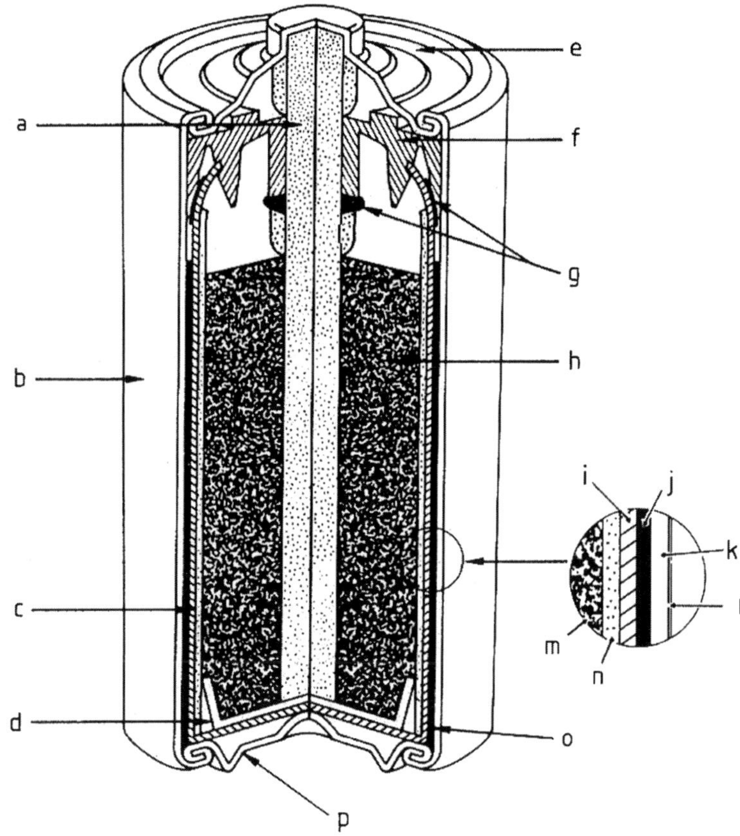

Figure 9.19 Leclanché cell. (a) Carbon electrode, (b) jacket-labeled polyethylene bonded tube, (c) separator coated paper, (d) paper cup, (e) metal cup (positive contact), (f) plastic closure, (g) sealant, (h) cathode mix (manganese dioxide, carbon, and electrolyte), (i) zinc can, (j) adhesive-backed barrier film, (k) paper, (l) label, (m) cathode mix (see h), (n) separator, (o) zinc can, and (p) metal bottom cover (negative contact). (Reproduced with permission from Ref. [25], © 1985, Wiley-VCH.)

$$2NiO(OH) + 2H_2O + 2e^- \rightarrow 2Ni(OH)_2 + 2OH^- \quad (9.24)$$

The cell voltage is 1.3 V. Eqs. (9.23) and (9.24) are formal equations; the exact processes in the nickel–cadmium battery are complex solid-state reactions. More details and other types of batteries are described in special literature.[25]

9.5.2 Lithium ion batteries

Energy storage in lithium ion batteries is based on the intercalation of lithium, either in a type of graphite (negative electrode) or in an oxide (positive electrode). Charging and discharging consist of the transport of lithium ions between the two storage media. The intercalation is connected with oxidation/reduction reactions. The principle process of a lithium ion battery is shown in Figure 9.22.

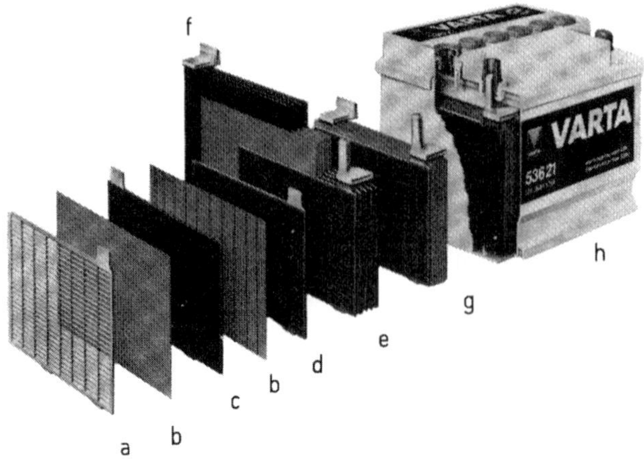

Figure 9.20 Components of a lead acid battery. (a) Cast lead alloy grid, (b) separator, (c) positive electrode, (d) negative plate, (e) negative plate group, (f) positive plate group, (g) plate block with separator, and (h) the complete battery. (Reproduced with permission from Ref. [25], © 1985, Wiley-VCH.)

In the development of primary and secondary lithium ion batteries a large variety of binary and ternary oxides, chalcogenides, etc. was investigated. Of special interest are oxides particles of nanosized dimensions. It is expected that these materials can be discharged and charged faster. The other reason for the trend to small particles is the large volume change connected with charging and discharging. Size changes for nanoparticles and nanoclusters are more flexible.

9.5.3 TiO_2-based photovoltaic cells

Under special conditions an illuminated titanium dioxide electrode generates a photocurrent that decomposes water into hydrogen and oxygen.[26] This discovery has led to the development of an electrochemical photocell, e.g., Grätzel.[27] The photoelectrons excited in the space-charge layer of a semiconductor electrode, e.g., a layer of TiO_2 particles, are transported through the external circuit to a counter electrode, e.g., a transparent indium-tin-oxide covered glass plate. At this electrode a suitable electron acceptor like J_2 is reduced to J^-, which then diffuses to the TiO_2 electrode and acts as a hole acceptor before it is reoxidized to J_2 by the photoholes that have accumulated at the surface. A schematic description of this mechanism is shown in Figure 9.23. The power of the cell can be represented in a power plot (Figure 9.24).

The band gap of TiO_2 does not overlap with the solar spectrum. Dyes that absorb the sunlight must be added to the electrolyte as sensitizers.[28] The photostability of these dyes is a serious problem. The liquid phase is another obstacle and there is now a trend to develop an all solid-state cell.[29–32] All materials must be optimized, especially the TiO_2. So far, nanosized particles have shown the best performance.[33,34]

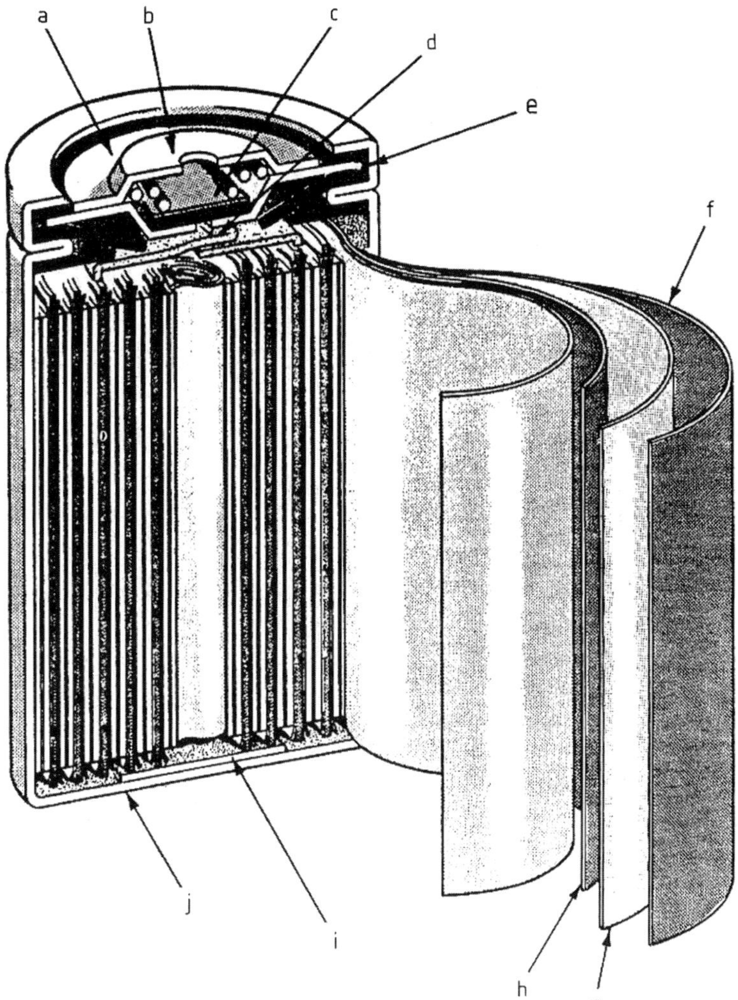

Figure 9.21 Nickel–cadmium rechargeable battery. (a) Cover, (b) positive contact, (c) resealable vent mechanism (in the event of a pressure build up), (d) positive tab welded to positive contact, (e) insulating seal ring, (f) negative plate, (g) separator, (h) positive plate, (i) negative tab, and (j) nickel-plated steel case. (Reproduced with permission from Ref. [25], © 1985, Wiley-VCH.)

9.5.4 Catalytic activity of oxide particles

Another important aspect of oxide particles is the catalytic properties of these materials. These properties have been used in the development of dimensionally stable electrodes (DSE) with a mixed oxide of RuO_2 and TiO_2 on Ti metal as a support. The role of the different oxides and improvement of the catalytic activity are currently under intensive investigation.[35] As in previous examples, new effects have already been observed with nanosized particles.[36]

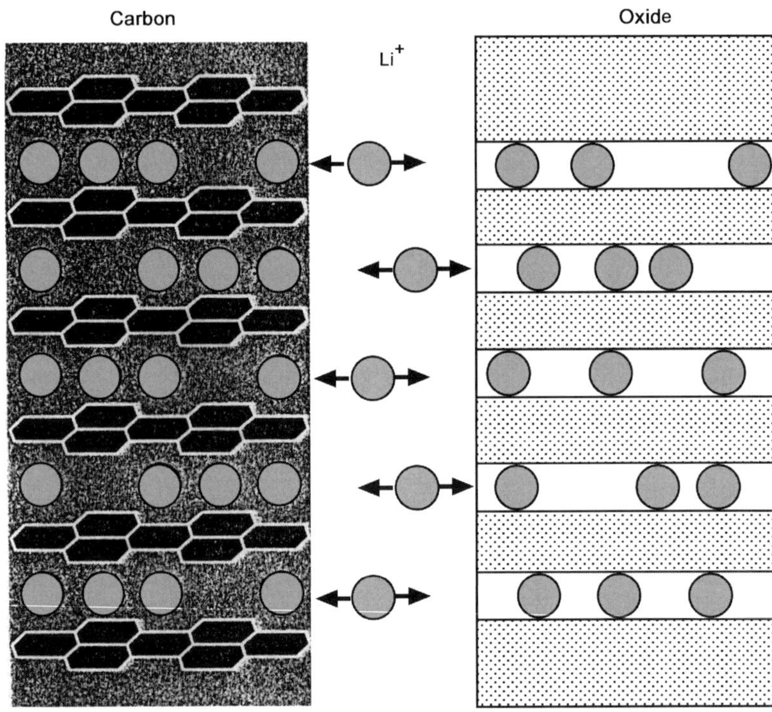

Figure 9.22 Rechargeable lithium ion battery (schematic presentation).

Another system, where catalytic active oxide particles might lead to a new development, is the polymer electrolyte membrane fuel cell (PEMFC). This type of fuel cell preferentially works with platinum and platinum alloy catalysts. The development of an effective oxide catalyst could solve some of the problems connected with the application of these systems.

9.6 OXIDE LAYERS

Oxide layers on metal surfaces are also important in electrochemistry.[37] Thin films on non-noble metals make the latter chemically inert. These passive films will be described in Chapter 10. On other metals, however, the so-called valve metals can be formed. For the formation of these films the metals are polarized to potentials of up to 100 V and above. Some examples are aluminum, titanium, tantalum, etc. In these cases the oxide behaves like a capacitor. If a potential is applied to the metal in an electrochemical cell against a counter electrode, a field gradient is built up through the oxide film. The strong field enables the transfer of ions through the layer by the high field mechanism. The oxide film grows until the ion transport through the film and film dissolution becomes stationary. The thickness of the film linearly depends on the cell voltage U.

Aluminum is a good example. The oxide film that is formed by anodic oxidation in oxalic acid protects the metal and prevents corrosion. The technical process is called

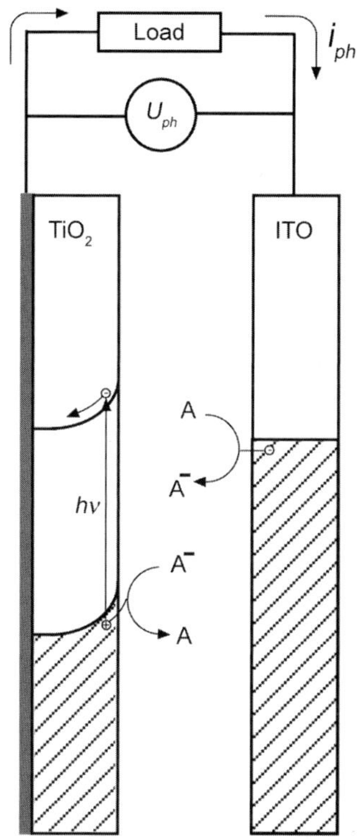

Figure 9.23 Photovoltaic cell. Separation of photoelectrons and photoholes in the space-charge region of the TiO_2 electrode generates a photocurrent to the counter electrode, e.g., a transparent indium-tin-oxide covered glass plate. Within the cell an A/A^- redox pair transports the current between the counter and semiconductor electrodes.

Eloxal process. These oxide films typically have a porous structure with a hexagonal pore arrangement. The pores can be used for coloring the metal surface and as templates for forming nanostructures. A high order can be achieved.[38]

9.7 ELECTROCHEMICAL DEPOSITION OF SEMICONDUCTORS

The development of vapor deposition techniques like chemical vapor deposition (CVP), plasma vapor deposition (PVD), or molecular beam epitaxy (MBE) have tempted electrochemists to prepare such films from a liquid phase with electrochemical reactions. Chemical processes can stimulate the film formation, e.g., the decomposition of precursors like sulfur compounds in the presence of metal ions like cadmium. Insoluble CdS is formed and a film is deposited by crystallization on a substrate.[39]

An electrochemical deposition on an electrode is the alternative to the chemical process. In some examples it is possible to deposit the semiconductor in one step, e.g. for Bi_2Te_3, a substance used in thermobatteries.[40]

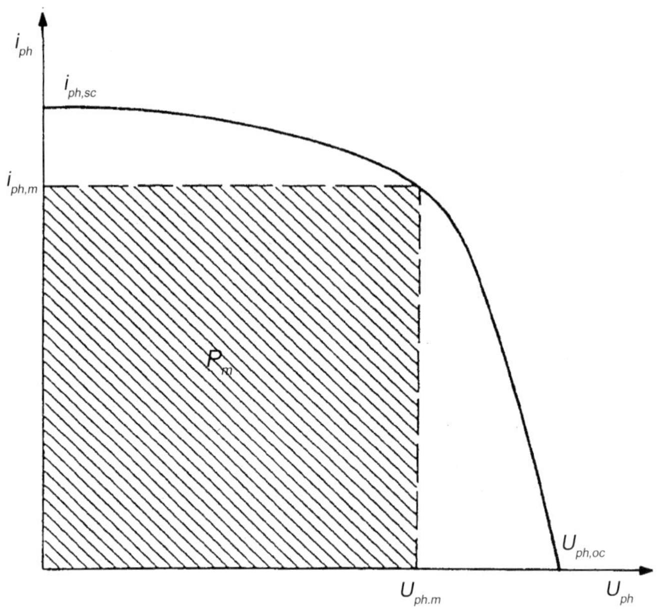

Figure 9.24 Power plot of a photovoltaic cell. $U_{\mathrm{ph,oc}}$ is the open-circuit cell voltage, $U_{\mathrm{ph,m}}$ the cell voltage at maximum power, $i_{\mathrm{ph,sc}}$ the short circuit photocurrent, $i_{\mathrm{ph,m}}$ the photocurrent at maximum power, and $P_{\mathrm{m}} = i_{\mathrm{ph,m}} U_{\mathrm{ph,m}}$ the power maximum.

A third approach is the layer-by-layer deposition of monolayers.[41] This process is a combination of two sequential underpotential deposition processes. In the first process a metal is deposited as UPD film. Then after rinsing, the electrolyte is changed, and by an anodic oxidation (a type of anodic underpotential deposition) the anionic component is deposited as a second layer on the primary UPD film. In both deposition processes epitaxial film growth is expected and has been partially confirmed.

A special thin-layer cell was developed for the process with a small electrolyte volume to make the electrolyte exchange as fast as possible. The deposition of the two components was done under time control in a computerized device.

REFERENCES

1. S.R. Morrison, Electrochemistry at Semiconductor and Oxidized Metal Electrodes, Plenum Press, New York, 1984.
2. R. Memming, Semiconductor Electrochemistry, Wiley-VCH, Weinheim, 2001.
3. N. Sato, Electrochemistry at Metal and Semiconductor Electrodes, Elsevier, Amsterdam, 1998.
4. S. Trasatti (Ed.), Electrodes of Conductive Metallic Oxides, Parts A and B, Elsevier, Amsterdam, 1980, 1981.
5. R. Memming, J. Electrochem. Soc., *116*, 785 (1969).

6. Yu.V. Pleskov, Yu.Ya. Gurevich, Semiconductor Photoelectrochemistry, Consultants Bureau, New York, 1986.
7. W. Plieth, G. Pfuhl, R. Matz, Photocurrent and Photopotential Transients of n- and p-GaAs and n- and p-InP, in N. Masuko, T. Osaka, Y. Fukunaka (Eds.), New Trends and Approaches in Electrochemical Technology, Kodansha Ltd., Tokyo, 1993, p. 235.
8. W. Lorenz, H. Handschuh, F. Bergmann, Chem. Phys., *215*, 157 (1997).
9. W.W. Gärtner, Phys. Rev., *116*, 84 (1959).
10. H. Gerischer, Electrochim. Acta, *35*, 1677 (1990).
11. L.M. Peter, J. Li, R. Peat, H.J. Lewerenz, W. Stumper, Electrochim. Acta, *35*, 1657 (1990).
12. J. Schefold, J. Electroanal. Chem., *341*, 111 (1992).
13. W. Plieth, C.A. Schiller, U. Rammelt, 57th Meeting of the International Society of Electrochemistry, Book of Abstracts, S7-P-70, 2006.
14. L.M. Peter, D. Vanmaeckelbergh, Time and Frequency Resolved Studies of Photoelectrochemical Kinetics, in R.C. Alkire, H. Gerischer, D.M. Kolb, C.W. Tobias (Eds.), Advances in Electrochemical Science and Engineering, Vol. 6, Wiley-VCH, Weinheim, 1999, p. 77.
15. H. Graefe, W. Plieth, Mater. Sci. Forum, *186*, 425 (1995).
16. W. Plieth, H.J. Rieger, Corrosion Sci., *29*, 267 (1989).
17. W. Plieth, H. Wetzenstein, Photoetching of III-V Semiconductors: Basic Photoelectrochemical Principles, in S.A. Campbell and H.J. Lewerenz (Eds.), Semiconductor Micromachining, Vol. 1, Wiley, Chichester, 1998, p. 93.
18. K. Bitterling, F. Willig, J. Electroanal. Chem., *204*, 211 (1986).
19. W. Plieth, C. Gutiérrez, G. Wilson, Pure Appl. Chem., *70*, 1395, 2409 (1998).
20. R.C. Alkire, D.M. Kolb, J. Lipkowski, P.N. Ross (Eds.), Advances in Electrochemical Science and Engineering, Vol. 9, Wiley-VCH, Weinheim, 2006.
21. L.J. Arsov, C. Kormann, W.J. Plieth, J. Electrochem. Soc., *138*, 2964 (1991); J. Raman Spectrosc., *22*, 573 (1991).
22. H.H. Strehblow, Elektroden, Elektrodenprozesse und Elektrochemie, in Bergmann-Schäfer (Ed.), Lehrbuch der Experimentalphysik, Bd. 5, Gase, Nanosysteme, Flüssigkeiten, 2nd Ed., DeGruyter, Berlin, 2006, p. 511.
23. D.A. Scherson, Mossbauer Spectroscopy, in H.A. Abruña (Ed.), Electrochemical Interfaces; Modern Techniques for *In situ* Interface Characterization, VCH, Weinheim, 1991.
24. P. Allongue, Scanning Tunneling Microscopy of Semiconductor Electrodes, in H. Gerischer, C.W. Tobias (Eds.), Advances in Electrochemical Science and Engineering, Vol. 4, VCH, Weinheim, 1995, p. 1.
25. D. Blomgren, B. Schumm, L.F. Urry, J. McBreen, D. Bernd, M. Kronenberg, in Ullmann's Encyclopedia of Industrial Chemistry, Batteries, Vol. A3, VCH, Weinheim, 1985, p. 343.
26. A. Fujishima, K. Honda, Bull. Chem. Soc. Jpn., *44*, 1148 (1971).
27. M. Grätzel, Nature, *414*, 338 (2001).
28. C.A. Kelly, G.J. Meyer, Coord. Chem. Rev., *211*, 295 (2001).
29. Y. Saito, T. Kitamura, Y. Wada, Sh. Yanagida, Synth. Met., *131*, 185 (2002).
30. D. Gebeyehu, C.J. Brabec, N.S. Sacriciftci, D. Vangeneugden, R. Kiebooms, D. Vanderzande, F. Kienberger, H. Schindler, Synth. Met., *125*, 279 (2002).
31. K. Murakoshi, R. Kogure, Y. Wada, Sh. Yanagida, Solar Energy Mater. Solar Cells, *55*, 113 (1998).
32. A.F. Nogueira, L. Micaroni, W.A. Gazotti, M.-A. De Paoli, Electrochem. Commun., *1*, 262 (1999).
33. L.M. Peter, E.A. Ponomarev, G. Franco, N.J. Shaw, Electrochim. Acta, *45*, 549 (1999).
34. U. Björkstén, J. Moser, M. Grätzel, Chem. Mater., *6*, 858 (1994).
35. S. Trasatti, Electrochim. Acta, *32*, 369 (1987); *45*, 2377 (2000).

36. J. Jirkovsky, M. Makarova, P. Krtil, Electrochem. Commun., *8*, 1417 (2006).
37. J.W. Diggle (Ed.), Oxides and Oxide Films, Vols. 1 and 2, Marcel Dekker, New York, 1972.
38. H. Masuda, F. Hasegwa, S. Ono, J. Electrochem. Soc., *144*, L127 (1997).
39. D. Lincot, M. Froment, H. Cachet, Chemical Deposition of Chalcogenide Thin Films from Solution, in R.C. Alkire, H. Gerischer, D.M. Kolb, C.W. Tobias (Eds.), Advances in Electrochemical Science and Engineering, Vol. 6, Wiley-VCH, Weinheim, 1999, p. 165.
40. K. Tittes, A. Bund, W. Plieth, A. Bentin, S. Paschen, M. Ploetner, H. Graefe, W.-J. Fischer, J. Solid State Electrochem., *7*, 714 (2003).
41. J.L. Stickney, Electrochemical Atomic Layer Epitaxy (EC-ALE): Nanoscale Control in the Electrodeposition of Compound Semiconductors, in R.C. Alkire, H. Gerischer, D.M. Kolb, C.W. Tobias (Eds.), Advances in Electrochemical Science and Technology, Vol. 7, Wiley-VCH, Weinheim, 2002, p. 1.

- 10 -

Corrosion and Corrosion Protection

10.1 CORROSION

In the previous chapters the deposition of materials of very different nature like metals, alloys, inter-metallic compounds, oxides, or semiconductors was described. The opposite of deposition is dissolution. Dissolution is the origin for corrosion. In the beginning of the process, surface defects like stains develop. With increasing progress of corrosion the functionality of the parts decreases and finally the parts get completely destroyed.

Corrosion is responsible for the permanent loss of economical values. Corrosion can affect all materials but the corrosion of metals and especially iron and steel is the most outstanding example of corrosive degradation and is called electrochemical corrosion.[1] In the case of iron and milled steel, the products of corrosion form an increasingly thicker surface layer of rust. Examples of this process were for many decades the corroding old cars. Other intensive corrosion damage can be found in the pipe systems of all kinds of cooling devices, where copper is often the material. Water pipes consisted for long time of lead, but are now usually of zinc coated steel. Corrosion products of lead or zinc deposit in the pipes and cause final degradation. Zinc-coated steel parts in an aggressive atmosphere show first zinc corrosion (white rust). Finally, the zinc layer is locally destroyed and corrosion of the steel starts (red rust). Different environments cause different types of corrosion. Thus one has to distinguish, e.g., corrosion in aqueous environment, corrosion in non-aqueous liquids, corrosion in melts, concrete, or soil, atmospheric corrosion, corrosion in hot gases, e.g., turbines.

Several forms of corrosion are observed and some examples are shown in Figure 10.1.

- Uniform corrosion.
- Non-uniform corrosion or local corrosion. One example is pitting corrosion, the formation of small round pits in the wall of sanitary pipes.
- Inter-granular or inter-crystalline corrosion.
- Stress corrosion and stress corrosion cracking.
- Crevice corrosion.
- Selective corrosion in alloys where one component dissolves and leaves a porous structure.

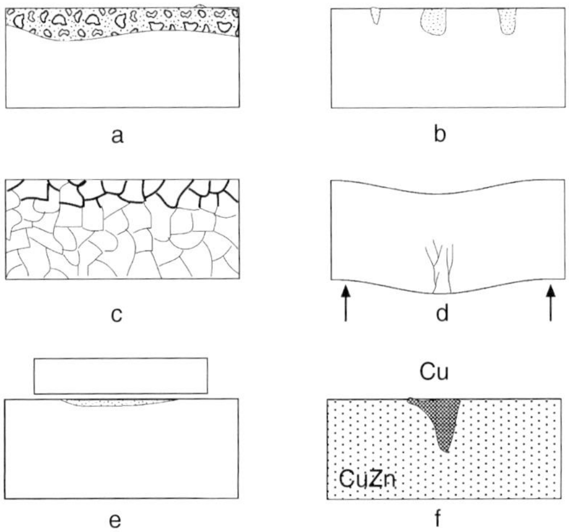

Figure 10.1 Examples of different forms of corrosion are: (a) uniform corrosion, (b) non-uniform corrosion, (c) inter-granular corrosion, (d) stress corrosion cracking, (e) crevice corrosion, and (f) selective corrosion.

Corrosion is one of the most intensively investigated phenomena. Therefore several textbooks exist on the market, which provide deeper insight into this matter.[2–4] In the following chapter, first the electrochemical processes connected with corrosion will be described. In the second part, the main means of protection will be discussed with emphasis on the electrochemically based protection mechanism.

10.1.1 Fundamental processes

The fundamentals of ion transfer across the double layer were described in Section 6.3. According to Eq. (6.43), the main process of electrochemical dissolution of a metal is described by the equation

$$\mathrm{Me} + v_L \mathrm{L} \rightarrow \mathrm{Me}^{z+} \cdot v_L \mathrm{L} + z\, e^- \qquad (10.1)$$

Similar to metal deposition the overall reaction can consist of partial reactions such as the separation from the kink site position to an ad-atom position followed by the transfer to the electrolyte. This will be described in more detail in the next section. To compensate the electron release a reduction process is necessary. In an acid environment (an acid solution) protons can be reduced

$$\mathrm{H}^+ + e^- \rightarrow \frac{1}{2}\mathrm{H}_2 \qquad (10.2)$$

The overall reaction can then be formulated as follows:

$$\mathrm{Me} + v_L \mathrm{L} + z\,\mathrm{H}^+ \rightarrow \mathrm{Me}^{z+} \cdot v_L \mathrm{L} + \frac{z}{2}\mathrm{H}_2 \qquad (10.3)$$

It is obvious that this reaction can only lead to metal dissolution, if the metal electrode potential is negative from the hydrogen electrode potential. This is the reason for the classification of metals into noble metals (the equilibrium potential is more positive than the standard hydrogen potential) and non-noble metals (the equilibrium potential is more negative than the standard hydrogen potential). The kinetic of the total process can be described by the Butler–Volmer equation for the two partial reactions.

$$i_{Me} = +zFk_{a,Me}c_L^{\nu_L}\exp\left(+\frac{\alpha_{a,Me}zF}{RT}E\right) - zFk_{c,Me}c_{Me^+}\exp\left(-\frac{\alpha_{c,Me}zF}{RT}E\right) \quad (10.4)$$

$$i_{H_2} = +Fk_{a,H_2}c_{H_2}\exp\left(+\frac{\alpha_{a,H_2}F}{RT}E\right) - Fk_{c,H_2}c_{H^+}\exp\left(-\frac{\alpha_{c,H_2}F}{RT}E\right) \quad (10.5)$$

Eq. (10.5) must turnover twice to compensate the metal dissolution. It was assumed in Eq. (10.4) that the electrochemical reaction order of the metal ion ligands is equal to the stoichiometric number. For both equations $\alpha_a + \alpha_c = 1$. If the difference between the metal electrode potential and the hydrogen electrode potential is larger than approximately 100 mV one can dismiss the backward reaction. The two remaining partial reactions can be combined in the following current–potential relationship

$$i = +zFk_{a,Me}c_L^{\nu_L}\exp\left(+\frac{\alpha_{a,Me}zF}{RT}E\right) - zFk_{c,H_2}c_{H^+}\exp\left(-\frac{\alpha_{c,H_2}F}{RT}E\right) \quad (10.6)$$

In the partial reactions $\alpha_a + \alpha_c \neq 1$. At the potential where the current densities of both partial reactions compensate each other, $i_{Me} = -i_{H_2}$, the overall current density is zero. This potential is referred to as free corrosion potential E_{cor}; the associated current density is called corrosion current density i_{cor}. The free corrosion potential is a mixed potential. This is shown in Figure 10.2.

Figure 10.2a is a linear presentation. The logarithmic presentation with the Tafel lines is shown in Figure 10.2b. The corrosion density i_{cor} is a measure for the corrosion rate r_{cor} and is the mass of dissolved metal in the time t per area unit:

$$r_{cor} = \frac{i_{cor}M}{zF}t \quad (10.7)$$

M is the molar mass of the metal.

Another possibility to oxidize the metal is the reduction of oxygen. The partial reaction is

$$\frac{1}{4}O_2 + \frac{1}{2}H_2O + e^- \rightarrow OH^- \quad (10.8)$$

The overall reaction of corrosion can then be formulated as follows:

$$Me + \nu_L L + \frac{z}{4}O_2 + \frac{z}{2}H_2O \rightarrow MeL^{z+} + z\,OH^- \quad (10.9)$$

The rate of the partial reaction of the oxygen reduction is given by the equation for the limiting oxygen diffusion current (bulk oxygen concentration c_{O_2}, diffusion layer

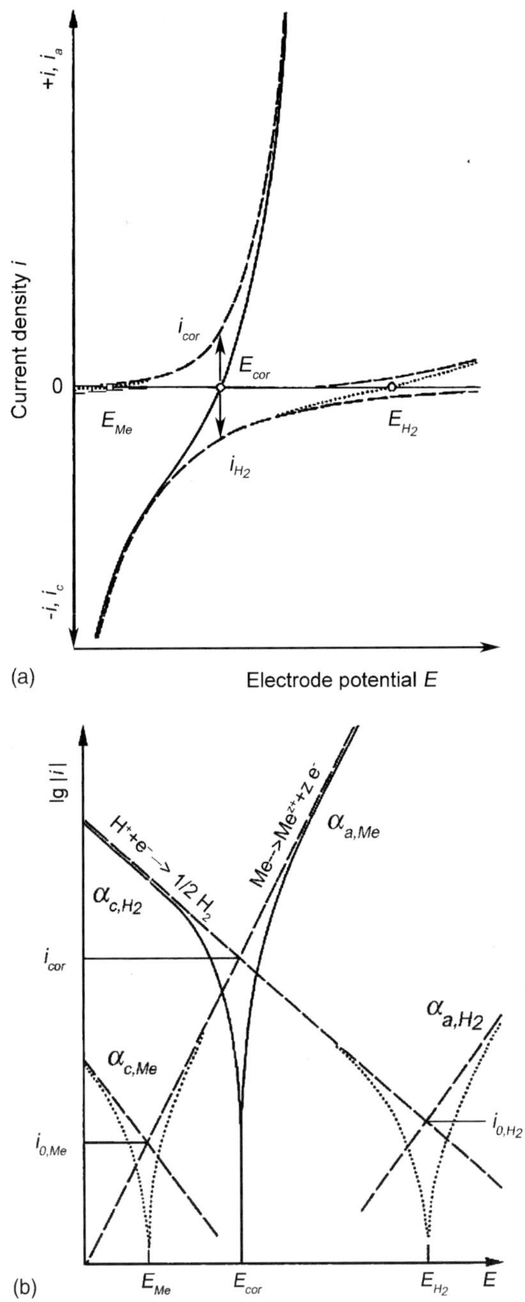

Figure 10.2 (a) Definition of the free corrosion potential E_{cor} (mixed potential) and the corrosion current density i_{cor}; linear current versus potential plot. Also shown are the Nernst potentials of the metal electrode and the hydrogen electrode. (b) Definition of the free corrosion potential E_{cor} (mixed potential) and the corrosion current density; the logarithm of current, log $|i|$ is plotted versus potential (Tafel plot) (p. 121). (Reproduced with permission from Ref. [2], © 2003, Springer-Verlag.)

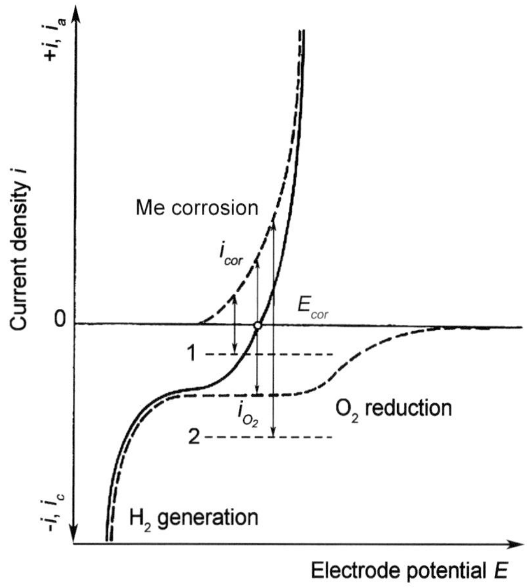

Figure 10.3 Corrosion in oxygen containing nearly neutral electrolyte. The corrosion rate decreases with decreasing oxygen concentration and the potential is shifted towards more negative potentials (1). Otherwise, with increasing oxygen reduction current the corrosion rate increases and the free corrosion potential shifts towards more positive values (2).

thickness δ)

$$i_{O_2} = -\frac{z}{4} FD \frac{c_{O_2}}{\delta} \tag{10.10}$$

D is the diffusion coefficient of oxygen. The rate of the partial reaction of oxygen concentration is usually of the order of mmol·dm^{-3} leading to diffusion-limited oxygen reduction currents. This limits the rate of corrosion as shown in Figure 10.3. The oxygen reduction current decreases with decreasing oxygen concentration. Keeping the oxygen concentration low is one way of corrosion limitation.

10.1.2 Mechanism of metal dissolution

As for deposition, the transition state of dissolution of a crystalline metal is the kink site position. From the kink site position there are two ways for separation (Figure 10.4).

(A) The transfer from the kink site position or step position, respectively to the ad-atom position (separation) followed by the transfer of the ad-atoms to the electrolyte (desorption). This can be described by the following two equations:

$$\mathrm{MeMe}^* + v_n\mathrm{L} \leftrightarrow \mathrm{Me}^* + \mathrm{Me}_{\mathrm{ad}}^{z_{\mathrm{ad}}} \cdot v_n\mathrm{L} + \Delta z_{\mathrm{sep}} e^- \text{ (Separation)} \tag{10.11}$$

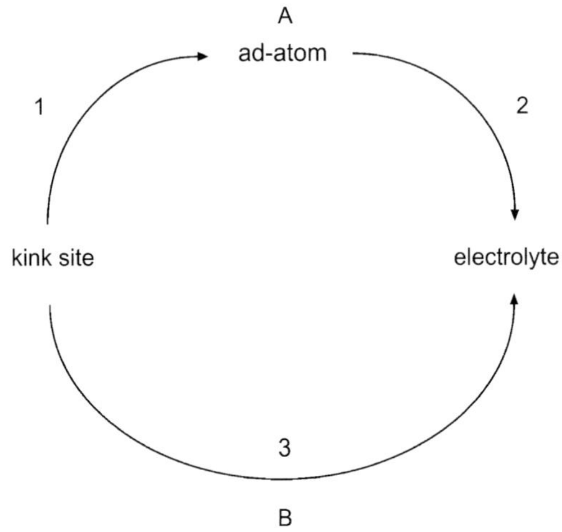

Figure 10.4 Mechanism of metal dissolution: (A) via ad-atom intermediates and (B) direct dissolution of kink atoms.

$$Me_{ad}^{z_{ad}} \cdot v_n L + (v_m - v_n)L \rightarrow Me_{ad}^{z+} \cdot v_m L + \Delta z_{des} e^- \text{ (Desorption)} \qquad (10.12)$$

The first equation describes the reproduction of the kink site position in the process of separation of an atom from the kink site position and formation of an ad-atom. The second step represents desorption of the ad-atom. The charge transfer is split. Following the description of partial charge transfer in Section 4.3 with the introduction of a partial charge transfer coefficient λ the charge transfer in the first step is $\Delta z_{sep} = \lambda z$, and the charge transfer in the second step $\Delta z_{ad} = (1 - \lambda)z$. The intermediate ad-atom can be a lower valent oxidation state stabilized as an anion complex.

If the first step, the separation, is rate determining (usually at higher potentials) the rate equation is

$$r_{sep} = k_{sep}[ksp]c_L^{v_n} \exp\left(+\frac{\alpha_{sep}\Delta z_{sep}F}{RT}E\right) \qquad (10.13)$$

The part of the potential drop across the double layer acting on the transition state of separation is described by α_{sep}. If the density of kink site positions $[ksp]$ in the first approximation is independent of the potential, the Tafel equation would be

$$\log|i| = \text{const} + v_n \log c_L + \frac{\alpha_{sep}\Delta z_{sep}F}{2.303RT}E \qquad (10.14)$$

With $\alpha_{sep}\Delta z_{sep}$ of the order of 0.5 the Tafel slope would be 118 mV.

If the second step, the desorption, is rate determining (for equilibrium conditions between atoms in kink positions and ad-atoms) a rate equation for ad-atom desorption, e.g., as derived in Chapter 4, Eq. (4.50) can be formulated:

$$r_{\text{des}} = k_{\text{des}} c_{\text{ad}} c_{\text{L}}^{(\nu_m - \nu_n)} \exp\left(+ \frac{\alpha_{\text{des}} zF}{RT} E \right) \tag{10.15}$$

Taking into account the equilibrium between the ad-atoms and the kink site positions a rather complex expression follows for the charge transfer coefficient α_{des}. The following Tafel equation

$$\log|i| = \text{const} + \nu_m \log c_{\text{L}} + \frac{\alpha_{\text{des}} zF}{2.303RT} E \tag{10.16}$$

describes several experimental results with a typical Tafel slope of 40 mV (25 °C)

(B) The direct dissolution, transfer of a charged ion from the kink site position to the electrolyte. In this case the rate equation for the direct anodic dissolution of kink site atoms would be

$$r_{\text{dis}} = k_{\text{dis}} [ksp] c_{\text{L}}^{\nu_m} \exp\left(+ \frac{\alpha_{\text{dis}} zF}{RT} E \right) \quad \text{(Direct dissolution)} \tag{10.17}$$

If the density of kink site positions $[ksp]$ is in a first approximation independent of the potential the Tafel equation is

$$\log i = \text{const} + \nu_m \log c_{\text{L}} + \frac{\alpha_{\text{dis}} zF}{2.303RT} E \tag{10.18}$$

With $\alpha_{\text{dis}} = 0.5$ and $z = 2$ the Tafel slope would be 59 mV.

10.1.3 Mechanisms of compensation reactions

Principally any reduction reaction, with an equilibrium potential more positive than the Nernst potential of the corroding metal, can compensate the metal dissolution. In practice, two reactions are of special importance, the reduction of hydrogen ions and the reduction of oxygen.

(i) *Hydrogen ion reduction*

The equilibrium potential of this reaction is $E_0 = 0\ \text{V}_{\text{SHE}}$ at pH = 0 and $E_0 = 0.826\ \text{V}_{\text{SHE}}$ at pH = 14. The overvoltage for the reduction of hydrogen is strongly dependent on the kind of metal and can reach more than 1 V.[5]

For the reduction of hydrogen ions two mechanisms are discussed.[6] According to the Volmer–Tafel mechanism the following two reactions take place:

$$\text{H}^+_{\text{solv}} + e^- \leftrightarrow \text{H}_{\text{ad}} \tag{10.19a}$$

$$\text{H}_{\text{ad}} + \text{H}_{\text{ad}} \leftrightarrow \text{H}_2 \tag{10.19b}$$

Alternatively, the Volmer–Heyrovsky mechanism is discussed.

$$H^+_{solv} + e^- \leftrightarrow H_{ad} \tag{10.19a}$$

$$H_{ad} + H^+_{solv} + e^- \leftrightarrow H_2 \tag{10.19c}$$

In both mechanisms the first step is the reduction of hydrogen ions and the formation of adsorbed hydrogen atoms. At higher pH values the direct reduction of water is possible. Metals with a strong affinity to hydrogen, like the platinum group metals, catalyze the reduction and the corrosion.

(ii) *Oxygen reduction*

The reduction of the ubiquitous oxygen is much more complex than the reduction of hydrogen. The reason is the strong bond in the oxygen molecule. The formal equilibrium potential of the reaction is $E_0 = 1.23\ V_{SHE}$ at pH $= 0$ and $E_0 = 0.40\ V_{SHE}$ at pH $= 14$.

To weaken the oxygen bond the molecule is adsorbed on the metal surface. The reduction might occur via several intermediate steps, including H_2O_2 to water according to the formal mechanism.

$$O_2 + 2\,H^+_{solv} + 2\,e^- \leftrightarrow H_2O_2 \tag{10.20a}$$

$$H_2O_2 + 2\,H^+_{solv} + 2\,e^- \leftrightarrow 2\,H_2O \tag{10.20b}$$

At higher pH values the hydrated hydrogen ions may be substituted by water and formation of hydroxyl ions.

10.1.4 Iron and steel

Iron is the most intensively investigated system. The so-called Bockris mechanism (Bockris, Drazic, Despic[6,7]) consists of the following steps:

$$FeFe^* + OH^- \leftrightarrow Fe^* + FeOH_{ad} + e^- \tag{10.21a}$$

$$FeOH_{ad} \rightarrow FeOH^+ + e^- \tag{10.21b}$$

$$FeOH^+ \leftrightarrow Fe^{2+} + OH^- \tag{10.21c}$$

The first step is ad-atom formation and reproduction of kink site positions denoted by FeFe* and Fe*. OH^- acts as a catalyst. The hydroxyl ions stabilize a one-valent ad-atom of iron. According to the authors, the second reaction step is rate determining. The authors observed a Tafel slope of 40 mV. The conclusion concerning the influence of the hydroxyl ions was based on the dependence of the Tafel lines on the pH value.

Heusler[8] determined the electrochemical reaction orders of the hydroxyl ions. With $\nu_{OH^-} = 2$, he formulated the following mechanism:

$$FeFe^* + OH^- \leftrightarrow Fe^* + FeOH_{ad} + e^- \tag{10.22a}$$

$$\mathrm{FeFe^* + OH^-_{ad} + FeOH_{ad} \rightarrow Fe^* + FeOH^+ + FeOH_{ad} + 2e^-} \tag{10.22b}$$

$$\mathrm{FeOH^+ + H^+ \leftrightarrow Fe^{2+} + H_2O} \tag{10.22c}$$

The equilibrium between kink site positions and ad-atoms is the same as in the Bockris mechanism. The Tafel equation of Heusler has the form:

$$i = 2Fk_{\mathrm{des}}[ksp]c^2_{\mathrm{OH^-}} \exp\left(+\frac{(1+2\alpha)F}{RT}E\right) \tag{10.23}$$

With $\alpha = 0.5$ Tafel lines with 29 mV slope are expected. This was confirmed by the experimental results.

10.1.5 Metallurgical aspects of iron and steel

For any analysis of corrosion results of a special metal, another factor of prime importance is the structure of the investigated metal. The metallurgical treatment must be taken into account. This complex nature of corrosion is best explained for the element iron. Iron can exist in two main modifications. Below 911 °C the stable modification is α-iron with a body centered cubic (bcc) structure, above 911 °C the stable modification is γ-iron, the face centered cubic (fcc) structure. A third modification is found at higher temperatures (δ-iron) for a bcc structure with different crystallographic parameters.

Iron always contains carbon. A part of the phase diagram of the iron/carbon system is shown in Figure 10.5. The carbon has different solubilities in the different iron modifications, which form mixed crystals (solid solutions). In α-iron the solubility is only 0.04% (ferrite) and in δ-iron the solubility is 0.36%. In the γ-modification with its fcc structure, carbon and iron form an intercalation lattice as a solid solution called austenite with the maximum solubility of 2.06% carbon at 1147 °C. Iron with more carbon is called cast iron. Iron with less than 2.06% carbon is called steel. During slow cooling of a melt (above 1147 °C) iron solidifies either as austenite (carbon content of the melt $<$4.3%) or as cementite (Fe_3C, carbon content of the melt $>$4.3%). At 1147 °C the melt solidifies in a eutectic mixture of both these phases called ledeburite.

At lower temperatures the austenitic phase decomposes into ferrite and cementite. The lowest point of stability of austenitic crystals is at a concentration of 0.8% carbon and a temperature of 723 °C. At this point the austenite transforms to a eutectoid mixture of ferrite with 0.02% carbon and cementite with 6.67% carbon. This eutectoid point of the solid solution is adequate to the eutectic point of a melt where the melt decomposes into a eutectic mixture. The eutectoid mixture of ferrite and cementite is called pearlite.

The process of heat treatment and different quenching rates of iron with carbon causes further structural differences. One product is martensite, an oversaturated (metastable) solid solution of carbon in ferrite. Martensite is obtained when the steel is heated to temperatures where austenitic steel is formed and then rapidly quenched.

This was a very short sketch of the metallurgical aspects of iron and steel. For details one must consult special literature.[10] All the different modifications of iron and steel have

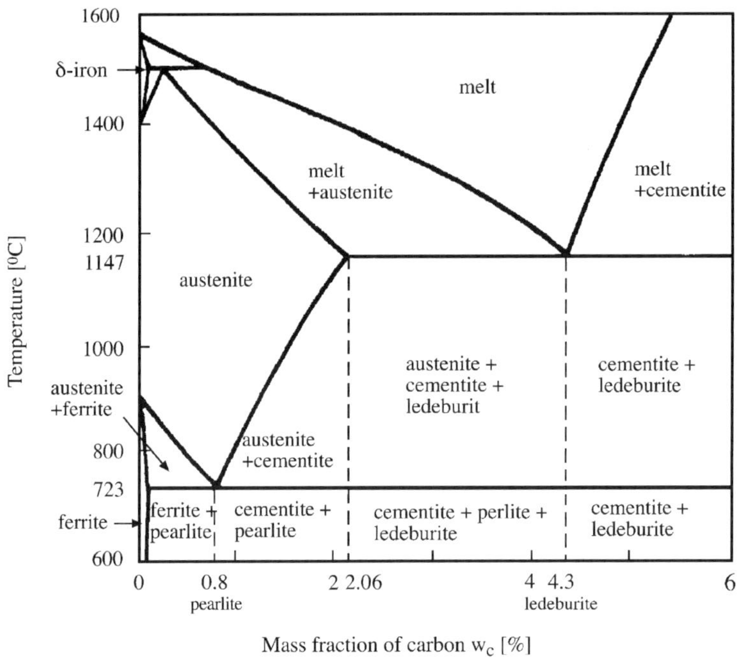

Figure 10.5 Phase-diagram of the iron carbon system at lower carbon concentration.[9]

different corrosion properties, which demonstrate the complex nature of corrosion research in practice.

10.1.6 Copper

Another reaction investigated in some detail was copper dissolution. It was assumed that anions like chloride stabilize the one-valent ad-atom. The old results by Bockris and Enyo[11] could then be described by the following two-step mechanism with Cu^+ as intermediate.

$$CuCu^* + X^- \leftrightarrow Cu^* + CuX_{ad} + e^- \tag{10.24a}$$

$$CuX_{ad} \rightarrow Cu^{2+} + X^- + e^- \tag{10.24b}$$

The authors assumed that the second reaction step was rate determining. The authors found a Tafel slope of 40 mV.

The mechanism changes in the presence of oxygen as a consequence of direct oxygen corrosion.

Magnussen *et al.*[12] recently investigated this system with scanning tunnelling spectroscopy. With a dynamic STM equipment Magnussen observed the step dynamics of mono-atomic islands on the surface of the (100) face of copper single crystals (Figure 10.6).

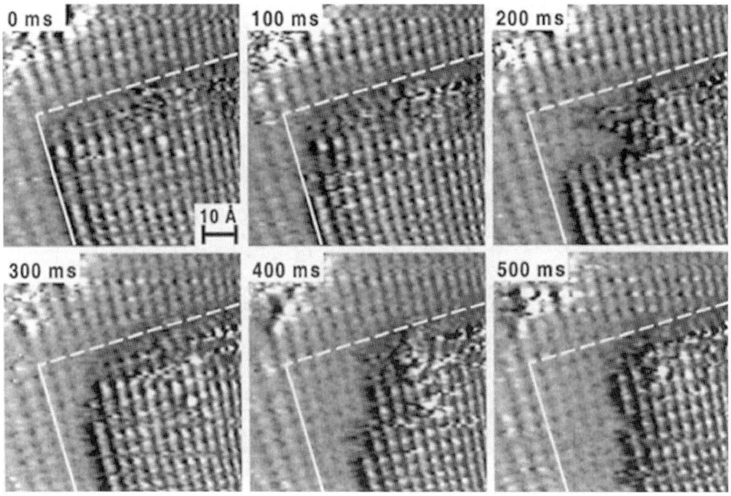

Figure 10.6 STM image of copper islands on the (100) face of a copper single crystal in 0.01 mol · dm^{-3} HCl; potential jump from −0.23 to +0.16 V; the dissolution of one step is observed, while the step rectangular to this step but with the same orientation is stable. The different stabilities could be explained by a Jahn–Teller effect of the surface lattice. (Reproduced with permission from Ref. [12], © 2001, Elsevier.)

From inspection of these pictures the different stability of steps with even the same orientation is remarkable. Since STM does not show ad-atoms, this method is of no help in confirming one or the other type of dissolution mechanism.

10.1.7 Zinc

A further example, which will be discussed, is zinc dissolution. Heusler investigated this system,[13] and two Tafel lines with a slope of 40 mV and 118 mV were observed (Figure 10.7).

As long as desorption (Eq. (10.12)) is rate determining a Tafel slope of 40 mV should be observed. When separation (Eq. (10.11)) becomes rate determining the Tafel slope should increase to a value around 118 mV. The experimental observations confirm these predictions.

10.1.8 Corrosion products

Corrosion is not finished with the formation of solvated metal ions. The metal ions can form secondary corrosion products by reaction with components on the metal surface. One typical reaction is the formation of hydroxides and oxides by reaction with water condensed on the surface in the humid atmosphere of a natural environment. In the presence of oxygen the metal ions can be oxidized. For example, oxygen can oxidize two-valent iron to three-valent iron. Two- and three-valent iron-oxide-hydroxides form the brown rust on the surface.

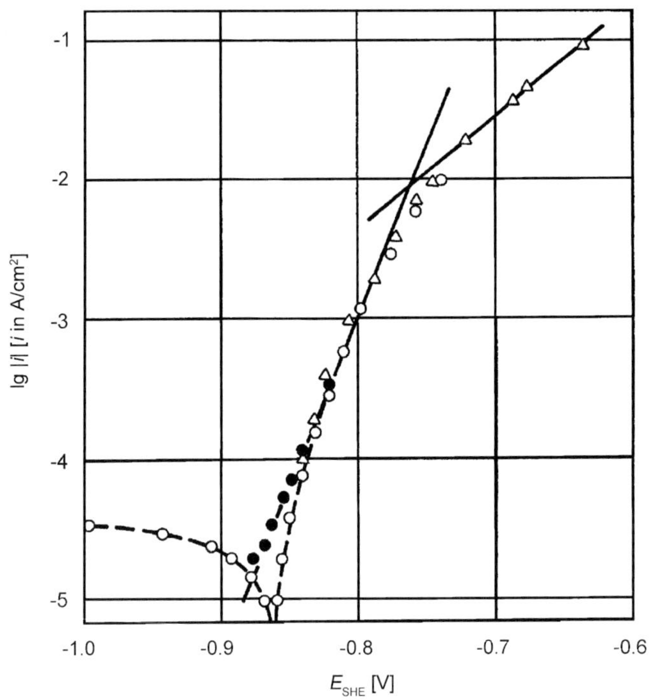

Figure 10.7 Zinc electrode in 1 mol·dm^{-3} $NaClO_4/HClO_4$, pH 3.25; two Tafel lines of dissolution were observed, one with a slope of 40 mV and the other with a slope of 118 mV.[13]

The reaction with CO_2 leads to oxy-carbonates. The reaction with sulfur can lead to sulfide layers, e.g., on silver ware. A very prominent example of secondary corrosion products is also the green patina on copper sheets. Details can be found in the special literature on this subject.[14,15]

10.1.9 Corrosion of alloys

Dissolution of alloys follows different principles. Dissolution of a freshly prepared surface can be described by the dissolution mechanism of a pure metal but now with different rate constants for the different alloy components (section General mechanisms). Because of the different corrosion rates a depletion of one component soon occurs. An intermediate region of different composition can develop (section Stationary dissolution conditions). In some cases the matrix of the nobler component is retained and a sponge-like structure develops. Again, this all concerns an oxide-free surface. The complications connected with oxide layer formation will be discussed in Section 10.2.

General mechanisms

The following mechanism is based on an alloy of two components A and B. In an approximate description on the surface four different kinds of kink site positions exist, AA, AB, BA, and BB. Three steps for each component describe the dissolution via ad-atoms.

For component A these equations are

$$AA^* \leftrightarrow A^* + A_{ad}^{z_{ad,A}} + \Delta z_{sep,A} e^- \tag{10.25a}$$

$$BA^* \leftrightarrow B^* + A_{ad}^{z_{ad,A}} + \Delta z_{sep,A} e^- \tag{10.25b}$$

$$A_{ad}^{z_{ad,A}} \rightarrow A^{z_A} + \Delta z_{dis,A} e^- \tag{10.25c}$$

The equations for the component B are

$$AB^* \leftrightarrow A^* + B_{ad}^{z_{ad,B}} + \Delta z_{sep,B} e^- \tag{10.26a}$$

$$BB^* \leftrightarrow B^* + B_{ad}^{z_{ad,B}} + \Delta z_{sep,B} e^- \tag{10.26b}$$

$$B_{ad}^{z_{ad,B}} \rightarrow B^{z_{ad,B}} + \Delta z_{dis,B} e^- \tag{10.26c}$$

At higher anodic potentials the direct dissolution from the kink site positions will become the rate-determining step described by the four equations

$$AA^* \rightarrow A^* + A^{z_A} + z_A e^- \tag{10.27a}$$

$$BA^* \rightarrow B^* + A^{z_A} + z_A e^- \tag{10.27b}$$

$$AB^* \rightarrow A^* + B^{z_B} + z_B e^- \tag{10.28a}$$

$$BB^* \rightarrow B^* + B^{z_B} + z_B e^- \tag{10.28b}$$

The rate equations for each component as well as the influence of solvation and complex formation can be formulated as was described for pure metals in Section 10.2.

Stationary dissolution conditions

At beginning corrosion the dissolution rates will be quite different depending on the surface composition after the alloy preparation. But with time the surface concentration will change and the partial corrosion rates will change. At longer times one could expect stable dissolution conditions not changing the alloy composition. This is expressed by the equation

$$\frac{r_A}{r_B} = \frac{x_A}{x_B} \tag{10.29}$$

Heusler[16] derived a special relation on the basis of this equation and with the following concept. The rate of dissolution of each component should be proportional to the mole fraction of this component in the surface x_A^s and x_B^s

$$r_A = k_A x_A^s \tag{10.30}$$

$$r_B = k_B x_B^s \tag{10.31}$$

with $x_A^s = 1 - x_B^s$. Combination of Eqs. (10.30) and (10.31) with Eq. (10.29) gives

$$\frac{k_A(1 - x_B^s)}{k_B x_B^s} = \frac{x_A}{x_B} \tag{10.32}$$

and after restructuring

$$\frac{1}{x_B^s} - 1 = \frac{x_A}{x_B}\frac{k_B}{k_A} \tag{10.33}$$

From this equation one derives the following two equations

$$r_B = k_B x_B^s = \frac{k_A k_B}{k_A x_B + k_B x_A} x_B \tag{10.34}$$

$$r_A = k_A x_A^s = \frac{k_A k_B}{k_A x_B + k_B x_A} x_A \tag{10.35}$$

The constant

$$K = \frac{k_A k_B}{k_A x_B + k_B x_A} \tag{10.36}$$

reflects the interaction of the components and the influence of the composition on the corrosion process.

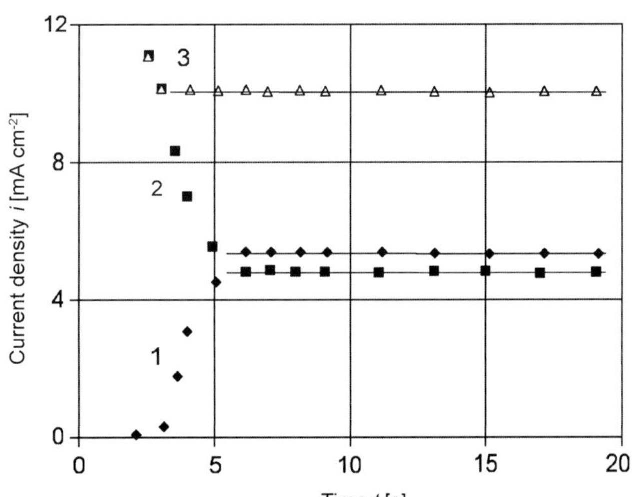

Figure 10.8 Time dependence of the partial current densities of Zn and Cu of α-brass ($CuZn_{30}$) in a chloride electrolyte, (1) partial current density of zinc dissolution, (2) partial current density of copper dissolution, and (3) total current density. (Reproduced with permission from Ref. [17], © 1980, Elsevier.)

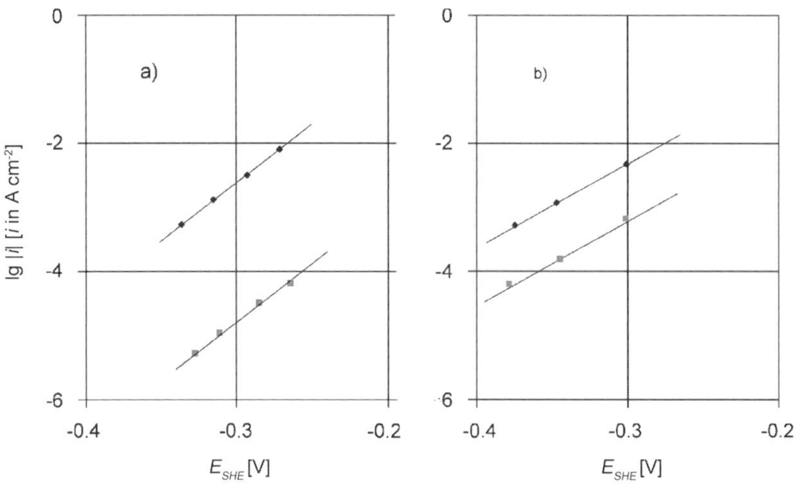

Figure 10.9 Potential dependence of the partial current densities of iron (spheres) and chromium (rectangles) dissolution from FeCr alloys in sulfuric acid; (a) $FeCr_{0.85}$, Tafel slope 40 mV and (b) $FeCr_{13.2}$, Tafel slope 100 mV. (Reproduced with permission from Ref. [17], © 1980, Elsevier.)

An example is the dissolution of Zn and Cu in CuZn alloys. The partial dissolution rates and its time dependence and final stationary value were determined by radiotracer methods (Kolotyrkin[17]). The results are shown in Figure 10.8.

With the stationary current densities from Figure 10.8 ($i_{Cu} = 5.3\ \mu A\ cm^{-2}$ and $i_{Zn} = 4.8\ \mu A\ cm^{-2}$) and assuming $z = 2$ for both metals $K_{Cu} = 3.9 \times 10^{-11}\ mol{\cdot}s^{-1}{\cdot}cm^{-2}$ and $K_{Zn} = 8.3 \times 10^{-11}\ mol{\cdot}s^{-1}{\cdot}cm^{-2}$. The origin for the difference needs further investigation.

In the beginning the zinc dissolution dominates and Cu dissolution is zero. But with time Zn dissolution rate decreases, Cu dissolution rate increases, and finally a stationary ratio equal to the alloy composition develops. In this situation an intermediate layer exist on the surface that regulates the zinc dissolution by the Zn diffusion through this layer. Otherwise, the great differences in the dissolution rates can also lead to a dezincation leaving a porous spongy Cu matrix.

Another example is the active dissolution of FeCr alloy (Figure 10.9). At low chromium content the observed Tafel slopes are similar to pure iron (40 mV). At higher chromium content the Tafel slopes are similar to pure chromium (100 mV). The surface layer changes from an iron rich to a chromium rich phase.

10.2 CORROSION PROTECTION

Since the invention of metal technology several thousand years ago, the prevention of metal corrosion has been a problem of fundamental practical importance. As the noble metals were corrosion resistant under normal conditions, the collection of knowledge of what could be done against corrosion really began with the invention of iron metallurgy by the Celtic tribes 3000 years ago.

Since then several main directions have developed:

- *Natural passivation.* Some non-noble metals show permanently and some under special oxidizing conditions the phenomenon of corrosion resistance. A special example is iron in contact with concentrated nitric acid, a phenomenon already discovered in 1797. Although iron dissolves in non-oxidizing acids, it is stable in nitric acid and shows a metal-like surface. The reason is the formation of a very thin protecting oxide film. This phenomenon will be described in more detail in the following chapters.
- *Inhibition.* Some molecules have the property of preventing corrosion or at least reducing the rate of corrosion. The molecules are called inhibitors. These molecules are used in closed systems like cooling aggregates.
- *Phosphatizing.* The surface of iron can be treated in such a manner that a thin layer is produced that does not fully protect the iron but inhibits the corrosion process and at the same time promotes adhesion of follow-up films. This is now a large-scale process in the car industry.
- *Conversion coatings.* Some metals can be treated by oxidizing agents, preferentially chromate solutions, to prepare an oxide film saturated by chromate ions. This is one of the most effective corrosion preventing treatments of metal surfaces and is applied to aluminum, copper, and zinc. Chromate ions are known for toxic properties. Political decisions have banned this surface treatment for the future. So far no alternative corrosion protecting treatment of similar effectiveness has been developed.
- *Surface coating with another metal with better resistance against corrosion.* The metal layer can be plated. Zinc plating is the most important example. But other methods are available to prepare a metal layer, e.g., mechanical plating, by dipping the parts into a liquid metal bath or spraying the metal.
- *Surface coating with a non-metallic inorganic material like enamel.* This coating is very effective as the parts can be used at higher temperatures but an isolating material covers the surfaces.
  For some metals an intrinsic inorganic coating can be produced by anodic oxidation and the formation of a thick surface oxide layer. One example is aluminum oxidation by the Eloxal process.
- *Surface coating with an organic material.* This is very effective but the metallic surface appearance is lost. The metallic conductivity is at least distorted. Furthermore, the coated parts cannot be used at higher temperatures. A scratch in the coating can have catastrophic effects because corrosion can proceed underneath the coating. In recent years a special organic coating has been described, the so-called intrinsically conducting polymers. These materials may have special corrosion protecting properties and are described later on in Chapter 11.
- *Cathodic protection.* If a metal is polarized to a potential that is sufficiently negative to its Nernst potential, the metal is protected. The polarization is possible by applying an external voltage. Another method is to form a galvanic element with a metal less noble then the metal to be protected. For iron a typical metal is magnesium. This method is used on a large scale in marine technology, in underground constructions, and in concrete reinforcements.

10.2.1 Passivity

Non-noble metals in dry air and at room temperature form a thin primary oxide film of different properties, structure, and thickness on the surface. If this oxide film is of low solubility, compact, and free of pores, then it is reducing the corrosion rate by orders of magnitude. This phenomenon is called passivity and the oxide film is called passive film. For some metals, like chromium, the passive film is very thin (some nanometers) and can become conducting. For another group of metals, like titanium, and aluminum, the so-called valve-metals, the oxide layer can get thicker (micrometers) and blocks further electrode processes.

Passivity of iron

A special example of a passive metal is iron. The primary oxide film on polycrystalline and pure iron (dry air, 20 °C) has a thickness of about 5 nm and is soluble in acid and neutral media. In strong nitric acid, an oxidizing medium, the usual dissolution of iron in acids is blocked by the formation of a passive film. In other acids this film can only be formed, if an anodic potential is applied. Polishing the electrode surface facilitates the formation of the passive film.

In Figure 10.10 the potentiostatic behavior of an iron electrode in sulfuric acid is shown. Beginning at zero current and scanning the potential in positive direction an anodic current is observed that increases with increasing potential. At a characteristic potential, the passivation potential with the passivation current density i_{max}, the current decreases by several orders of magnitude. The following potential region is called the passive region. In this region the current density of approximately 5 μA cm^{-2} remains constant (independent of

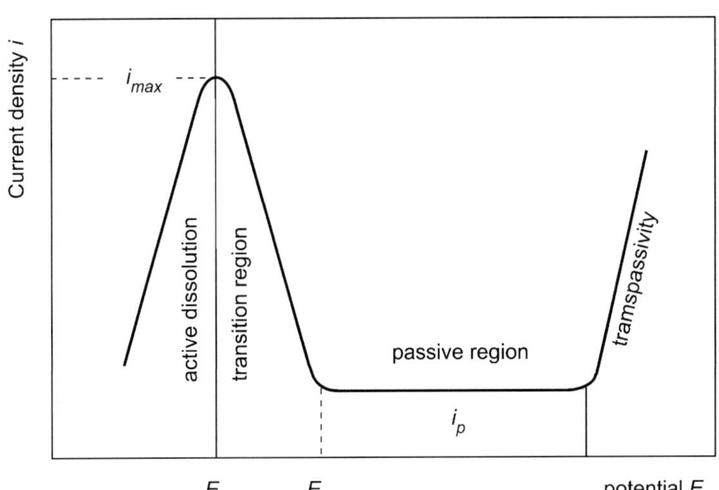

Figure 10.10 Current–potential plot of an iron electrode in an acid electrolyte (schematic representation). E_p passivation potential, i_{max} passivation current density, E_F Flade potential, and i_p passive current density.

the potential). At higher electrode potential, which is called the transpassive region, a new current increase starts. In this region the passive oxide can be oxidized connected with an increase in corrosion rate or oxygen evolution starts.

In the transition region between active dissolution and passive behavior a transformation process from a film of adsorbed iron hydroxide species into the passive oxide film takes place.[18]. The potential when this transformation is completed is called the Flade potential E_F. For iron the transformation process may be formulated schematically by the equation:

$$2\ Fe(OH)_{2,ad} \rightarrow Fe_2O_3 + H_2O + 2\ H^+ + 2\ e^- \tag{10.37}$$

In acid solution (pH 1–6) this process is pH dependent according to the equation (for 25 °C)

$$E_{Fl} = (0.58 - 0.059\ pH)\ V \tag{10.38}$$

Beginning the potential scan in the passive region and scanning in cathodic direction a steep decrease of the current density is observed at the Flade potential caused by the dissolution of the passive film. The same potential is observed if the passive current is switched off and before the potential jumps to the active region. This effect is called switch-off activation or self-activation. The activation time depends on the thickness and the dissolution rate of the oxide film. Therefore the Flade potential is also called activation potential.

Thus the stability of the passive film depends on two parameters, the electrode potential and the pH value. Pourbaix[19] developed special diagrams of stability regions of oxides on metal surfaces as function of electrode potential and pH value. The diagrams were calculated from thermodynamic equilibrium values for selected reactions between the metal and aqueous electrolyte. A Pourbaix diagram for iron is shown as example in Figure 10.11 (Kaesche[20]).

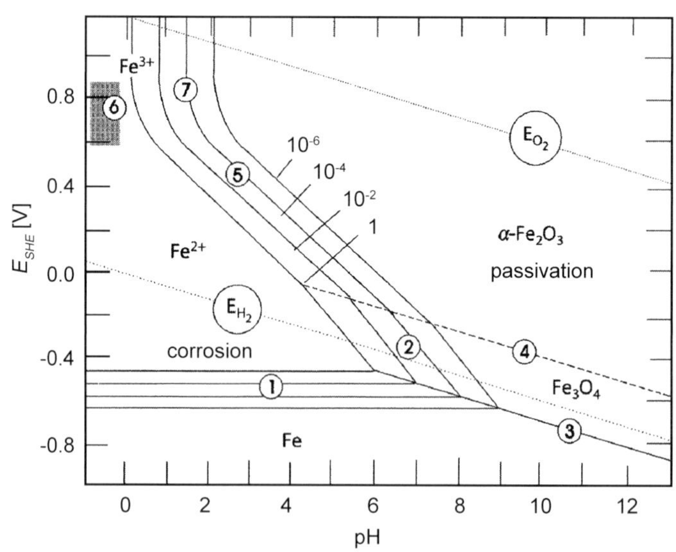

Figure 10.11 Pourbaix diagram of the iron–water system at 25 °C (after Kaesche[20]). The parameters mean the activities of the dissolved ions in mol·dm^{-3}. (Reproduced with permission from Ref. [2], © 2003, Springer-Verlag.)

The reactions of the different equilibria described in Figure 10.11 are the following:

(1) $Fe \leftrightarrow Fe^{2+} + 2\,e^{-}$

(2) $3\,Fe^{2+} + 4\,H_2O \leftrightarrow Fe_3O_4 + 8\,H^{+} + 2\,e^{-}$

(3) $3\,Fe + 4\,H_2O \leftrightarrow Fe_3O_4 + 8\,H^{+} + 8\,e^{-}$

(4) $2\,Fe_3O_4 + H_2O \leftrightarrow 3\,Fe_2O_3 + 2\,H^{+} + 2\,e^{-}$

(5) $2\,Fe^{2+} + 3\,H_2O \leftrightarrow Fe_2O_3 + 6\,H^{+} + 2\,e^{-}$

(6) $Fe^{2+} \leftrightarrow Fe^{3+} + e^{-}$

(7) $2\,Fe^{3+} + 3\,H_2O \leftrightarrow Fe_2O_3 + 6\,H^{+}$

Additionally, the oxygen and the hydrogen reactions are included. The diagram is characteristic for the formal procedure of the calculation. Not all possible reactions were considered in the diagram, e.g., higher valent Fe ions.

For iron, nickel, and chromium the experimental values of passivation and the thermodynamic data in the Pourbaix diagram are quite different because additional kinetic effects determine the regions of passivity and active dissolution for these metals. For other metals, e.g., zinc and copper the correspondence between the Pourbaix diagram and regions of stability and corrosion is better.

The chemistry and structure of the passive film was the subject of many investigations. The film is amorphous. Vetter[21] and Göhr and Lange[22] suggested a duplex film structure (Figure 10.12). Analytical investigations described the film as an inner film of Fe_3O_4 and an outer film of Fe_2O_3. However, electronic equilibrium is established on the metal/oxide interface. Therefore, on the metal/oxide side the metal must be in the lowest stable valence state, which is FeO. Oxygen ions are in equilibrium with the electrolyte on the oxide/electrolyte interface as described by the equation

$$H_2O \leftrightarrow O^{2-}_{ox} + 2\,H^{+} + 2\,e^{-} \qquad (10.39)$$

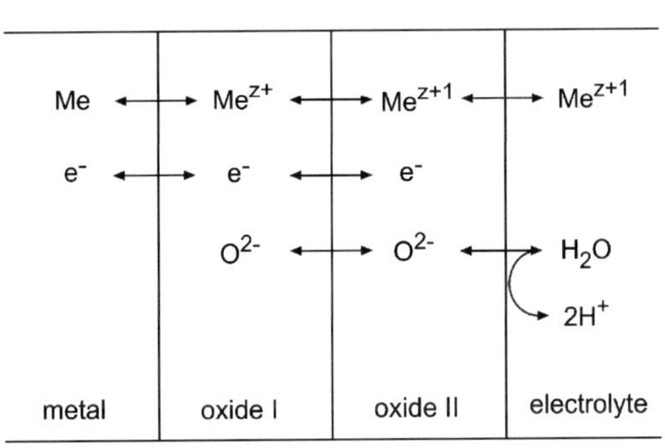

Figure 10.12 Duplex structure of the passive film of a metal, phase equilibriums, and transport processes in the film.

The oxidation state on this side depends on the potential. In the passive region of iron this is the three-valent iron.

The duplex structure shown in Figure 10.12 is a schematic representation and there can be a gradient of the metal ion valence in the film, e.g., an intermediate Fe_3O_4 structure. Furthermore, the outer film may have a partial hydroxide character.

By ellipsometry it was found that the thickness is a linear function of the potential (Vetter and Gorn[23], Sato[24]). Sato's results are shown in Figure 10.13.

A possible interpretation could be as follows. At a potential E_1 and a potential drop across the oxide of $\Delta\varphi_1$ for stationary conditions the thickness is δ_1. The corrosion rate is constant. The field strength in the film is $\Delta\varphi_1/\delta_1$. If the potential is switched to E_2 the field strengths increase and the corrosion rate increases, which leads to a growth of thickness. If the field strength in the film is $\Delta\varphi_2/\delta_2$, the corrosion current is back to its stationary value. It follows that with increasing thickness the field strength remains constant (Figure 10.14). The field strength is of the order of 10^6–10^7 V cm^{-1}. The ion transport in the oxide depends on the field strength $\Delta\varphi/\delta$ according to the equation

$$i = i_0 \exp\left(\beta \frac{\Delta\varphi}{\delta}\right)$$

The constant field strength is the reason for the potential independence of the corrosion current.

A very important step in the description of the properties of passive films was the description of these films as semiconductors.[26–28] The electronic properties of the passive oxide depend on the properties of the valence band and the conduction band. The oxide can behave as an n-semiconductor or p-semiconductor.

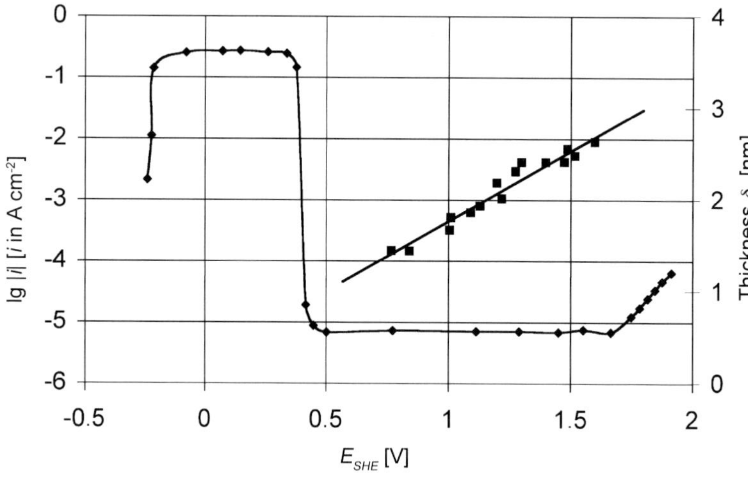

Figure 10.13 Thickness of a passive film of iron as a function of the potential. Comparison with the current–potential plot. (Reproduced with permission from Ref. [25], © 1980, Elsevier.)

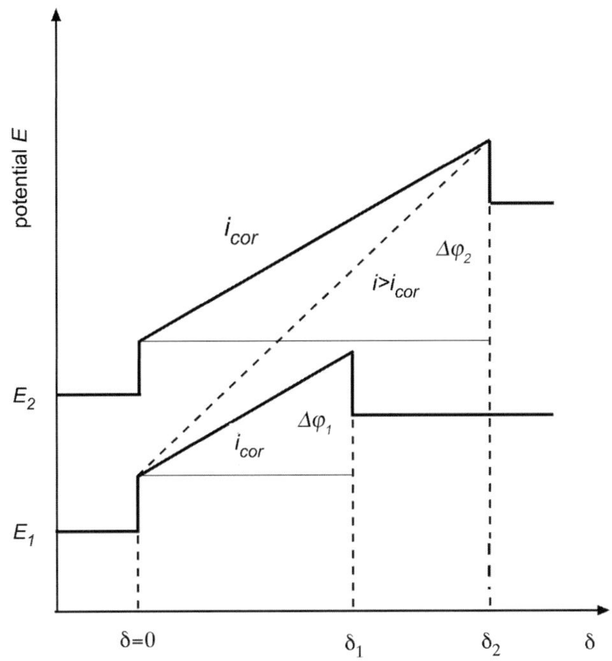

Figure 10.14 Dependence of the thickness of the passive film on iron on the potential. The field strength in the film is $\Delta\varphi_1/\delta_1$. If the potential jumps from E_1 to E_2 the field strengths increase as well as the film growth. During the time of film growth from thickness δ_1 to δ_2, a current higher than the corrosion current flows and decreases with time until the field strength is back to its old value.

The investigation of the semiconducting properties is possible by the typical methods of semiconductor electrochemistry as was described in Chapter 9. The band structure of the passive film of iron is shown in Figure 10.15.

The flat band potential can be associated with the Flade potential. The Flade potential of 0.58 V is 0.18 V more positive than the flat band potential. A possible explanation for the difference is the contact potential at the metal/oxide interface. The potential gradient in the oxide determines the bend bending. The region of passivity is terminated if by anodic polarization the Fermi level passes the top level of the valence band.

Passivity of nickel and chromium

Similar current–potential diagrams were obtained for nickel. The current–potential curve for Ni is shown in Figure 10.16. The passive layer of Ni behaves like a p-semiconductor. At 0.85 V NiO is oxidized to Ni_2O_3 and is thus connected with a steady increase in corrosion rate (Figure 10.16). At higher potentials a further oxidation to NiO_2 with a new decrease in current is observed. Additionally oxygen evolution starts.

Chromium is an example with an extremely thin passive layer and a much smaller corrosion current. At 1.0 V the passive film is oxidized to chromate. This is the region of transpassive dissolution (Figure 10.17).

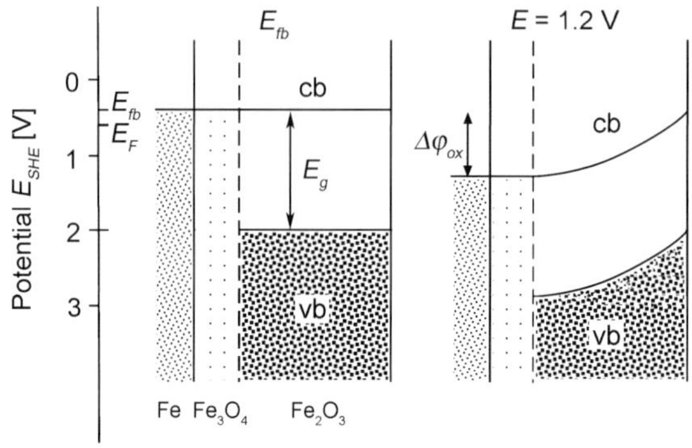

Figure 10.15 Semiconductor model for the passive film of iron, with a band structure at the flat band potential (0.4 V_{SHE} at pH = 0) and at 1.2 V_{SHE}. The band gap is 1.6 V. With anodic polarization a potential gradient across the oxide layer $\Delta\varphi$ is built up, which is the origin of the transport of iron ions through the oxide.

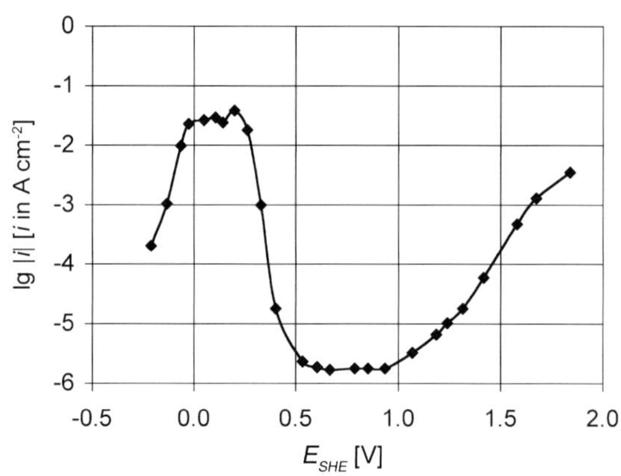

Figure 10.16 Current–potential diagram and passivation of Ni.[29,30]

On nickel and chromium the passive film is formed at normal atmospheric conditions. If protected by this film the metals behave like noble metals.

Passivity of alloyed steel

Metallurgical aspects of alloyed steel. The alloying of the different modifications of iron with other metals leads to a shear infinite number of materials all with different corrosion properties. The alloying element stabilizes one or the other modification of steel. In a rough scheme one can distinguish the following main routes:

– *Ferritic steels*
  These are chromium steels with more than 13% of chromium and a low carbon content (less than 0.08%). The chromium content stabilizes the ferritic modification. Ferritic steels are magnetic and are highly resistant against corrosion.

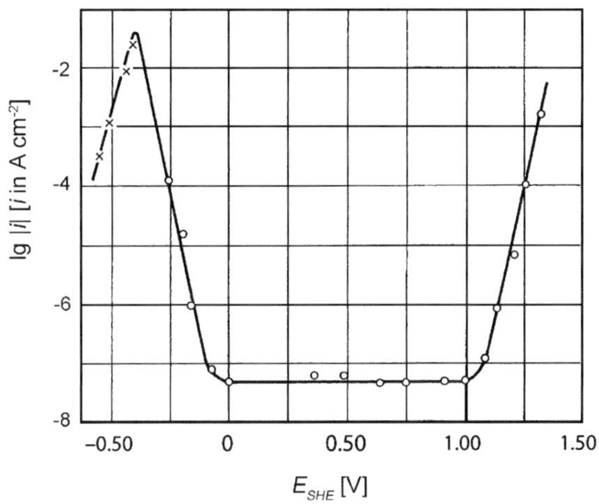

Figure 10.17 Current–potential diagram and passivation of Cr.[31]

- *Austenitic steels*
  These steels contain chromium (12–25%) and nickel (8–25%). The high nickel content stabilizes the austenitic modification. After a heat treatment of the region of austenite transformation and rapid cooling the steels remain in the austenitic form. They are non-magnetic. The most famous example is the18/8 chromium–nickel steel.
  The corrosion resistance is improved if compared with ferritic steels but the steels are sensitive to inter-crystalline corrosion as a result of chromium carbide precipitation at grain boundaries. Low carbon content or additional alloying metals like titanium, niobium, and tantalum, can suppress this sensitivity.
- *Ferritic–austenitic steels*
  Steels with high chromium content but lesser nickel content (4–6%) are a combination of ferritic and austenitic modification. Their chemical stability is similar to ferritic steels but they are not so sensitive to inter-crystalline corrosion.
- *Martensitic steels*
  Steels with a chromium content of 13–17% but with a carbon content of 0.2–0.4% can be hardened by a heat treatment forming the martensitic structure, an oversaturated solution of carbon in ferritic crystals. The hardening is also improving the corrosion resistance.

Even more materials can be obtained by varying the heat treatments. Each modification has different corrosion properties and the heat treatment can also modify the corrosion properties.

Selected examples of the passivity of alloyed steel. The extreme corrosion resistance is best demonstrated for iron–chromium alloys. As an example in Figure 10.18 passivity plots of these alloys are shown.

The protection mechanism is based on a modification of the composition of the passive film. The metals with a more stable oxide are enriched in the passive layer and become dominating. With an increased concentration of the metal with the higher corrosion resistance the protection is increased and a saturation value is finally achieved.

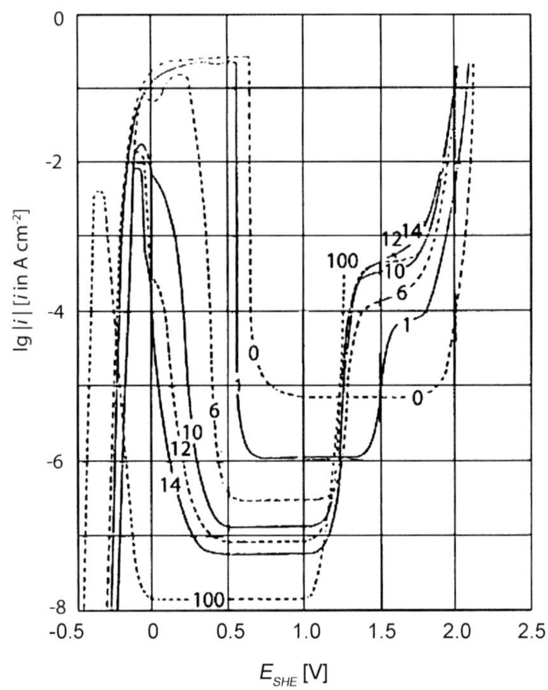

Figure 10.18 Current–potential plots of iron–chromium alloys in 0.5 mol·dm^{-3} H_2SO_4. (The figures in the diagram are weight percent chromium.) (Reproduced with permission from Ref. [32], © 1989, Elsevier.)

Passivity breakdown: pitting corrosion

Iron and steel in the presence of aggressive anions like chloride ions show the phenomenon of local breakdown of the passive film. On pipes, vessels, etc., semi-spherical pits develop on the surface, which penetrate the walls and destroy the parts with time. The omnipresence of chloride ions makes pitting corrosion a very general and dangerous phenomenon. The process has been intensively investigated.[33,34]

The major problem of the process is the large potential difference between the active surface in the pit and the passive film at the edge of the pit (Figure 10.19).

The development of pits starts with a crack or a hole of atomic dimension in the passive film caused, e.g., by tensions or by local chemical dissolution of the film. Permanent pitting corrosion can start above a critical potential and a critical concentration of the chloride ions. Above these critical values repassivation is prevented by the adsorption of the aggressive anions in the crack or the hole. The small dimensions of the crack or hole stabilize the large potential drop between active and passive surface.

In the second phase of pit growth corrosion products of dissolved metal ions and chloride anions are deposited in the hole. This film further stabilizes the potential drop and provides the conditions for an electrochemical polishing of the surface in the hole. This finally creates the larger semi-spherical forms of the pits.

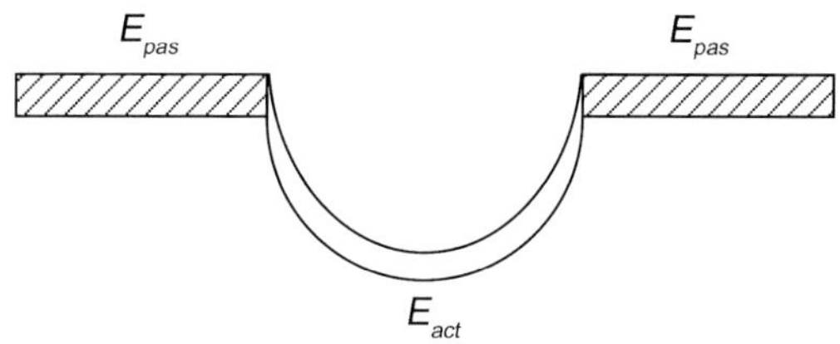

Figure 10.19 Model of pitting corrosion. A salt film stabilizes the potential difference between active iron surface in the pit and the passive surface.

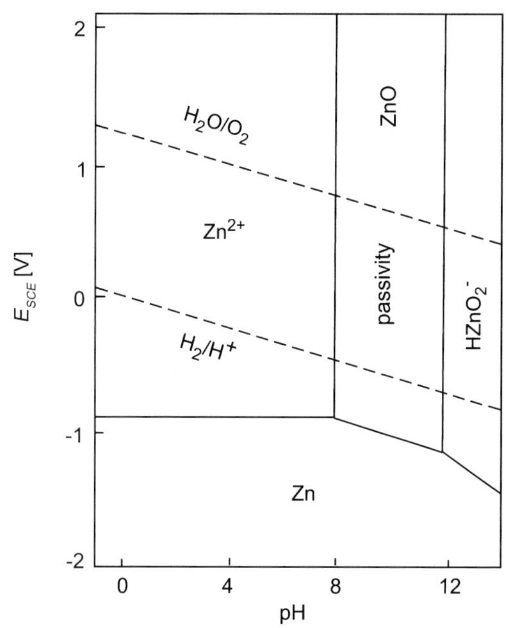

Figure 10.20 Pourbaix diagram of zinc.[35]

Passivity of zinc

Zinc is an example of a metal with an amphoteric oxide layer. The Pourbaix diagram is shown in Figure 10.20. The oxide is stable between pH 8 and pH 12. A natural oxide layer can be formed on the zinc surface but the usual procedure is a surface treatment in an oxidizing solution. An example is the chromatizing described in Section 10.2.5. With this treatment a mixed oxide of the formal composition $ZnO{\cdot}CrO(OH)$ is formed. In the case of chromate treatment the oxide film contains chromate ions. The chromate ions are responsible for the self-healing properties of these films.

Because of the ban of chromate for environmental reasons, alternative ways of protection of zinc are being looked for. In addition to different oxidizing treatments, zinc alloys have been developed. ZnNi, ZnFe, and ZnCo were some of the first examples. New trends are ZnMn and ZnMg. The corrosion protection of the alloys is probably connected with a change of the semiconducting properties of the oxides. The alloying of the less noble metal shifts and enlarges the band gap inhibiting the oxygen reduction on these surfaces.[36]

Metals with insulating passive films (valve metals)

While the previous examples were limited in the anodic polarization potential either by transpassive dissolution or by oxygen evolution valve metals can be polarized to potentials of up to 100 V and above. Examples are aluminum, titanium, tantalum, hafnium, and zirconium. Formation characterization and properties of these oxides were treated in Chapter 9.

10.2.2 Cathodic protection

The protection by passivation is based on the anodic formation of a passive film. A metal can also be protected by polarization to a negative (cathodic) potential where the dissolution is thermodynamically prevented. The negative polarization can be achieved by the application of an external cathodic current. Typical protected constructions are all kind of pipelines, marine constructions, or iron and steel embedded in concrete. The first step in the development of an external cathodic protection device is the determination of the necessary power, current density, and distance of the contact elements by measurement of the resistance of the environmental material of the metal construction to be protected. Then the necessary anodes, e.g., rods of ferrosilicon or magnetite in a coke bed are connected to the positive pole and the protected construction to the negative pole of the power source (Figure 10.21).

A second possibility is the generation of a local element by combination of the part to be protected with a less noble metal, preferentially Mg. The magnesium dissolves and protects the iron.

10.2.3 Corrosion inhibition

Inhibition is a method of corrosion protection by molecules that are adsorbed on the metal surface and reduce the rate of either metal dissolution (anodic inhibition) or the rate of the

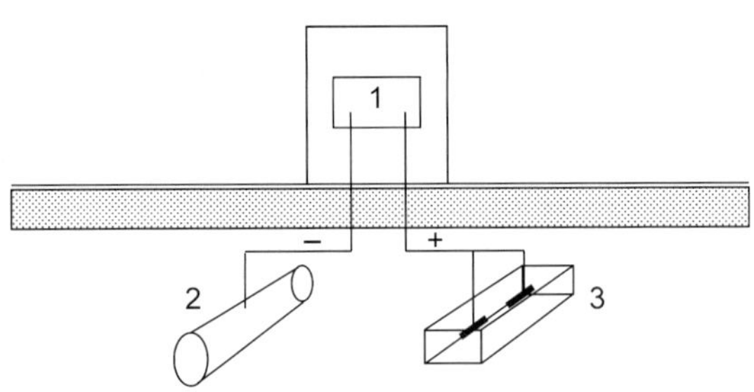

Figure 10.21 Schematic description of a cathodic protection device. Power source (1), steel construction (2), and inert anodes (3).

counter reaction, e.g., oxygen reduction (cathodic inhibition). Connected with the inhibition is a shift of the free corrosion potential. This shift is towards more positive potentials for anodic inhibitors and towards more negative potentials for cathodic inhibitors (Figure 10.22).

The adsorption of the inhibitor on the metal surface follows similar rules as were discussed for deposition additives in Chapter 9.

The selection and use of corrosion inhibitors is an important field of applied corrosion research and details on the formulation of inhibitor systems are private. In publications only general information can be found.[37]

Different inhibitors are used for different metals. Besides iron and unalloyed steel, copper aluminum, and zinc have to be protected. One problem is that a system often contains metal combinations and inhibitors for two or even more metals must be combined. Most inhibitors are used in cooling systems. There are open as well as closed systems. Each system demands different measures and inhibitors.

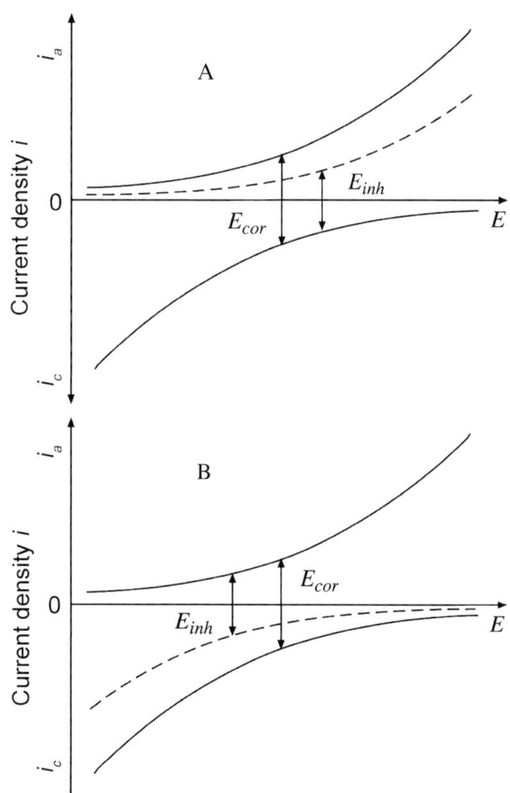

Figure 10.22 Current–potential diagram for anodic (A) and cathodic (B) inhibition of corrosion. In the first case (A) the free corrosion potential is shifted towards more positive values. In (B) the corrosion potential is shifted towards more negative potentials compared to the inhibitor free electrolyte.

Some examples of typical combinations of inhibitors in closed systems are

- Sodium tetraborate (Borax) is a simple measure to control the pH of cooling water, usually in closed systems.
- Sodium benzoate and sodium nitrite are used for combinations of unalloyed steel with copper and aluminum.
- Triethanolammonium phosphate and sodium mercaptobenzothiazole for steel and copper.
- Sebacic acid ester and benzotriazole or tolyltriazole for combinations of various metals.

In open systems, more inorganic substances are used to prevent oxidation of organic inhibitors.

- Chromates were used for a long time but are now substituted by less toxic compounds.
- Molybdates.
- Zinc phosphonates and phosphates.
- Combinations of phosphonates and polymers.
- Polyphosphates and polyphosphate silicates.

10.2.4 Phosphatizing

A more intensive treatment of the surface of iron, unalloyed steel, zinc magnesium, and aluminum is phosphatizing. This involves providing a film of phosphates (film forming phosphatizing) that protects against immediate corrosion and provides an undercoat of follow-up coating procedures. The phosphatizing solution contains phosphates and several metal ions of Zn, Fe, and others in diluted phosphoric acid. Etching of the metal surface, especially near the grain boundaries, is the primary process. In the following equations the phosphatizing of iron is used as an example.

$$Fe + 2\,H^+ \rightarrow Fe^{2+} + H_2$$

The pH is shifted to higher pH values by this process, which influences the dissociation equilibrium of the phosphate:

$$PO_4^{3-} + H^+ \rightarrow HPO_4^{2-}$$

Then the iron ions and the zinc ions in the electrolyte form an insoluble zinc–iron phosphate, which is schematically described by the following equation:

$$Fe^{2+} + Zn^{2+} + 2\,HPO_4^{2-} \rightarrow ZnFe(HPO_4)_2$$

By a preliminary activation step the surface can be modified and a very fine grain structure of the phosphate film is achieved. Usually by combinations of three or more metal cations, e.g., Zn and Ni or Zn and Mn the film properties are further improved.

10.2.5 Chromatizing

Chromatizing is an oxidative treatment that forms an oxide film or increases the thickness of an oxide film and incorporates chromate ions. The process is applied to zinc, aluminum, copper, and also to phosphatized steel. The reaction mechanism of chromatizing in an acidic dichromate solution can be schematically described by the following equations:

$$Zn \rightarrow Zn^{2+} + 2e^-$$

$$Cr_2O_7^{2-} + 8\,H^+ + 6\,e^- \rightarrow 2\,CrO(OH) + 3\,H_2O$$

$$Zn^{2+} + H_2O \rightarrow ZnO + 2\,H^+$$

The oxide film grown in this process has a thickness of 0.5 µm. It is a film of a zinc oxide/hydroxide containing different chromium oxides (CrO(OH), Cr_2O_3, Cr_2O_5, CrO_3) and zinc chromate ($ZnCrO_4$). The chromate ions are rather mobile.[38] If a zinc layer on a steel substrate gets a scratch somewhere on the zinc, the chromate will diffuse to this scratch and repassivate the surface (the self-healing mechanism).[39]

The toxic properties of chromate have led to the political demand to develop an alternative to chromatizing. One possibility is to treat the surface with hexafluorotitanate or hexafluorozirconate. Then a film containing the metal oxide mixed with titanium dioxide or zirconium dioxide covers the surface. One possible substitute for chromatizing is a treatment with a three valent chromium bath.

10.2.6 Corrosion protection by surface coatings

The final solution for corrosion protection is the coating of the surface with a film of an inert substance. This can be a layer of a metal with better protection properties than the bulk material. One example is the covering of iron or unalloyed steel by a zinc layer with a thickness of some micrometers. Zinc coating is possible by dipping into a hot zinc bath or by electroplating. The zinc layer must be further treated for sufficient corrosion protection. This can be chromatizing as described in the previous chapter or by a newly developed process free of chromate.

Another example is chromium plating on steel parts. The typical process consists of a primary copper layer, a secondary nickel layer, and the final chromium plating.

Nickel-plating itself is used for corrosion protection without additional chromium. Usually two nickel layers are plated from different plating baths. A nickel layer of dull finish covers a bright nickel layer. The nickel layers have slightly different potentials, which develop some cathodic protection.

An inorganic coating with a non-metallic material is an enamel layer. Enamel is a glasslike film. The metallic part is covered by the enamel mass, usually a liquid suspension, by dipping into the suspension or by spraying. After drying the part is heated to 800 or 900 °C.

The third possibility for corrosion protection is a coating with an organic polymer. This can be a thick rubber-like material, a polymer, or just by painting. New materials are being developed that provide improved corrosion protection. One example is intrinsically conducting polymers, which will be described in Chapter 11.

REFERENCES

1. The nomenclature in this chapter is based on the IUPAC recommendation; K.E. Heusler, D. Landolt, S. Trasatti, Pure Appl. Chem., *61*, 19 (1989).
2. H. Kaesche, Corrosion of Metals, Springer, New York, 2003.
3. L.L. Shreir, R.A. Jarman, G.T. Burstein (Eds.), Corrosion, 3rd ed., Vol. 2, Butterworth-Heinemann, Oxford, 1994.
4. P. Marcus, J. Oudar (Eds.), Corrosion Mechanism in Theory and Practice, Marcel Dekker, New York, 1995.
5. K.J. Vetter, Elektrochemical Kinetics, Academic Press, New York, 1967, p. 539.
6. J.O'M. Bockris, D. Drazic, A.R. Despic, Electrochim. Acta, *4*, 325 (1961).
7. A. Despic, Deposition and Dissolution of Metals and Alloys. Part B: Mechanisms, Kinetics, Texture, and Morphology, in B. Conway, J.O'M. Bockris, E. Yeager, S.U.M. Khan, R.E. White (Eds.), Comprehensive Treatise of Electrochemistry, Vol. 7, Plenum Press, New York, 1983, p. 451.
8. K.E. Heusler, Iron, in A.J. Bard (Ed.), Encyclopedia of Electrochemistry of the Elements, Vol. 9, Part A, Marcel Dekker, New York, 1982.
9. T.B. Massalski, H. Okamoto, H. Subramanian, L. Kasprzak, Binary alloys Phase Diagrams, 2nd Ed, ASM International, Metals Park, OH, 1991.
10. F.J. Humphreys, M. Hatherly, Recrystallization and Related Annealing Phenomena, Elsevier, Amsterdam, 1995.
11. J.O'M. Bockris, M. Enyo, Trans. Faraday Soc., *58*, 1187 (1962).
12. O.M. Magnussen, L. Zitzler, B. Gleich, R.J. Behm, Electrochim. Acta, *46*, 3725 (2001).
13. L. Gaiser, K.E. Heusler, Electrochim. Acta, *15*, 161 (1970).
14. V. Kucera, E. Mattson, Atmospheric Corrosion, in F. Mansfeld (Ed.), Corrosion Mechanisms, Marcel Dekker, New York, 1987.
15. W. Grauer, Werkstoffe und Korr., *31*, 870 (1980); *32*, 113 (1981).
16. K.E. Heusler, Corrosion Sci., *39*, 1177 (1997).
17. Y.M. Kolotyrkin, Electrochim. Acta, *25*, 89 (1980).
18. G.L. Griffin, J. Electrochem. Soc., *131*, 18 (1984).
19. M. Pourbaix, Atlas d'eqilibres électrochimique a 25 ^{0}C, Gautier-Villars, Paris, 1963.
20. H. Kaesche, Corrosion of Metals, Springer, New York, 2003, p. 45.
21. K.J. Vetter, J. Electrochem Soc., *110*, 5597 (1963); Electrochim. Acta, *16*, 1923 (1971).
22. H. Göhr, E. Lange, Z. Elektrochem. Ber. Bunsenges. Phys. Chem., *61*, 1291 (1957).
23. F. Gorn, K.J. Vetter , Z Physikal Chem. NF, *77*, 317 (1972); K.J. Vetter, F. Gorn, Z Physikal Chem. NF, *86*, 113 (1973).
24. N. Sato, K. Kudo, T, Noda, Z Physikal Chem. NF, *98*, 271 (1975).
25. N. Sato, Corrosion Sci., *31*, 1 (1990).
26. U. Stimming, J.W. Schultze, Ber. Bunsenges. Phys. Chem., *80*, 1297 (1976); Electrochim. Acta, *24*, 859 (1979).
27. N. Sato, J. Electrochem. Soc., *129*, 255 (1982).
28. H. Gerischer, Corrosion Sci., *31*, 81 (1990).
29. K.J. Vetter, K. Arnold, Z. Elektrochem., *64*, 240, 407 (1962).
30. N. Sato, G. Okamoto, J. Electrochem. Soc., *110*, 703 (1963).
31. Ya. M. Kolotyrkin, Z. Elektrochem., *62*, 700 (1958).

32. R. Kirchheim, B. Heine, H. Fischmeister, H. Hofmann, H. Knote, U. Stolz, Corrosion Sci., *29*, 899 (1989).
33. H. Böhni, Localized Corrosion, in F. Mansfeld (Ed.), Corrosion Mechanisms, Marrcel Dekker, New York, 1987, p. 285.
34. H. Strehblow, Mechanism of Pitting Corrosion, in Corrosion Mechanism in Theory and Practice, P. Marcus, J. Oudar (Eds.), Marcel Dekker, New York, 1995, p. 201.
35. G. Wranglen, An Introduction to Corrosion and Protection of Metals, Butler & Tunner Ltd., Stockholm, 1972, p. 262.
36. R. Hausbrand, M. Stratmann, M. Rohwerder, Steel Res., *74*, 453 (2003).
37. V.S. Shastry, Corrosion Inhibitors, Principles and Applications, Wiley, Chichester, 1998.
38. L. Xia, R.L. McCreery, J. Electrochem. Soc., *145*, 3083 (1998).
39. J. Chao, G. Frankel, R.L. McCreery, J. Electrochem. Soc., *145*, 2258 (1998).

– 11 –

Intrinsically Conducting Polymers

With the discovery of polyacetylene and the process of oxidative doping in the 1970s of the last century,[1] a new class of polymers was introduced with conducting properties of the same order of magnitude as was usually found for metals. For this reason the name synthetic metals was suggested.[2] Conducting polymers have been synthesized before by mixing polymers with metal or graphite powders. But in the case of this new class of materials, the conductivity was an intrinsic property. Therefore, the name intrinsically conducting polymers (ICPs) was recommended. Many aspects of these new materials have been summarized in a growing amount of review literature.[3–5]

The metal-like property of these polymers is based on their chemical nature, which consists of chains of conjugated double bonds. If these polymers are oxidized, they become electrical conducting. In the neutral state they can have properties like an inorganic semiconductor. This has now become as important as the metal-like conductivity.[6] One example is the development of an "organic field effect transistor" (OFET).

Some examples of ICPs are shown in Figure 11.1. The typical oxidation–reduction process between neutral (semiconducting) and oxidized (metal-like conducting) state is shown for polythiophene in Figure 11.2.

Conducting properties can also be achieved by reduction of the neutral state. An example is the poly(dibutoxyphenylenevinylene).[7] A platinum electrode coated with polymer in acetonitrile showed the transition between neutral and oxidized state at +1 V versus a platinum quasi reference electrode and the transition between neutral and reduced state at –1.8 V. The cyclic voltammogram is shown in Figure 11.22b as an example to determine characteristic semiconductor properties from cyclic voltammograms. The reduced state has received much less attention than the oxidized state.

According to Figure 11.2 the oxidation process is reversible as it is possible to switch between the oxidized and neutral states. Otherwise, oxidation and reduction of conducting polymers is a very complex process.[8] Therefore, the reversibility is limited and depends on the timescale. In a periodic oxidation–reduction cycle, only quasi-stable oxidized and reduced polymers can be expected. During synthesis or in the process of oxidation the chain of the conjugated polymer is positively charged. The charging can reach a degree of up to 25–30%. Anions are intercalated into the polymer structure to compensate the positive charges of the polymer chain. In the following reduction the anions can be ejected or cations can be intercalated. The anion or cation transport can be connected with solvent

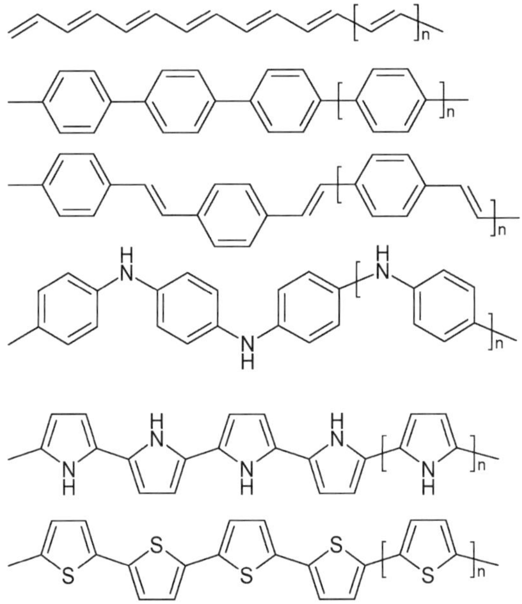

Figure 11.1 Some typical intrinsically conducting polymers (ICPs) are PA, trans-polyacetylene; PPP, polyparaphenyle; PPV, polyparaphenylenvinylene; PANI, polyaniline; Ppy, polypyrrole; and PT, polythiophene.

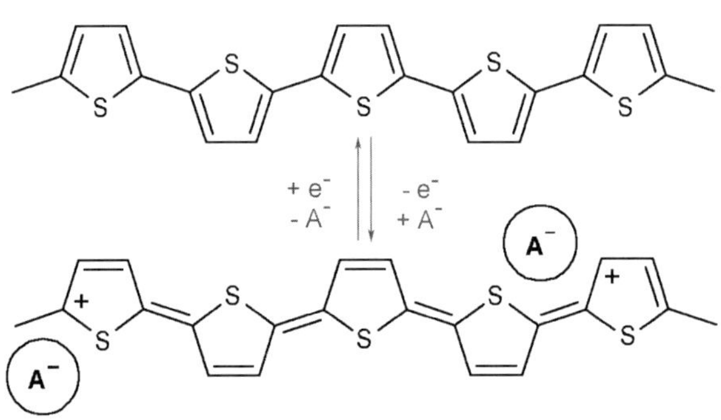

Figure 11.2 Oxidation–reduction of intrinsically conducting polythiophene with switching between neutral (semiconducting) and oxidized (metal-like conducting) states.

transport and with structural changes of the polymer. Thus, chemical relaxation processes accompany the charging and discharging. These structural fluctuations are a characteristic feature of these materials and a reason for the limited reversibility of the redox process.

Other characteristic properties are connected with the different oxidation states. ICPs are hydrophilic in the oxidized state and hydrophobic in the neutral state. The specific volume of the oxidized state is usually larger than the specific volume of the neutral state. The many new properties lead to suggestions for a variety of applications. One of the most remarkable has been the development of an organic battery.

11.1 CHEMICAL SYNTHESIS

Conducting polymers can be synthesized by following conventional polymer synthesis routes. One can transfer the Ziegler–Natta synthesis of polyethylene to polyacetylene.[9] The Ziegler–Natta catalyst is solved in large excess in an inert solvent with acetylene streaming over the surface. A black film of polyacetylene is formed on the liquid surface.

Figure 11.3 shows an example of polycondensation with a Grignard reaction. Treating a solution of 2-bromthienyl-3-alkyl-5-magnesiumbromide in the presence of [1,3-bis (diphenylphosphino)propane]nickel(II)-chloride as a catalyst leads to formation of poly-3-alkylthiophene and magnesium bromide (Kumada–McCullough process).[10]

Polycondensation can be used for the synthesis of poly-3-alkylthiophenes with high regioselectivity. The condensation can either lead to a head–tail, head–head, or tail–tail combination (Figure 11.4).

Oxypolymerization is especially important. Several oxidizing agents are known but $FeCl_3$ in chloroform or acetonitrile is one that is very often used.[11] The formal reaction equation is shown in Figure 11.5.

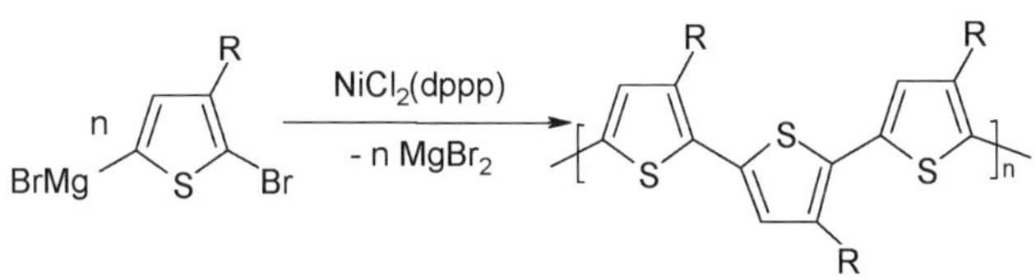

Figure 11.3 Synthesis of poly-3-alkylthiophene by polycondensation. Catalyst: [1,3-bis(diphenylphosphino)propane]nickel(II)-chloride.

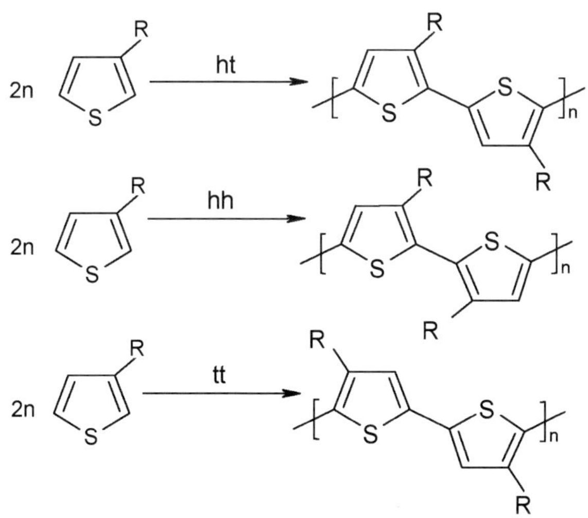

Figure 11.4 Different combinations of 3-alkylthiophene groups such as ht, head–tail; hh, head–head; and tt, tail–tail combination.

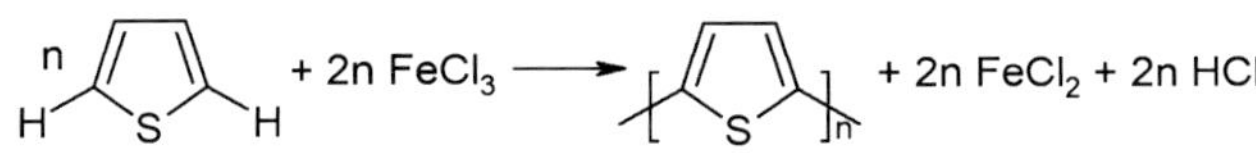

Figure 11.5 Oxypolymerization of thiophene with $FeCl_3$.

The $FeCl_3$ must be used in large excess. The process consists of oxidation of the monomer, formation of a radical cation, dimerization of two radical cations, and growths of a polymer chain. The mechanism is similar to the electrochemical oxidation that will be described in the next section. For the electron transfer between the iron(III) ions and monomer or polymer molecules a complex formation was postulated.

The growing polymers finally become insoluble and form solid particles, which limits the length of the chain but results in the oxidized form of the polymer.

This oxidized form contains anions that compensate the positive charges on the polymer chain and solvent molecules. Type, size, and charge of the anions strongly influence the polymer structure. With small anions like perchlorate or large voluminous anions like tosylsulfonate, polymers of different structure and properties can be synthesized.

The product of the chemical synthesis for polyacetylene was a freestanding film. In other cases, the product can be a powder of particles of different shapes and sizes from micrometer down to nanometer dimension.

11.2 ELECTROCHEMICAL SYNTHESIS AND SURFACE FILM FORMATION

Oxypolymerization can also be performed by anodic oxidation. In this case the product of the synthesis is a surface film of conducting polymers.

The electrochemical deposition of conducting polymers has been intensively investigated for the last few decades.[12] The mechanism of polymerization can be described by the formation of radical cations followed by (a) coupling between radical cations, as suggested by Genies *et al.*[13] and (b) reaction of a radical cation with a neutral molecule, e.g., Koßmehl *et al.*[14] The majority of experimental facts support the radical cation coupling. In Figure 11.6 radical cation coupling is shown in detail for polythiophene.

The mechanism consists of the following main steps:

1. The oxidation of monomers to radical cations.
2. Coupling between two radical cations in the α-position.
3. Dimer formation by the loss of two protons (stabilization step).
4. Oxidation of the dimer.
5. Trimer formation by reaction of dimer and radical cation.

The process continues with oxidation, coupling, and deprotonation steps until the final polymer chain length is reached.

The complex nature of the electrode surface has a large influence on all these steps. Some examples are

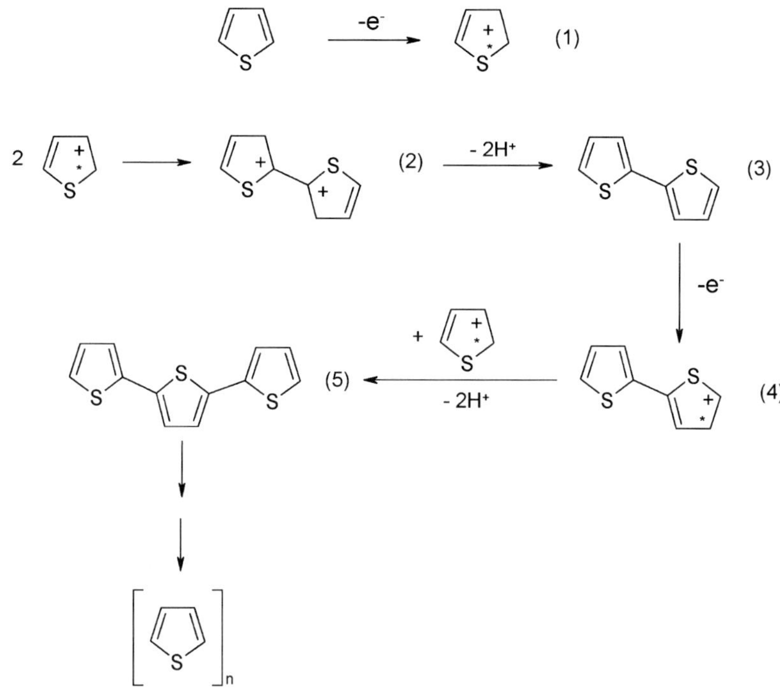

Figure 11.6 Radical cation coupling mechanism for polythiophene.

- The electrode material will influence the process, i.e., the chemical nature of the electrode material as well as the crystallographic structure.
- The solvent molecules, the cations and anions of the supporting electrolyte, and the monomer molecules can adsorb on the electrode surface with a strong influence on the partial reactions.
- The character of the inner part of the double layer can vary from hydrophobic to hydrophilic; the field strength in this area can vary due to ion adsorption.
- The orientation of the adsorbed molecules will influence local reactivity.
- The anions that are built-in into the film have a strong influence like in the chemical synthesis. The interaction of the anions with the electrode surface supports this influence.
- The properties of the electrode surface are governed by the potential. Therefore, the formation potential is a strong influence. The oxidation potential of the monomers is usually more positive than the oxidation potential of the growing polymer chain. Because the potentials should not be higher than the potential necessary for a polymer growths (to avoid damage of the polymer by over-oxidation), the potential program for the electrosynthesis of ICPs should be carefully selected.

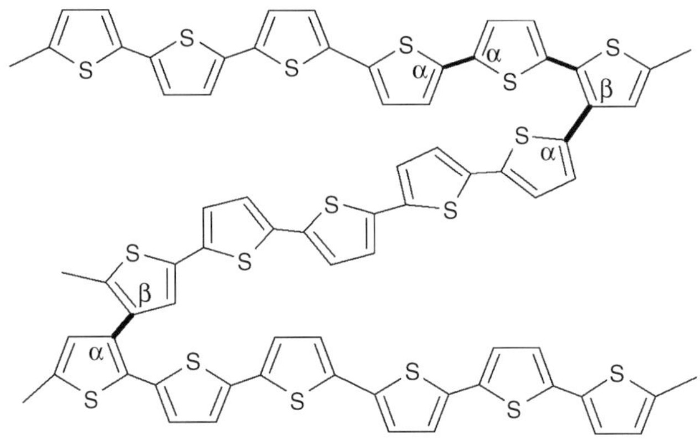

Figure 11.7 Network polymers formed by α–β coupling.

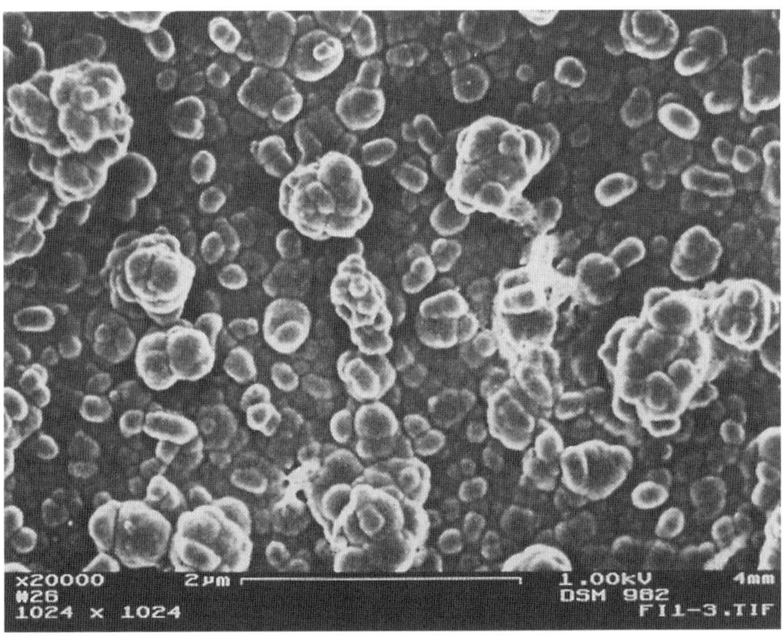

Figure 11.8 Cauliflower structure of a thicker film of polythiophene.

So far, a formation of linear polymer chains has been assumed. Although coupling in α–α position is most likely, it is also possible between α–β positions, whereupon a network is generated (Figure 11.7).

The film formed on the surface usually has an irregular structure of growing spheres of different sizes and is therefore sometimes called a cauliflower structure, e.g., Figure 11.8.

Otherwise, it has been found[15] that the film has a compact structure of coherent grains just after the start of the deposition (Figure 11.9). The different conductivity of the different grains will then lead to the cauliflower structure of the thicker films.

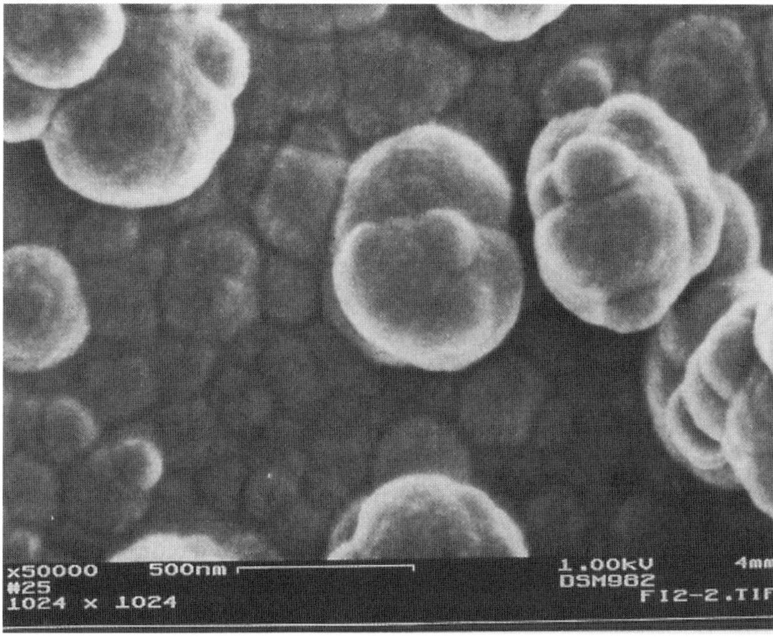

Figure 11.9 Structure of a polythiophene film of few monolayers. On the substrate a compact layer with shell structure is formed. The following film has a porous structure.

The complete film has a duplex structure of a thin compact inner film and a porous outer part. This is important to understand the properties of these films.

11.3 FILM FORMATION WITH ADHESION PROMOTERS

Electrochemical film formation is restricted to conducting materials. On inert metals a stable and adherent film may form. On corrosive metals like iron, oxidation of the metal prevents formation of a stable film. Films on non-conducting materials can be formed by chemical oxidation but often suffer from insufficient adhesion.

Films on these materials can be formed if adhesion promoters are used.[16,17] With adhesion promoters it is possible to deposit films of conducting polymers on insulators, semiconductors with highly polished surfaces, and corrosive metals. An adhesion promoter consists of an adhesion group, e.g., chloro- or hydroxy silane groups, phosphonic acid groups, which can bond to the surface of the solid, spacer group, and a monomer molecule as a head group.

For the preparation of a polymer film on a solid, e.g., a silicon wafer, the solid is dipped into the solution of the adhesion promoter like chloroform for approximately 1 h. The adhesion promoter can be expected to form a monomolecular film on the surface of the solid. The structure of this monomolecular film is similar to that of a self-assembling monolayer (SAM). After rinsing the solid specimen is brought into a solution of the monomer

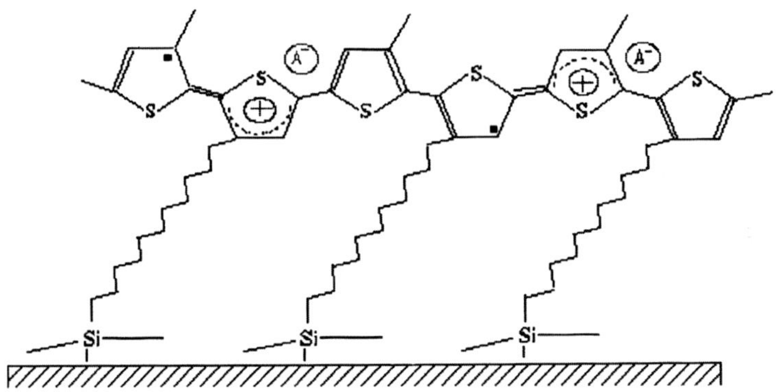

Figure 11.10 Structure of a polymer film with adhesion promoter.

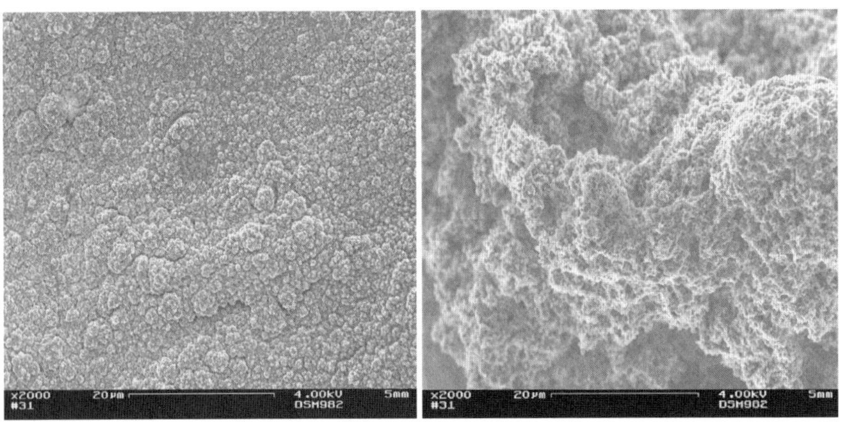

A B

Figure 11.11 Comparison of films formed (A) with and (B) without an adhesion promoter layer.

and an oxidizing agency, e.g., thiophene with $FeCl_3$. Then the film is grown with covalent bonds between the adhesion promoter molecules and polymer network (Figure 11.10).

The SAM-like layer induces a structure of the polymer film that is different from films formed without the adhesion promoter (Figure 11.11).

11.4 ION TRANSPORT DURING OXIDATION–REDUCTION

The characteristic feature of conducting polymers is the switching between oxidized and neutral state or, in the case of the electrical properties, between metal-like conducting and semiconducting or ion-conducting states. It was pointed out that this is a complex process with several partial reactions. The separation of the partial reactions is a prime goal of investigations. In this context electrochemical quartz crystal microbalance (QCMB) investigations have been especially successfully applied.[18–23] A short introduction of the method was given in Section 4.2.3. The Sauerbrey equation relates the frequency shift Δf of the resonance frequency f_0 of a metal-coated quartz crystal during electrochemical formation or oxidation of a film of a conducting polymer to the mass change Δm of the film.[24]

$$\Delta f = -C_{\mathrm{SB}} \frac{\Delta m}{A} \tag{11.1}$$

The Sauerbrey constant C_{SB} for a 10 MHz AT-quartz is 226.01 $\mathrm{Hz \cdot cm^2 \cdot \mu g^{-1}}$ (AT describes the type of cut, which is like a slice from a quartz crystal (Section 4.2.3)). It was mentioned that the application of the simple Sauerbrey equation is limited to rigid films. Polymer films can have viscoelastic properties or can develop a high surface roughness. In these cases the mechanical impedance Z_{M} contains terms that take into account viscoelasticity and surface roughness, and the resonance frequency shift becomes a complex quantity.

$$\Delta f^* = \Delta f + j\frac{\Delta w}{2} \tag{11.2}$$

The complex term is determined by the half width of the resonance frequency plot Δw. Then it is necessary to make a complete analysis of the resonance behavior of the quartz crystal with a network analyzer or an impedance bridge.[25–30] The difference between the rigid film and a viscoelastic film was shown in Figure 4.17. Recently, a semi-quantitative relation between viscoelastic and roughness properties and the complex frequency shift could be derived.[31]

11.4.1 Analyzing oxidation–reduction cycles using QCMB

An example of an ICP is poly(-3,4-ethylenedioxythiophene) with the short name PEDOT. The polymer can be prepared in acetonitrile or in aqueous electrolytes. The polymer films formed have quite different oxidation–reduction properties. For the film prepared in dry acetonitrile, the change of the half width of the quartz resonance frequency in the oxidation–reduction cycle is negligible (Figure 11.12).[32,33]

Because of the small damping such a film can be used to study anion exchange. The mass change can be calculated with Eq. (11.1). Such a mass change during oxidation and reduction of the film is shown in Figure 11.13.

If the current density is recorded together with the frequency shift, the charge change ΔQ in the redox cycle can be determined. The cyclic voltammogram in a tetra-butyl-ammonium-perchlorate/acetonitrile electrolyte is shown in Figure 11.14.

The region of determination of the charge change (potentials between 0.2 and 0.6 V) is marked in the figure. The plot of mass change versus charge change for 3,4-ethylenedioxythiophene for three different electrolytes in acetonitrile is shown in Figure 11.15.

From the slope of this plot the apparent molar mass can be determined by multiplying the slope with Faraday's constant

$$M_{\mathrm{app}} = zF\frac{\Delta m}{\Delta Q} \tag{11.3}$$

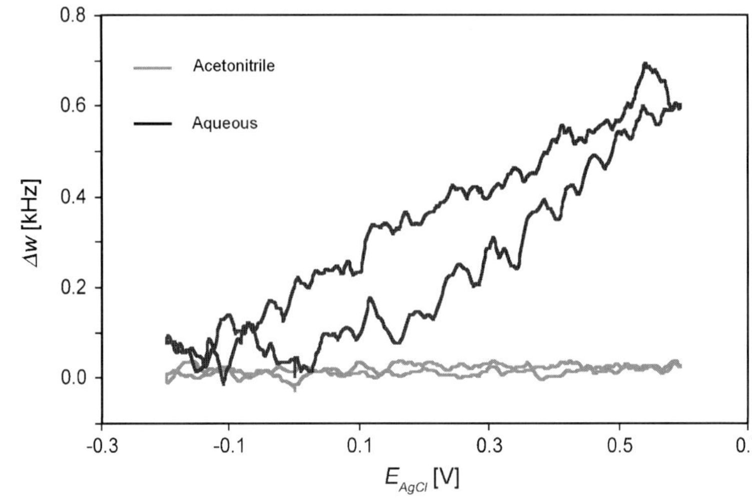

Figure 11.12 Half width of the resonance frequency Δw as function of the potential of a poly (-3,4-ethylenedioxythiophene) film grown in dry acetonitrile or water, respectively containing 0.1 mol·dm^{-3} 3,4-ethylenedioxythiophene and 0.1 mol·dm^{-3} $LiClO_4$.

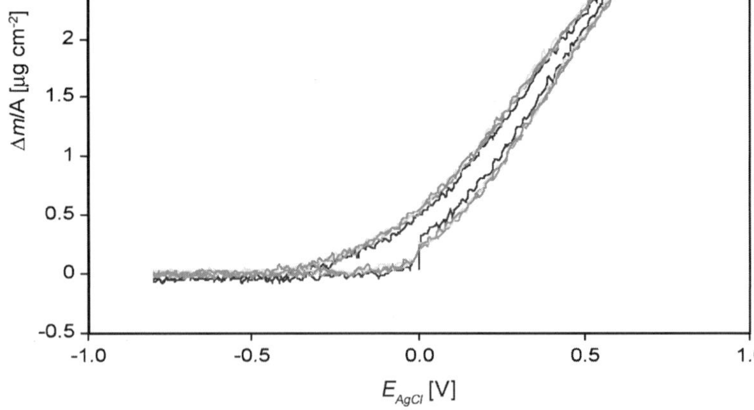

Figure 11.13 Mass change per electrode area $\Delta m/A$ during an oxidation–reduction cycle of poly (-3,4-ethylenedioxythiophene) in 0.1 mol·dm^{-3} $NaClO_4$/acetonitrile.

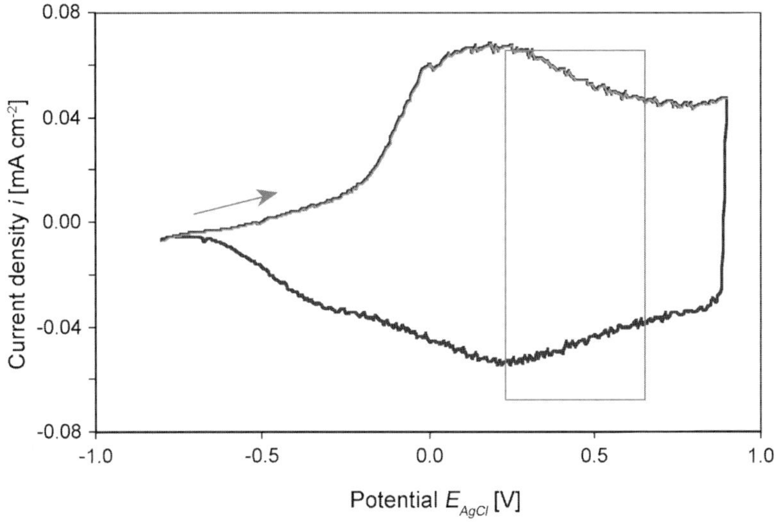

Figure 11.14 Cyclic voltammogram of an oxidation–reduction cycle of poly(-3,4-ethylenedioxythiophene) in 0.1 mol·dm^{-3} tetra-butyl-ammonium-perchlorate/acetonitrile, scan rate 10 mV·s^{-1}. The frame shows the region of data evaluation.

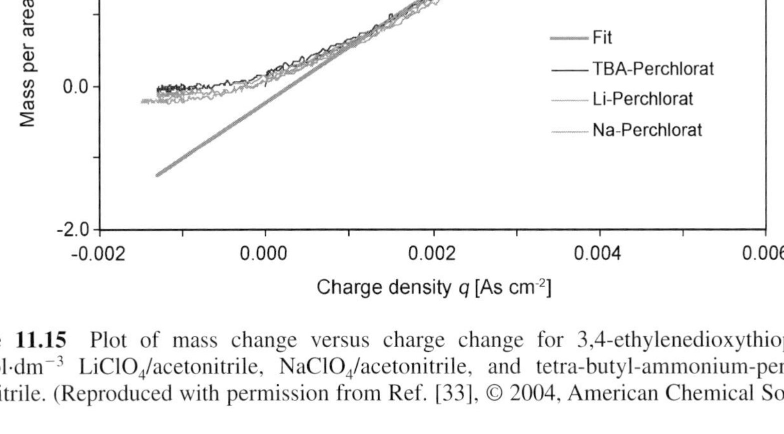

Figure 11.15 Plot of mass change versus charge change for 3,4-ethylenedioxythiophene in 0.1 mol·dm^{-3} $LiClO_4$/acetonitrile, $NaClO_4$/acetonitrile, and tetra-butyl-ammonium-perchlorate/acetonitrile. (Reproduced with permission from Ref. [33], © 2004, American Chemical Society.)

The oxidation–reduction process consists of several transport processes between polymer and electrolyte: (i) the exchange of anions, (ii) the exchange of solvent molecules accompanying the anion transport, (iii) the exchange of cations, and (iv) the exchange of solvent molecules accompanying the cation transport. Therefore, M_{app} consists of several terms.

$$M_{app} = M_{an} + \alpha M_{solv} + \beta M_{cat} \tag{11.4}$$

For some polymers only anion exchange is observed. Then $\beta = 0$ and α, associated with the exchange of solvent molecules with one anion, can be determined. For poly(-3, 4-ethylenedioxythiophene) in $LiClO_4$/acetonitrile $\alpha = -1$ was found. This means one perchlorate anion substitutes one acetonitrile molecule. The experiments are also possible in aqueous electrolytes. Results for different anions in aqueous electrolytes are shown in Table 11.1.[32,34]

The slope of the plot in Figure 11.15 can vary with the potential. This means the apparent mass is a function of the potential.[32,35]

Another example for the complex behavior of anion, cation, and solvent transport is the polypyrrole. Usually, anion, cation, and solvent transport can be observed, if this polymer is cycled in aqueous 0.1 mol·dm^{-3} KCl electrolyte. How to separate the contributions of anion and solvent flux was described above. Additional techniques to separate the different contributions of ion and solvent fluxes are described in the literature.[36,37]

There are ways to enforce either an anion or cation exchange by

- Only cation exchange, which can be enforced by incorporation of immobile anions in the polymer during the synthesis, e.g., polystyrene sulfonate. In this case a strong change of Δw accompanies the frequency change Δf and the simple Sauerbrey equation cannot be applied. Moreover, values of Δf and Δw do not return to the initial values. Such behavior can be explained by a "break-in" effect.[38]
- Only anion exchange, which should be observed if the polymer is cycled in a system with a large cation. Such a situation was found for the cycling of polypyrrolemolybdate in tetrabutylammonium bromide.[34]

Table 11.1

Anion and solvent exchange in poly-3,4-ethylenedioxythiophene during oxidation and reduction in aqueous electrolytes in the potential range from 0.2 to 0.6 V versus AgCl (sat. KCl). The coefficient α in Eq. (11.4) represents the number of solvent molecules exchanged together with one anion. Negative values stand for replacement. The polymer was synthesized in acetonitrile

Anion	V_{anion} (Å)	$\Delta m/\Delta Q$ (10^{-4} g·C^{-1})	M_{app} (g·mol^{-1})	α
Perchlorate	50	6.9	63	−2
Nitrate	20	4.7	45	−1
Toluenesulfonate	250	8.8	85	−5
Sulfate	18	6.9	130	+2
Citrate	490	4.7	169	−3

11.5 ELECTRICAL AND OPTICAL FILM PROPERTIES

In addition to oxidation and reduction there is a change in the electrical properties between neutral state and metal-like conducting behavior. For investigation of the electrical properties electrochemical impedance spectroscopy is used. The principles of this method were described in Chapters 4 and 5.

11.5.1 Impedance of conducting polymers

The film grown on an electrode surface has a duplex structure with a thin, compact first layer that is directly on the electrode surface and a porous second layer contacting the electrolyte. An equivalent circuit can be used to represent the electrical properties of this film. The components of an equivalent circuit can be determined by impedance spectroscopy. Therefore, this method has become one of the key methods for the characterization of conducting polymers.

The development of an adequate equivalent circuit has been controversially discussed in the literature. Gabrielli *et al.* considered the polymer primarily as a non-porous layer.[39] Transport processes in the polymer matrix dominated the impedance. Vorotyntsev *et al.* developed a model that took into account the electron transfer at the metal–polymer interface, transport of charge carriers in the film, and ion transfer at the polymer–electrolyte interface[40] (Figure 11.16).

There has also been some discussion about the electron and ion transport contributions in the film.[41] Popkirov *et al.* made experiments by growing a polymer film between two neighboring platinum electrodes. They were able to show that the electron conductivity dominated the charge transport and the conductivity in the polymer.[42] The contribution by ion diffusion was only 5%.

Another problem was that experimental results showed a dependence of the impedance data on the film thickness. For example, Komura *et al.*,[43] Albary *et al.*,[44] and Fletcher took

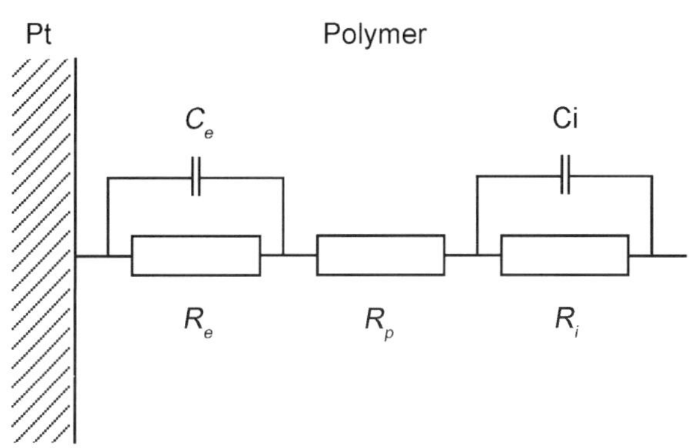

Figure 11.16 Impedance model of Vorotyntsev *et al.*[40] R_e and C_e represent the properties of the metal/polymer interface, R_i and C_i the properties of the metal/electrolyte interface, and R_p the resistance of the bulk polymer.

into account the obviously porous nature of the film by applying transmission line models.[45] In the transmission line model electronic and ionic conduction lines are connected by Faraday capacitances representing the redox properties.

In experiments covering a larger potential region, from the oxidized state until the complete neutral state, a new resonance circuit was found not described by the transmission line model. A new model was suggested by Pickup *et al.*,[46] which was used and modified later by Rammelt and Plieth *et al.*[47,48] This model is corroborated by the duplex film structure (Figure 11.9). A compact layer on the metal/polymer interface with neutral state properties in the neutral state and double-layer properties in the oxidized state describes the compact polymer film; the transmission line model represents the porous part (Figure 11.17).

The analysis of this model is simplified by the experimental observation that in the impedance spectrum of conducting polymers three frequency regions can be distinguished. Impedance spectra as a function of the electrode potential, represented as Bode plots, are shown in Figure 11.18. The compact film on the metal–polymer interface determines the high-frequency region. It can be neglected in the oxidized state. The transport processes in the porous film part dominate the medium frequency range together with the Faraday capacitance. Separation into ionic and electronic properties is usually not possible. Finally, structural changes of longer relaxation time determine the low-frequency region.

The high frequency part of the impedance spectra can be used for the calculation of space-layer capacitance C_{SP} and space-layer resistance R_{SP}. These data are shown in Figure 11.19.

In the diagram the cyclic voltammogram of the polymer film is also shown. At positive potentials the film is metal conducting with high capacitance values and low resistance values. Capacitance decreases and resistance increases as the polymer becomes reduced. With the reduction the anions are expelled. The film becomes neutral state.

It was mentioned that the properties of the polymer film depend on the type of anion. In Figure 11.20, another RC versus potential diagram is shown for a polypyrrole film prepared by anodic oxidation in a polystyrene-sulfonate electrolyte. The large ion cannot

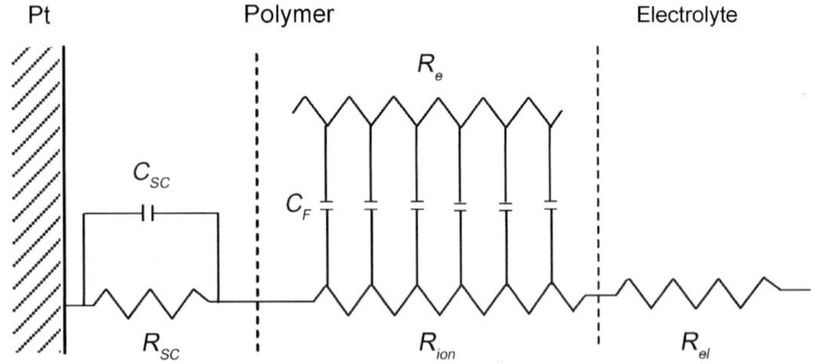

Figure 11.17 Equivalent circuit of the conducting polymer film on an inert electrode surface. The model assumes a compact film on the metal surface with semiconducting properties in the neutral state (R_{SC} and C_{SC}) and a porous film towards the electrolyte with electron resistance R_e and ion resistance R_{ion} and a Faraday capacitance C_F. R_{el} is the electrolyte resistance.

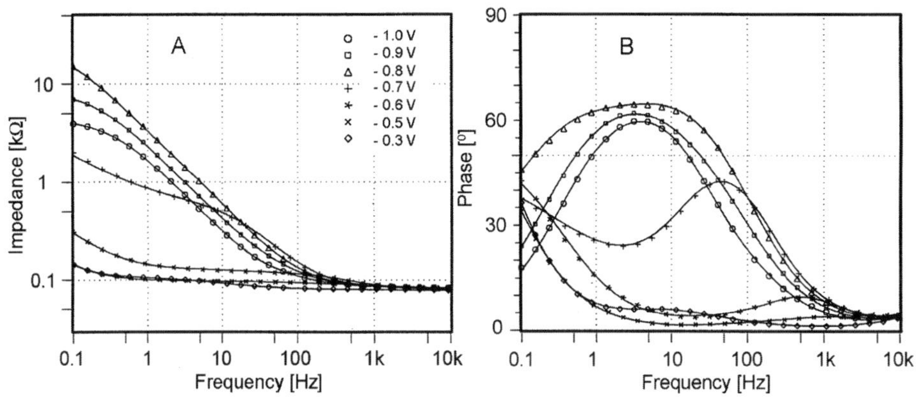

Figure 11.18 Impedance data measured for polypyrrole as function of the potential. The polymer film was prepared by anodic oxidation in a perchlorate electrolyte. Bode representation: (A) impedance versus frequency and (B) phase angle versus frequency.

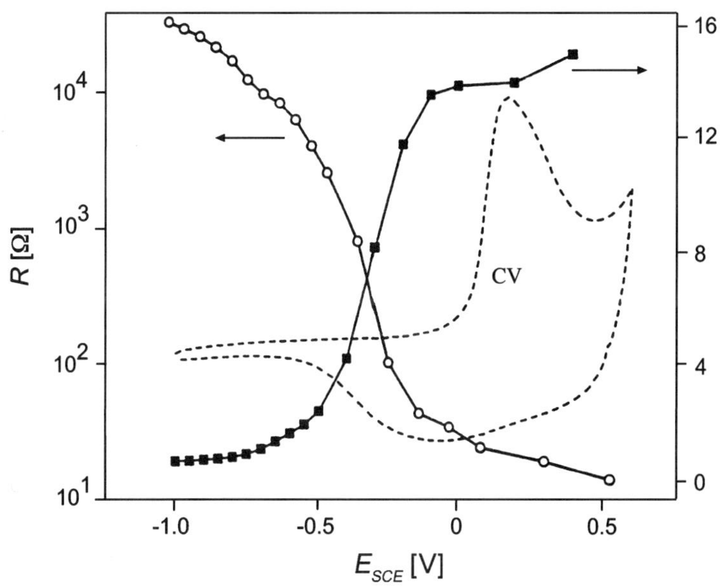

Figure 11.19 High frequency part of capacitance and resistance of a polypyrrole film as function of the potential. The film was prepared by anodic oxidation in a perchlorate electrolyte. Additionally, the cyclic voltammogram is shown. The film has metal-like properties at positive potentials ($E > 0$ V) and neutral state properties at negative potentials ($E < -0.5$ V).

be expelled. The polymer matrix is reduced and the compensation of the charge is only possible by built-in cations. The film does not show neutral state properties. In an intermediate potential region high-frequency capacitance increases and high-frequency resistance decreases with the potential but then capacitance increases and resistance decreases with further reduction.

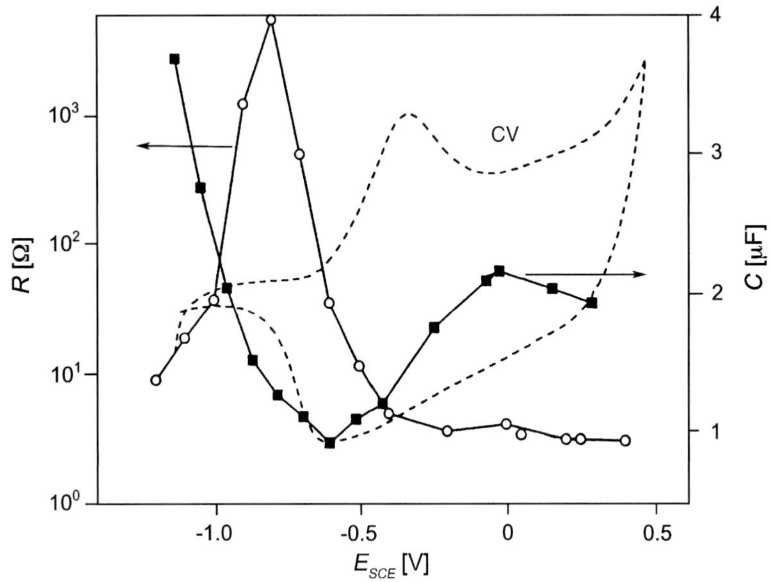

Figure 11.20 High frequency part of capacitance and resistance of a polypyrrole film as function of the potential. The film was prepared by anodic oxidation in a polystyrene-sulfonate electrolyte. Additionally, the cyclic voltammogram is shown, whereby the film becomes electron conducting at more positive potentials ($E > -0.5$ V) and ion conducting at negative potentials ($E < -0.8$ V).

11.5.2 Neutral state properties

Space-charge capacitance and space-charge resistance of polypyrrole as a function of the potential were shown in Figure 11.19. Similar results can be obtained for polythiophene. The potential dependence of the capacitance values in the cathodic region is typical for a p-type semiconductor. Then one should observe a linear dependence of C^{-2} on the potential (Mott–Schottky plot). Such a plot is shown for polythiophene in Figure 11.21.

The linear dependence confirms the semiconductor model for the neutral state of polythiophene.

A hysteresis is observed which shows the relaxation of the polymer matrix if the material is switched between the oxidized and neutral states. The flat band potential for the potential scan in negative direction is 0.46 V. The flat band potential for the scan in positive direction is 0.61 V. The average value of the acceptor density determined from the slope of the Mott–Schottky plot is 6.5×10^{18} cm^{-3}.

Data about the band structure, e.g., the band gap can also be obtained from current potential diagrams. The complete voltammogram including anodic oxidation and cathodic reduction of the neutral form must be known for such an analysis. A schematic representation is given in Figure 11.22a. An example is the cyclic voltammogram shown in Figure 11.22b.

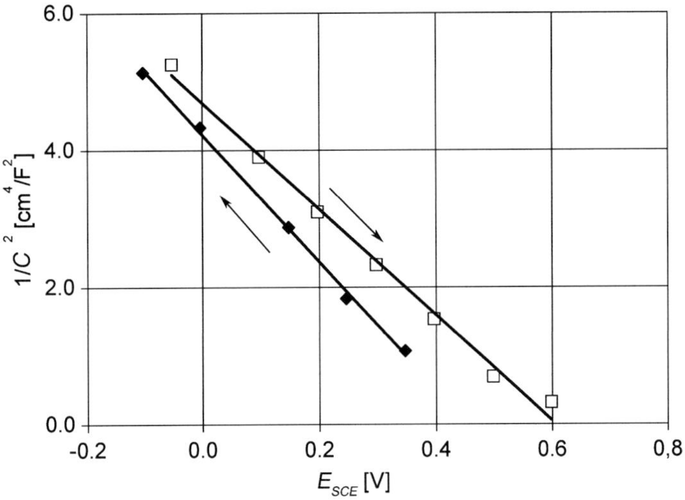

Figure 11.21 Mott–Schottky plot for polythiophene. Electrolyte 0.1 mol·dm^{-3} $LiBF_4/H_2O$, potentials versus saturated calomel electrode (SCE). The polythiophene films were prepared by electrochemical oxidation in acetonitrile.

The described use of electrochemical data for comparison with band properties determined by other methods is sometimes disturbed by the additional reactions possible in an electrochemical cell, like hydrogen and oxygen evolution.

Data of the band gap for some ICPs are summarized in Table 11.2.

11.5.3 Photoelectrochemical properties

Photoelectrochemical measurements (introduced in Chapter 9) corroborate and extend capacitance measurements. Photocurrent spectra of poly-bithiophene are shown in Figure 11.23. The spectra are shown for different solvents. The highest photocurrents (largest quantum-chemical efficiency) were observed in acetonitrile, which decreased with an increasing amount of water.

The dependence of the photocurrent on the potential (at 2.5 eV) is shown in Figure 11.24. The measurements confirm the flat band potential determined from capacitance measurements.

11.5.4 Polaron–bipolaron model of conducting polymers

The conductivity of the conducting polymers in the oxidized state is not as easy to explain. The main feature of the conducting polymers is the chain structure of alternating double bonds. The structure can change from a benzoic to quinoidic form. For the most important conducting polymers like polythiophene, polypyrrole, and polyaniline, the two structures

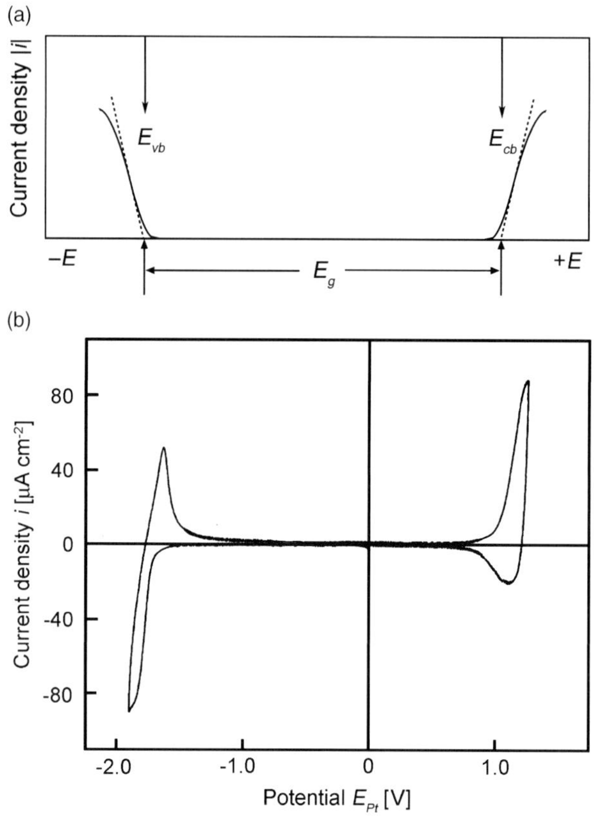

Figure 11.22 (a) Determination of data of neutral state properties of a polymer from a voltammogram. The absolute value of the current density is plotted. The band gap can be approximately determined from the extrapolation of the current increase to the zero line. Band edges of conduction band and valence band can be approximately determined, if the physical potential scale is used (Chapter 3). The standard hydrogen potential has an approximate value of 4.5 eV. (b) Cyclic voltammogram of a platinum electrode coated with poly(dibutoxyphenylenevinylene) in acetonitrile with 0.1 mol·dm^{-3} tetrabutylammoniumtetrafluoroborate. A platinum quasi reference electrode was used, the sweep rate was 100 mV s^{-1}. An electrochemical band gap of 2.78 eV is determined. (Reproduced with permission from Ref. [7], © 2004, American Chemical Society.)

Table 11.2

Band gaps for some ICPs: (i),[49] (ii),[50] (iii),[51] (iv),[52] (v),[53] and (vi)[54]

Polymer	Band gap E_g (eV)	Source
Polyaniline	2.3	Disilvestro and Haas (i)
Polypyrrole	2.2	Inoue and Yamase (ii)
Polythiophene	1.8–2.0	Paramunage *et al.* (iii)
Poly(3-methyl)-thiophene	1.9	Lukkari *et al.* (iv)
PEDOT	1.5	Girotto *et al.* (v)
Poly-paraphenylene	2.2	Hebestreit (vi)

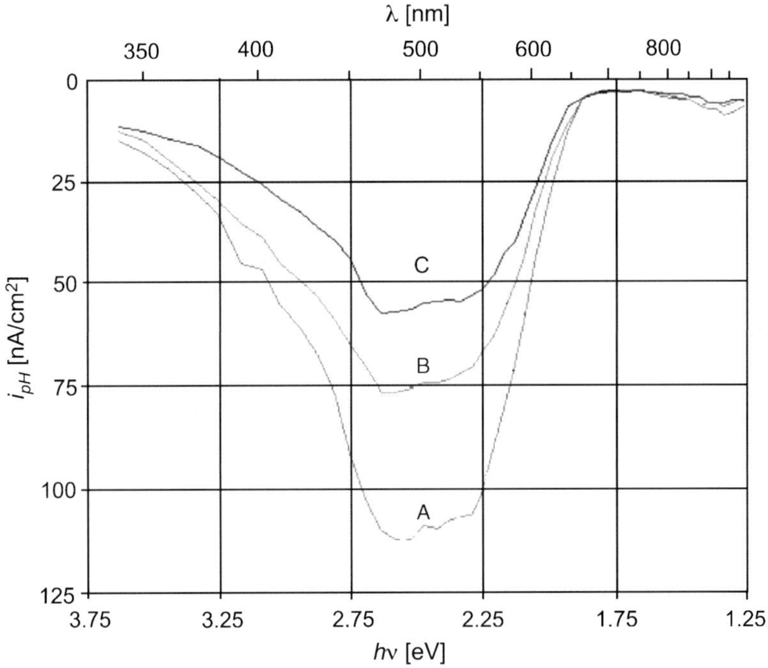

Figure 11.23 Photocurrent spectra of poly-bithiophene in the UV/vis region, $E = +0.2\ V_{SCE}$. (A) in acetonitrile, (B) H_2O/acetonotrile 60/40, and (C) H_2O, 0.1 mol·dm^{-3} $LiBF_4$.

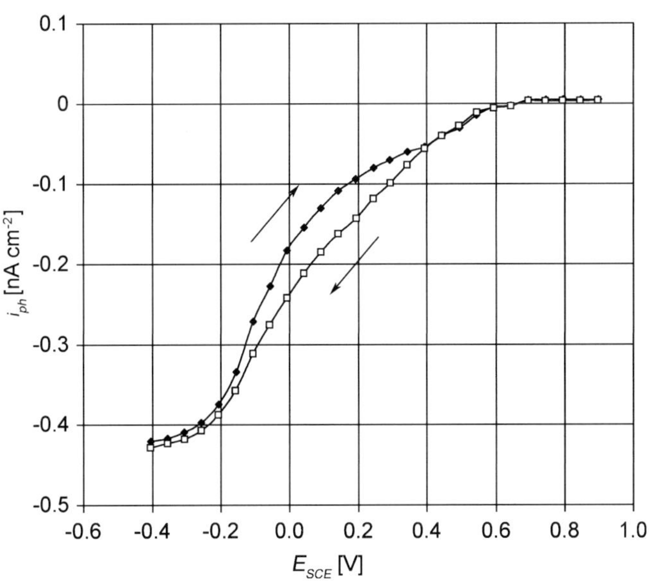

Figure 11.24 Dependence of the photocurrent of a poly-bithiophene film on the potential (at 2.5 eV), 0.1 mol·dm^{-3} $LiBF_4$, $\lambda = 500$ nm, and light intensity 120 mW·cm^{-2}.

have different energies (non-degenerate ground state, Heger[55]). The polaron–bipolaron model was suggested to explain electronic properties.[56] Polaron and bipolaron are radical cation and bi-cation states of the polymer (Figure 11.25).

It is assumed that with oxidation new bands form within the band gap, which could explain the metal-like conductivity. The energetic details of these bands are basically obtained from spectro-electrochemical measurements (Section 11.5.5). The polaron–bipolaron band model for polypyrrole is shown in Figure 11.26 as an example.

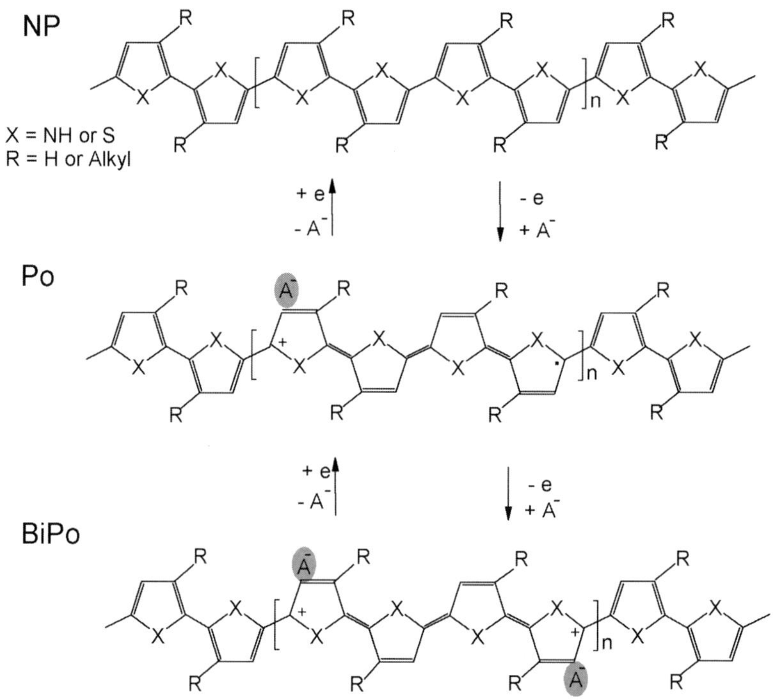

Figure 11.25 Neutral polymer (NP), polaron (Po, radical cation), and bipolaron (BiPo, bi-cation) of conducting polymers.

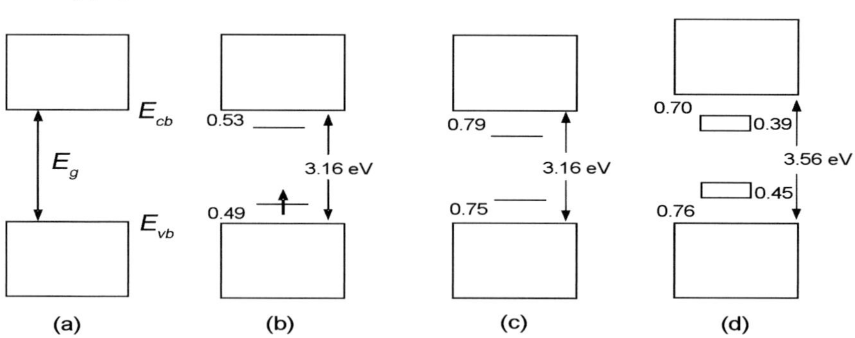

Figure 11.26 Polaron–bipolaron band model for polypyrrole. (a) Neutral state, (b) polaron, (c) bipolaron, and (d) development of the bipolaron band with increasing degree of oxidation.

Further details about the oxidized state were obtained from experiments with hexathienyl by Garnier *et al.*[57] Formation of dimers was observed in these experiments with oligomers. Heinze *et al.* found evidence for the formation of a σ-dimer.[58] Dunsch *et al.* were able to experimentally confirm the formation of π-dimers.[59]

11.5.5 Spectro-electrochemical methods

The development of *in situ* spectroscopic methods for electrochemical applications was a major step to extend the experimental arsenal of electrochemistry. A description and comparison of the different methods can be found in the literature.[60,61]

The methods can principally be divided into two classes to investigate the properties of (a) reflected light and (b) diffuse or direction-oriented scattered light.

The experimental problem of reflectance spectroscopy is the usually small effect of a thin surface film on the reflected light. The change of the intensity during reflection is measured in UV/vis reflectance absorption spectroscopy and IR reflectance absorption spectroscopy. In the case of IR reflectance absorption spectroscopy, another problem is the absorption of the electrolyte. This has led to the development of thin-layer cells.

If working with polarized light, one can measure the change of the polarization parameters of the reflected light. This method is called ellipsometry and was intensively used for studying the thickness of passive films. Examples were shown in Chapter 10.

The measurement of scattered light is done with Raman and fluorescence spectroscopy. The frequency shift of laser light is measured. Examples were shown in Chapter 7.

Especially useful for the investigation of ICPs and the polaron–bipolaron state was a combination of UV/vis spectroscopy and electron-spin-resonance (ESR) spectroscopy. The cell developed by Petr, Dunsch, and Neudeck[62] is an example for the experimental verification and is shown in Figure 11.27.

The central element of the cell is an LIGA structure that serves as a working electrode. The optical spectroscopy is made through two quartz rods. In the LIGA structure the light is reflected at a grazing angle of incidence. The LIGA structure is placed between two quartz tubes in the resonator of the ESR spectrometer. Counter and reference electrodes are placed outside the resonator of the ESR spectrometer. Examples of investigations on polypyrrole are the papers by Neudeck, Dunsch *et al.*[63,64]

11.6 COPOLYMERIZATION

So far, polymers of only one type of monomer molecules have been discussed. It is possible to carry out the preparation procedure with mixtures of different monomers in which case formation of copolymers is possible. Copolymerization opens ways for new materials.

When two monomers can react in the polymerization process this can either be in a regularly alternating way, whereby A is followed by B and B is followed by A. The product is an alternating copolymer (Figure 11.28a). Otherwise, the monomers can react in a statistical manner so that the product is a random copolymer (Figure 11.28b). In a third

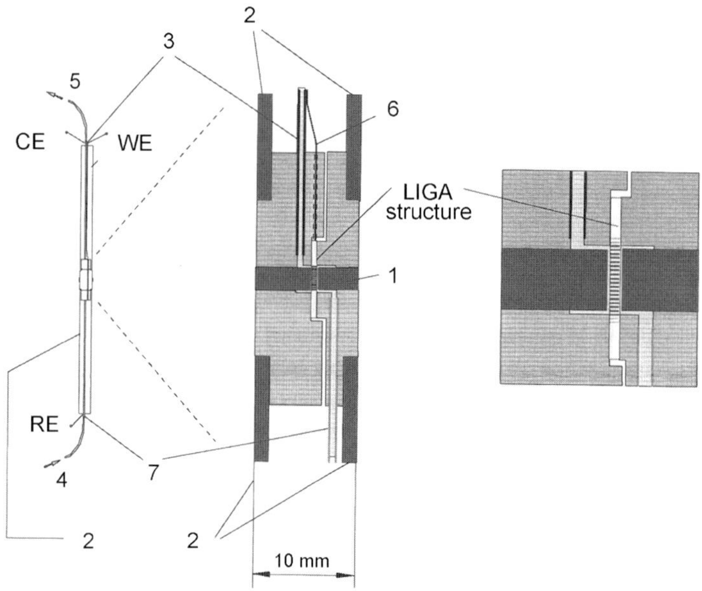

Figure 11.27 Spectro-electrochemical cell for simultaneous measurements of ESR and UV/vis spectra. (1) Quartz rod, (2) quartz pipes, (3) platinum tube, (4) electrolyte in, (5) electrolyte out, (6) Ag wire; contact to LIGA structure, and (7) Ag tube; WE, RE, and CE are working, reference, and counter electrodes (Petr, Dunsch, Neudeck).[62]

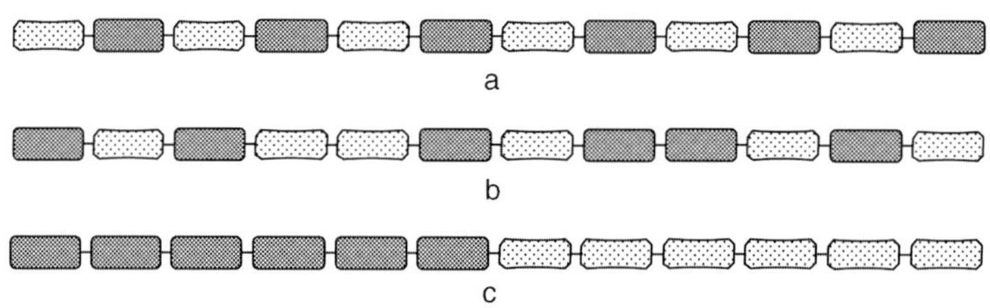

Figure 11.28 Alternating (a), random (b), and block copolymers (c).

case where reaction between A and B is the exception, the product could be a block copolymer of polymer A and polymer B. This may be compared with a eutectic mixture (Figure 11.28c).

The formation of copolymers from a mixture of monomers depends on several requirements:

- The oxidation potential of the monomer should be similar. In Table 11.3 monomers and their oxidation potentials are shown.
- The reactivities of the radical cations formed in the primary oxidation process determine the type of copolymer. If the reactivity of A with B is much larger than for between A and A or B and B, an alternating copolymer is formed. If, on the other hand,

Table 11.3

Oxidation potentials of different monomers and oligomers: (i),[65] (ii),[66] (iii),[67] and (iv)[68]

Monomer	Oxidation potential (E_{ox}/V)	Reference electrode
Aniline	1.4	SCE (i)
Pyrrole	1.2	SCE (i)
3,4-Ethylendioxythiophene	1.39	Ferrocene: +0.352 $V_{Ag/AgCl}$/ CH_2Cl_2 (ii)
2-(3-Thienyl)-ethylacetate	1.4	$Ag/AgClO_4/CH_3CN$ (iii)
3-Methylthiophene	1.86	SCE (i)
Thiophene	2.07	SCE (i)
3-Bromothiophene	2.1	SCE (i)
3-Cyanothiophene	2.46	SCE (i)
3-Nitrothiophene	2.69	$Ag/AgClO_4/CH_3CN$ (iv)
Bithiophene	1.31	SCE (i)
Terthienyl	1.05	SCE (i)
Pentathienyl	0.43	SCE (i)

the reactivity between A and A and between B and B is larger than between B and B, a block copolymer will be formed. Similar reactivity for all combinations of A and B gives a random copolymer.

– Reactivity is a property of the monomer but it is also influenced by the experimental conditions. Therefore, the further conditions should be considered such as the solvent and the composition of the electrolyte, influence of additives, and temperature.

The experimental syntheses of copolymers and monopolymers follow the same routes. Both chemical and electrochemical oxidations are possible. Electrochemical oxidation has the advantage of a continuously variable oxidation potential. The electrochemical oxidation can be made either potentiostatically at a specified oxidation potential or galvanostatically. Then the current density determines the potential.

Step methods were applied to compensate differences in monomer oxidation potential as well as the reactivity's potential.[69]

11.6.1 Mechanism of copolymerization

In Section 11.2, the mechanism of polymerization by coupling radical cations was described. This mechanism can be applied to oxidative copolymerization.[70–72] The radical cations can react with radical cations or with neutral molecules,[73,74] but there is experimental evidence that reaction is only possible between radical cations.[75]

The mechanism consists of three steps. In the first step the monomers are oxidized

$$A \rightarrow A^{+*} + e^- \tag{11.5}$$

$$B \rightarrow B^{+*} + e^- \tag{11.6}$$

The second step is coupling of the radical cations with the growing radical cation polymer chains

$$R - A^{+*} + A^{+*} \xrightarrow{k_{AA}} R - A^{+}A^{+*} \quad (11.7a)$$

$$R - A^{+*} + B^{+*} \xrightarrow{k_{AB}} R - A^{+}B^{+*} \quad (11.7b)$$

$$R - B^{+*} + A^{+*} \xrightarrow{k_{BA}} R - B^{+}A^{+*} \quad (11.7c)$$

$$R - B^{+*} + B^{+*} \xrightarrow{k_{BB}} R - B^{+}B^{+*} \quad (11.7d)$$

The third step is the proton release

$$R - A^{+}A^{+*} \rightarrow R - AA + 2H^{+} \quad (11.8a)$$

$$R - A^{+}B^{+*} \rightarrow R - AB + 2H^{+} \quad (11.8b)$$

$$R - B^{+}A^{+*} \rightarrow R - BA + 2H^{+} \quad (11.8c)$$

$$R - B^{+}B^{+*} \rightarrow R - BB + 2H^{+} \quad (11.8d)$$

The third step is usually fast. For the continuation of polymerization an oxidation step must follow the proton release. One can combine second step, proton release, and subsequent oxidation. The resulting equation simultaneously describes both the chain growth and new chain oxidation

$$R - A^{+*} + A^{+*} \xrightarrow{k_{AA}} R - AA^{+*} + 2H^{+} + e^{-} \quad (11.9a)$$

$$R - A^{+*} + B^{+*} \xrightarrow{k_{AB}} R - AB^{+*} + 2H^{+} + e^{-} \quad (11.9b)$$

$$R - B^{+*} + A^{+*} \xrightarrow{k_{BA}} R - BA^{+*} + 2H^{+} + e^{-} \quad (11.9c)$$

$$R - B^{+*} + B^{+*} \xrightarrow{k_{BB}} R - BB^{+*} + 2H^{+} + e^{-} \quad (11.9d)$$

The end of the chain growths is achieved if another inert molecule terminates the process.

The Markov chain theory can be applied on this reaction scheme (Chapter 7). The reactivity of the radical cations A^{+*} and B^{+*} with a polymer chain are described by rate equations

$$r_{AA} = k_{AA}\left[R - A^{+*}\right]\left[A^{+*}\right] \quad (11.10a)$$

$$r_{AB} = k_{AB}\left[R - A^{+*}\right]\left[B^{+*}\right] \quad (11.10b)$$

$$r_{BA} = k_{BA}\left[R - B^{+*}\right]\left[A^{+*}\right] \tag{11.10c}$$

$$r_{BB} = k_{BB}\left[R - B^{+*}\right]\left[B^{+*}\right] \tag{11.10d}$$

The concentration of radical cations is usually proportional to the monomer concentration in the electrolyte.

$$\left[A^{+*}\right] \propto c_A \tag{11.11}$$

$$\left[B^{+*}\right] \propto c_B \tag{11.12}$$

Probabilities for the reaction of the chain with A or B can then be formulated (Chapter 7 and literature given there).

$$p_{AA} = \frac{r_{AA}}{r_{AA} + r_{AB}} \tag{11.13a}$$

$$p_{AB} = \frac{r_{AB}}{r_{AA} + r_{AB}} = 1 - p_{AA} \tag{11.13b}$$

$$p_{BA} = \frac{r_{BA}}{r_{BA} + r_{BB}} \tag{11.13c}$$

$$p_{BB} = \frac{r_{BB}}{r_{BA} + r_{BB}} = 1 - p_{BA} \tag{11.13d}$$

An equation for the composition of the polymer (defined as the ratio of the number of monomers in the chain, N_A/N_B) can thus be derived

$$\frac{N_A}{N_B} = \frac{p_{BA}}{p_{AB}} = \frac{c_A}{c_B}\frac{(k_{AA}/k_{AB})c_A + c_B}{c_A + (k_{BB}/k_{BA})c_B} \tag{11.14}$$

One can substitute the ratio of monomer units in the polymer chain by v and the ratio of the monomer concentration in the electrolyte by w.

$$N_A/N_B = v \qquad c_A/c_B = w \tag{11.15}$$

The ratio of the rate constants in Eq. (11.14) can be substituted by r_1 and r_2.

$$\frac{k_{AA}}{k_{AB}} = r_1 \qquad \frac{k_{BB}}{k_{BA}} = r_2 \tag{11.16}$$

By introducing these substitutions into Eq. (11.14) and rearranging the equation, two linear equations can be derived

$$w\left(1-\frac{1}{v}\right)=r_1\frac{w^2}{v}-r_2 \quad (11.17a)$$

$$\frac{v-1}{w}=-r_2\frac{v}{w^2}+r_1 \quad (11.17b)$$

Eq. (11.14) is known as the copolymerization equation and was first derived by Mayo and Lewis on a kinetic basis.[76] A derivation on a probability basis was later published.[77,78] Feinman and Ross derived the linear equations.[79]

The equation was applied to 2,2′-bithiophene-pyrrole copolymerization.[70] The result plotted in the linear form, Eq. (11.17), is shown in Figure 11.29. These authors confirmed the copolymerization equation. Recently, Dung[80] *et al.* investigated the same system with similar results.

With some classical polymerization reactions, e.g., styrene with methyl metacrylate, deviations were observed, e.g., by Fukuda *et al.*,[81] and O'Driscoll and Reilly,[82] whereby a penultimate unit effect was discussed and corrections introduced. This is expected on the basis of the Markov chain model because in this model only the last added unit controls the next step. As the correction method of O'Driscoll also took errors in the variables into account, it is called the error in variables method (EVM).[83]

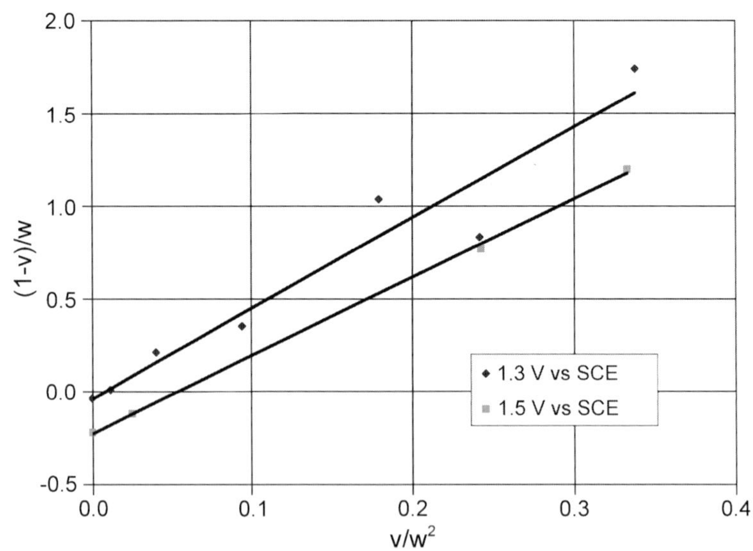

Figure 11.29 22′-bithiophene-pyrrole copolymerization for two polymerization potentials, analyzed according to Eq. (11.17), $v = N_{Py}/N_{BTh}$, $w = c_{Py}/c_{BTh}$, $r_1 = k_{PyPy}/k_{PyBTh} = 4.6$, and $r_1 = k_{BThBTh}/k_{BThPy} = 0.1$.[70]

11.6.2 Structure analysis of copolymers

The determination of the type of copolymers formed in the preparation process is an important characterization step. One method of characterization is cyclic voltammetry as shown, e.g., by Yohannes *et al.*[84] Cyclic voltammograms of copolymerization of pyrrole and bithiophene is shown in Figure 11.30.

The homogeneous redox peaks of the copolymers are a first hint that copolymers were formed.

Another method to study the polymer structure of a copolymerization process is mass spectroscopy.[86] Ionization of polymer fragments, which can be obtained, e.g., by an extraction process from the polymer, is possible through several methods. One method is matrix-assisted laser desorption ionization coupled with time-of-flight mass separation, commonly called MALDI-TOF. Another method is electrospray ionization mass spectroscopy. The solution of fragments is sprayed in a high electric field. The droplets are charged and after evaporation the remaining charged fragments can also be separated in the mass spectrometer. The final list of fragments A_xB_y obtained is interpreted as an image of the composition of the polymer chains.

The most advanced method to determine the polymer structure is solid-state NMR spectroscopy. For the details of this method one must refer to the special literature.[87,88] As an example, the application of double quantum magical angel spinning (DQ MAS) NMR spectroscopy to determine the structure of copolymers of 3-methylthiophene and ethyl-3-thiophene acetate (ETA) will be described.[89]

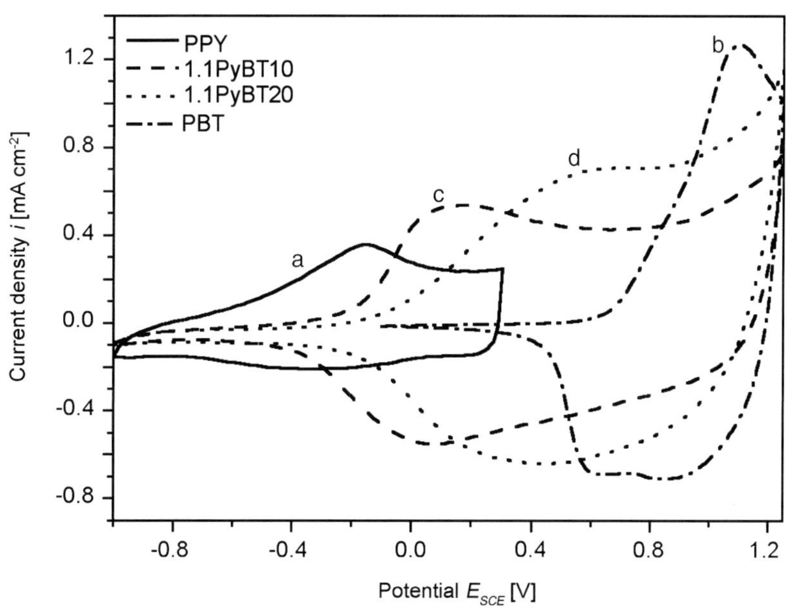

Figure 11.30 Cyclic voltammograms of (a) polypyrrole, (b) poly-bithiophene, (c) the copolymer formed at 1.1 V versus SCE in acetonitrile, monomer ratio Py:BTH = 1:10, and (d) the copolymer formed at 1.1 V versus SCE in acetonitrile, monomer ratio Py:BTH = 1:20.[85]

Figure 11.31 shows the one-dimensional (1D) ^{1}H NMR spectra of the homopolymers poly-(3-methylthiophene) (PMT) and poly-(ethyl-3-thiophene acetate) (PETA). Figure 11.32 shows the 1D ^{1}H NMR spectra of the corresponding copolymer MtETa15, the number indicating the ratio PMT:PETA = 1:15 during preparation of the copolymer. For the assignment of the peaks to certain type of protons in the polymer chain, the letters shown in Figure 11.33 label the protons. The peak assignments were made by comparison of the intensities following the principle methods of NMR spectroscopy. Peak assignments and relative intensities are summarized in Table 11.4. In the assignment of the peaks to the protons in the spectrum of the copolymer MtETa15, the possible peak from b-type protons is not observed.

The data in Table 11.4 show that the peak positions of the homopolymers are shifted in the spectrum of the copolymer. There is also the overlap of the a- and c-type protons in the spectrum of the copolymer. This indicates that the copolymer is a random copolymer rather than a block copolymer. The other argument for a random copolymer is the missing peak from b-type protons in the spectrum of the copolymer.

But the real information on the structure is provided by two-dimensional double quantum ^{1}H NMR (2D DQ ^{1}H NMR) experiments. This method gives information about double quantum coherence through dipole coupling, which is a strong indication of spatial proximity. In general, there is no distinction between intra- and inter-chain interactions.

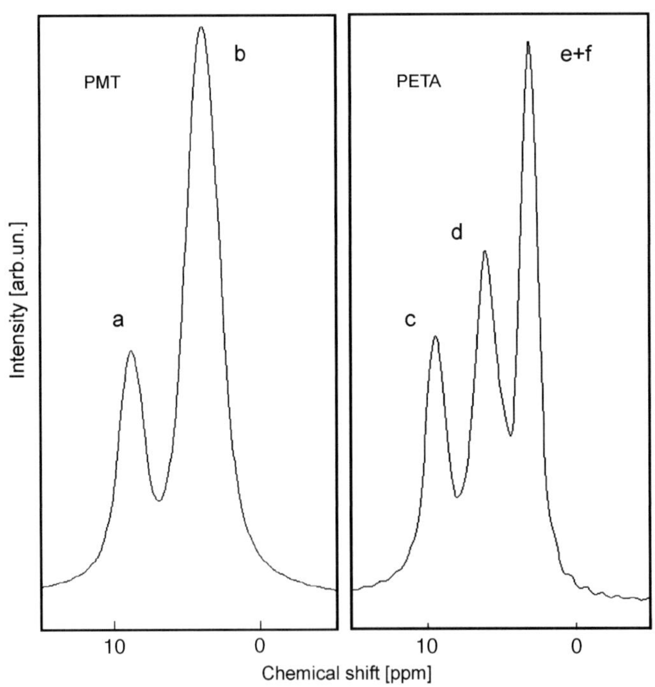

Figure 11.31 1D ^{1}H NMR spectra of poly-(3-methylthiophene) (PMT) and poly-(ethyl-3-thiophene acetate) (PETA).

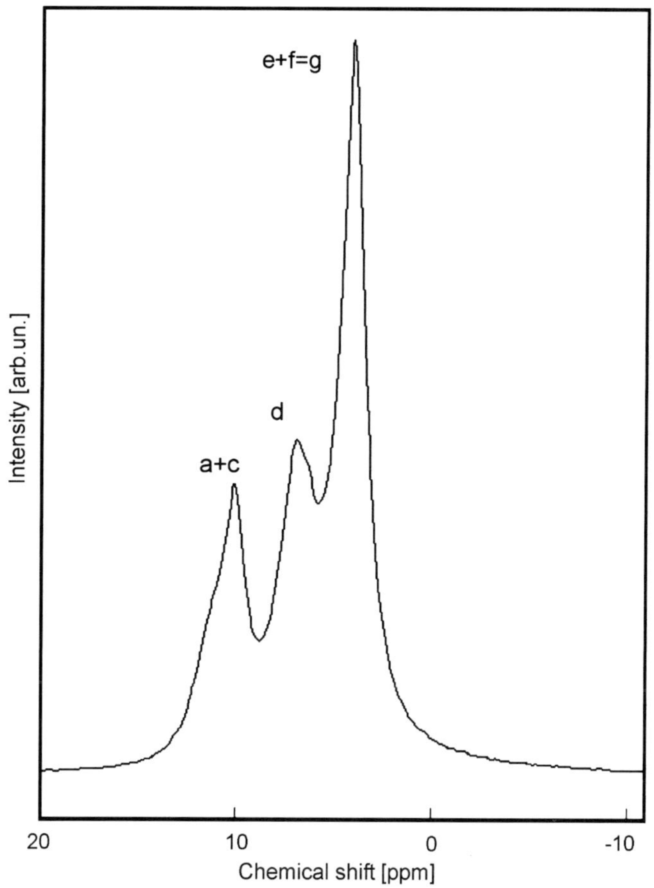

Figure 11.32 1D ^{1}H NMR spectra of the corresponding copolymer MtETa15.

Figure 11.34 shows the 2D DQ ^{1}H NMR spectrum and Figure 11.35 the 1D off-diagonal spectrum along the bar in Figure 11.34. The spectrum at the off-diagonal bar shows the strong interaction between a- and b-type protons. This shows that a- and b-type protons are in close proximity in the polymer.

Figure 11.36 shows the 2D DQ ^{1}H NMR spectrum of PETA and Figure 11.37 shows the 1D off-diagonal spectra along the lines (1), (2), and (3) in Figure 11.36. There is a moderate interaction between c- and d-type protons in Figure 11.37 (1) and a strong interaction between d- and (e+f)-type protons in Figure 11.37 (3). There is a weak interaction between c- and (e+f)-type protons in Figure 11.37 (2). The center of the off-diagonal spectrum in Figure 11.37 (2) has a high density. The reason is a strong overlap of the d- and (e+f)-type protons.

Figure 11.38 shows the 2D DQ ^{1}H NMR spectra of the copolymer MtETa15 and Figure 11.39 the 1D off-diagonal spectra along the lines (1), (2), and (3) in Figure 11.38.

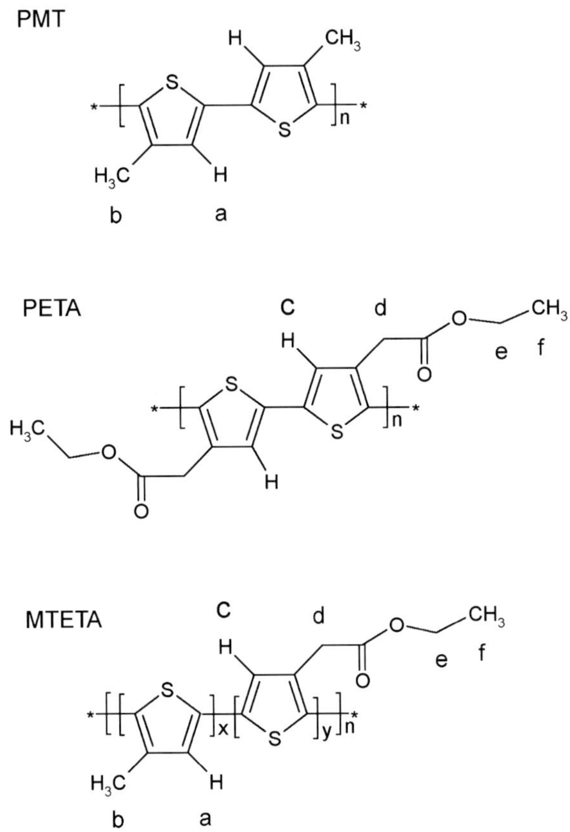

Figure 11.33 Repeating units of the pure polymers PMT, PETA, and of the corresponding copolymer MtETa15; labeling of the different protons.

Table 11.4

Peak assignment and relative peak intensities of poly-(3-methylthiophene) (PMT), poly-(ethyl-3-thiophene acetate) (PETA), and of the corresponding copolymer MtETa15

Sample	Peak	Chemical shift (ppm)	Peak intensity (%)
PMT	a	8.8	25
	b	3.9	75
PETA	c	9.5	25
	d	6.1	35
	e+f = g	3.2	40
MTETA	a+c	10.1	17
	d	7.0	31
	e+f = g	4.1	52
	b	not observed	not observed

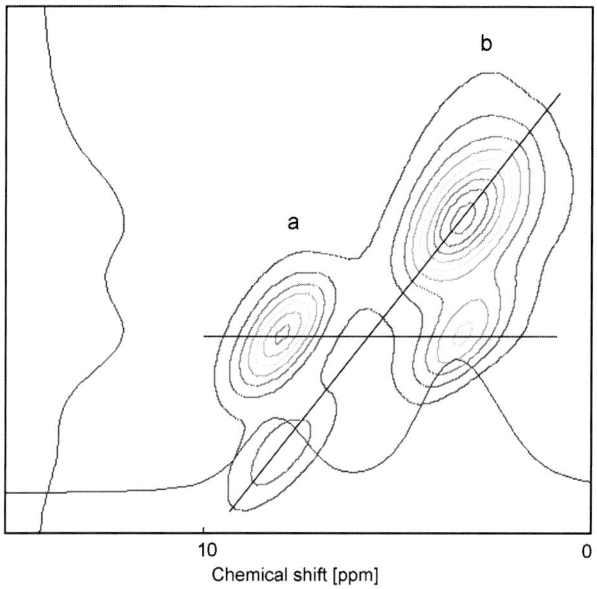

Figure 11.34 2D DQ ^1H NMR spectrum of poly-(3-methylthiophene). There is a strong interaction between a- and b-type protons which indicates the fact that a- and b-type protons are close to each other in the sense of space.

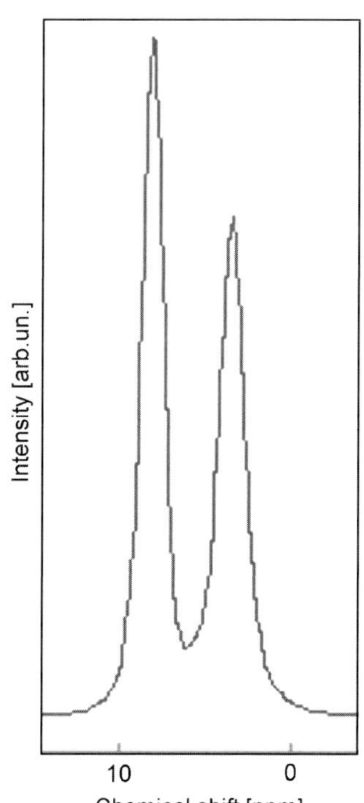

Figure 11.35 1D spectrum at the off-diagonal bar of Figure 11.34.

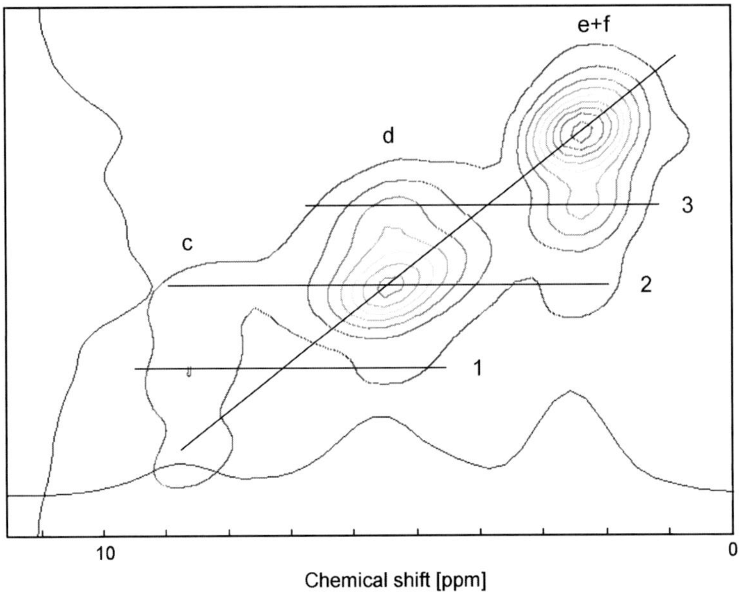

Figure 11.36 2D DQ ^{1}H NMR spectrum of poly-(ethyl-3-thiophene acetate) (PETA).

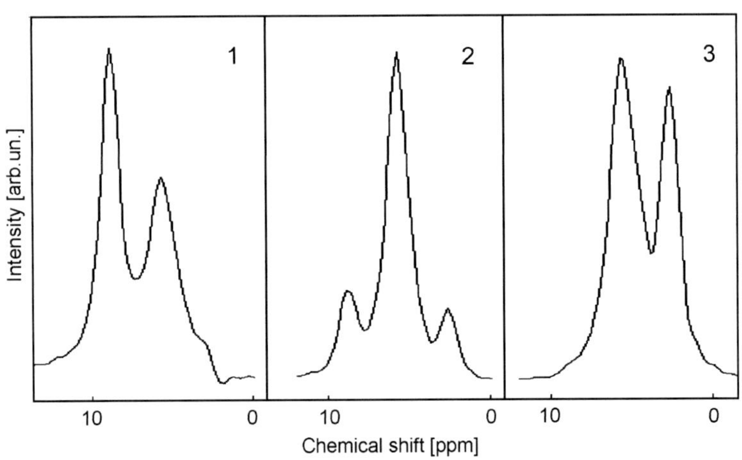

Figure 11.37 1D off-diagonal spectra along the bars (1), (2), and (3) of Figure 11.36.

There is a moderate interaction between (a+c)- and d-type protons in Figure 11.39 (1) and a strong interaction between d- and (e+f)-type protons in Figure 11.39 (3). There is a weak interaction between (a+c)- and (e+f)-type protons, and similarly to pure PETA, there is an overlap of the strong interaction between d- and (e+f)-type protons seen in Figure 11.39 (2).

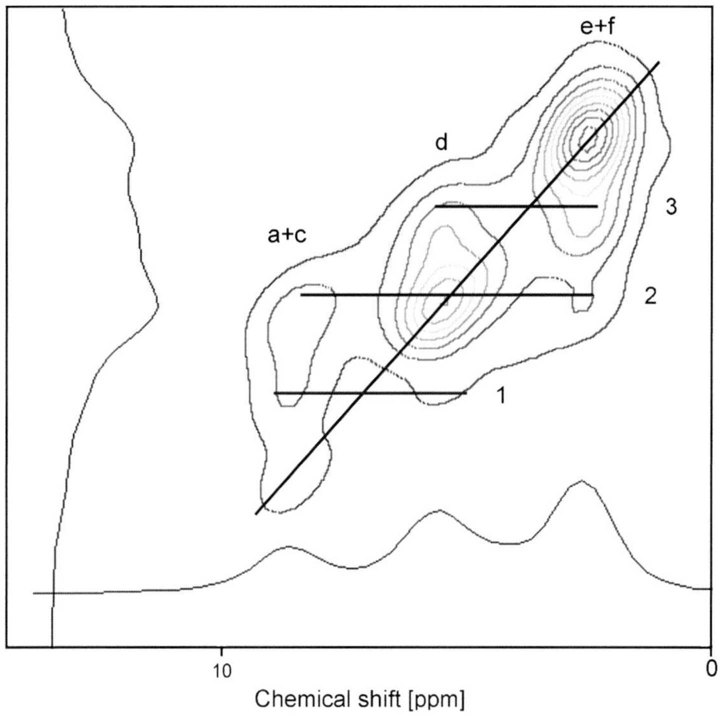

Figure 11.38 2D DQ ^{1}H NMR spectra of the corresponding copolymer MtETa15.

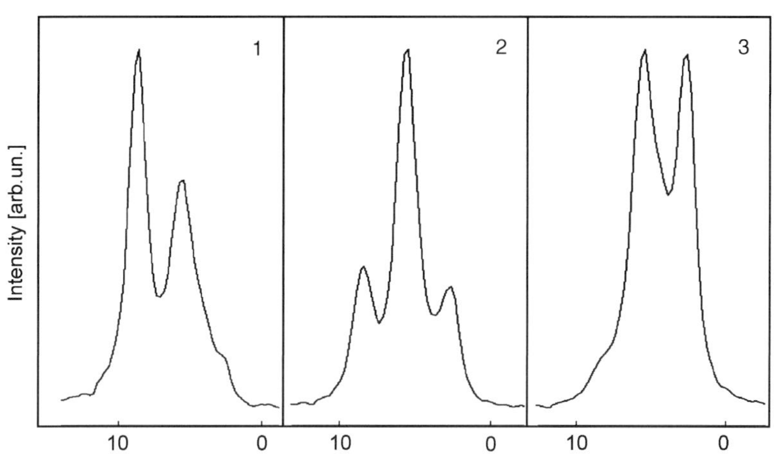

Figure 11.39 1D off-diagonal spectra along the bars (1), (2), and (3) of Figure 11.38.

One should note that the interaction between (a+c)- and (e+f)-type protons can be expected to be stronger if the copolymers have an alternating structure. Especially, a- and (e+f)-type protons should be closer in space to enhance the relative intensities of the peaks in the spectra shown in Figure 11.38. Therefore, the copolymers are random copolymers, which is consistent with the shift of the peak positions in the 1D spectra from homopolymers to copolymers.

When comparing the 2D ^{1}H NMR spectrum of pure PETA with those of the copolymer MtETa15, it should be emphasized that the increase in the intensity of anisotrospic interactions from the pure PETA (see Figure 11.36) to the copolymer MtETa15 (Figure 11.38) is explained by the increase of the 3-methylthiophene content in the functionalized copolymer chains, resulting from the more compact structure with a closer proximity between protons either in space or in bonding network.

The spectra are not ideal because the structure of these polymers is very irregular. Nevertheless, the spectra provide extremely good evidence for a structure of pairs of methyl-3-thiophene (MT) and ethyl-3-thiophene acetate (ETA). Moreover, the results show that the acetate group is still active, which is a prerequisite for using these polymers as a basis for complex modifications, e.g., with enzymes and is in agreement with the results of Welzel *et al.*[90]

11.6.3 Properties of copolymers

Copolymers are synthesized in order to hopefully combine the properties of the components. In the case of a statistical mixture of the monomers one can expect the properties to vary in a continuous manner by changing the ratio between the two components.

This kind of behavior was already seen in the cyclic voltammograms in Figure 11.30. The cyclic voltammograms showed that the oxidation potential could be continuously changed. A similar continuous shift is expected and could be found for the band structure, if the compounds are characterized using the methods described in Section 11.5.

Another reason to prepare copolymers is to provide special chemical functions for a conducting polymer. An example is the copolymer of MT and ETA. Combining ETA with MT provides conducting polymers with the full functionality of the acetate group. This group can now be used for further modification of the polymer layer. The problems of polymerization of 3-substituted thiophene have been known for some time, e.g., Bäuerle *et al.*[91] The substitution group can sterically hinder the polymerization process. Mixing the substituted monomers with simple monomers can be one way to incorporate the problematic monomers into copolymers. Experiments to form copolymers of pyrrole and metal salenes are a similar example.[92]

11.7 CORROSION PROTECTION BY INTRINSICALLY CONDUCTING POLYMERS

For a couple of years now there has been an ongoing discussion about how effective ICPs are in corrosion protection. The first argument to consider electrochemically prepared thin films of ICPs as corrosion protective coatings came from the positive value of the redox potential of these materials. The surface can be expected to be noble if the substrate surface is covered by a film of ICP.[93–95] While this is known to be true for protective metal films, it is not possible for the porous films of polyaniline, polypyrrole, etc., to protect an iron or steel surface. Through the pores the less noble substrate would corrode even faster because a local element would form with the pores acting as anodes.

A second attempt to explain the corrosion protection of polyaniline was the formulation of a mechanism with a catalytic reduction of oxygen and a simultaneous formation of a passivating metal oxide film on the substrate.[96–99] Later it was shown that this mechanism is not supported by experimental investigations.[100]

Obviously, there are still many open questions about the corrosion protective properties of ICPs.

11.7.1 Film formation on non-noble metals

The electrochemical film formation by oxypolymerization was discussed for inert electrodes like Pt or Au. A stable and adherent film can be formed on inert metals. But on metals like iron undergoing anodic dissolution during the anodic oxidation process the preparation is more complicated.

In case of iron or steel, anodic or chemical oxidation is accompanied by dissolution of the metal. The oxidation potential of most monomers is in the region of metal dissolution. Therefore, strong corrosion inhibits film formation. Formation of passivating oxide films, which might be possible at a higher potential, will also be delayed or inhibited.

In the literature several ways to avoid or minimize the corrosion have been described.

Beck *et al.* used oxalic acid to form a layer of iron oxalate to prevent the dissolution and support the formation of the polymer layer.[101,102] Ferreira[103] etched the steel surface with a dilute nitric acid to form an iron nitride layer, which inhibited the iron dissolution.

For iron and steel parts, a different film preparation was developed[104] that was based on adhesion promoters as described in Section 11.3. After formation of the SAM layer in a solution containing the adhesion promoter, the electrode was transferred into a cell filled with a solution of the monomer molecules and a supporting electrolyte. Then an anodic current pulse was applied for some seconds or minutes. A film of polythiophene formed on the iron or steel parts with a thickness depending on the pulse time. Oxidation of the iron or steel was avoided by this procedure. The film had a compact structure, as was shown by SEM pictures. The adhesion promoter film provided excellent protection against delaminating.

For some applications the formation of compact films can be substituted by dispersing polymer pigments in a polymer coating.[105] Instead of pigments for purely conducting polymers, modified pigments have been developed that are based on a core-shell concept.[106]

11.7.2 Kinetic experiments of corrosion protection

A mechanism describing corrosion protection must specify the partial reaction of the corrosion process, which will be inhibited. This can either be the anodic metal dissolution or the cathodic process, which in most cases is oxygen reduction. A long-discussed mechanism dating back to Wessling[107] is based on the assumed passivation of the metal surface by the positive redox potential of the conducting polymer, which in this case is polyaniline.

It was first thought that the conducting polymer oxidized the metal and initiated the passivation process. Later it was assumed that oxygen reoxidized the reduced ICP.

To understand the different reactions taking place in a film of ICPs formed on an iron or steel substrate, an experiment was described where the reaction on the iron surface and the reaction in the polymer film could be studied separately. The principal electrode configuration is shown in Figure 11.40.[100] The conducting polymer film was formed on a platinum electrode. An insulating acryl resin film covered the main part of a separate iron

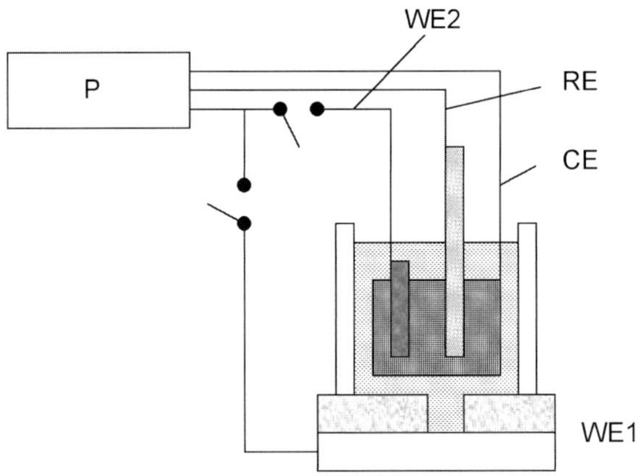

Figure 11.40 Electrochemical cell for a separate investigation of the reactions on an iron electrode and on a conducting polymer film electrode. WE1, working iron/steel electrode; WE2, polythiophene covered platinum electrode; CE, counter electrode; RE, reference electrode; and P, potentiostat.

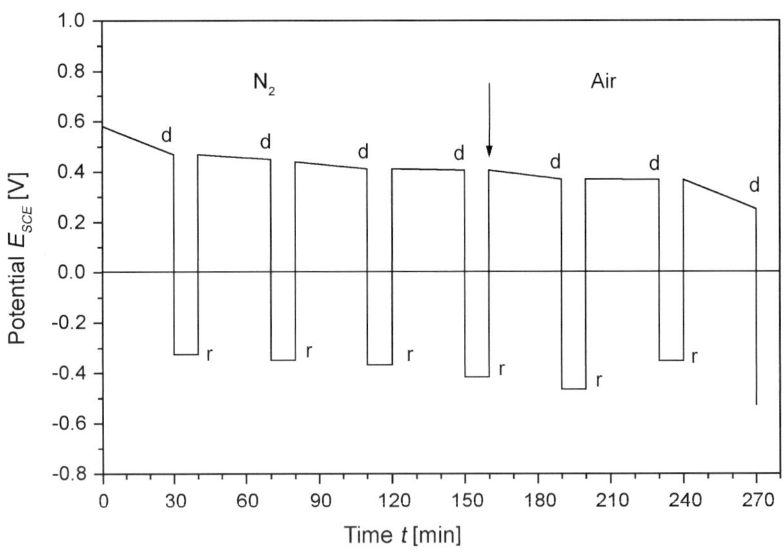

Figure 11.41 Open-circuit potential of a mild steel electrode connected to a polythiophene electrode: (d) disconnected and (r) reconnected.

electrode; the iron electrode was only exposed to the electrolyte in a small scratch. The behavior of both electrodes was investigated under all suitable combinations. Additionally, a switch could connect the platinum and iron electrodes.

The experiment provided the following information:

- The conducting polymer polarizes the iron electrode towards its own redox potential. The potential of the iron electrode under connected and disconnected electrochemical conditions is shown in Figure 11.41. Otherwise, if connected, the iron electrode remains in an active state.
- In nitrogen atmosphere the conducting polymer cannot oxidize the iron. The polymer is reduced and the iron in the scratch oxidized only in oxygen atmosphere.
- There are no results supporting reoxidation of the conducting polymer by oxygen. After a while the ICP is reduced and stays in the neutral states.

Similar results were obtained with polypyrrole.[108] No protection of the iron or mild steel by just the ICP was observed.

11.7.3 Role of anions for a possible corrosion protection of conducting polymers

A remarkable property of ICPs on iron or mild steel is the shift of the electrode potential to more positive values. A mixed potential is established by a coupling (in oxygen atmosphere) to a permanent reduction of the conducting polymer. But if conducting polymers

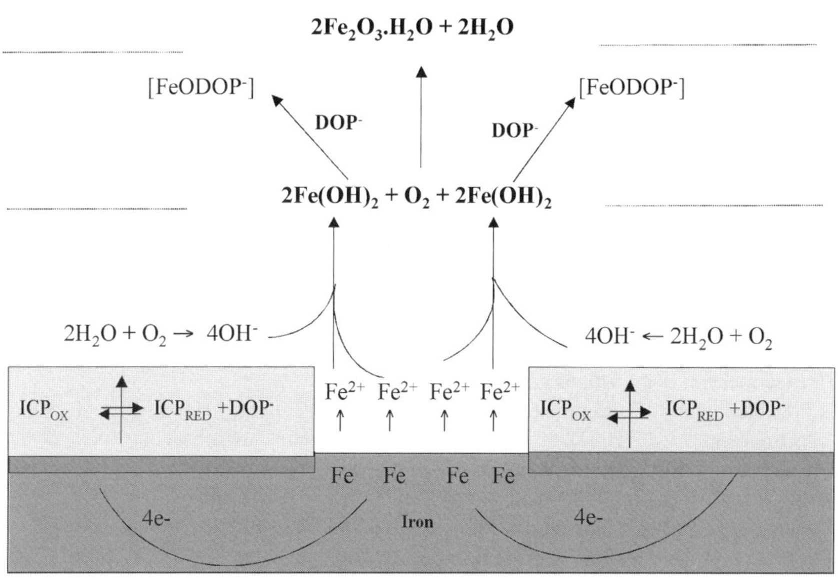

Figure 11.42 Mechanism of corrosion protection by conducting polymers and anion (DOP^-) release. The mechanism was suggested by Kinlen *et al.*[109,110]

are to protect metals against corrosion, there must be an additional effect, which has been recently identified as anion release. Therefore, conducting polymers can only provide corrosion protection if the polymer film contains anions that support passivation. The principal mechanism is shown in Figure 11.42.

This mechanism is supported by a variety of experimental results. The molybdate ion was intensively investigated and found to be incorporable into the polypyrrole films.[111]

Corrosion protection of conducting polymers is also involved with an inhibition of the oxygen reduction.[112] Oxygen reduction and the formation of OH anions were the reasons for the delaminating of protective films around film defects. A very successful method to investigate delaminating is provided by the Kelvin probe connected with surface potential measurements as developed by Stratman, Rohwerder *et al.*[113]

REFERENCES

1. A.G. MacDiarmid, A.J. Heeger, H. Shirakawa, Angew. Chem., *113*, 2643 (2001).
2. A.J. Heeger, Synth. Met., *55–57*, 3471 (1993).
3. T.A. Skotheim, R.J. Elsenbaumer, J.R. Reynolds, Handbook of Conducting Polymers, Marcel Dekker, New York, 1998.
4. P. Chandrasekhar, Conducting Polymers, Fundamentals and Applications—A Practical Approach, Kluwer Academic Publishers, Boston, London, 1999.
5. J. Heinze, Conducting Polymers, in A.J. Bard, M. Stratmann (Eds.), Encyclopedia of Electrochemistry, Vol. 8, H.J. Schäfer (Ed.), Organic Electrochemistry, Wiley-VCH, Weinheim, 2004, p. 605.
6. G. Hadziioannou, P.F. van Hutten, Semiconducting Polymers: Chemistry, Physics and Engineering, Wiley-VCH, Weinheim, 2000.
7. U. Janakiraman, K. Doblhofer, C. Fischmeister, A.B. Holmes, J. Phys. Chem. B, *108*, 14368 (2004).
8. K. Doblhofer, Thin film on electrodes: A physicochemical approach, in J. Lipkowski, P.N. Ross (Eds.), The Electrochemistry of Novel Materials, VCH, Weinheim, 1994, p. 141.
9. H. Shirakawa, S. Ikeda, Polymer J., *2*, 231 (1971).
10. D. Fichou (Ed.), Handbook of Oligo- and Polythiophenes, Wiley-VCH, Weinheim, 1999, p. 9.
11. T. Olinga, B. Francois, Synth. Met., *69*, 297 (1995).
12. J. Heinze, Topics in Current Chemistry: Electrochemistry IV, in E. Steckhan (Ed.), Vol. 152, Springer, Berlin, 1990, p.1.
13. E.M. Genies, G. Bidan, A.F. Diaz, J. Electroanal. Chem., *149*, 101 (1983).
14. G. Koßmehl, G. Chatzitheodorou, Macromol. Chem. Rapid Commun., *2*, 551 (1982).
15. A. Fikus, R. Pelz, E. Kern, W. Plieth, Electrochim. Acta, *43*, 2233 (1998).
16. A. Fikus, W. Plieth, D. Appelhans, D. Ferse, H.-J. Adler, F.-J. Schmidt, B. Adolphi, J. Electrochem. Soc., *146*, 4522 (1999).
17. C.M. Intelmann, U. Rammelt, W. Plieth, X. Cai, E. Jähne, H.-J. Adler, J. Solid State Electrochem., *11*, 1, (2007).
18. M. Skompska, Electrochim. Acta, *45*, 3841 (2000).
19. D.B. Wurm, Y.-T. Kim, Langmuir, *16*, 4533 (2000).
20. C. Visy, J. Kankare, E. Kriván, Electrochim. Acta, *45*, 3851 (2000).
21. W. Paik, I.-H. Yeo, H. Suh, Y. Kim, E. Song, Electrochim. Acta, *45*, 3833 (2000).
22. G. Inzelt, Electrochim. Acta, *45*, 3865 (2000).
23. V. Syritzki, A. Öpik, O. Forsen, Electrochim. Acta, *48*, 1409 (2003).
24. G. Sauerbrey, Z. Phys., *155*, 206 (1959).

25. A.L. Kipling, M. Thompson, Anal. Chem., *62*, 1514 (1990).
26. A. Glidle, A.R. Hillman, S. Bruckenstein, J. Electroanal. Chem., *318*, 411 (1991).
27. S.J. Martin, V.E. Granstaff, G.C. Frye, Anal. Chem., *63*, 2272 (1991).
28. M. Rodahl, F. Höök, A. Krozer, P. Brzezinski, B. Kasemo, Rev. Sci. Instrum., *66*, 3924 (1995).
29. A. Bund, G. Schwitzgebel, Electrochim. Acta, *45*, 3703 (2000).
30. J. Schröder, R. Borngräber, R. Lucklum, P. Hauptmann, Rev. Sci. Instrum., *72*, 2750 (2001).
31. A. Bund, M. Schneider, J. Electrochem. Soc., *149*, E331 (2002).
32. S. Neudeck, Diplom-Thesis, Dresden, 2004.
33. A. Bund, S. Neudeck, J. Phys. Chem. B., *107*, 6743 (2004).
34. W. Plieth, A. Bund, U. Rammelt, S. Neudeck, L. Duc, Electrochim. Acta, *51*, 2372 (2006).
35. S. El Sana, C. Gabrielli, S. Perrot, Russ. J. Electrochem., *40*, 267 (2004).
36. M.J. Henderson, A.R. Hillman, W.E. Vieil, C. Lopez, J. Electroanal. Chem., *458*, 241 (1998).
37. H.M. French, M.J. Henderson, A.R. Hillman, E. Vieil, J. Electroanal. Chem., *500*, 192 (2001).
38. E. Kriván, C. Visy, J. Kankare, Electrochim. Acta, *50*, 3851 (2005).
39. C. Gabrielli, O. Haas, H. Takenouti, J. Appl. Electrochem., *17*, 82 (1987).
40. M.A. Vorotyntsev, L.I. Dhaikhin, M.D. Levi, J. Electroanal. Chem., *364*, 37 (1994).
41. B.W. Johnson, D.C. Read, P. Christensen, A. Hamnett, R.D. Armstrong, J. Electroanal. Chem., *364*, 103 (1994).
42. G.S. Popkirov, E. Barsukov, R.N. Schindler, Electrochim. Acta, *40*, 1857 (1995).
43. T. Komura, N. Kitani, K. Takahashi, Bull. Chem. Soc. Jpn., *67*, 2669 (1994).
44. W.J. Albery, C.M. Elliott, A.R. Mount, J. Electroanal. Chem., *288*, 15 (1990).
45. S. Fletcher, J. Chem. Soc. Faraday Trans., *89*, 311 (1993).
46. X. Ren, P.G. Pickup, J. Electroanal. Chem., *420*, 251 (1997).
47. A. Fikus, U. Rammelt, W. Plieth, Electrochim. Acta, *44*, 2025 (1990).
48. U. Rammelt, S. Bischoff, M. El-Desouki, R. Schulze, W. Plieth, L. Dunsch, J. Solid State Electrochem., *3*, 406 (1999).
49. J. Disilvestro, O. Haas, Electrochim. Acta, *36*, 361 (1991).
50. T. Inoue, T. Yamase, Bull. Chem. Soc. Jpn., *56*, 985 (1983).
51. D. Paramunage, M. Tomkievic, D.S. Ginley, J. Electrochem. Soc., *134*, 1384 (1987).
52. J. Lukkari, M. Alanko, V. Pitkänen, K. Kleemola, J. Kankare, J. Phys. Chem., *91*, 8525 (1994).
53. E.M. Girotto, W.A. Gazotti, M.-A. DiPaoli, J. Phys. Chem. B, *104*, 6124 (2000).
54. N. Hebestreit, Thesis, Technische Universität, Dresden, 2004.
55. A.J. Heeger, Angew. Chem. Int. Ed., *40*, 2591 (2001).
56. J.L. Brédas, G.B. Street, Acc. Chem. Res., *18*, 309 (1985).
57. D. Fichou, B. Xu, G. Horowitz, F. Garnier, Synth. Met., *41–43*, 463 (1991).
58. J. Heinze, H. John, M. Dietrich, P. Tschunky, Synth. Met., *119*, 49 (2001).
59. A. Merz, J. Kronberger, L. Dunsch, A. Neudeck, A. Petr, L. Parkanyi, Angew. Chem. Int. Ed., *38*, 1442 (1999).
60. W. Plieth, C. Gutiérrez, G. Wilson, Pure Appl. Chem., *70*, 1395, 2409 (1998).
61. C. Gutiérrez, C. Melendrez (Eds.), Spectroscopic and Diffraction Techniques in Interfacial Electrochemistry, Kluwer Academic Publishers, Dordrecht, 1990.
62. A. Petr, L. Dunsch, A. Neudeck, J. Electroanal. Chem., *412*, 153 (1996).
63. P. Rapta, A. Neudeck, A. Petr, L. Dunsch, J. Chem. Soc. Faraday Trans., *94*, 3625 (1998).
64. A. Neudeck, A. Petr, L. Dunsch, Synth. Met., *107*, 143 (1999).
65. H. Lund, M. Baizer (Ed.), Organic Electrochemistry—An Introduction and a Guide, 3rd Ed., Marcel Dekker Inc., New York, 1991, p. 1377.
66. M. Dietrich, J. Heinze, J. Electroanal. Chem., *369*, 87 (1994).
67. F. Li, W.J. Albery, Electrochim. Acta, *37*, 393 (1992).
68. R.J. Waltmann, J. Bargon, Can. J. Chem., *64*, 76 (1986).

69. P.H. Audebert, A. Neudeck, L. Dunsch, P. Audebert, P. Capdevielle, M. Maumy, J. Electroanal. Chem., *470*, 77 (1999).
70. B.L. Funt, E.M. Peters, J.D. Van Dyke, J. Polym. Sci. A: Polym. Chem., *24*, 1529 (1986).
71. S.V. Lowen, J.D. Van Dyke, J. Polym. Sci. A: Polym. Chem., *28*, 451 (1990).
72. E.M. Peters, J.D. Van Dyke, J. Polym. Sci. A: Polym. Chem., *29*, 1379 (1991).
73. I. Rodriguez, M.L. Macros, J. Gonzalez-Velasco, Electrochim. Acta, *32*, 1181 (1987).
74. Y. Wei, J. Tian, D. Glahn, J. Phys. Chem., *97*, 12842 (1993).
75. H.-P. Welzel, G. Kossmehl, G. Engelmann, W.-D. Hunnius, W. Plieth, J. Solid State Electrochem., *5*, 141 (2001).
76. F.R. Mayo, F.M. Lewis, J. Am. Chem. Soc., *66*, 1594 (1944).
77. H.W. Melville, B. Noble, W.F. Watson, J. Polym. Sci., *2*, 229 (1947).
78. G. Goldfinger, T. Kane, J. Polym. Sci., *3*, 462 (1948).
79. M. Fineman, S.D. Ross, J. Polym. Sci., *5*, 259 (1950).
80. X.-D. Dang, Thesis, TU Dresden (2006).
81. T. Fukuda, Y.-O. Ma, H. Inagaki, Makromol. Chem., Rapid Commun., *8*, 495 (1987).
82. K.F. O'Driscoll, P.M. Reilly, Makromol. Chem., Makromol. Symp., *10/11*, 355 (1987).
83. T.P. Davis, K.F. O'Driscoll, M.C. Piton, M.A. Winnik, Macromolecules, *23*, 2113 (1990).
84. T. Yohannes, J.C. Carlberg, O. Inganäs, T. Solomon, Synth. Met., *88*, 15 (1999).
85. X.-D. Dang, C.-M. Intelmann, U. Rammelt, W. Plieth, J. Solid State Electrochem., *8*, 727 (2004).
86. X.-D. Dang, C.-M. Intelmann, U. Rammelt, W. Plieth, J. Solid State Electrochem., *9*, 706 (2005).
87. M.J. Duer, Introduction to Solid State NMR Spectroscopy, Blackwell Science Ltd., Oxford, 2004.
88. S.P. Brown, H.W. Spiess, Chem. Rev., *100*, 4125 (2001).
89. X.-D. Dang, S. Ok, U. Scheler, W. Plieth, Unpublished; See: X.-D. Dang, Thesis, TU Dresden (2006).
90. H.-P. Welzel, G. Koßmehl, G. Engelmann, J.P. Hunius, W. Plieth, J. Solid State Electrochem., *5*, 141 (2001).
91. P. Bäuerle, S. Scheib, Adv. Mater., *5*, 848 (1993).
92. P.H. Aubert, A. Neudeck, L. Dunsch, P. Audebert, P. Capdevielle, M. Maumy, J. Electroanal. Chem., *470*, 77 (1999).
93. M. Fahlman, S. Jasty, A.J. Epstein, Synth. Met., *85*, 1323 (1997).
94. M.C. Bernard, A. Hugot-LeGoff, S. Joiret, N.N. Dinh, N.N. Toan, J. Electrochem. Soc., *146*, 995 (1999).
95. H.N.T. Le, B. Garcia, C. Deslouis, Q.L. Xuan, Electrochim. Acta, *46*, 4259 (2001).
96. H. Wessling, Adv. Mater., *6*, 226 (1994).
97. H. Wessling, J. Posdorfer, Electrochim. Acta, *44*, 2139 (1999).
98. A. Mirmoshseni, A. Oladegaragoze, Synth. Met., *114*, 105 (2000).
99. J.L. Camelet, J.C. Lacroix, S. Aeiyach, P.C. Lacaze, J. Electroanal. Chem., *445*, 117 (1998).
100. U. Rammelt, P.T. Nguyen, W. Plieth, Electrochim. Acta, *48*, 1257 (2003).
101. U. Barsch, F. Beck, Synth. Met., *55–57*, 1638 (1993).
102. F. Beck, R. Michaelis, F. Schloten, B. Zinger, Electrochim. Acta, *39*, 229 (1994).
103. C.A. Ferreira, S. Aeiyach, J.J. Aaron, P.C. Lacaze, Electrochim. Acta, *41*, 1801 (1994).
104. U. Rammelt, P.T. Nguyen, W. Plieth, Electrochim. Acta, *46*, 4251 (2001).
105. B. Wessling, J. Phys. II France, *6*, 385 (1996).
106. N. Hebestreit, J. Hofmann, U. Rammelt, W. Plieth, Electrochim. Acta, *48*, 1779 (2003).
107. B. Wessling, Metallic properties of conductive polymers due to dispersion, in H.S. Nalva (Ed.), Handbook of Conducting Molecules and Polymers, Vol. 3, Wiley, New York, 1997, p. 497.
108. L.M. Lee, Thesis, Technische Universität, Dresden, 2006.

109. P.J. Kinlen, V. Menon, Y. Ding, J. Electrochem. Soc., *146*, 3690 (1999).
110. P.J. Kinlen, Y. Ding, D.C. Silvermann, Corrosion, *58*, 490 (2002).
111. U. Rammelt, L.M. Duc, W. Plieth, J. Appl. Electrochem., *35*, 1225 (2005).
112. G. Paliwoda-Porebska, M. Rohwerder, M. Stratmann, U. Rammelt, L.M. Duc, W. Plieth, J. Solid State Electrochem., *10*, 730 (2006).
113. R. Hausbrand, M. Stratmann. M. Rohwerder, Steel Res., *74*, 453 (2003).

– 12 –

Nanoelectrochemistry

12.1 GOING TO ATOMIC DIMENSIONS

New instrumentation and devices for gaining insight into the structure of matter on the atomic level has opened the door to this fascinating nanoworld. Research and development of nanoscience and nanotechnology in electrochemistry follow several main streams. One of them is the investigation of the surface with atomic resolution, which has been described throughout this book. A number of reviews exist about this surface nanoelectrochemistry.[1]

Another area of study has been the preparation of atomic-level surface topographies assembled from clusters of nanostructures like nanodots, nanowires, nanoarrays, and nanoelectrodes, etc.[2]

Additionally, supramolecular carbon nanostructures have been made from nanoballs, -rods, and -tubes.[3] New materials can be formed with the different nanosized structures but preparation of their materials and properties is somewhat different than for traditional materials. For example, nanocrystalline metals are much harder.[4] These materials and unusual properties combine to create composites, for example, from the co-deposition of particles with metals in a plating process or in compositionally modulated multi-layers. Also of interest are core-shell particles, which consist of an inorganic core and a conducting polymer shell, and the traditional composites with larger subunits. Ways leading to the preparation of such nanocomposites is an interesting topic and will be discussed in this chapter.

12.2 CO-DEPOSITION

Co-deposition describes the embedding of particles in a metal matrix during a plating process. The ideal structure of such materials is shown in Figure 12.1. Spherical particles of different sizes are embedded in a homogeneous metal matrix. Another form of particles is flakes. Co-deposition of mica flakes with Zn^{2+} is shown as an example in Figure 12.2.

The principles have been known since the 1960s but the mechanism of co-deposition and the conditions for optimized particle incorporation are still unclear. In Table 12.1 some examples of systems mentioned in the literature are given.

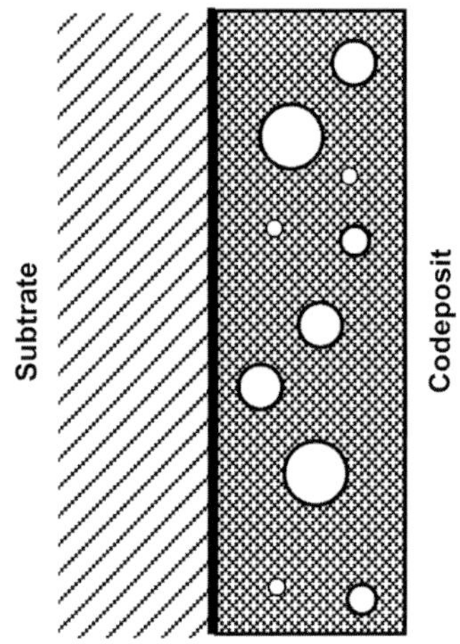

Figure 12.1 Idealized picture of a composite layer formed by metal plating.

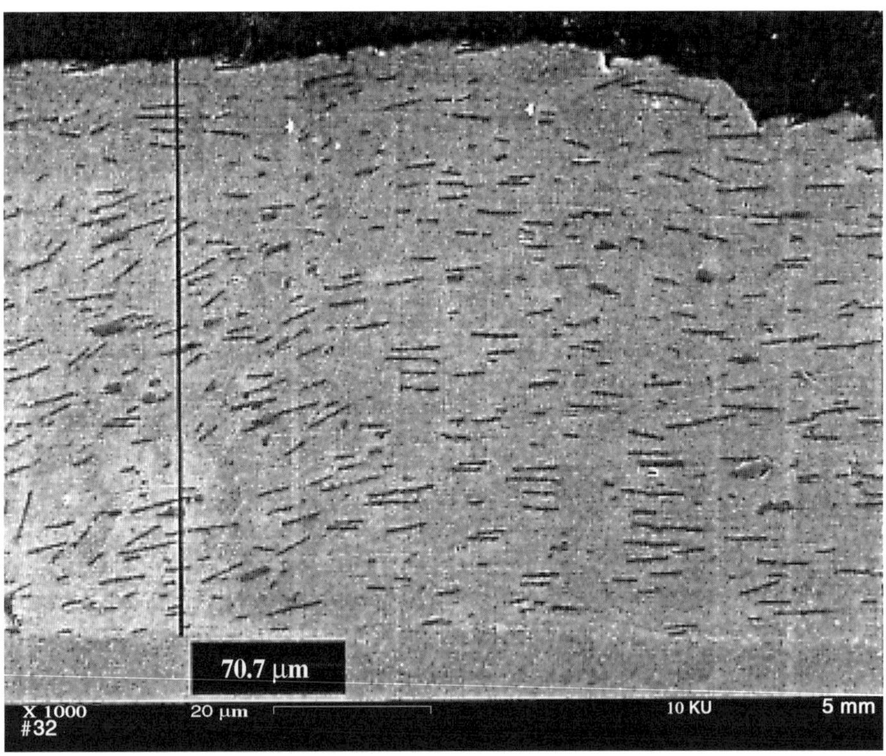

Figure 12.2 Co-deposition of mica flakes in a zinc matrix.[5]

Table 12.1

Some of the particle–metal matrix combinations and applications that are known so far

Metal matrix	Particles	Application	Industrial branch
Ni	SiC	Wear resistant	Car industry
Ni	PTFE	Friction control	
Ni	Al_2O_3	Dull nickel	
Ni	Micro capsules	Different applications	
Ni	$NiFeS_2$	Catalytic properties	
Co	Cr_2C_3	Temperature resistant	Air industry
CoNi	Cr_2C_3	Temperature resistant	Air industry
Ag	Al_2O_3, nanodiamond	Wear resistant	Electronic industry
Au	Al_2O_3, nanodiamond	Wear resistant	Electronic industry
Zn	SiO_2	Corrosion resistance	

12.2.1 Particle dispersions

For co-deposition the particles must be dispersed in the plating bath. Either the dispersion can be made in a separate bath that is mixed later on with the plating bath or the particles can be directly added to the plating bath. In both cases, in the formation of a stable dispersion there must be no separation of the particles and little or no agglomeration. The theoretical background for a stable particle solution is provided by the DLVO theory (Derjaguin–Landau–Verwey–Overbeek theory[6]). In a stable dispersion electrostatic repulsion forces must compensate van der Waals and other attraction forces. Electrostatic forces are dependent upon the charge on the particle surface and dominate at larger distances between the particles, which creates a larger dilution. Van der Waals forces dominate at a low particle charge and smaller particle distance. For sufficiently large electrostatic repulsion forces, the sum of both charges has a maximum, which is shown in Figure 12.3.

The maximum is an activation barrier that prevents or reduces agglomeration. For a stable dispersion the particles must have a large charge and a high activation barrier. An experimental parameter proportional to this charge is the zeta potential. Therefore, the determination of the zeta potential is an important experimental procedure in characterizing the particles for co-deposition.

12.2.2 Determination of the zeta potential

One method for the measurement of the zeta potential is based on the measurement of the relative electrophoretic motion of particles and electrolytes in an electric field. To measure the velocity of the particles laser doppler anemometry can be used. A typical apparatus is shown in Figure 12.4.

The measurements are usually made in a diluted electrolyte (e.g., 10^{-3} $mol \cdot dm^{-3}$ aqueous KCl) with an extensive diffuse double layer around the particle (Figure 12.5).

By an applied electrical field, the particle, which is positively charged in the example shown in Figure 12.5, moves in the direction of the negative electrode, while the compensating

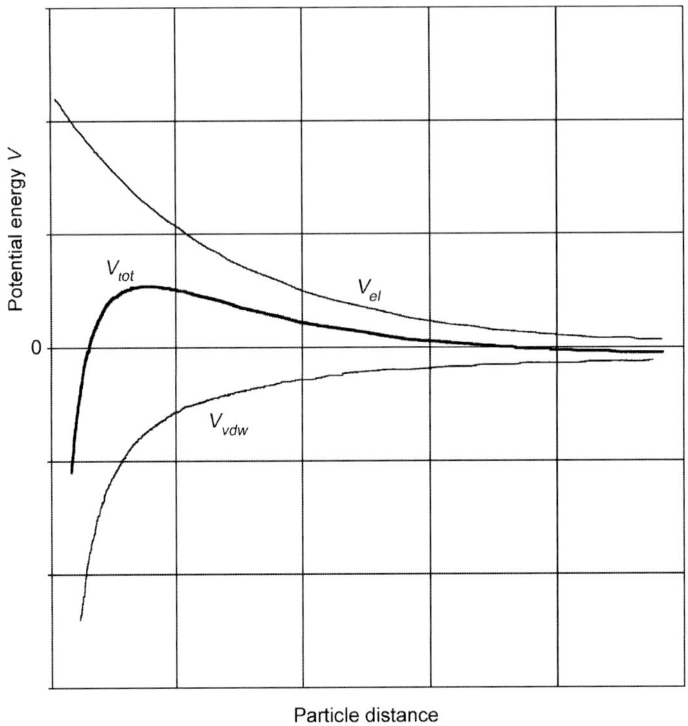

Figure 12.3 Interaction energies in colloidal dispersions. V_{el} electrostatic repulsion force depending on particle charge, V_{vdW} van der Waals attraction force, and V_{tot} sum of electrostatic and van der Waals forces. The maximum is the activation energy preventing coagulation.

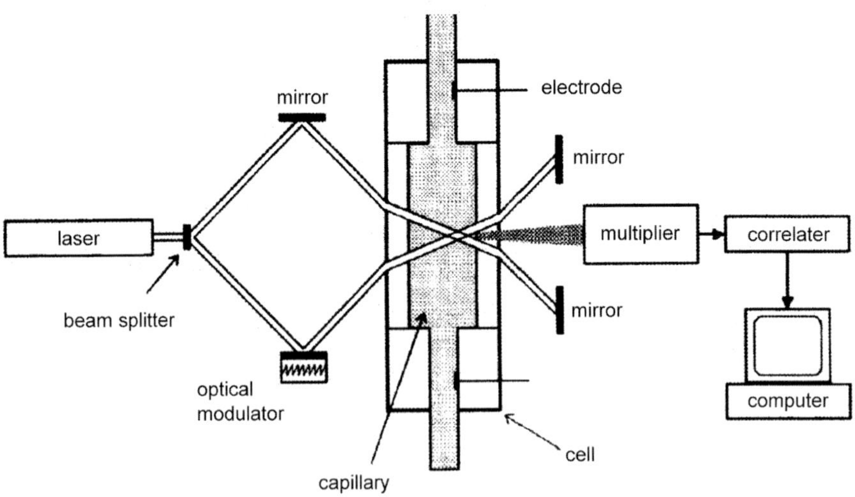

Figure 12.4 Apparatus for the measurement of zeta potentials.[7]

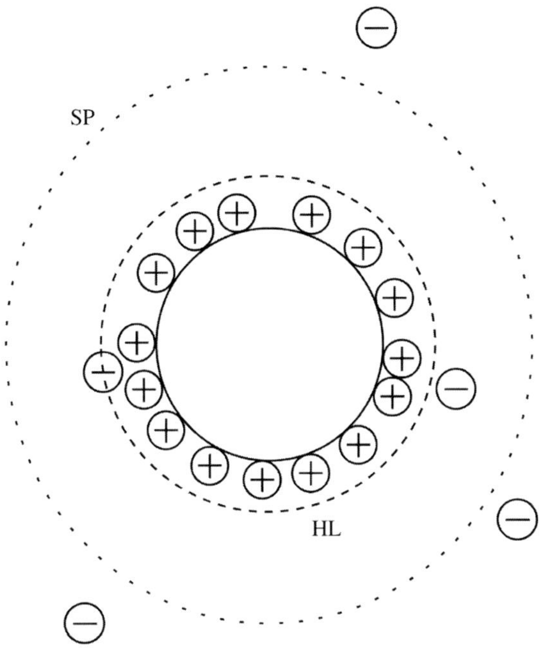

Figure 12.5 Particle with surface charge (consisting of an intrinsic particle charge and charge of specifically adsorbed ions), Helmholtz layer HL, part of the diffuse double layer and the shear plane SP of the zeta-potential measurement (schematic representation).

negative charge in the diffuse double layer moves towards the positive electrode. The potential at the shear plane between particle and electrolyte is the zeta potential determined by this method. Figure 12.5 shows that the zeta potential includes an intrinsic particle charge, charge of specifically adsorbed ions, and charge in the Helmholtz layer. The potential can be calculated from the measurement of the particle velocity according to the Smoluchowski equation

$$\zeta = -\frac{\eta}{\varepsilon\varepsilon_0} u_E \tag{12.1}$$

η is the viscosity, u_E the electrophoretic mobility, and $\varepsilon\varepsilon_0$ the product of relative permittivity and vacuum permittivity. With the viscosity $\eta(H_2O) = 0.894\,\text{mPa s}$ and the relative permittivity $\varepsilon(H_2O) = 78.54$, the equation between zeta potential in mV and the electrophoretic mobility u_E in μm s^{-1} per V cm^{-1} is

$$\zeta = -12.87 u_E \tag{12.2}$$

For small spherical particles and dilute electrolytes the zeta potential is larger by the factor 1.5

$$\zeta = -1.5 \frac{\eta}{\varepsilon\varepsilon_0} u_E \tag{12.3}$$

12.2.3 Factors influencing zeta potential and particle properties

The charge of the particle will be influenced by several experimental parameters. This is the chemical nature of the particle. Typical inorganic particles used for co-deposition are oxides and carbides. The surface charge of these particles depends on the chemical composition. The surface charge also depends on pH of the electrolyte. The characteristic pH for when the zeta potential is zero is the isoelectric point. Isoelectric points of some particles used for co-deposition are shown in Table 12.2.

The zeta potential also depends on the pretreatment of the particles. This is shown for TiO_2 particles in Figure 12.6 where the zeta potential of two TiO_2 samples is shown as a function of the pH.

The surface charge of these particles is also a function of the electrolyte composition. The following electrolyte components can change the surface potential:

(i) The pH of the electrolyte can change, the protonation or deprotonation of the surface leads to more positive or negative charges.
(ii) Electrolyte ions can be adsorbed on the particle surface. Particularly interesting has been the adsorption of the metal cations to be deposited. Mechanistic models of co-deposition have been developed that suggest the formation of a first metal layer on the particles by reduction of the adsorbed ions when the particle comes into contact with the metal surface.
(iii) Inert particles can adsorb on the surface, especially in the case of surface-active organics. One characteristic feature of surface-active substances is the change of the surface tension (surface energy).
(iv) The size and form of the particles are important. Study of the incorporation of nanoparticles has shown that size has a large influence on particle incorporation. The same can be said about the shape. For example, it is thought that non-spherical particles may stick better to surface irregularities.

Table 12.2

Isoelectric points of different particle materials[8–10]

Material	pH of isoelectric point	Remarks
Al_2O_3	2	
Diamond	<1	
Fe_2O_3	8	Haematite
Mica	2	
PbO	10	
SiO_2	2	
SnO	6	
TiO_2	3.5	Anatase
TiO_2	4.5	Rutile
Y_2O_3	5.5	
ZrO_2	4	

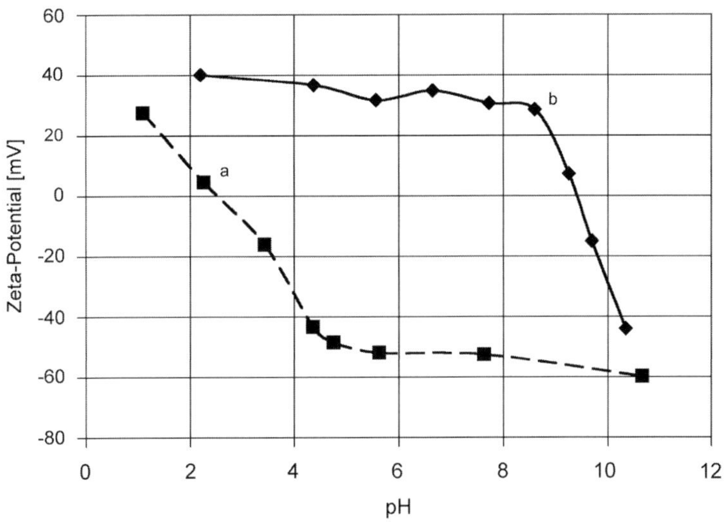

Figure 12.6 Zeta potential of TiO_2 particles measured in 10^{-3} mol·dm^{-3} KCl as a function of the pH. (a) Anatase and (b) Bayertitan R KB4.

12.2.4 Properties of the metal surface

The co-deposition of particles demands a pronounced adhesion of particles on the metal surface. Therefore, not only the properties of the particles are important. The same detailed analysis of surface properties is also necessary for the metal surface. The following properties should be considered:

(i) *Surface charge*. It was pointed out that the particles must have a surface charge to avoid agglomeration in the electrolyte and to form a stable colloidal solution. The particle surface charge in combination with the surface charge of the electrode can influence the adhesion process. The surface charge of the electrode is determined by the relative position of the deposition potential to the potential of zero charge (pzc). If the deposition process occurs at potentials negative to the pzc, the metal surface will be negatively charged. This means the compensating charge in the electrolyte part of the double layer will be positive. The influence of the charge will be discussed in Section 12.2.6. The determination of the potential of zero charge was described in Chapter 4.

(ii) *Adsorption of electrolyte components and the particle surface are equally important.* A specific adsorption of ions can change the position of the potential of zero charge (Esin–Markov effect). Adsorption of surface-active organics will influence the hydrophobic or hydrophilic character of the surface and also change the surface tension.

(iii) *Surface morphology*. This property is significant for the adhesion process of particles on the surface. A rough surface will provide better possibilities for surface adhesion of particles than a very smooth surface. The surface topography will also influence the surface energy.

12.2.5 Process parameters influencing the incorporation

The incorporation of particles will also be influenced by the experimental process parameters. The most typical ones are:

(i) *Current density*. The rate of metal ion discharge and the deposition potential are both determined by the current density. The current distribution in the plating cell and around the particles should also be taken into account. Electrolyte resistance near and below the particle changes. Celis and Fransaer[11] have calculated the shielding effect of an isolating particle.
(ii) *Convection in the plating cell*. The transport of particles to the cathode is mainly determined by convection. A diffusion layer is built up near the surface. But for larger particles, the diffusion layer is as thick as for the particle diameter. This means the most important forces acting on a particle are hydrodynamic (Section 12.2.6 and Figure 12.9). Convection can remove particles from a surface position. Therefore, an optimized convection is an important factor in co-deposition.
(iii) *Temperature*. As usual, temperature can influence the co-deposition process.
(iv) *Particle concentration*. The amount of particles adsorbed on the electrode surface depends on the particle concentration. However, as will be shown later, adsorption of particles is important but not a guarantee for particle incorporation.

12.2.6 Mechanistic models

Several models were developed to describe and understand the complex nature of co-deposition. Some examples will be briefly described in this chapter.

Guglielmi model

This model was one of the first models developed. It described the incorporation of particles by a two-step adsorption process of particles (Figure 12.7). In the first step a weak agglomeration of particles on the surface was assumed and in the second a kinetically controlled irreversible adsorption was the final deposition process.

Guglielmi applied a Langmuir-type adsorption concept and a Butler–Volmer equation to formulate the final deposition process. But his final equation for the volume percent of co-deposited particles x_V is rather empirical. It can be written in the following form:

$$x_V = \frac{Kc_V}{1 + Kc_V} \frac{zF \cdot \rho_M \cdot v_0}{M \cdot i} \left(\frac{i}{i_0}\right)^{\beta} \tag{12.4}$$

In this equation c_V is the volume percent of particles in the electrolyte, K a Langmuir-type adsorption constant, ρ_M the density of the deposited metal (g cm^{-3}), v_0 a system constant (cm s^{-1}), zF the Coulomb charge of the deposited metal ions (in C mol^{-1}), M the molar mass of the metal atoms (g mol^{-1}), i the deposition current density (A cm^{-2}), i_0 the exchange current density of metal deposition, and β another system constant. This equation starts with an adsorption isotherm and then compares an empirical weight

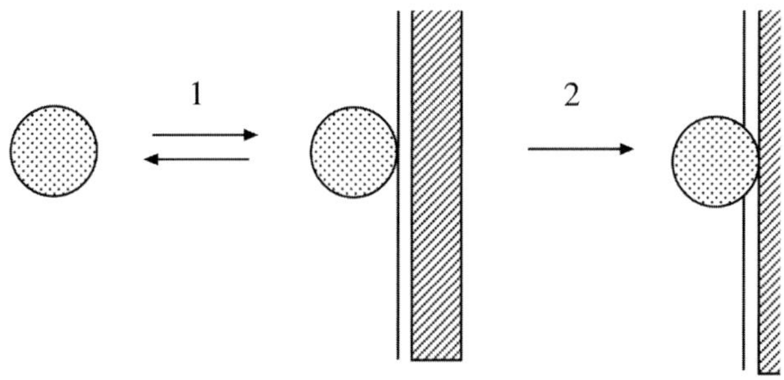

Figure 12.7 Guglielmi model of co-deposition. (1) Physisorption and (2) chemisorption followed by incorporation.[12]

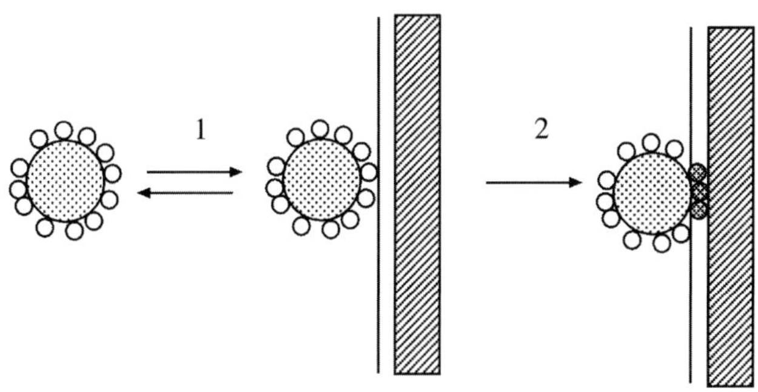

Figure 12.8 Kariapper–Foster model of co-deposition whereby metal ions are adsorbed on the particles. The first step describes physisorption. In the second, part of the adsorbed metal ions get reduced and form a "mechanical" bond with the metal surface.

of the deposited metal with the true weight of metal deposited per square centimeter and finally introduces a current density factor reduced to exchange current density with an additional empirical exponent β.

Kariapper–Foster model[13]

Kariapper and Foster used the Guglielmi model but introduced the adsorption of metal ions on the particle surface. First of all, the particles had to be positively charged for incorporation. Contact between the particle and the metal surface could be expected to then reduce part of the adsorbed cations. The authors also postulated the formation of a "mechanical bond" between particles and metal surface (Figure 12.8).

The authors added other empirical factors to the Guglielmi equation such as a "mechanical" factor that depended on convection forces, among other things. The cation adsorption model was the dominating concept for the description of co-deposition for several years thereafter.

Fransaer–Buelens–Celis–Roos contributions

Celis, Roos, and Buerlen[14] used the concept of metal ion adsorption on the particles for the derivation of an equation for the particle concentration in the metal matrix. The partial reduction of metal ions on the particle was considered the dominating effect for particle incorporation. An empirical probability function was formulated.

The mass percent of particles in the metal matrix x_m was defined by the following equation:

$$x_m = \frac{\Delta m_P}{\Delta m_{Me} + \Delta m_P} 100 \tag{12.5}$$

Δm_{Me} is the deposited metal mass per area unit (cm^2) and time (s) given by Faradays equation

$$\Delta m_{Me} = \frac{M \cdot i}{zF} \tag{12.6}$$

In this equation M is the molar mass of the metal, i the deposition current density of the metal, and zF the equivalent charge of the deposited metal ions.

For Δm_P, the deposited particle mass per area unit (cm^2) and time (s), the following equation was given

$$\Delta m_P = \frac{4\pi}{3} r_P^3 \cdot \rho_P \cdot N_P \cdot P \tag{12.7}$$

In this equation $(4\pi/3)r_P^3$ is the volume of one particle, ρ_P the specific mass of the particles, and N_P the number of particles passing the diffusion layer per area unit and time. P is the mentioned probability of particle incorporation. N_P as well as P are empirical factors.

A subsequent paper of Fransaer, Celis, and Roos[15] described the different forces acting on the particles during the deposition process while trying to find an equation for the trajectory of the particle in front of the electrode. The calculation of a trajectory of a particle is possible if all the forces acting on the particle in front of the electrode are taken into account. This was done as completely as possible. The forces acting on the particle on the surface are shown in Figure 12.9.

The number of particles adsorbed on the metal surface depended on hydrodynamic shear forces. These forces were simulated by a rotating disc electrode and then compared with experiments studying polystyrene particle co-deposition with copper. One result was a calculation of the adhesion force of the particles on the copper surface as a function of the current density showing a maximum between 1 and 2 A dm^{-2} (Figure12.10).

The maximum coincided with a minimum of the distance between particle and copper surface calculated from capacitance measurements. Maximum and minimum also coincided with the potential of zero charge of the copper electrode. To explain maximum and minimum, a new force was postulated that depended on the reorientation of the solvation shell around particles and electrode surface. The orientation of the solvent molecules on the electrode surface should have a minimum at the pzc. On the particle side, formation of a close contact with the metal surface would be easier for hydrophobic particles with only a weak interaction with solvent molecules. A maximum co-deposition should be observed for hydrophobic particles at the pzc. A number of papers published experimental results that support this concept.[16–17] Dedeloudis and Fransaer[18] even measured the attraction of hydrophilic and hydrophobic particles to a metal surface by placing the particles on the tip

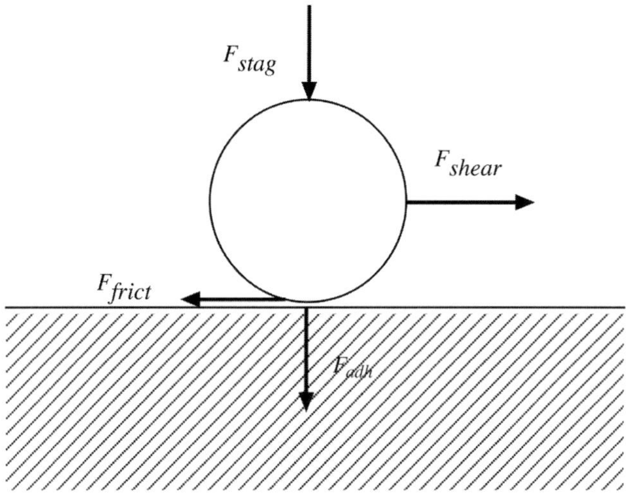

Figure 12.9 Forces acting on a particle adsorbed on an electrode surface; F_{adh} adhesion forces, F_{frict} friction forces, F_{shear} shear forces, and F_{stag} stagnation forces. (Reproduced with permission from Ref. [15], © 1992, The Electrochemical Society.)

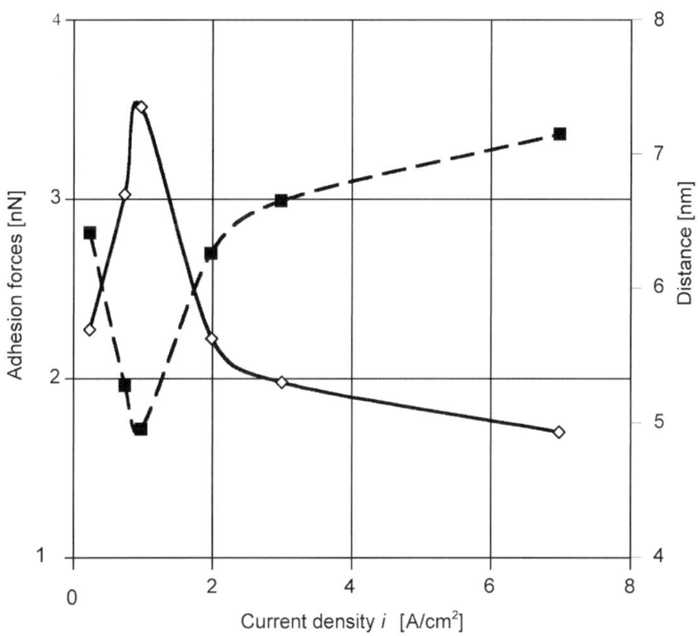

Figure 12.10 Adhesion force (◇) and equilibrium distance (■) between particle and electrode surface determined by rotating disc experiments with polystyrene particle co-deposition and copper. (Reproduced with permission from Ref. [15], © 1992, The Electrochemical Society.)

of an atomic force microscope. From all these experiments it can be concluded that the hydrophobic character of particles has an influence on incorporation.

Some results, however, may be restricted to the special conditions of polyethylene particle co-deposition with copper. For example, deposition at other metals like Zn occurs at potentials that are far more negative from the potential of zero charge. In this case no minimum of the hydration force can be expected. Nevertheless, a maximum was observed in the particle concentration versus current density plot.[5]

Role of the electrostatic interaction

Usually it is assumed that only positively charged particles can be incorporated because the charge of the cathode is negative. Positive particle charge was considered a prerequisite for incorporation. But it was found for the incorporation of nanoparticles of Al_2O_3 and nanodiamonds in an Au matrix that negatively charged particles had been incorporated in this case (Wünsche, Bund, Plieth[19]). The zeta potential of Al_2O_3 particles is shown as a function of the pH value in Figure 12.11. In the KCl electrolyte used for the measurements of the zeta potential the values are positive; the zero value is observed above pH 9. But when citrate ions are added or when the electrolyte used for the Au deposition is added, the zeta potential is negative, at least at the pH of the deposition electrolyte (pH = 5.8).

The zeta potential of nanodiamonds is always negative. Therefore, the electrostatic forces between the particles and the metal surface must be considered in more detail.

The particles have a characteristic double-layer structure that is determined by the intrinsic surface charge, specifically adsorbed ions and ions compensating the surface

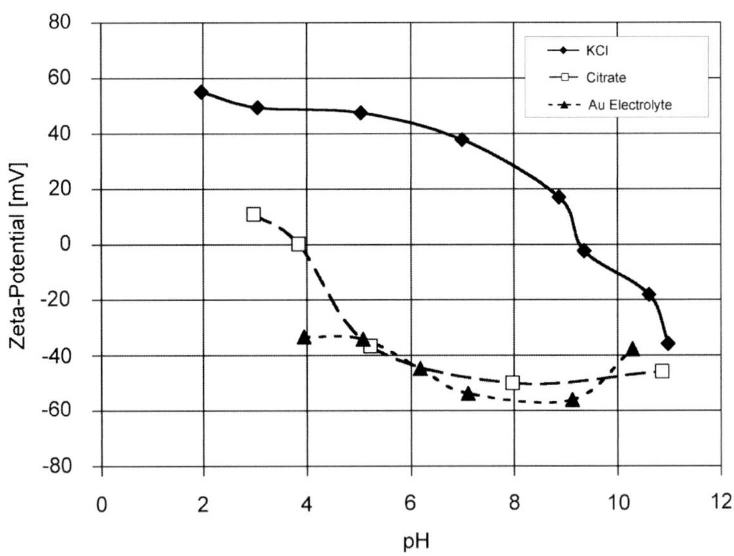

Figure 12.11 Zeta potential of Al_2O_3 particles, concentration 0.2 $g \cdot dm^{-3}$. (a) 10^{-3} $mol \cdot dm^{-3}$ KCl (pH adjustment with HCl or KOH), (b) 10^{-3} $mol \cdot dm^{-3}$ KCl + 10^{-3} $mol \cdot dm^{-3}$ citric acid, and (c) 10^{-3} $mol \cdot dm^{-3}$ KCl + 5 $mol \cdot dm^{-3}$ Au plating electrolyte.[19]

charge in the electrolytic part of the electrochemical double layer. The zeta potential represents the surface charge of the particles, charge of specifically adsorbed ions, as well as the charge in the Helmholtz layer. A positive zeta potential means a negative charge for the diffuse double layer and vice versa. This was shown in Figure 12.5.

On the metal side the charge will be determined by the position of the electrode potential relative to the potential of zero charge (pzc). The potential of zero charge, determined by the capacitance minimum of the electrode in a dilute inert electrolyte (typical $< 10^{-3}$ mol·dm^{-3} electrolyte concentration), is always the potential, when the metallic part of the double layer has no free charge. Specifically adsorbed ions add a dipole potential and can therefore shift the potential of zero charge. At potentials that are negative from the pzc the surface is negatively charged. In the diffused double layer positive charge is accumulated. But at the electrolyte concentrations used in the plating bath, additional ions can be adsorbed or can accumulate in the Helmholtz layer and change the sign of the charge of the metal surface seen by the particles. The situation is sketched in Figure 12.12. This figure shows that ions with a negative zeta potential can be attracted from the metal surface by electrostatic forces.

Valdes model[20]

Valdes suggested a very special model. He took into account the experimental observation that the particle concentration in the deposited metal as a function of current density went

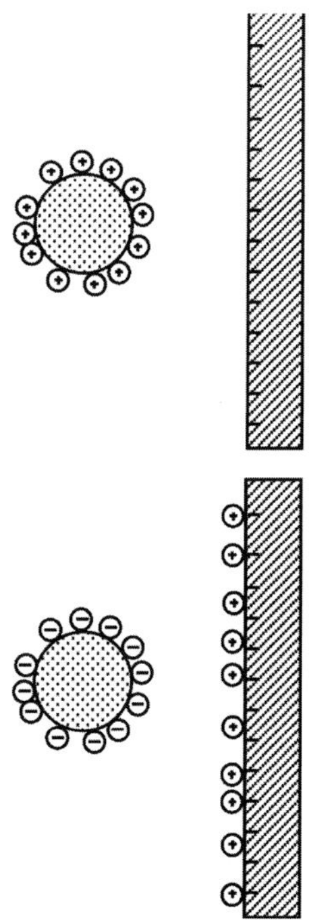

Figure 12.12 Electrostatic interaction between particles and electrode surface. Particles can be charged positively and negatively. For attraction of particles by the electrode surface the effective charge of the electrode surface must be considered.

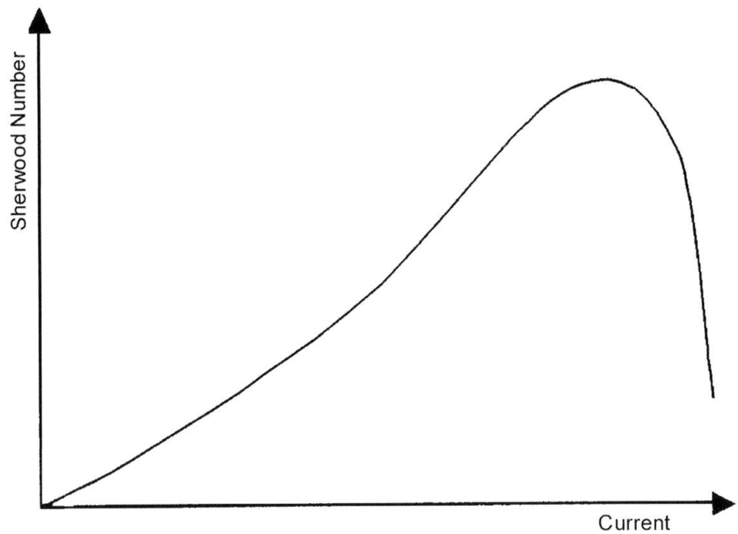

Figure 12.13 Particle Sherwood number as a function of the current density in the model of Valdes.

through a maximum.[21] In the Fransaer–Celis–Roos model this was explained by the maximum adsorption of particles on the surface at the potential of zero charge. Valdes developed a model of particle migration from places of low electrolyte concentration to places of high electrolyte concentration. With increasing current density the metal ion concentration in the diffusion layer should decrease and the particles should migrate into the bulk electrolyte. A Sherwood number was defined by going through a maximum with increasing current density (Figure 12.13). A maximum of particles was adsorbed at the maximum of the Sherwood number.

12.2.7 General concepts for the development of a model

The process of co-deposition of particles consists of two consecutive reactions:

(i) *Particle transport and adsorption on the metal surface*. Particles that come into the vicinity of the surface due to hydrodynamic forces or by diffusion or migration may touch the metal surface and, if adhesion is strong enough, stay for what is called the residence time of particles τ_p on the surface. The adsorption process can be described by the following rate equation:

$$r_{p,ad} = k_{p,ad} N_{p,el} \left(N_{p,max} - N_{p,ad} \right) \tag{12.8}$$

$N_{p,el}$ is the particle density in the electrolyte; $N_{p,max} - N_{p,ad}$ is the free surface area. The rate constant of particle adsorption $k_{p,ad}$ depends on the energy barrier for reorganization of solvent molecules around the particles and surface.

After adsorption the adhesion forces between particle and surface will keep the particle adsorbed on the surface until hydrodynamic forces or thermal motion are strong enough to overcome the adsorption barrier. This can be described by the equation

$$r_{\mathrm{p,sep}} = k_{\mathrm{p,sep}} N_{\mathrm{p,ad}} \tag{12.9}$$

The reciprocal value of the rate constant of separation can be defined as the residence time of co-deposition

$$\tau_{\mathrm{p}} = \frac{1}{k_{\mathrm{p,sep}}} \tag{12.10}$$

At equilibrium $r_{\mathrm{p,ad}} = r_{\mathrm{p,sep}}$ and a Langmuir type of adsorption, isotherm is obtained.

Proportionality between surface density and electrolyte can be formulated for small coverage

$$N_{\mathrm{p,ad}} \propto N_{\mathrm{p,el}} \tag{12.11}$$

(ii) *Metal deposition.* Parallel to the stay of the particle on the surface for the residence time the metal goes ahead with the deposition rate in proportion to the current density. If the metal has time enough to build up a wall of newly formed metal around the particle of a critical thickness δ_{crit}, the particle will become incorporated. The critical thickness, which is an empirical factor, is a fraction of the particle diameter

$$\delta_{\mathrm{crit}} = x_{\mathrm{crit}} d_{\mathrm{p}} \tag{12.12}$$

This is shown in Figure 12.14A.

The thickness that grows during the residence time is determined by the current density and is given by the equation

$$\delta = \frac{i \cdot \tau_{\mathrm{p}} \cdot M}{z \cdot F \cdot \rho} \tag{12.13}$$

M is the molar mass of the metal, z the charge of the metal ions, F Faradays constant, and ρ the density. The particle density in the growing metal phase is proportional to the thickness growth and therefore proportional to the product of current density and particle residence time.

The current density can influence the residence time. This is either possible by the change of the potential with increasing current or by forces acting on the particle.

The growth of the metal layer around the particle is the main process of incorporation. The surface tension balance between particle, metal, and electrolyte determines whether the metal grows around the particle. The situation can be described by the classical equation

$$\sigma_{\mathrm{me}} = \sigma_{\mathrm{mp}} + \sigma_{\mathrm{ep}} \cos\Theta \tag{12.14}$$

σ_{me} is the surface tension between metal surface and electrolyte, σ_{mp} the surface tension between metal and particle, and σ_{ep} the surface energy between particle and electrolyte.

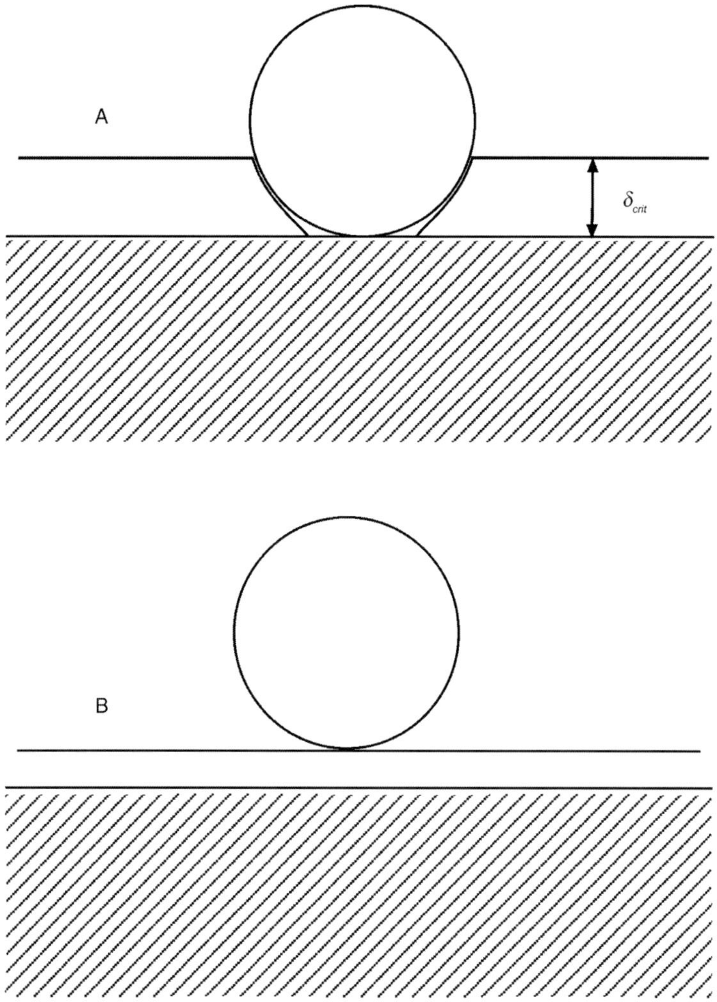

Figure 12.14 Growth of the metal layer on the metal-electrolyte surface with adsorbed particles. (A) A metal layer grows around the particle, (B) a metal layer grows beneath the particle that is floating on the metal layer.

The angle Θ is the angle between the surface tension vectors $\sigma_{mp} / \sigma_{ep}$. This is shown in Figure 12.15.

The surface charge of the metal-electrolyte interface will influence the surface tension, which makes the angle between the metal and particle surfaces depend on the potential.

$$\cos\Theta(E) = \frac{\sigma_{me}(E) - \sigma_{mp}}{\sigma_{ep}} \tag{12.15}$$

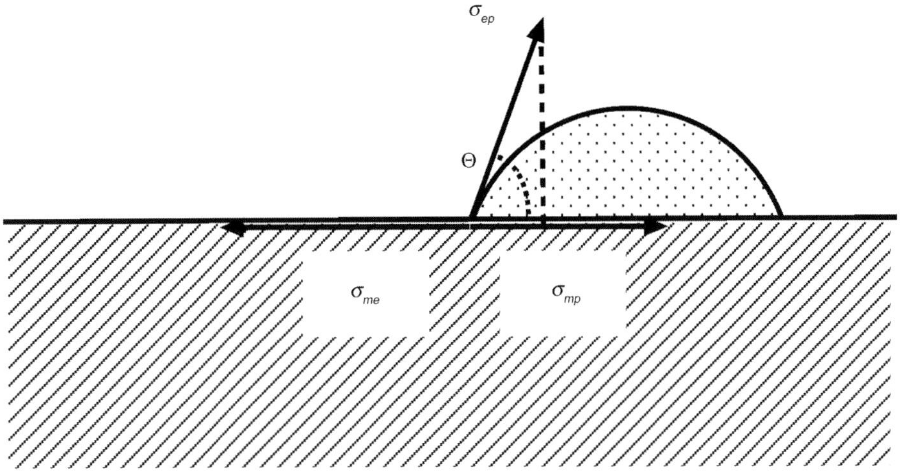

Figure 12.15 Surface energies between particle–metal–electrolyte phases. In equilibrium, the forces must balance each other.

The cosine of the angle Θ is a probability function for incorporation. The particle will be incorporated, if $\cos\Theta = 1$. The probability for incorporation decreases, if $\cos\Theta < 1$. The probability is zero for $\cos\Theta \leq 0$.

Summarizing, the probability can have the following values

$$P = 1 \qquad \Theta = 0^\circ$$
$$P = 0 \qquad \Theta \geqslant 90^\circ$$

The surface tension $\sigma_{me}(E)$ depends on the potential and has a maximum at the potential of zero charge. A first approximation of the potential dependence of the surface tension can be described by the following equation:

$$\sigma(E) = \sigma_{max} - \frac{C}{2}(E - E_{pzc})^2 \tag{12.16}$$

σ_{max} is the surface tension at the maximum of the surface-tension-potential diagram (electro-capillary maximum). The probability for incorporation is potential dependant

$$P(E) = \cos\Theta(E) = \frac{\sigma_{max} - \sigma_{mp}}{\sigma_{ep}} - \frac{C}{2\sigma_{ep}}\left(E - E_{pzc}\right)^2 \tag{12.17}$$

One can write an equation for the particle density in the co-deposition layer $N_{p,co}$. The amount of particles incorporated into the metal matrix is proportional to (i) the concentration in the electrolyte $N_{p,el}$ (assuming there is proportionality between adsorbed particles and particles in the electrolyte), (ii) by the product $\tau_p \cdot i$ (the charge deposited during the residence time of the particles), and (iii) by the probability function $P(E)$ of particle incorporation.

$$N_{p,co} = k_{co} \cdot P(E) \cdot \tau_p \cdot i \cdot N_{p,el} \tag{12.18}$$

Substituting $P(E)$ by Eq. (12.17) one obtains

$$N_{p,co} = k_1 \cdot \tau_p \cdot i \cdot N_{el} - k_2 \cdot \tau_p \cdot i \cdot (E - E_{pzc})^2 \cdot N_{el} \tag{12.19}$$

E is a function of the current density getting more negative with increasing current density. Therefore, the second term increases faster with increasing current density than the first term. This predicts a maximum in the dependence of the particle concentration on the current density.

Another way to explain this maximum could provide the possible dependence of the residence time on the current density.

12.2.8 Examples

Much of the focus in recent years on the application of co-deposits has been on particles with increased wear resistance. The deposition of SiC with Ni in car engines is a classic example.

One of the recently investigated dispersion coatings is the co-deposition of microcapsules in nickel. The microcapsules of an organic material can be filled with a variety of surface-active substances. For example, substances that diminish friction can be incorporated. This may be an inorganic material like MoS_2 or organic lubricants. An example of microcapsules deposited with Ni (Dietz[22]) is shown in Figure 12.16.

One classical metal that is deposited as a corrosion protective layer is zinc. Zinc treated with chromic acid shows superior corrosion protection. Attempts to substitute chromate

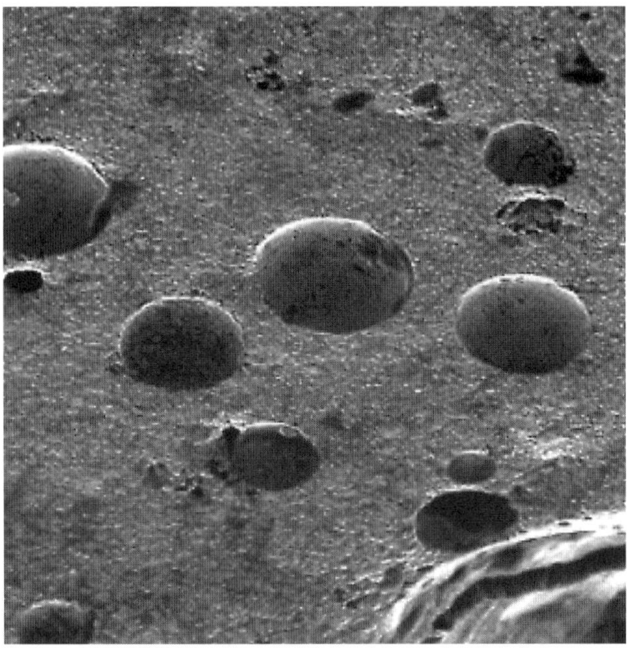

Figure 12.16 Co-deposition of microcapsules with nickel (Dietz). (Reproduced with permission from Ref. [22], © 2005, Eugen G. Leuze Verlag.)

treatment of zinc surfaces have led to new investigations in zinc deposition, i.e., the co-deposition of zinc and SiO_2 particles, which was shown in Figure 12.2.

Development of these new applications is still on going and one can expect a variety of new materials in the near future.

12.3 COMPOSITIONALLY MODULATED MULTI-LAYERS

The deposition of compositionally modulated multi-layers is another type of composite materials. First, experiments with copper–nickel layers showed an improvement of mechanical properties.[23,24] The interest in compositionally modulated multi-layers intensified with the discovery of the giant magneto resistance effect (GMR) by Grünberg and by Fert in the late 1980s.[25] It was observed that combinations of very thin ferromagnetic metal layers separated by non-magnetic metal layers in a magnetic field showed a very large change in the resistance. Since then such stacks of layers have been intensively studied and the preparation of the layers has also been tried by electrochemical methods. So far electrochemical preparation cannot compete with preparation by vapor deposition. Research should be intensified to improve the electrochemical techniques of multi-layer preparation.

12.3.1 Plating of multi-layers

For the preparation of multi-layers two methods can be applied (i) the single bath technique and (ii) the dual bath technique.

(i) *Single bath technique.* The electrolyte can be used for the deposition of two metals and therefore contains ions of both metals. The deposition potential of the two metals must differ. The deposition of layers of different metals can then be achieved by varying the potential or current. Variation of the layer composition is possible if a noble component is deposited together with a non-noble one. Examples are the deposition of copper–nickel, cobalt–platinum, and cobalt–gold layers.[26]

Mostly pulse plating with current pulses is applied for the deposition. Examples of pulse forms are shown in Figure 12.17.

In Figure 12.17A a simple double pulse program is shown. In Figure 12.17B a third pulse that has zero current or even reversed current is added between the two deposition pulses. This pulse is applied to stop the non-noble metal deposition. In the low current pulse only the noble component will be deposited at higher currents and at more cathodic potential the non-noble component will be deposited together with the noble component. The concentration of the noble component must be so low that the amount of noble metal co-deposited with the non-noble component is negligible. This shows that the plating conditions are very restrictive. For this reason the dual bath technique could be an alternative for electrochemical deposition of a special metal combination.

(ii) *Dual bath technique.*[27] Plating of a multi-layer from different electrolytes seems to be easier. The samples must only be transported between two plating baths. But the transport and rinsing make realization in larger scale applications difficult. Only a few examples have been published. A technique for rinsing is shown in Figure 12.18.

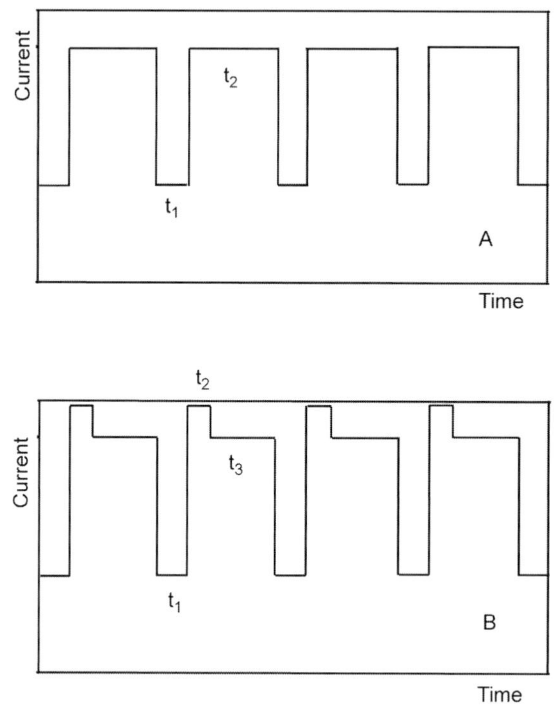

Figure 12.17 Different pulse forms in a single bath plating of compositionally modulated multi-layers. (A) Double pulse program with deposition of the less noble component for time t_1 and deposition of the more noble component for time t_2 and (B) an additional pulse with zero current to stop deposition of less noble metals.

12.3.2 Examples of multi-layers

An SEM image of the cross section of a cobalt–gold layer is shown in Figure 12.19 (Zielonka).[26] The current density varied between 50 A dm^{-2} (deposition of an alloy of 69% gold and 31% cobalt) and 0.8 A dm^{-2} (deposition of an alloy of 26% gold and 74% cobalt). Sodium hypophosphite was added to the electrolyte to reduce tensions and achieve the parallel layer's orientation.

For the most important application in GMR layers, the small rippling in the example shown in Figure 12.19 is not acceptable.[28] Another problem is the alloy character of the layers obtained by the single bath plating. Therefore, more experiments including ones to refine the dual bath technique are necessary in this area.

12.4 CORE-SHELL COMPOSITES

Electrically conducting polymers provide a new way to prepare conducting composites. It was shown in Chapter 11 how to prepare films of conducting polymers on non-conducting materials using adhesion promoters. The surface of the polymer became conducting. With the

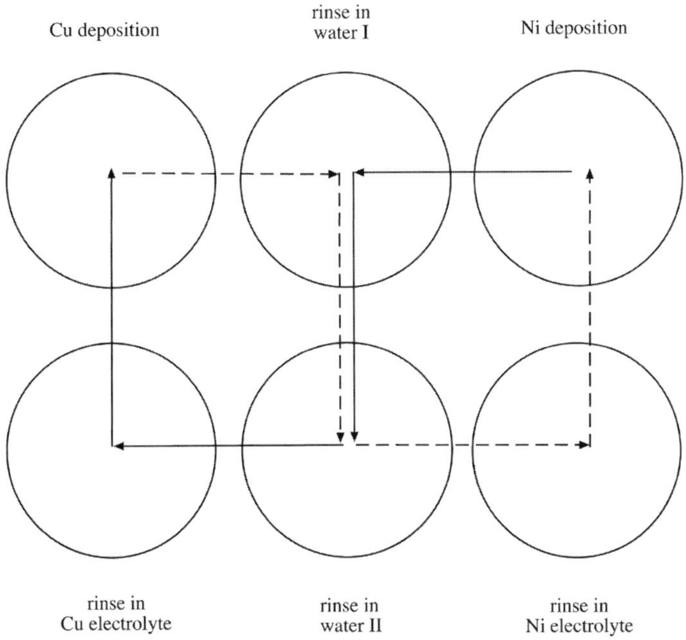

Figure 12.18 Rinsing procedure applied in a dual bath plating of copper–nickel. After deposition the samples are rinsed twice in water and in the electrolyte before the next layer is deposited.

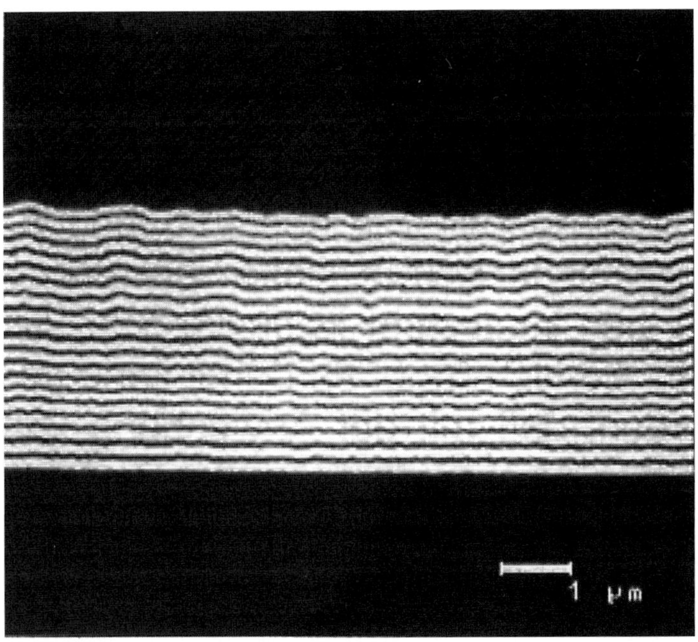

Figure 12.19 SEM image of a cobalt–gold deposition of compositionally modulated multi-layer (Zielonka). (Reproduced with permission from Ref. [26], © 1997, Eugen G. Leuze Verlag.)

same method it was possible to prepare films with a variety of inorganic particles. A new development allowed formation of such composites without the adhesion promoter. Oxide particles are typical materials for this procedure and have particle diameters that go down to nanometer size. A summary of the literature is available.[29]

12.4.1 Preparation procedure

The chemical preparation of conducting polymers by oxidative coupling leads to particles of conducting polymers in the micrometer and nanometer range. If this process is carried out in presence of inorganic particles, a thin film of conducting polymers is formed on the particles. Irregularities of the particles act as nucleation centers for film growth. The film is ultrathin and is usually a couple of nanometers thick.

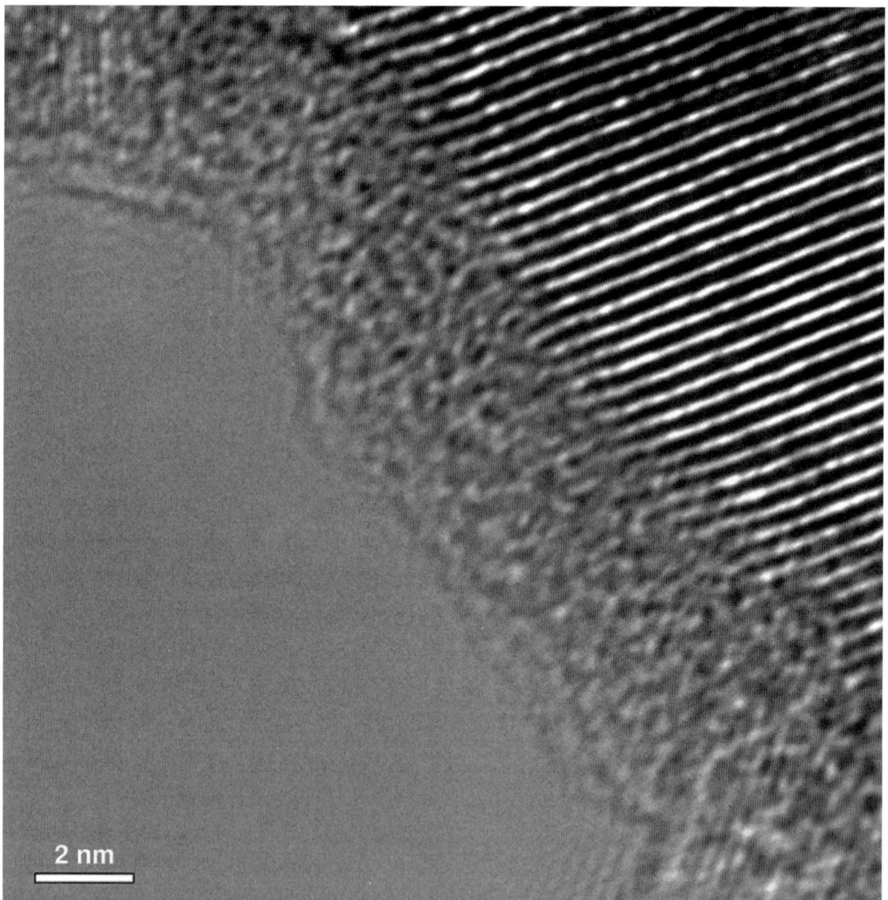

Figure 12.20 TEM image of core-shell particle. TiO_2 nanoparticle with polythiophene shell. The transition from the crystalline TiO_2 to the amorphous polythiophene is shown. Shell thickness is 4 nm.

12.4.2 Particle characterization: applications

An example of formation of a conducting polymer shell around the inorganic kernel is shown in Figure 12.20. The amorphous polymer shell is in direct contact with the crystalline structure of the kernel.

Further characterization is possible by impedance spectroscopy, photo-electrochemical measurements, and even simple cyclic voltammetry as described in previous chapters.

For applications of such particles film formation is important. Electrophoretic deposition is a possible method for forming compact particle films.[30]

In summary, the combination of oxide particles with conducting polymers forms a promising class of materials. They are more stable than purely conducting polymer particles. The dark and intensive color of these purely conducting polymer particles is compensated; the non-conducting oxide powders become conducting. In the reduced state of the conducting polymer the particles show photocurrents whereby the sign of the photocurrents depends on the wavelength.

REFERENCES

1. W.J. Lorenz, W. Plieth (Eds.), Electrochemical Nanotechnology, Wiley, Weinheim, 1998.
2. G. Staikov (Ed.), Electrocrystallization in Nanotechnology, Wiley, Weinheim, 2006.
3. (a) J. Seetharamappa, Sh. Yellappa, F. D'Souza, Interface, *15*, 23 (2006); (b) F. D'Souza, Interface, *15*, 27 (2006); (c) N. Martin, N. Solladié, J.-F. Nierengarten, Interface, *15*, 29 (2006); (d) L. Dunsch, Sh. Yang, Interface, *15*, 34 (2006); (e) B. Weissmann, Sh. Subramoney, Interface, *15*, 42 (2006); (f) T. Hasobe, Sh. Fukusumi, P.V. Kamat, Interface, *15*, 47 (2006).
4. S. Zein El Abedin, E.M. Moustafa, R. Hempelmann, H. Natter, F. Endres, Chem. Phys. Chem., *7*, 1535 (2006).
5. M. Azizi, W. Schneider, W. Plieth, J. Solid State Electrochem., *9*, 429 (2005).
6. E.J.W. Verwey, J. Th. Overbeek, Theory of the Stability of Lyophobic Colloids, Elsevier, Amsterdam, 1948.
7. R.H. Müller, Zeta-Potential und Partikelladung in der Laborpraxis, Wissenschaftliche Verlagsgesellschaft, Stuttgart 1996.
8. M. Azizi, Thesis, Technische Universität Dresden, 2003.
9. F. Wuensche, Diplomarbeit, Technische Universität Dresden, 2003.
10. H. Sahrhage, Thesis, Technische Universität Dresden, 2006.
11. J.P. Celis, J. Fransaer, Galvanotechnik, *88*, 2229 (1997).
12. N. Guglielmi, J. Electrochem. Soc., *119*, 1009 (1972).
13. A.M.J. Kariapper, J. Foster, Trans. Inst. Met. Finish., *52*, 87 (1972).
14. J.P. Celis, J.R. Roos, C. Buelens, J. Electrochem. Soc., *134*, 1402 (1987).
15. J. Fransaer, J.P. Celis, J.R. Roos, J. Electrochem. Soc., *139*, 413 (1992).
16. V. Terzieva, J.P. Celis, J. Fransaer, J. Electrochem. Soc., *147*, 198 (2000).
17. P. Nowak, R.P. Socha, M. Kaicheva, J. Fransaer, J.P. Celis, Z. Stoinov, J. Appl. Electrochem., *30*, 429 (2000).
18. C. Dedeloudis, J. Fransaer, Langmuir, *20*, 11030 (2004).
19. F. Wünsche, A. Bund, W. Plieth, J. Solid State Electrochem., *8*, 209 (2004).
20. J.L. Valdes, J. Electrochem. Soc., *134*, 223C (1987).
21. C. White, J. Foster, Trans Inst. Met Finish., *56*, 92 (1978); *59*, 8 (1981).

22. A. Dietz, Galvanotechnik, *96*, 1627 (2005).
23. A.R. Despic, V.D. Jovic, S. Spaic, J. Electrochem. Soc., *136*, 1651 (1989).
24. D.M. Tench, J.A. White, J. Electrochem. Soc., *137*, 3061 (1990).
25. P.A. Grünberg, Acta Mater., *48*, 239 (2000).
26. A. Zielonka, Galvanotechnik, *88*, 1122 (1997).
27. A.S.M.A. Haseeb, J.P. Celis, J.R. Roos, J. Electrochem. Soc., *141*, 230 (1994).
28. I. Bakonyi, L. Peter, Z. Rolik, K. Kiss-Szabo, Z. Kupay, J. Toth, L.F. Kiss, J. Padar, Phys. Rev. B, *70*, 54427 (2004).
29. N. Hebestreit, J. Hofmann, U. Rammelt, W. Plieth, Electrochim. Acta, *48*, 1779 (2003).
30. Qu.-T. Vu, M. Pavlik, N. Hebestreit, J. Pfleger, U. Rammelt, W. Plieth, Electrochim. Acta, *51*, 1117 (2005).

Index